T0328126

POLYGENERATION WITH POLYSTORAGE

POLYGENERATION WITH POLYSTORAGE

For Chemical and Energy Hubs

Edited by

KAVEH RAJAB KHALILPOUR

ACADEMIC PRESS

An imprint of Elsevier

Academic Press is an imprint of Elsevier
125 London Wall, London EC2Y 5AS, United Kingdom
525 B Street, Suite 1650, San Diego, CA 92101, United States
50 Hampshire Street, 5th Floor, Cambridge, MA 02139, United States
The Boulevard, Langford Lane, Kidlington, Oxford OX5 1GB, United Kingdom

Notices
Knowledge and best practice in this field are constantly changing. As new research and experience broaden our understanding, changes in research methods, professional practices, or medical treatment may become necessary.

Practitioners and researchers must always rely on their own experience and knowledge in evaluating and using any information, methods, compounds, or experiments described herein. In using such information or methods they should be mindful of their own safety and the safety of others, including parties for whom they have a professional responsibility.

To the fullest extent of the law, neither the Publisher nor the authors, contributors, or editors, assume any liability for any injury and/or damage to persons or property as a matter of products liability, negligence or otherwise, or from any use or operation of any methods, products, instructions, or ideas contained in the material herein.

Library of Congress Cataloging-in-Publication Data
A catalog record for this book is available from the Library of Congress

British Library Cataloguing-in-Publication Data
A catalogue record for this book is available from the British Library

ISBN 978-0-12-813306-4

For information on all Academic Press publications visit our website at https://www.elsevier.com/books-and-journals

  Working together to grow libraries in developing countries

www.elsevier.com • www.bookaid.org

Publisher: Jonathan Simpson
Acquisition Editor: Lisa Reading
Editorial Project Manager: Thomas Van Der Ploeg
Production Project Manager: Kamesh Ramajogi
Cover Designer: Christian J. Bilbow

Typeset by SPi Global, India

CONTENTS

12. Biorefinery Polyutilization Systems: Production of Green Transportation Fuels From Biomass   373

Pankaj Kumar, Mohan Varkolu, Swarnalatha Mailaram,
Alekhya Kunamalla, Sunil K. Maity

13. Energy-Water Nexus: Renewable-Integrated Hybridized Desalination Systems   409

Hesamoddin Rabiee, Kaveh Rajab Khalilpour, John M. Betts, Nigel Tapper

14. Renewable Energy Integration in Combined Cooling, Heating, and Power (CCHP) Processes   459

Dia Milani

CONTRIBUTORS

Zainul Abdin
Faculty of Information Technology, Monash University, Melbourne, VIC, Australia

Christos Agrafiotis
Institute of Solar Research, Deutsches Zentrum für Luft- und Raumfahrt/German Aerospace Center (DLR), Linder Höhe, Köln, Germany

John M. Betts
Faculty of Information Technology, Monash University, Melbourne, VIC, Australia

Hamid Ghanbari
Technology Specialist, Chemical Engineering, COMSOL Inc., Stockholm, Sweden

James Hinkley
Sustainable Energy Systems, School of Engineering and Computer Science, Faculty of Engineering, Victoria University of Wellington, New Zealand

Faezeh Karimi
Faculty of Engineering and Information Technology, The University of Sydney, Sydney, NSW, Australia

Kaveh Rajab Khalilpour
Faculty of Engineering and Information Technology, Monash University, Melbourne, VIC, Australia

Pankaj Kumar
Department of Chemical Engineering, Indian Institute of Technology Hyderabad, Sangareddy, India

Alekhya Kunamalla
Department of Chemical Engineering, Indian Institute of Technology Hyderabad, Sangareddy, India

Swarnalatha Mailaram
Department of Chemical Engineering, Indian Institute of Technology Hyderabad, Sangareddy, India

Sunil K. Maity
Department of Chemical Engineering, Indian Institute of Technology Hyderabad, Sangareddy, India

Dia Milani
CSIRO Energy Centre, Newcastle, NSW, Australia

Joel Parraga
School of Chemical and Biomolecular Engineering, University of Sydney, Sydney, NSW, Australia

Hesamoddin Rabiee
Faculty of Information Technology, Monash University, Melbourne, VIC, Australia

Ahmad Rafiee
Cardiff School of Engineering, Cardiff University, Cardiff, United Kingdom

Nigel Tapper
School of Earth Atmosphere & Environment, Monash University, Melbourne, VIC, Australia

Mohan Varkolu
Department of Chemical Engineering, Indian Institute of Technology Hyderabad, Sangareddy, India

Anthony Vassallo
School of Chemical and Biomolecular Engineering, University of Sydney, Sydney, NSW, Australia

PREFACE

This book is about the decentralization of the energy network and the associated challenges and opportunities. In this context, it addresses the historical and future trends in energy network diversification with polygeneration and polystorage systems, at both macro- and micro-level.

Before the Industrial Revolution, food, water, and energy supply chains were decentralized and scattered. Producers were consumers of their products (farms, agriculture, etc.), and the redundancies were supplied to neighborhood community markets. The efficient utilization of wind and water power in the 18th century led to a transition from hand production methods to new manufacturing processes. This change was slow, as water power was geographically limited. However, it turned into an Industrial Revolution as steam engines emerged. Steam power led to the rapid transformation of traditional production systems (food, textiles, iron, chemicals, transportation, etc.) to modern factory systems.

The rational consequence of industrialization could be the attention devoted to the "economy of scale" for bipartisan benefit of higher income for the supplier and lower prices for the consumer. This led in the 19th century to the Second Industrial Revolution for large centralized factories, beginning with the mass manufacture of steel and expanding to all other industries. Internal combustion engines were also emerged during this period, leading not only to the birth of automotive industries but also to the need for liquid fuels and consequently the birth of the modern petroleum industry, with multiple products beyond mere Kerosene for lighting.

Various technoeconomic reasons made liquid fuels the most desired form of energy source; consequently, crude oil turned into "black gold," and the 20th century became the "century of oil". There were severe geopolitical consequences as crude oil became the origin of several political conflicts and wars around the world, a situation that has continued to our time. Technologically, however, crude oil contributed notably to the scientific development. Crude oil refineries could be considered the most complex polygeneration systems in the history of human civilization. Many new sciences and technologies have been born or inspired by refining industries. For instance, several state-of-the-art system control and optimization theories have been developed in efforts applied to the operation of refineries. The HSE (Health, Safety and Environment) science has originated from, and developed by, this industry and its associates.

Over time, the chemical and energy industries have grown and become more complex and sophisticated. The more industries have been able to separate and recover the many components in black gold, the more subsequent processes have been born. For instance, following the development of refineries, the petrochemical industry was born to further process refinery products for value-added chemicals. As a consequence, polymer science and technology emerged in the 20th century, which enables infinite product synthesis.

Having said this, while the downstream fossil fuel industry has made substantial advances in the production of value-added chemicals, fossil-fuel-based power generation (accounting for more than 40% of consumed fossil fuels) has not experienced a marked transformation in production concepts. Until now, coal- or oil-based Rankine cycles and gas-based Brayton cycles have been the main methods of power generation, both of which are considered the major contributors to climate change. There have been some efforts for combined power and chemical generation, which have led to the development of pyrolysis and integrated gasification combined cycle processes, also referred to as coal refinery. However, with climate change threats and limited emission budgets, there is an international effort to move away from fossil fuels toward renewables.

The future society is now envisaged as one with "100%-renewable," "net-zero emission," or even "net-negative emission" systems. These targets cross the "decentralization" path. Today, more than a century since the centralization wave, we have learned that overcentralization, at least in energy and electricity networks, has drawn the overall system into suboptimality. This has led to the birth of decentralization concepts. In recent years, several often closely defined concepts have been introduced for decentralized energy networks. These include "microgrid," "mesogrid," "nanogrid," "minigrid," "energy internet," "community energy network," "social energy network," "peer-to-peer energy network," and "virtual power plant". This book focuses on the opportunities and challenges associated with decentralization of the energy network, addressing the following issues:

> *Resource storage for improved flexibility*: Before the industrialization, humans and all living beings lived in harmony with the nature. They had learned to adapt to nature's variability through resource storage in oversupply seasons. There are elaborate discussions in this book that, with humans having damaged nature cycles, the only way forward for us is adjustment of our lifestyle and industrial practices within nature's

constraints. This can be achieved through behavioral change and flexible manufacturing in which energy storage plays a critical role. Several chapters offer solutions for the flexible operation of small-scale and industrial-scale systems based on renewable resources.

➢ *Integration and connectedness*: Everything in the universe is integrated. The biosphere and atmosphere within which we live have reached the existing equilibrium over billions of years. Any disturbance in any element will take a certain time, depending on its magnitude, until the entire system reaches a new equilibrium. The Industrial Revolution empowered humans to take mastery and control over nature, leading to today's disintegrated Earth with its damaged biosphere, vast deforestation, polluted and acidified oceans, and energy-food-water struggles. Nevertheless, humans have succeeded in developing a sociotechnically connected world. Although decentralization of the energy network is necessary, the act of decentralization requires us to consider integration with nature cycles as well as with other sociotechnical networks. Of particular importance is the case of the energy-water-food nexus. Some chapters in the book discuss mechanisms for energy network integration and interconnection with other networks.

➢ *Hybridization and diversification*: Historically, energy and, specially, electricity generation processes have been through relatively simple processes such as combustion with very low efficiencies. Now, attention is being directed toward diversified energy generation systems, including all types of renewable energy and also conventional technologies. Hence, the energy industry is moving toward hybrid systems, which are able to process multiple feeds (solar irradiation, wind speed, biomass, geothermal, coal, oil, gas, wastes, etc.) and produce multiple products (e.g., power and chemicals). This indeed complicates the energy production system, especially in the face of intermittent and difficult-to-predict renewable resources. A consequence of high renewables penetration is the need for various energy storage technologies. Then comes the complexity of integrating hybrid storage and hybrid generation systems. Almost all the chapters in this book address polygeneration and polystorage system development and associated challenges in various industries, including chemical, oil and gas production, iron and steelmaking, and water desalination. Some chapters also address the sustainable utilization of carbonaceous fuels as alternatives to combustion. The critical roles of renewables-based fuels and various energy vectors such as hydrogen are also elaborated.

In recent years (2010s), there has been some hyperemotionality about the role of photovoltaic (PV) and battery technologies to solve the entire energy network challenges. The scientific community remembers speculative terms such as "greenhouse mafia" or "fossil-fuel mafia" and should be cautious about the development of any form of "renewable mafia" stereotype. Innovation is limitless and it behooves academia and industry to support all technologies, even those in the early phase of their learning curves. The main motto of this book is that *our future energy network will be a colorful mix of diverse energy generation and storage technologies, integrated and interconnected with other networks.*

➤ *Moving back to prosumption and closed-loop consumption*: Before the Industrial Revolution, humans were prosumers (concurrently producer and consumer). The Industrial Revolution transformed the lifestyle by centralizing production systems in order to benefit from the economy of scale. This resulted in the development of complex supply chain systems to link mass producers to the now-distant customers. With distributed energy sources, such as photovoltaic cells, this paradigm is shifting back to pre-Industrial Revolution periods, so that households (or generally small-scale energy consumers) can generate energy locally with clean energies, reducing their demand for distant fossil-fuel-based power plants. Thus, we are entering the era of prosumers and do-it-yourself in the context of energy systems. Although this is undoubtedly a positive step forward toward efficiency and closing the waste cycle, it creates several technical challenges that need to be addressed. For instance, the one-directional energy network is being converted into a bidirectional-flow network, with several associated challenges. Some chapters in the book discuss mechanisms such as prosumer cooperation for reducing network security risks. Furthermore, some chapters emphasize the critical role of closed-cycle manufacturing and waste minimisation through 3R (reduce, reuse, recycle) or 6R (Reduce, Reuse, Recover, Redesign, Remanufacture, and Recycle) approaches. In this context, elaborative discussions are provided on the importance of fostering the refurbishment and remanufacturing industries. One chapter is also dedicated to the critical issue of CO_2 recycling and utilisation.

Though the book is continuous, every chapter has its identity, enabling interested readers to study individual chapters. While appreciating the contributing authors, I also thank several people without whose help this book could not have been developed. First and foremost, I thank Thomas van der

Ploeg and Lisa Reading from Elsevier for their support throughout the development of this book. Kamesh Ramajogi and Indhumathi Mani work on the book edition and pagination was phenomenal. I express my appreciation of Prof. Manos Varvarigos from Monash University for his valuable inputs for several chapters. Dr. Igor Skryabin, Prof. Wojciech Lipinski, and Prof. Ken Baldwin from the Energy Change Institute of the Australian National University provided constructive feedbacks. Colleagues at Monash Infrastructure (MI) and Monash Energy and Materials Institute (MEMSI) offered continuous support while the book was in preparation. Writing requires mental and environmental peace. I am grateful to the staff of Revesby Workers Club in Sydney, where I wrote some major portion of the book, for their warm support and writer-quality coffees. Last but not least, I dedicate this work to my beloved little Daniel Khalilpour, who I missed accompanying to playgrounds while I was preparing this text for his sake and the sake of all children of today and tomorrow, deserving of a safe planet.

Sincerely,
Kaveh Rajab Khalilpour

CHAPTER 1

Moving Forward to the Past, With Adaptation and Flexibility: The Special Role of Resource Storage

Kaveh Rajab Khalilpour
Faculty of Engineering and Information Technology, Monash University, Melbourne, VIC, Australia

Abstract

This chapter describes three paradigms in human-nature interactions with respect to the role of "resource storage." The first paradigm is pre-Industrial Revolution, when humans had learned adaptation to the nature variability through storage of resources at oversupply periods. The second paradigm is post-Industrial Revolution, when access to reliable source of fossil fuels reduced the dependence on the nature and led to base-load industries and lifestyles, which were less and less in harmony with the nature's cycles. The new paradigm, after notably damaging the biosphere, is the "move forward to the past" with adjusting the lifestyle and the industrial practices with the nature's constraints through flexible manufacturing and consumption behavioral change in which energy storage and waste recycle play critical roles.

Keywords: Adaptation, Climate change mitigation, Renewable energies, Energy storage, Flexible manufacturing systems, Flexibility

1 THREE PARADIGMS IN HUMAN RELATIONS WITH NATURE

1.1 Paradigm 1: Pre-Industrial Revolution: Adaptation and Flexibility

The climate at any point of time and any location on Earth is variable. This is due to several reasons, the most important being the position from the sun, which is not located at the center of the Earth's orbital plane. Furthermore, the Earth's axis of rotation has a 23.5° tilt angle relative to the orbital plane. These, along with other reasons, have led to the development of various ecosystems and climate patterns. To simplify, humans have clustered the weather patterns into two (tropical regions) or more categories called seasons.

Polygeneration with Polystorage
https://doi.org/10.1016/B978-0-12-813306-4.00001-X

1

The variability or seasonality of the nature is probably the first challenge of life for any living being on this planet. Abraham Maslow's hierarchy of needs identifies physiological needs as the most basic necessity of human life [1]. Some such needs include water, food, energy, and shelter, the availability of which is dependent on nature's seasonality. In striving for survival, living beings through evolution have learned the inefficiency of solitude for dealing with the dynamics of nature. As such, they have learned the notion of living together by building cooperative groups and communities [2]. Consequently, civilizations have arisen in regions having rivers with fertile lands in the vicinity.

In communities, human (as well as other intelligent living beings) learned various adaptation mechanisms for survival (see Fig. 1). One approach, perhaps a passive one, for some living beings is hibernation. The other tactics include cyclic or permanent migrations, as well as storage of essential goods during seasons of abundance.

Storage is undoubtedly a tightly embodied feature of nature. Early humans did not need to look far to discover the storage concept: it was embodied in their bodies. A human baby is born with 10% of weight as fat stored in the upper back. This energy, which is used during the first three days after birth, guarantees the baby's survival against surrounding threats (see Fig. 2). Early humankind also observed weight gain with food overconsumption and weight loss during hunger periods. As such, they discovered

Hibernation   **Storage at oversupply times**   **Food caching**

Nomads   **Animal migration**

Fig. 1 Various adaptation and survival mechanisms for living beings against nature's variability.

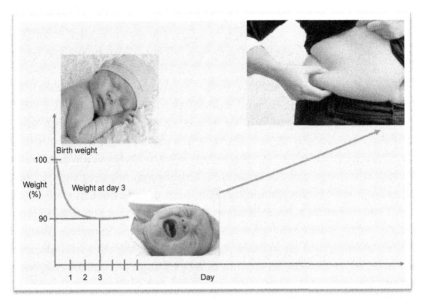

Fig. 2 Energy storage embodiment in the human body as a survival mechanism.

various ways of storing essential goods during oversupply periods for use in shortage seasons. They divided the year into various seasons and for each season they defined certain actions for production and storage of necessity goods. While in tropical zones, there were two seasons based on rain quantity, there were other geographies where the communities were dividing the year into more seasons (e.g., six seasons in Australian Aboriginal communities [3]), each with certain action items.

Some storage examples employed by our ancestors include *food storage:* drying and salting (dehumidification), and underwater (cold) storage; *energy storage:* wood, olive oil, beeswax; *water storage:* wells and qanat development. Community members also learned to exchange goods among each other and trade was developed.

Improved human adaptation and flexibility in extreme nature conditions led to reduced fatalities and increased population. The obvious consequence of population increase is the reduction of resources per capita and the need for expansion of territory in the search for more resources. The outcome was wars, which could define the fraction of a community who had to migrate, something that still continues in our time. Subsequently, the communities spread around and distribution of societies developed across the world, each with unique resources and limitations based on their geography.

Communities also found smarter means of access to necessities such as water, even at locations distant from surface waters. For instance, they learned to make wells, and later developed more intelligent solutions such as qanats (Fig. 3).

Early isolated communities were truly prosumers (producer-consumer), as they had to rely only on their own production. As communities were scattered, intercommunity trade was developed. In fact, the early supply chain networks were developed for balancing commodity supply and demand. A community could trade its overproduced commodity with some other essentials from neighbor communities. This intercommunity supply chain further improved human flexibility against natural constraints.

As time passed, human social networks were further integrated and long-distance supply chain networks were developed. An example is the Silk Road connecting the East to the West (see Fig. 4). Along the way, currencies were invented to facilitate easy commodity exchange; all these interactivities led to the growth of civilization despite wars and disasters.

The consequence of global interconnection was access to an extended demand market. This motivated communities with rich resources to develop new production and supply strategies for maximizing their benefit from their on-demand resources. Despite all collective creativities, however, productivity

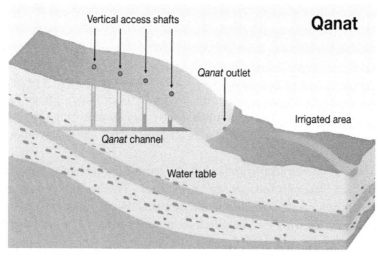

Fig. 3 Schematic of a qanat, a smart approach to create a channel from water tables at higher elevation and directing water to the ground at lower elevations. *(Image: Courtesy of GRID-Arendal.)*

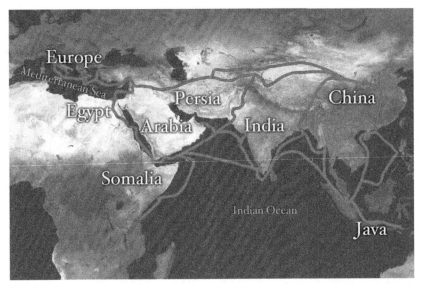

Fig. 4 The Silk Road: the ancient supply chain network and trade routes (land and sea routes shown with different colors).

did not increase beyond ox-driven farming and transport. At best, wind or water mills were developed. This continued until the Industrial Revolution.

1.2 Paradigm 2: Post-Industrial Revolution: From Flexibility to Baseload

As discussed in the previous section, before the Industrial Revolution, humans led a life in harmony with nature. Through evolution and thousands of years of civilization, humankind adapted to nature's seasonality and extreme conditions and learned to live flexibly with climate dynamics. The 18th century, however, was a turning point in human behavior and interaction with the nature. The technological ignition came from the development of new types of efficiently engineered water wheels and windmills by John Smeaton (1724–92) in the mid-18th century [4].

The efficient utilization of wind and especially water power led to a transition from hand-production methods to new manufacturing processes over the late 18th and early 19th centuries. Despite their notable contribution to industrial development, water and wind powers were geographically limited. However, this manufacturing reform became the well-known *Industrial Revolution* when steam engines emerged and dissolved the limitations of water and wind power. Steam power led to fast transformation of traditional production systems (food, textiles, iron, chemicals, transportation, etc.) to modern factory systems [5].

The rational consequence of industrialization could be the attention to the *economy of scale* in order to minimize the levelized cost of production. This led to the "Second Industrial Revolution" in the 19th century through the development of large centralized factories. This began with the mass manufacture of steel and expanded to all other industries.

Steam engines were in fact the main step in moving away from harmony with nature. Both wind and water are nature's renewable resources. But steam engines needed carbon-based fuels. Biomass was the only renewable carbon-based fuel, which mankind had used before for living. No doubt steam engines contributed notably to the deforestation crisis. Nevertheless, wood was not considered an ideal industrialization fuel due to several constraints including geographic limitations, seasonality, volume intensity, and difficulty of transportation. These led to the emergence of coal, oil, and gas fuels. The evolution of the primary energy mix, over the last two centuries, is illustrated in Fig. 5. Until the Industrial Revolution, wood had been the major energy source. Then coal was explored and gradually became the dominant source of energy.

Crude oil was a known fluid for thousands of years (named variously across the world) and had scattered applications, such as asphalting roads,

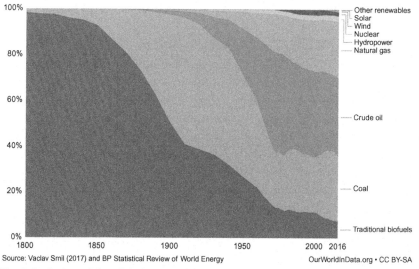

Source: Vaclav Smil (2017) and BP Statistical Review of World Energy   OurWorldInData.org • CC BY-SA

Fig. 5 Evolution of the global primary energy mix from 1800 to the present. *(Image: Courtesy of Our World in Data.)*

illumination, or even medical uses. In 1847, while the exploration of coal was reaching its climax, the Scottish chemist James Young could distil paraffins from coal and crude oil and produce various liquid fuels. Very soon, the oil industry developed, with the most successful company of the time being John Rockefeller's Standard Oil Company, from which today's Exon-Mobil, Chevron, British Petroleum (BP), and Marathon originated. Interestingly, for the first three or four decades, the oil industry focused on kerosene production for illumination. But today's oil industry was born with the development of liquid-fuel-based internal combustion engines [6]. Suddenly, oil refining diversified production from kerosene only to kerosene, gasoline, and diesel, and later evolved into today's multiproduct refineries with wide range of products from gas to LPG, gasoline, jet oil, diesel, fuel oil, asphalt, lubricants, coke, solvents, and more.

Briefly, the massive energy demand post the Industrial Revolution directed competition toward oil exploration and recovery from the late 19th century. With the gradual exploration of crude oil resources and the development of refining technology, oil became the dominant primary energy source in the second half of the 20th century (Fig. 5). The energy intensity and ease of transportation of oil compared to coal made crude oil the most desired energy source in the 20th century, although the geographic separation of supply sources and demand regions caused significant geopolitical issues across the world [7]. It is due to this significant industrial and sociopolitical impact that the 20th century is sometimes called the "century of oil."

The low cost of fossil fuels did not foster development of clean energy sources. Even among fossil fuels, the cleanest fuel, natural gas, is still the least utilized energy source due to its relatively higher price. One interesting historical fact relates to the automotive industry. Most of us may think that the automotive industry began with fossil-fuel-based internal combustion engines. But surprisingly, electric vehicles (EV) were dominant automobile products in the market in the late 19th and early 20th centuries, with a peak in 1913 (see Fig. 6) [8]. After that, Henry Ford's success in efficient and cheap internal combustion engine cars doomed the EV market to such an extent that today we can hardly believe its history. Automobiles and the general transportation sector is today the main customer of the petroleum industry, contributing significantly to almost a quarter of the world's CO_2 emissions [9].

The early 1900s also witnessed a tipping point in the world's human population due to many reasons, the most significant being the invention of the

Fig. 6 Edison with an electric car in 1913 (Electric vehicles were dominant automobiles of the market until 1913, but afterwards cheap internal combustion engines took over and the EV market almost vanished).

ammonia production process called Haber-Bosch, from hydrogen and nitrogen. This process has in some ways contributed to the growth of civilization on Earth [10]. Ammonia is not an abundant chemical and is found in trace quantities in nature, from nitrogenous animal and vegetable matter. Traditional agriculture was able to receive a limited amount of nitrogen from organic wastes. Without the Haber-Bosch process, and without access to a synthetic source of nitrogen, humans could produce only half as much as the current agriculture products [10] and it would need four times more land [11]. Synthetic fertilizers enabled humans to significantly improve agriculture productivity and this is considered one of the key drivers of population increase from around 1.8 billion [12] during the time of the invention of the Haber-Bosch process (1913) to more than quadruple of that today (7.6 billion in 2017) (see Fig. 7).

In summary, until the Industrial Revolution, human lives were in harmony with nature's dynamics and almost all consumptions came from renewable resources. Thereafter, however, humans developed an artificial feeling of victory over nature's variable but renewable resources. Instead of relying on fluctuating water, wind, and solar resources, it was now possible to explore fossil fuels from the Earth's deposit and to develop reliable baseload production systems almost disconnected from seasonal variations. The fast population growth of the 20th century accelerated the demand

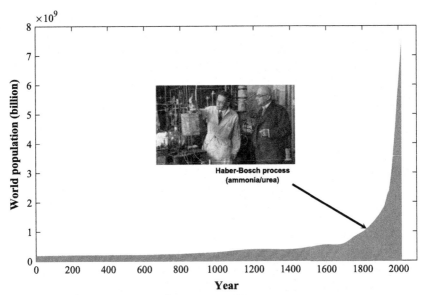

Fig. 7 World population during the years 0–2017.

for energy and thus fossil fuel extraction. This, however, came at a significant cost to the environment and, with less than a century's leadtime, to humans. Added to this, deforestation, land misuse, and release of various chemicals to the environment caused significant environmental problems.

Environmental Dutch disease: Our behavior with regard to fossil fuel exploration can be analogized to the famous "Dutch disease." According to the *Financial Times*, "Dutch disease is the negative impact on an economy of anything that gives rise to a sharp inflow of foreign currency, such as the discovery of large oil reserves. The currency inflows lead to currency appreciation, making the country's other products less price competitive on the export market. It also leads to higher levels of cheap imports and can lead to deindustrialization as industries apart from resource exploitation are moved to cheaper locations" [13]. It is now well understood that countries should remove redundant currencies (e.g., deposition in reserve banks) to avoid currency appreciation and subsequent inflation. With this phenomenon, we can now explain what we have done with our environment over the last century. Let us assume carbon or CO_2 as currency, the Earth's underground as a reserve bank, and the environment as the local economy. The Earth had deposited redundant carbon currency over billions of years and had developed a stable biogeochemical cycle with the atmosphere and beyond (identical to a stable economy). We opened the Earth's reserve bank and continuously injected its CO_2 currencies into our environment. What we observe today is an environmental Dutch disease, where overinjection of CO_2 currency has caused CO_2 inflation to such an extent that there is a significant threat of full environment (economy) collapse. We have created this crisis while we could consider fossil fuels as a wealth of nature for critical demand periods (under extreme conditions) experienced by current or future generations.

2 ENVIRONMENTAL CHANGE CRISIS

Over billions of years, the Earth had developed almost closed biogeochemical cycles within itself, with the atmosphere, and beyond. From the Industrial Revolution, however, the Earth entered a new era called "Anthropocene," in which humans became the main drivers of environmental changes [14, 15]. Johan Rockstrom and his colleagues from Oxfam have identified nine key Earth system processes [16], as shown in Fig. 8.

Humankind has opened these cycles within almost a century (see, e.g., Fig. 9). Probably the most significant threat has been through the emission of pollutants, which occurred in several ways, especially by releasing buried carbons (in fossil form) for energy, and reducing carbon sinks (deforestation) for food and other needs. Next, we focus on climate change as a result of fossil fuel emissions.

2.1 Brief Historical Background on Climate Change Crisis

Today, climate change is attributed to the emission of certain so-called anthropogenic greenhouse gases (GHGs) including carbon dioxide, methane, nitrous oxide, hydrofluorocarbons (HFC), perfluorocarbons (PFC), and sulfur hexafluoride (SF6). Naturally occurring GHGs in the atmosphere include water vapor (36%–70%), carbon dioxide (9%–26%), methane (4%–9%), and ozone (3%–7%) [17]. Since the Industrial Revolution, the emission of anthropogenic GHGs has steadily increased. According to the Fifth Assessment Report of the Intergovernmental Panel on Climate Change (IPCC), "human influence on the climate system is clear" and the recent anthropogenic GHGs emissions are the highest in the history [18]. Among GHGs, carbon dioxide (CO_2) is the most important in terms of quantity emitted. Fossil fuels account for about three-quarters of the increase in anthropogenic CO_2 emissions, while the remainder is attributed to land-use changes such as deforestation [19].

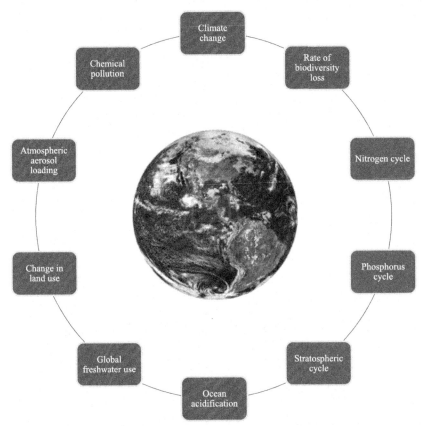

Fig. 8 Nine key environmental systems required to be controlled [16].

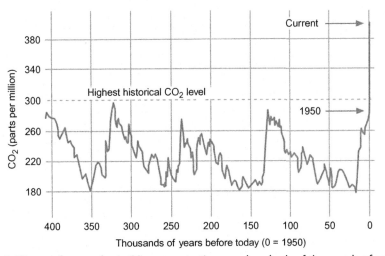

Fig. 9 History of atmospheric CO_2 concentration over hundreds of thousands of years (*Image from NASA's Global Climate Change website.*)

These emissions are considered to have caused an imbalance in the thermodynamics of the Earth and its atmosphere. The thermodynamic consequences of extra GHG emissions can be categorized in two classes: (1) absorption in oceans and (2) accumulation in the atmosphere.

Absorption of GHGs in oceans—ocean acidification: The average pH of oceans was about 8.2 for a few million years, until the times of the Industrial Revolution, from when it has steadily declined due to the extra absorption of CO_2 [20]. According to Sabine et al. [21], oceans took up around 48% of anthropogenic CO_2 emissions during 1800–1994 (57% during 1958–2009 [22]). If all the anthropogenic CO_2 were accumulated in the atmosphere, the current CO_2 concentration could be much higher than it is now. Therefore, CO_2 absorption by oceans has played the role of a natural sink for the mitigation of anthropogenic GHGs. But, with rapid acidification of the oceans, it is noticed that marine biogeochemistry (e.g., calcification process) has been affected in various ways. According to Turley et al. [20], the ultimate positive or negative impact is still uncertain, but evidence of a negative impact on some marine species is rapidly building [23].

Accumulation of GHGs in the atmosphere—global warming: GHGs can absorb and emit infrared radiation and thus warm the Earth's lower atmosphere and surface. According to historical data by NASA (the U.S. National Aeronautics and Space Administration), the average concentration of atmospheric CO_2 was 280 ppm in the years 1000–1800, the centuries before the Industrial Revolution. In 1900 the concentration was 296 ppm and the average temperature was 13.92°C. This value reached 311 ppm (13.84°C) by 1950. Though the value was high, it was still not high enough for scientists to clarify this as a phenomenon rather than noise. In the second half of the 20th century, however, the data clearly showed a trend with values breaking the historical trend (see Fig. 9). It became 331 ppm in 1975 and 369 by the year 2000 and in early 2015 the value passed 400 pm.

It is estimated that if the growth in consumption of fossil fuels continues at the current rate, the average global temperature will rise about 3.5°C by 2035 [24] and about 5–6°C by the end of this century [25]. This is regarded as a major concern that will cause serious natural disasters. Here are some of the very significant possible consequences:

➢ Sea level rise, mainly due to cryosphere melting. The sea level has already risen about 20 cm from the 1880 level and is estimated to rise between 0.5 and 1 m by the end of this century [23].

➢ Thermodynamically, when the temperature increases, the solubility of a gas decreases in liquids. Therefore, the CO_2 uptake by oceans may reduce as a result of the increase of ocean temperatures. This could result in a notable unexpected elevation in the atmospheric CO_2. Likewise, the temperature may cause ocean deoxygenation, which is harmful for nutrient cycling and thus for the life of marine habitats [26].

➤ Some potential GHGs are in stable liquid or solid form in the current thermodynamic state of the Earth. An example is natural gas hydrates (NGH) beneath the oceans. The total amount of these potential sources of GHGs is estimated to be much higher than the total GHGs of the atmosphere. The threat is that the temperature rise could result in vaporization and release of these gases. Even a low percentage release may be disastrous.

Avoiding such GHG emissions is therefore seen as imperative, and governments have recognized the need to adopt a key objective of curbing extreme weather, higher temperatures, worsening droughts and floods, and rising sea levels. In the next section, the global actions taken toward mitigation of GHGs thus far are discussed.

2.2 Global Action Against Climate Change

CO_2 is a colorless and odorless gas, vital for life. It is expectable and acceptable, therefore, that initially there was no concern about the overemission of this clean gas. It was in 1824 that Joseph Fourier discovered the greenhouse effect stating that "The temperature [of the Earth] can be augmented by the interposition of the atmosphere, because heat in the state of light finds less resistance in penetrating the air, than in re-passing into the air when converted into non-luminous heat." The greenhouse effect theory helped Svante Arrhenius to investigate the impact of increased atmospheric CO_2 on Earth's surface temperature using physical chemistry principles. In 1896, he concluded that "Human-caused CO_2 emissions, from fossil-fuel burning and other combustion processes, are large enough to cause global warming" [27]. This made him the first person who predicted global warming. The critical requirement after this was data-driven proves. In 1938, Guy Stewart Callendar studied data records from 147 weather stations around the world. He observed both temperatures and CO_2 concentration rise since Arrhenius's day. He concluded that "CO_2 has caused global warming." This is known today as "Callendar effect" [28]. Two decades later, in 1958, Charles David Keeling was able to get the US government support to start still-continuing measurements of atmospheric CO_2 at two locations, Hawaii and Antarctica, the recording which continues until now with data reports in the form of so-called "Keeling curve". These data could be considered as the first undisputable proof that CO_2 concentrations are rising. Despite such evidence, until the late 20th century, there was ongoing debate whether anthropogenic carbon dioxide warming was at noise level within natural climate variability.

The main global action against climate change began with the ozone depletion issue. In 1973, Rowland and Molina justified that CFC molecules had the potential to diffuse into the stratosphere, over time being broken to chlorine atoms by UV radiation. Then the chlorine has the potential to break down ozone (O_3) [29]. This argument, which later brought the Nobel Prize to the two scientists, ignited a global threat followed by international negotiations to stop emitting all ozone-depleting substances (ODS). Finally, in 1987, a treaty well known as the "Montreal protocol" was produced and came into force from 1989. This protocol is claimed to be the most successful international agreement in the history of United Nations (UN), as all members of the UN adhered to it and it could notably reduce ODS emission within a decade.

The gradual global warming stimulated discussions about the mitigation of GHGs. In 1981, James Hansen and his colleagues from the NASA published a data-driven paper concluding that anthropogenic CO_2 emissions would cause global warming more severe than previous predictions [30]. They projected that anthropogenic CO_2 warming would "emerge from the noise level of natural climate variability by the end of the century," which is exactly what happened (see Fig. 9). In 1988, one year after the successful Montreal protocol, when political decision makers were pleased about their achievement, the Intergovernmental Panel on Climate Change (IPCC) was formed to collate and assess evidence on climate change. In the same year, James Hansen found an opportunity to show evidence about the GHG threat in the US congress. These led to worldwide awareness of a global warming crisis [31]. Ultimately, in the UN's Earth Summit held in June 1992 in Brazil, a treaty known as the United Nations Framework Convention on Climate Change (UNFCCC) was produced, with the aim of "stabilization of greenhouse gas concentrations in the atmosphere at a level that would prevent dangerous anthropogenic interference with the climate system." The original treaty was legally nonbinding, with no mandatory limits on the emission of GHGs for individual nations and no enforcement provisions. It came into force from 1994 when more than 50 countries had signed the UNFCCC. The participating countries established an annual Conference of the Parties (COP), to assess progress in dealing with climate change. In 1997, in the COP-3 in Kyoto Japan, a protocol with legal bind-ings was developed that later became famous as the Kyoto Protocol. From February 2005, the protocol came into force by obligating industrialized countries, addressed as Annex I, to reduce their emission of GHGs by about 5.2% from the level of 1990 during 2008–2012 [32]. The Kyoto protocol was a notable achievement toward reducing the emission of GHGs and it was expected to be as successful as the Montreal protocol on the issue of ozone depletion.

A few years after COP-3, discussions began on the so-called post-Kyoto or post-2012 roadmap. Sadly, two decades from the beginning of the 21st century, we can say that still there is no sound protocol on climate change mitigation. Obviously, the reason lies in the significant capital investment requirements for any CCM option and thus the difficulty of cost distribution among the participant countries.

2.3 Practical Solutions for Climate Change Crisis

Carbon-based fossil fuels today comprise more than 85% of global energy resources [33] and their demand is increasing at a rapid rate. As shown in Fig. 10, the total energy consumption in 1970 was 4.91 billion tonnes oil equivalent (Btoe) (equiv 205.6 exajoule, EJ), but this value shot up to

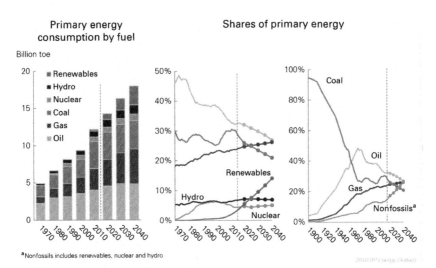

ᵃNonfossils includes renewables, nuclear and hydro

Fig. 10 World energy consumption by fuel type—historical and projected data. *(Image: Courtesy of British Petroleum. BP energy outlook. British Petroleum, London; 2018.)*

9.39 and 13.28 Btoe in 2000 and 2016, respectively. It is projected that by 2030 the global energy demand will be a whopping 16.32 Btoe, and will reach almost 18 Btoe by 2040 (see Fig. 10) [34]. In 2040, despite the increase in energy demand, the share of nonfossil energies including renewables (wind, solar, geothermal, biomass, and biofuels), nuclear, and hydro will exceed a quarter of the energy mix (26.0%). Oil (~26.9%), gas (26.2%), and coal (20.9%) will each take around one quarter of the demand (see right-hand graph of Fig. 10).

The most important issue is the fact that the increase in demand is almost completely attributable to developing countries. Energy is the fourth necessity of human life, after food, water, and shelter. According to the International Energy Agency, currently 1.4 billion people lack access to electricity and about 2.7 billion people are still using biomass for cooking. Strikingly, these values will not change notably anywhere in the near future (considering population rise) and in 2030 there will still be 1.2 billion people with no access to electricity [24]. Therefore, the fairness in development of the basic needs in the underdeveloped regions translates to an increased energy demand. This increasing demand rate is not expected to cease in future unless—and this is unlikely—these developments slow down. Therefore, a global paradox exists today: we require more energy for development and reducing inequalities, yet the GHGs and other pollutants generated from the energy sources have the potential to ruin our society.

In summary, there are two related crises ahead: (1) an energy supply crisis, along with an increasing trend in energy demand and shrinkage of some conventional energy sources such as oil, and (2) a climate change crisis, considering the fact that the majority of our current demand (>85%) relies on fossil fuels. Therefore, if the consumption of these fuels continues, a serious climate change crisis is ahead.

After a few decades of studies, the sustainable energy policy for mitigating the forthcoming crises is considered to have three pillars: (1) efficiency improvement, (2) replacing fossil fuels with clean energy sources such as renewables, biofuels, and nuclear energy, and (3) CO_2 capture, storage, and utilization (CCSU) [35, 36] (Fig. 11).

Efficiency improvement is certainly an important step toward a clean and sustainable environment. The advantage of this approach is its contribution to both mitigation of energy demand (by consumption reduction), and achievement of CCM objectives (by the consequent emission reductions). Improvement of the efficiency or demand-side efficiency of industrial

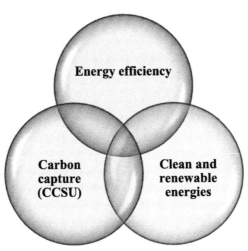

Fig. 11 Three key technological pillars of climate change mitigation (CCSU, carbon capture, storage, and utilization).

processes, e.g., building design and insulation, appliances, and vehicles, plays a key role in reducing global energy consumption. However, the problem is the huge capital cost required for establishing the bases. Cost is also the key barrier for other options. Renewable energies are undoubtedly superior energy sources compared to fossil fuels, due to their minimal associated emissions of CO_2 and other pollutants. Until the recent decade, however, most renewable technologies had not reached competitiveness with fossil fuels.

The idea of carbon capture, storage, and utilization (CCSU) derives from the proposition that, if we cannot stop using fossil fuels, at least attempts should be made to capture the CO_2 emitted from large fossil-fuel consuming sources such as power plants and steel plants. The captured CO_2 can then be either reused or sent by pipeline to be stored underground or beneath the oceans rather than emitted to the atmosphere. CCS is therefore viewed as a medium-term solution and a bridge from the current fossil-fuel-based energy system to one that has near-zero carbon emissions. For some time, there was a belief that there is today no credible pathway toward stringent GHG stabilization targets without implementation of CCS technologies in power plants [37]. However, no cost-effective CCSU solution has yet been introduced to be practically implemented across the world, and interest in this option is seemingly in decline [38].

3 PARADIGM 3: RECENTLY—ADAPTATION AND FLEXIBILITY

It was discussed earlier in this chapter that Rockstrom and his colleagues from Oxfam identified nine key environmental systems, shown in Fig. 8. They developed a "planetary boundaries" concept by identifying quantitative values as environmental ceiling for each of the nine processes [16]. On this basis, Raworth [39] developed a doughnut for sustainable development with eleven social boundaries at its core and nine environmental boundaries at the outer layer. As shown in Fig. 12, between these two boundaries there is a so-called safe and just space for humanity, which is a socially just and environmentally safe space to flourish.

Now, it might be easier to explain why climate change negotiations have not so far led to outstanding solutions. This lies in the fact that the three climate change mitigation options (efficiency, clean energies, and CCSU of Fig. 11) are generally costly and face numerous socioeconomic barriers. In other words, countries have not been able to find a "safe and just space" for climate change mitigation, because most solutions cannot satisfy the social boundaries. For instance, the immediate consequence of closing emission-intensive industries such as power and steelmaking plants is job loss, with its own social aftermath.

In the absence of socially just technologies, we have continued to violate environmental boundaries. Over the last decade, there have been efforts to change the "target-and-timetable" approaches of CCM to "budget" approaches [23]. This is claimed to give flexibility to governments to build their clean infrastructure and then start reducing their emissions at a faster pace. Consequently, toward the end of the second decade of the 21st century, we still face short- to medium-term uncertainty in international political decision making. It seems that the main practical solutions in the medium term are (i) clean and renewable energies such as PV and wind, (ii) "stringent policies" for industries in the form of carbon tax or renewable energy intensives, and (iii) the removal of energy subsidies for end users. The consequence of such passive strategies has been the release of over 500 Gtonnes-CO_2 during 2000–2017 [40]. According to the IPCCC and the United Nations Environment Programme (UNEP), if the moderate goal of capping the temperature rise at 2°C (450 ppm CO_{2e}) is to be achieved, the total carbon dioxide emission budget since the late 19th century (when the emission rate began to the increase) should not exceed 3670 Gtonnes-CO_2. According to the UNEP, starting from 2012 only 1000 Gtonnes-CO_2

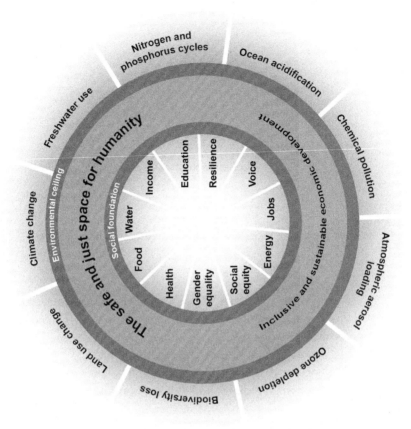

Fig. 12 Illustration of sustainable development doughnut with 11 inner social boundaries and nine outer environmental boundaries. *(Image courtesy of Raworth K. A safe and just space for humanity: can we live within the doughnut. oxfam policy and practice: climate change and resilience, vol. 8, Oxfam, Oxford; 2012. p. 1–26.)*

emission budget is left, meaning that "to stay within the 2°C limit, global carbon neutrality will need to be achieved sometime between 2055 and 2070… and total global greenhouse gas emissions need to shrink to net-zero sometime between 2080 and 2100" [41].

We should note that today there is no magical solution to the climate change problem, as most available methods are costly and consumers have to pay for them. *The more we move forward in time, the more we conclude that renewable energies and behavioral change are the main steps forward.*

Today, behavioral change is necessary for both industries and individuals. As shown in Fig. 13, there are three main pillars for behavioral change:

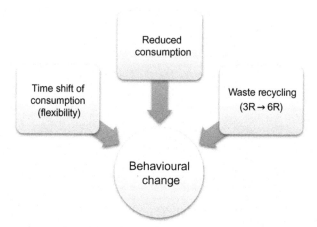

Fig. 13 Three main pillars of behavioral change with respect to energy usage.

consumption reduction, time shift of consumption, and waste recycling. There is no doubt that demand reduction (whether through efficient appliances or compromises in lifestyle through reduced usage) has a direct impact on in fossil fuel reduction and thus on the environment.

The next behavioral change is adaptation and flexibility to the time of resource availability, which requires time shifts of consumption. In the field of flexible manufacturing systems (FMS), there are several definitions for flexibility, some of which are listed in Fig. 14 [42, 43]. Some common forms of flexibility are "operation flexibility," "production flexibility," "expansion flexibility," and "volume flexibility." Perhaps the best known of these is "market flexibility," which is traditionally defined as operating a process based on dynamic market signals. These definitions should now be expanded by the addition of "nature flexibility," which considers the availability of renewable resources.

In short, flexibility can be defined as "the ability to adapt to changes." A longer definition, also with consideration of the environment, could be "The ability to respond to potential changes, whether endogenous or exogenous, certain or uncertain, in a timely and cost-effective manner, and with minimal environmental impact." In simple market language, it can be expressed as "operating a system based on renewable energy resources variability and market signals."

We are obliged, at least on energy-related issues, to "move fast to the past" with adaptation to nature's renewable but variable resources. This

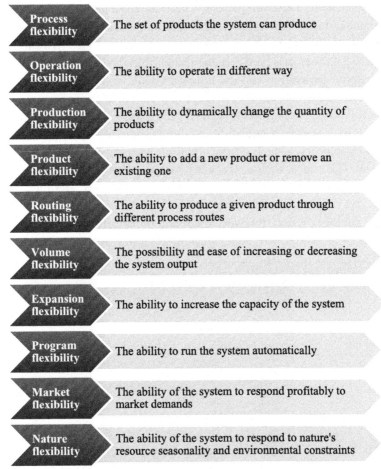

Process flexibility   The set of products the system can produce

Operation flexibility   The ability to operate in different way

Production flexibility   The ability to dynamically change the quantity of products

Product flexibility   The ability to add a new product or remove an existing one

Routing flexibility   The ability to produce a given product through different process routes

Volume flexibility   The possibility and ease of increasing or decreasing the system output

Expansion flexibility   The ability to increase the capacity of the system

Program flexibility   The ability to run the system automatically

Market flexibility   The ability of the system to respond profitably to market demands

Nature flexibility   The ability of the system to respond to nature's resource seasonality and environmental constraints

Fig. 14 Various forms of "flexibility" in flexible manufacturing systems (FMS) [42–46].

adaptation requires a significant industrial reform from baseload operation to flexible operation.

From the end-user's point of view, our lifestyle should be also adapted to renewable energy resources. We have begun installing photovoltaic cells on the rooftops of our houses and, with the declining trend in PV prices, such adaptations will have widespread use around the world, even in developing countries. The next step should be to adapt our energy consumption patterns with PV cells (or other renewable technology) output. For instance, instead of running energy-intensive appliances in the evening, we could program them to operate in the middle of the day during solar energy availability. Alternatively, surplus renewable energy can be stored in energy storage devices to be used in demand periods. This is the third paradigm in our individual and industrial interaction with nature, which has just begun and might be the only approach to operate our lives within safe environmental and just social boundaries and save the planet from the serious threat, which now faces our generation and the future residents of the Earth.

The third pillar of behavioral change in Fig. 13 is waste management by closed loop waste generation through recycling rather than through the current open loop approach, which has visibly changed the appearance of our lands and waters, let alone its critical environmental impacts. While the concept of 3Rs (Reduce, Reuse, Recycle) has been known and has been the foundation of urban design in several locations, there are arguments that it is not yet a fully closed-loop approach. A recent alternative proposal is the 6R (reduce, reuse, recover, redesign, remanufacture, and recycle) approach [47], which can better guarantee closing the waste cycle (see Fig. 15).

Fig. 15 The 6R (reduce, reuse, recover, redesign, remanufacture, and recycle) concept for a closed-loop product life-cycle system.

Such reform requires complex product manufacturing and supply chain systems to add the three new Rs (recover, redesign, and remanufacture) to conventional thinking [48]. But they can lead to a better world for its inhabitants.

4 CONCLUSION

This chapter provided an overview of the civilization journey that humans have made in interaction with the nature. The consequences of anthropogenic disturbance to the natural cycles were elaborated. It was then discussed that the only way forward is adjustment of our lifestyle and the industrial practices with the nature's constraints through consumption behavioral change and flexible manufacturing in which energy storage and waste recycle play critical roles.

REFERENCES

[1] Maslow AH. A theory of human motivation. Psychol Rev 1943;50:370–96.
[2] Hawkley LC, Cacioppo JT. Loneliness matters: a theoretical and empirical review of consequences and mechanisms. Ann Behav Med 2010;40:218–27.
[3] Bodkin F, Robertson L. D'harawal climate and natural resources. Sussex Inlet, NSW: Envirobook; 2013.
[4] Smeaton JXVIII. An experimental enquiry concerning the natural powers of water and wind to turn mils, and other machines, depending on a circular motion. Philos Trans 1759;51:100–74.
[5] Landes DS. The unbound prometheus: technological change and industrial development in Western Europe from 1750 to the present. Cambridge: Cambridge University Press; 2003.
[6] Singer CJ, Williams TI. A history of technology. Oxford: Clarendon Press; 1978.
[7] Yergin D. The prize: the epic quest for oil, money & power. New York: Free Press; 2011.
[8] Thomson C. The fascinating evolution of the electric car. Tech Insider; 2015.
[9] IEA. CO2 emissions from fuel combustion. Paris: International Energy Agency; 2015.
[10] Smil V. Detonator of the population explosion. Nature 1999;400:415.
[11] Smil V. Nitrogen cycle and world food production. World Agric 2011;2:9–13.
[12] Angus M. Development centre studies the world economy volume 1: a millennial perspective and volume 2: historical statistics: volume 1: a millennial perspective and volume 2: historical statistics. OECD Publishing; 2006.
[13] Financial-Times. Definition of Dutch disease. 2018.
[14] Crutzen PJ, Stoermer EF. The 'Anthropocene'. IGBP Newsl 2000;41:17–8.
[15] Crutzen PJ. Geology of mankind. Nature 2002;415:23.
[16] Rockström J, Steffen W, Noone K, Persson Å, Chapin Iii FS, Lambin EF, et al. A safe operating space for humanity. Nature 2009;461:472.
[17] Philander SG. Encyclopedia of global warming and climate change. SAGE Publications; 2008.
[18] IPCC. Climate change 2014: impacts, adaptation, and vulnerability. Part A: global and sectoral aspects. In: Field CB, Barros VR, Dokken DJ, Mach KJ, Mastrandrea MD,

Bilir TE, Chatterjee M, Ebi KL, Estrada YO, Genova RC, Girma B, Kissel ES, Levy AN, MacCracken S, Mastrandrea PR, White LL, editors. Contribution of working group II to the fifth assessment report of the intergovernmental panel on climate change. Cambridge and New York, NY: Cambridge University Press; 2014.

[19] Houghton JT, IPCC Working Group I. Climate change 2001: the scientific basis: contribution of working group I to the third assessment report of the intergovernmental panel on climate change. Cambridge and New York: Cambridge University Press; 2001.

[20] Turley C, Blackford J, Widdicombe S, Lowe D, Nightingale P. Reviewing the impact of increased atmospheric CO2 on oceanic pH and the marine ecosystem. In: Schnellnhuber HJ, Cramer WP, editors. Avoiding dangerous climate change. Cambridge and New York: Cambridge University Press; 2006.

[21] Sabine CL, Feely RA, Gruber N, Key RM, Lee K, Bullister JL, et al. The oceanic sink for anthropogenic CO2. Science 2004;305:367–71.

[22] Le Quere C, Raupach MR, Canadell JG, Marland G, Bopp L, Ciais P, et al. Trends in the sources and sinks of carbon dioxide. Nat Geosci 2009;2:831–6.

[23] Steffen W. Steffen W, editor. The critical decade: climate science, risks and responses. Canberra: The Climate Commission; 2011. p. 69.

[24] IEA. World energy outlook 2010: executive summary. Paris: International Energy Agency; 2010.

[25] Jha A. Global temperatures could rise 6C by end of century, say scientists. Guardian 2009;.

[26] Breitburg D, Levin LA, Oschlies A, Grégoire M, Chavez FP, Conley DJ, et al. Declining oxygen in the global ocean and coastal waters. Science 2018;359(6371)eaam7240. https://doi.org/10.1126/science.aam7240.

[27] Maslin M. Global warning: a very short introduction. Oxford: OUP Oxford; 2008.

[28] Fleming J. The callendar effect: the life and work of guy stewart callendar (1898-1964). Boston: American Meteorological Society; 2013.

[29] Molina MJ, Rowland FS. Stratospheric sink for chlorofluoromethanes: chlorine atom-catalysed destruction of ozone. Nature 1974;249:810.

[30] Hansen J, Johnson D, Lacis A, Lebedeff S, Lee P, Rind D, et al. Climate impact of increasing atmospheric carbon dioxide. Science 1981;213:957–66.

[31] Kerr RA. Hansen vs. the World on the Greenhouse Threat. Science 1989;244:1041–3.

[32] UNFCCC. Kyoto protocol. Kyoto: United Nations; 1998.

[33] BP. BP statistical review of world energy. London: British Petroleum; 2017.

[34] BP. BP energy outlook. London: British Petroleum; 20182018.

[35] Damm DL, Fedorov AG. Conceptual study of distributed CO2 capture and the sustainable carbon economy. Energ Convers Manage 2008;49:1674–83.

[36] IEA. World energy outlook 2009. Editioni—Climate Change ExcerptParis: IEA; 2009.

[37] Deutch JM, Moniz EJ. Summary for policy makers, In: Retro-Fitting of Coal-Fired Power Plants for CO_2 Emissions Reductions SymposiumMassachusetts Institute of Technology; 2009.

[38] Karimi F, Khalilpour R. Evolution of carbon capture and storage research: trends of international collaborations and knowledge maps. Int J Greenhouse Gas Control 2015;37:362–76.

[39] Raworth K. A safe and just space for humanity: can we live within the doughnut. In: Oxfam policy and practice: climate change and resilience. vol. 8. Oxford: Oxfam; 2012. p. 1–26.

[40] US-Energy-Information-Administration. Annual energy outlook 2017 with projections to 2050: CreateSpace Independent Publishing Platform; 2017.

[41] UNEP. The emissions gap report 2014. Nairobi: United Nations Environment Programme (UNEP); 2014.

[42] Browne J, Dubois D, Rathmill K, Sethi S, Stecke K. Classification of flexible manufacturing systems. The FMS Magazine. 2.

[43] Sethi AK, Sethi SP. Flexibility in manufacturing: a survey. Int J Flex Manuf Syst 1990;2:289–328.

[44] Nigel S. The flexibility of manufacturing systems. Int J Oper Prod Man 1987;7:35–45.

[45] Shewchuk JP, Moodie CL. Definition and classification of manufacturing flexibility types and measures. Int J Flex Manuf Syst 1998;10:325–49.

[46] Gupta YP, Goyal S. Flexibility of manufacturing systems: concepts and measurements. Eur J Oper Res 1989;43:119–35.

[47] Jaafar IH, Venkatachalam A, Joshi K, Ungureanu AC, De Silva N, Rouch KE, et al. Product design for sustainability: a new assessment methodology and case studies. In: Environmentally conscious mechanical design. 2007 (Chapter 2 in this book).

[48] Jayal AD, Badurdeen F, Dillon OW, Jawahir IS. Sustainable manufacturing: modeling and optimization challenges at the product, process and system levels. CIRP J Manuf Sci Technol 2010;2:144–52.

CHAPTER 2

The Nexus Era: Toward an Integrated, Interconnected, Decentralized, and Prosumer Future

Kaveh Rajab Khalilpour
Faculty of Engineering and Information Technology, Monash University, Melbourne, VIC, Australia

Abstract

In this chapter, four network factors including integration, connectedness, decentralization, and prosumption are discussed as as measures of sustainable development. We analyze nature and the human condition with respect to these four factors before and after the Industrial Revolutions, and we explain the sustainable way forward for a prosperous future.

Keywords: Nexus analysis, Integration, Connectedness, Decentralization, Prosumer, Energy vector, Waste recycle, Global grid, Energy-Water-Food nexus

1 INTRODUCTION

Networks are embodied features of nature's complex systems. Just within our bodies, there are several multiscale network structures from proteins to genes, neurons, cells, and blood circulatory systems. All creatures are also consciously engaged in some networks, starting right from birth with their family tree. On a global scale, all the physical and social networks interact and build the biosphere network. We here identify the four network features of integration, connectedness, decentralization, and prosumption as critical factors for a sustainable society, industry, and biosphere (Fig. 1). Now, each of these features is defined.

1.1 Connectedness

Connectedness is defined as "the state of being joined or linked." It seems there is a drive in nature for living beings to seek connection. Evolutionary scientists explain that, in striving for survival, living beings through

Polygeneration with Polystorage
https://doi.org/10.1016/B978-0-12-813306-4.00002-1

27

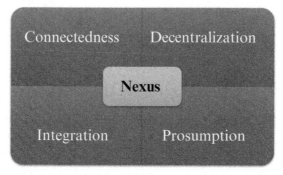

Fig. 1 The "nexus" era: the requirement of future society and industry to be decentralized, integrated, interconnected, and to demonstrate prosumption.

evolution have learned that solitude is inefficient for dealing with the dynamics of nature. They have learned the notion of living together by building cooperative groups and communities [1]. The growth of civilization is directly linked to the degree of connectedness. Beginning with primitive tribes, it took hundreds of thousands of years until humans slowly developed communication skills such as language and writing [2], which then facilitated building larger communities, villages, cities, countries, and in our time, the so-called global village.

Today's global village is undoubtedly indebted to the scientific and Industrial Revolutions as, without that progress, physical and social connectedness was impossible. Before the Industrial Revolution, humans and anything related to them were scattered and distributed across the world. Over the last few thousand years, during the Bronze and Iron ages, humans were able to make strong progress in civilization. They created cities, invented language writing (a necessity for managing life in cities) [3], and developed crosscommunity trade, which subsequently led to the birth of supply chain pathways (e.g., the Silk Road) and currencies. However, all these progressions were extremely slow by today's standards, as the manufacturing and transportation were still mostly oxen-driven. Thus the level of connectedness was limited to such an extent that the Earth's surface had not been discovered fully, let alone the discovery of macro-, micro-, and nanoworlds. For example, the continent of America and the country of Australia were discovered only during the scientific revolution.

The invention of steam engine ignited a chain of developments from transportation systems (including road, rail, sea, and later air), to complex supply chains for food, clothing, steel, chemicals, and oil refining. Gradually, the supply mechanisms for necessity commodities were transformed into networks such as water pipeline, gas pipeline, and electricity grid. The need for communication systems faster than postage gave birth to telegraphs, cable phones, wireless telecommunication, and now global connectedness through the internet.

Social and technological connectedness is probably the most outstanding achievement of all living beings on Earth and one of the positive outcomes of the Industrial Revolution. Connectedness improves the efficiency of physical networks by reducing communication time and reducing wastes through supply-demand balancing. Connectedness also improves the performance of social networks, with minimization of the power misconduct that is generally a feature of an isolated (atomized) community.

In summary, global connectedness has resulted in better technology performance and improved civilization. Not only should it be preserved, but also it should flourish as the main human heritage.

1.2 Decentralization

The terms centralization and decentralization both seem to have been born in the space of politics and government structure, in the era of the French revolution [4]. In the second half of the 20th century, however, those terms gained widespread attention. Probably the most symbolic statements in this regard are "whenever something is wrong, something is too big" by Leopold Kohr [5] and "small is beautiful" by Ernst Friedrich Schumacher [6]. Later, the terms entered into the public sphere when two popular futurist authors, Alvin Toffler [7, 8] and John Naisbitt [9], elaborated on decentralization in their best-selling books.

This term decentralization has also found widespread use in industrial manufacturing. From the beginning of human civilization until the 18th century, manufacturing system development was mainly bound to ox power and, when available, water wheel or wind mills. Steam engines, however, empowered humans and led to the development of modern manufacturing systems. The second Industrial Revolution occurred around the fundamental idea of technological centralization with the goal of utilizing "economy of scale" for increased revenue and more affordable products. Furthermore, moving often-noisy production systems out of sight could contribute to a better lifestyle. Around two centuries after the centralization wave, we now have learned, at least in the context of energy and electricity networks, that overcentralization moves the network away from optimality. The phenomenon was nicely summarized by Toffler when he wrote:

> The Second Wave Society is industrial and based on mass production, mass distribution, mass consumption, mass education, mass media, mass recreation, mass entertainment, and weapons of mass destruction. You combine those things with

*standardization, centralization, concentration, and synchronization, and you wind
up with a style of organization we call bureaucracy.*

Toffler [8]

Therefore, assuring the maximum benefit of social and industrial connectedness requires us to consider centralization in any part of a network as a risk,
socially and technologically. In network theory, there are evolving research
areas including "networks of networks," "interconnected networks," and
"multilayer networks," which address the right levels of centralization-
decentralization for reliable and robust societies and industries, which are
decentralized but connected (see Fig. 2).

As an example, in the energy sector, concepts such as "microgrid,"
"energy hub," "community energy network," and "social energy network"
are proposed for shifting the conventional low-efficiency and centralized
electricity network into a decentralized network of energy hubs. This shift
restructures the electricity network toward delivering reliable and affordable
energy with minimum loss and environmental impacts. Nevertheless, at the
same time of energy network decentralisation, there are substaintial research
activities on building the so-called "global grid" or "supergrid" which interconnects the electricity networks of almost all continents (see Chapter 6).

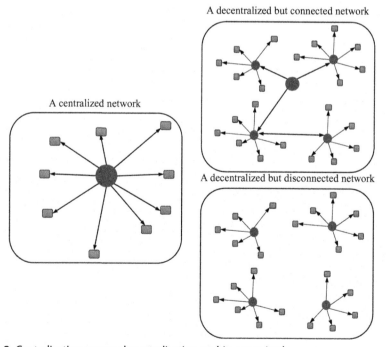

Fig. 2 Centralization versus decentralization and its magnitude.

1.3 Integration

Everything in the universe is integrated. The biosphere and atmosphere within which we live have reached the existing equilibrium over billions of years. Given any disturbance in any element, it will take a certain amount of time, depending on the magnitude of the disturbance, until the entire system reaches a new equilibrium. Johan Rockstrom and his colleagues from Oxfam have identified nine key Earth system processes: climate change, rate of biodiversity loss, nitrogen cycle, phosphorus cycle, stratospheric cycle, ocean acidification, global freshwater use, change in land use, atmospheric aerosol loading, and chemical pollution [10] (see also Fig. 3).

Since the Industrial Revolution, humans have become the main drivers of environmental changes and have imposed disturbances on these equilibriums to such an extent that some of the cycles were opened. As Fig. 4

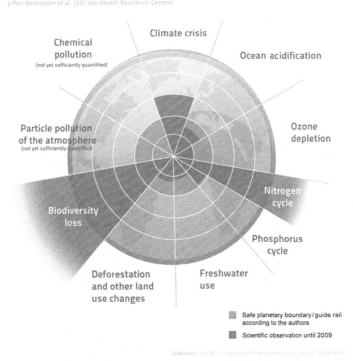

Fig. 3 Nine key Earth system processes, the planetary boundaries (beyond which severe circumstances are consequential), and the current status of each system [11]. *(From Wikimedia Commons, https://en.wikipedia.org/wiki/Planetary_boundaries#/media/ File:Planetary_Boundaries.png.)*

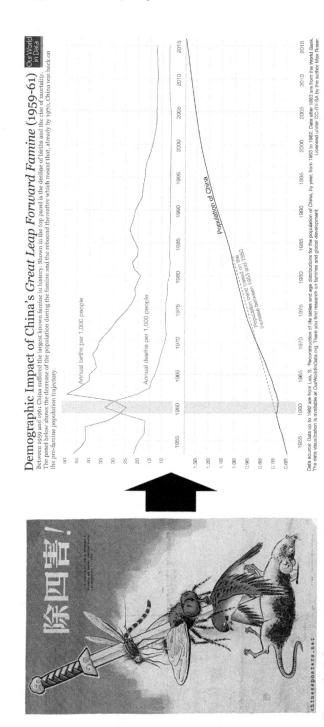

Fig. 4 The left image illustrates China's Four Pests Campaign for the extermination of rats, flies, mosquitoes, and sparrows, which caused a severe ecological imbalance leading to reduced agricultural productivity, and consequently the Great Chinese Famine leading to death (right graph) of tens of millions of people from hunger. *(Right image: Courtesy of Our World in Data.)*

illustrates, the Earth's biodiversity and nitrogen cycles have been significantly damaged, and in recent years, atmospheric CO_2 concentration has also passed from noise level to create climate change crisis [12]. The so-called Callendar effect refers to these integration phenomena, stating that increased CO_2 concentration passes the disturbance to the Earth by warming it [13, 14].

We have discussed some examples on global scales. However, the dynamics of integration are multiscale. An example of national-level ecosystem integration is the story of China's Four Pests Campaign. Following China's aspirational Great Leap Forward program for industrialization (1958–1962), the government identified four pests to exterminate: rats, flies, mosquitoes, and sparrows. The resulting campaign caused a severe ecological imbalance to the extent that in less than 2 years, farms and forests were filled with bugs and locusts that had formerly been eaten by sparrows. The resulting damage to the farms' products was much more severe than that caused previously by sparrows feeding on some seeds [15, 16]. Fig. 5 illustrates the consequence of China's Four Pests Campaign on the country's population during the late 1950s and early 1960s. This ecological imbalance intensified the Great Chinese Famine, leading to the death of 20–45 million people from hunger [17].

Another example, among numerous others elaborated in Andrew Goudie's valuable book [18], is the Australian rabbit crisis. European wild rabbits were brought to Australia by the first fleet of the migrants, in the late 18th century, mainly for food purposes. A few decades later, in the early 19th century, they were also grown for hunting. But they bred very quickly, to such an extent that their population reached 10 billion by 1920 [18]. Though

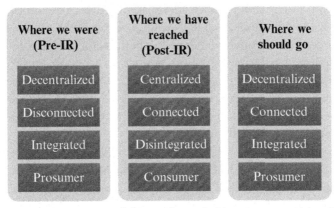

Fig. 5 Human society journey and the way forward with respect to social and industrial decentralization and interconnection, integration with biosphere processes, and demand management (*IR*, Industrial Revolution).

the numbers later declined to 200 million, they significantly affected Australian agriculture productivity, destroyed several spices, caused soil erosion, and obviously incurred a significant cost for their control [19].

In summary, all nine Earth system processes are in fact interconnected. Any disturbance in one system will not only propagate within that system but will also transfer to other Earth systems. We have shared some evidence of the high level of integration in the biosphere and the atmosphere and have shown the extent of catastrophes a disturbance can cause. To the Earth systems, we can add anthropological systems such as industrial, economic, and social systems, which are again interconnected within and with the Earth systems (see Fig. 6).

1.4 Prosumer

The word "prosumer" is derived from the words "producer" and "consumer." In today's supply chain network, the producer or supplier is generally at one end of the network, and the consumer or demand side is at the other end. Often there is limited communication between the two entities, though in the last few decades the necessity of this linkage has been recognized and research fields such as "demand-side management (DSM)" have been born. Nevertheless, though DSM enables the creation of bidirectional "information flow" between the two entities, the physical flow between them is still one directional, i.e., the demand side is a consumer and the supply side is a producer. In the preindustrial world, however, the gap between producer and consumer was narrow.

If we go back to early civilizations, there was no gap between their supply and demand. They were truly prosumers (producer-consumer), as they had to rely only on their own production. Isolation is indeed the other extreme of network centralization. In all its meanings, isolation ("atomized society" in today's definition) causes vulnerability, both technically and socially. In any isolated network, members have limited or no access to the outside world and, consequently, they must satisfy all needs by themselves. Over time, as humans developed to communities of two, three, and more, resource exchange mechanisms were also developed and the production-consumption gap started to grow.

Gradually, when tribes, villages, cities, and civilizations developed, intercommunity exchange mechanisms were also further developed. In fact, the early supply chain networks were developed to balance commodity supply and demand. A community could trade its overproduced commodity with

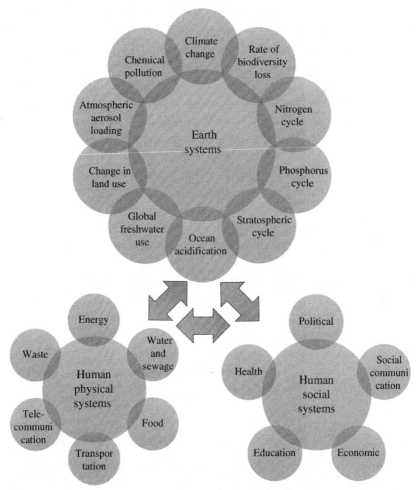

Fig. 6 Illustration of the integration and interconnection of Earth systems with anthropogenic (social and physical) systems.

some other essentials from neighbor communities. This intercommunity supply chain further improved human survivability against natural constraints. As time passed, human social networks were further integrated, and long-distance supply chain networks were developed. An example is the Silk Road connecting the East to the West. Along the way, currencies were invented to facilitate easy commodity exchange; all these interactivities led to the growth of civilization despite wars and disasters. But until the Industrial Revolution, the gaps between producers and suppliers were still minimal; generally, producers and suppliers were connected. The Industrial Revolution and the consequent centralization separated the consumer from the producer. Around two centuries after the centralization wave, today we have learned that overcentralization, at least in energy and electricity networks, moves the network away from optimality. Electricity networks not only account for the nontrivial amount of energy losses during transmission and distribution, but also for the huge network infrastructure development capital account for the major part of the energy bill [20].

Thus, the other arm and complement of decentralization is the "prosumer rise" and "do-it-yourself" movements, which attempt to reduce the producer-consumer distance to a sociotechnically optimal level by giving part of the production tasks to the consumer. An obvious example includes the IKEA-type movement, which, among many other benefits, has notably reduced furniture transportation costs (reduced emissions) through modularization and volume reduction of products and passing the assembly task to the consumer. Maybe the most important example of prosumer rise is found in the energy network, where households, commercials, and industries invest in their own in-house generation technologies such as photovoltaic cells, instead of relying fully on energy network supply [21]. Users can even supply their surplus energy to the grid, reminiscent of the time their ancestors supplied their surplus dairy product to local markets.

2 DESIGNING OUR NEXT SOCIOTECHNICAL NETWORKS

2.1 Three Phases in Human and Sociotechnical Networks

In the previous section, the interaction of each of the four features of human sociotechnical network development with natural networks was elaborated. Now, if we look at all four features combined, it becomes evident that there have been two distinct historical paradigms, with the Industrial Revolution era being the buffer period. Recognition that neither of these paradigms has been optimal implies the need for a move to a new paradigm. These three paradigms are illustrated in Fig. 5 and discussed briefly next.

2.2 Pre-IR: Disconnected, Decentralized, Integrated, Prosumer

Before the Industrial Revolution, while humans had trivial dominance over nature, the biosphere and atmospheric systems operated in full integration achieved over billions of years. Humans and anything related to them were scattered across the world, though they had a history of slow and oxen-driven endeavors to develop interconnections among themselves for social and trade purposes. The associated disconnection and isolation placed them in vulnerable situations against life challenges. This disconnection and the constraints of small-size communities necessitated living a prosumer life.

2.3 Post-IR: Connected, Centralized, Disintegrated, Consumer

Before the Industrial Revolution, humans and other living beings lacked the power to influence the Earth's systems and the biosphere, and atmosphere systems were in their natural condition. The Industrial Revolution empowered humans to assume mastery and control over nature. In less than two centuries, they moved some of the systems away from their equilibriums and opened cycles, which had been closed for billions of years (see Fig. 4). This disintegration even applied to human-made sociotechnical networks. Consequently, we have today a disintegrated Earth with the damaged biosphere, vast deforestation, polluted and acidified oceans, and energy-food-water struggles. With all the damage that humans have imposed on nature, we have succeeded in developing a sociotechnically connected world. Supply chain networks are also radically centralized, resulting in supply chain systems with efficiency concerns.

2.4 The Way Forward: Connected, Decentralized, Integrated, Prosumer

Connectedness as our civilization heritage should be kept empowered. Supply chain systems need to be decentralized to a techno-enviro-socio-economically optimal level. Industrialization converted the old prosumers to consumers. Along with decentralization, the prosumer concept needs to be remobilized and all human activities should be integrated within and with the Earth systems (see Fig. 6). This book [22] has provided several chapters explaining the role of diversified energy generation systems and energy storage systems in integration with chemical and energy industries to improve the sustainability and harmony of these systems with nature's processes. In the following section, some integration issues are also separately elaborated.

3 SOME NEXUS ISSUES FOR CHEMICAL AND ENERGY INDUSTRIES

3.1 From Petroleum Refineries to Integrated Energy and Chemical Polygeneration Complexes

After sun, hydropower, and wind, carbonaceous resources have been the main source of energy for humans throughout the history of civilization. Learning to burn wood triggered a revolution in all aspects of early human lifestyle, from diet to comfort and safety. Although there are scattered historical evidence of fossil fuel usage (coal, oil, and gas) across the world, systematic use of coal has a history close to a millennium. Widespread use of oil and gas has a much shorter history, starting with James Young (1811–1883) who invented crude oil refining in the 1840s.

Crude oil refineries could be addressed as the first and (probably still) the most complex production systems in the history of human civilization. Many new sciences and technologies have been born or inspired by refining industries. Over time, the chemical and energy industries have grown and become more complex and sophisticated. The more the industries have been able to separate and synthesize the hundreds of components in the so-called black gold, the more subsequent processes have been born. For instance, following the development of refineries, petrochemical processes were developed to further process refinery products for value-added chemicals. Consequently, the petrochemical complex and the polymerization industry was born in the 20th century, enabling infinite product synthesis.

Though the major fraction of refinery products has been used as fuels, at least this industry has advanced the production of value-added chemicals. However, the coal and natural gas industries have not been notably advanced and these two fossil fuel types have been primarily combusted for power generation. Consequently, fossil fuels are the major contributors to climate change due to the emission of CO_2 and other harmful chemicals.

Fig. 7 illustrates various pathways for carbonaceous fuels including fossil (natural gas, crude oil, and coal) and biomass. Until now, coal- or oil-based Rankine cycles and gas-based Brayton cycles have been the main methods of power generation, both of which are considered major contributors to climate change. It is no wonder that, in the absence of environmental policies, combustion-based processes have been the most economical pathways for biomass and fossil fuel energy utilization. But there is not much CO_2 budget left for the 2°C target, and a change in this approach is inevitable. Indeed, the approach must change if we want an integrated and sustainable biosphere

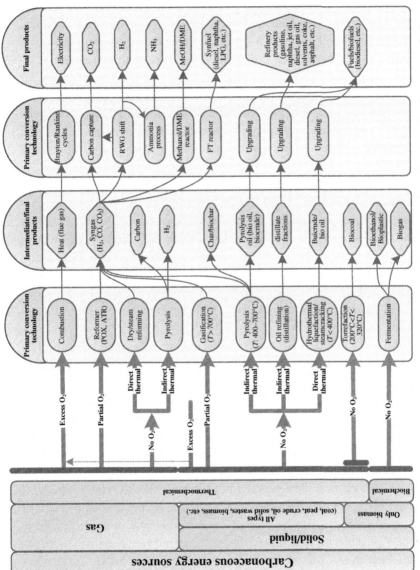

Fig. 7 Beyond simple combustion, toward diversified and more sustainable pathways for utilization of carbonaceous fuels.

and atmosphere. The way forward is the development of flexible and emission-neutral processes for combined power and chemical productions, as summarized in Fig. 7.

Gaseous fuels include natural gas, biogas, and synthetic natural gas (mainly composed of methane and some light hydrocarbons such as ethane and propane). The conventional approach for energy utilization from gaseous resources has been direct combustion with excess air to generate thermal energy. Alternatively, this energy can be recovered in turbines for power generation (e.g., the open loop Brayton cycle). The remaining energy of effluent gas can be further utilized for steam (and then power) generation using the Rankine cycle. If CO_2 emission removal is required, the exiting flue gas should be put through a carbon capture process to remove the CO_2 [23].

An alternative route for gaseous resource utilization is partial oxidation in a process called reformer [2, 24]. A third approach for the utilization of gaseous fuels is anaerobic (no oxygen). The initial reactions are endothermic, and the thermal energy is supplied either directly or indirectly. The anaerobic direct-thermal process is still considered a reformer because this process also generates syngas through the direct supply of high-temperature steam to the fuel chamber. The anaerobic indirect-thermal process, called pyrolysis, dissociates gases into their primary components, largely carbon and hydrogen [3].

Liquid and solid fuels are either fossil-based or renewable biomass. The utilization technologies for these fuels are (1) thermochemical and (2) biochemical. While the first category can process both organic and inorganic fuels, biochemical processes are used only for biomass. Like gaseous fuels, thermochemical processes can be classified into three groups based on the amount of oxygen used. With excess air, liquid and solid fuels can be directly combusted for heating or power generation (through Rankine cycles). They can also be gasified with partial oxygen levels in high-temperature (>700°C) reactors called gasifiers. The product of gasification is also syngas, though it has lower H_2/CO ratios than gaseous fuels. Thermochemical conversion of solid/liquid fuels can also be carried out anaerobically through pyrolysis, with two possible modes, direct thermal and indirect thermal. Indirect-thermal pyrolysis occurs at lower temperatures than gasification (400–700°C), and its products contain less gas, liquids (bio-oil), and solids (char/biochar). Biofuels such as biodiesel are produced through this process. In direct-thermal pyrolysis, superheated water is directly supplied to the feed in the reactor. This process is often called "hydrous pyrolysis," "hydrothermal liquefaction," or "steam cracking." Another anaerobic indirect-thermal pyrolysis, specific for biomass, is called "torrefaction" or "mild pyrolysis," due to the reactions occurring at lower temperatures (200–320°C). The advantage of this process is that the product, called "torrefied biomass" or "biocoal," is denser, with improved energy intensity for efficient transportation or subsequent processing.

Alternatively, organic materials can be converted biochemically through fermentation (digestion) to products such as biogas, bioethanol, and bioplastics. In fact, analogous to crude oil refineries, today biorefinery concepts are being developed, which, rather than limiting biomass utilization to combustion, attempts to convert them to value-added chemicals, as shown in Fig. 8.

3.2 The Era of Energy Hubs With Flexible Energy Vectors

Several definitions and classifications in the energy sector, such as renewable/nonrenewable, primary/secondary, and energy carriers warrant explanation. We here provide a brief discussion about some of these classifications, hoping they can lead us smoothly to the flexible energy vector discussion.

3.2.1 Renewable/Nonrenewable

While our minds know how to associate energy types with renewables and nonrenewables, the real definition might be complex. From an accurate scientific view, the definition of renewable/nonrenewable is difficult. Based on the following two different assumptions, it can be argued that (1) all energy sources are nonrenewable or (2) all energy sources are renewable:

(1) All energy sources are nonrenewable

The Sun is the main source of energy on Earth. We know that if we don't use solar energy today, we can use it tomorrow. There is no historical evidence of any disconnection of the solar energy supply on Earth. Therefore, we call solar energy and its associates, such as wind, renewables. But in reality, our Sun is composed of a limited quantity of hydrogen, though this quantity is significantly large. It currently fuses about 600 million tons of hydrogen into 596 million tons of helium every second [25]. The other 4 million tons of the matter is converted into energy, of which a small fraction is received on Earth. Though the sun has enough hydrogen to fuse stably at least for the next 5 billion years, in reality, it is nonrenewable and will die one day. Therefore, our so-called renewables are not really indefinitely renewable.

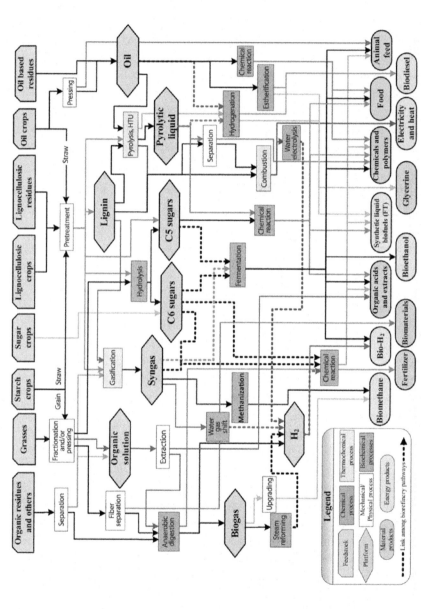

Fig. 8 The International Energy Agency's classification of biorefinery processing pathways. (*Image reproduced with permission from Francesco C, et al. Toward a common classification approach for biorefinery systems. Biofuels Bioprod Biorefin 2009;3(5):534–46.*)

(2) All energy sources are renewable

Another stance is within the boundaries of the biosphere and atmosphere. It can be argued that all energy sources are renewable as they are from nature to nature. The difference between various energy types is just the timeframe required for their renewal. It takes 1 day for solar energy renewal, some weeks or seasons for biomass renewal, and much longer for fossil fuel renewal. In fact, fossil fuels are the energy that nature has synthesized and stored over millions of years, and it will keep doing that as long as it exists. With this line of thinking, we can argue that we impose the term "nonrenewable" to some energy resources only because of their longer renewal period than that of the human lifecycle. With this argument, all energy sources are renewable, though some are clean and some not.

3.2.2 Primary and Secondary Energy

There are different definitions of primary and secondary energies. Probably the best definition for primary energy is the "energy embodied in natural resources (e.g., coal, crude oil, sunlight, uranium) that has not undergone any anthropogenic conversion or transformation" [26]. The best definition for secondary energy also might be that of the United Nations: "secondary energy should be used to designate all sources of energy that results from the transformation of primary sources." In a much more simple way, any energy source that is naturally available for direct use is primary, and those which require additional processing are secondary.

3.2.3 Energy Carrier

According to the International Organization for Standardization (ISO) document number 13600:1997(E), an energy carrier is a "substance or a phenomenon that can be used to produce mechanical work or heat or to operate chemical or physical processes" [27]. Thus this definition not only covers primary and secondary energies but also encompasses other media, which can hold energy. Water is not an energy source. But it can be elevated to hold potential energy for later use. Therefore, despite hydro being neither a primary nor a secondary energy source, it is an energy carrier with a notable contribution to our daily energy supply.

3.2.4 Energy Vectors

A critical issue in the energy supply chain, especially in future energy systems with highly variable renewable uptake, is coordination between supply and demand variabilities. This condition requires "flexibility" in the energy supply chain in terms of having enough "lead" or "lag" in supply time, creating the need for "energy vectors," a less popular term in the energy sector.

According to Orecchini [28] "a tool that allows the transportation and/ or storage of [a quantity of] energy, [in space and time], is called energy vector." Another definition could be an energy vector is a multipurpose and flexible energy carrier which can be primary or produced from other energies, and is transportable with lead and lag time, both directly and with conversion to other energy carriers.

The suitable energy vectors for today's complex energy supply chains are those with the capability of transformability, transportability, and storability (Fig. 9). Such energy vectors are obviously required to have high energy density while being safe, sustainable, and economically feasible. This definition explains why the world has been so reliant on fossil fuels. Crude oil, coal, and natural gas have been the three top energy vectors over the last century. Especially, the energy density and ease of transportation of oil, compared to coal and gas, led to crude oil becoming the most desired energy source in the 20th century, to the extent that the 20th century is often called the "century of oil" [29].

The challenge with renewable energy resources is either that they are nonenergy vectors or when they are energy vectors, they are not flexible. For instance, hydro, one of the best and so far dominant renewable energy sources, cannot be categorized as an energy vector as water cannot flow in space (though can flow in time). Biomass is an energy vector, but the low energy density of solid biomass makes its distant transportation infeasible unless the energy is converted to other forms such as electricity. Geothermal energy is an excellent energy vector with flexibility in time lead/lag, but its transportation cannot be considered for large distances unless the energy is converted to other forms. These limitations lead to the electron energy vector.

Electron energy vector. Generally, most renewable resources, such as solar irradiation, wind speed, and ocean energies, are not energy vectors as they cannot flow in space. They require conversion to secondary energies. The electron has played the role of the best energy vector for these renewable energy sectors. Photovoltaic (PV) cells, wind turbines, and other renewable technologies can pass the primary renewable energy to electrons, and then the energized electrons can move almost spontaneously in time and space through the electricity network. The critical challenge of the electron energy vector is, however, electron storage (to provide lead/lag time).

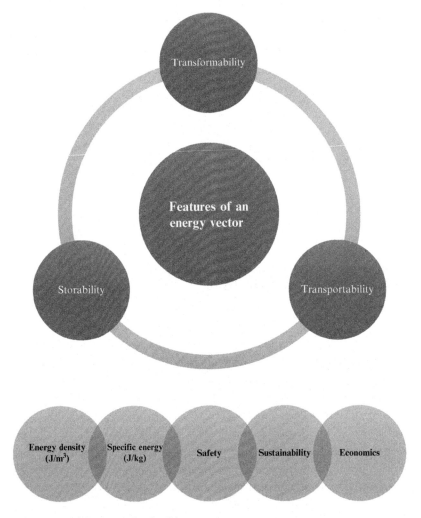

Fig. 9 Required features of a flexible energy vector.

Hydrogen energy vector: The relatively high cost of electron storage devices (batteries, flywheels, etc.) hinders long-term electron storage. Thus, although the electron is the best energy vector for direct energy supply to very distant locations, there are requirements for complementary energy vectors with the capability of long-term energy storage (longer lead and lag times) to assure reliability of the supply-demand balance. This need evokes chemical energy carriers overarched with hydrogen (Chapter 4). Put simply, primary renewable energy sources can be used to split water and produce H_2. Then, hydrogen has the superb capability to flow flexibly in time and space across the entire energy network. It can be stored in various forms (e.g., compressed or liquefied tanks) to offer lead and lag time. It can be used in fuel cells to generate electricity and thermal energy. It can even be converted to other energy forms such as ammonia or synthetic natural gas, both of which are also energy vectors.

Toward energy hubs with polygeneration and polystorage systems: We have elaborated on the critical role of electrons and hydrogen in future energy networks. Given the diversity of renewable resources in various geographies and also the development of diverse energy generation and storage systems, the future energy network will be composed of distributed energy hubs with colorful polygeneration and polystorage systems, with various energy vectors assuring reliability of energy supply.

3.3 Integration of Physical Networks and the Energy-Water-Food Nexus

Theorists have argued that complex networks pose the most significant challenge of the current century [30]. Modern society deals increasingly with complex networks, especially infrastructure. These networks are often interlinked and tightly bound up with our daily lives. We rely on water and energy networks for basic needs and transportation and communication networks for work and social life. Added to this, over the last decade the development of social networks, empowered by smartphones, has been integrated with the critical infrastructure networks, and these networks have played notable roles in improving public welfare using more reliable infrastructures while also posing new integration challenges.

Despite the fundamental similarity of networks, historically there has been very limited transdisciplinary research on critical networks. For instance, telecommunication networks have been studied in the discipline of Electrical and Computer Engineering, whereas transportation science has been hosted in Civil Engineering communities. The consequence is the development of various networks, which are internally integrated but externally disintegrated. There is a strong push for this trend to be changed. While it was possible for early generations of networks to be tackled to some extent independently, the level of interdependency has been increasing over time (see Fig. 10). We are now entering the era of "networks of networks" in which several networks, from infrastructure to social, are tightly interlinked. Topics such as "interconnected networks," "interdependent networks," "multilevel

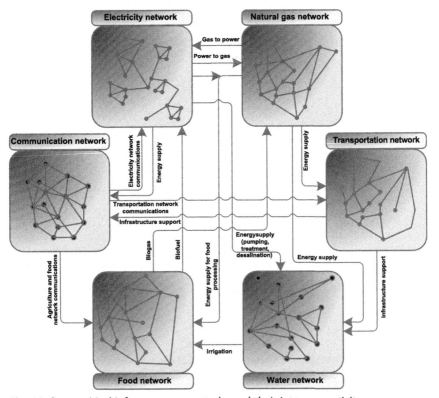

Fig. 10 Some critical infrastructure networks and their interconnectivity.

networks," "multilayer networks," and "multiplex networks" are finding increasing attention in the space of *critical infrastructure resilience* management.

One of the subsumed problems in this context is the topic of the energy-water-food nexus. The energy and water networks are interdependent, agriculture and food supply chains being tightly dependent on water availability for irrigation and subsequent food manufacturing processes. Water and energy networks are also integrated; the energy industry is water dependent for utility and feed (hydropower). Its effluents also affect water resources. On the other hand, the water processing industry is reliant on energy for water treatment and transportation.

The interdependency of energy and food networks is a contemporary dilemma. The food supply chain is energy intensive and reliant on energy network. On the other hand, some food resources can be used for energy (fuel) production. Over recent decades, there has been significant deforestation and land use change around the world, especially for palm oil tree

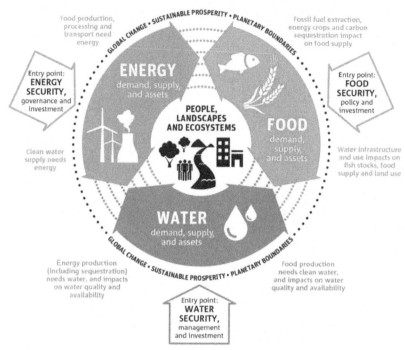

Fig. 11 The energy-water-food nexus: schematics of the interdependencies across the three networks. *(Image: Courtesy of the United Nations.)*

plantation, with the aim of biofuel reduction [31]. One of the resultant emerging dilemmas of our time is the preference conflict between energy and food. The answer to this is clearly complex, requiring a holistic nexus analysis approach to consider the interactions with all the sociotechnical and environmental processes (see Fig. 11). The main driving institution behind this nexus analysis is the United Nations, and it seems the problem of energy-water-food nexus was coined by this institution in 1984 [32].

3.4 Waste Management: From Open-Cycle Production to Closed Cycle With 6R

Perhaps the most anthropogenic damage to the environment has been through open-cycle production and consumption systems. We earlier explained how the centralization of supply chain networks along with other factors created distance between producers and consumers. A consequence of that distance was the life-cycle breakage of the specific product. Petrochemical complexes produced water bottles; not only did consumers have no use for them, but also they were not degradable by nature. The need for more food increased fertilizer use, which subsequently polluted water resources. The explored crude oil has also some water and gas content. The low economic value of these two components led to the discharge of the water to oceans and flaring (or venting) of the associated gas [33, 34]. The so-called produced water contains components such as BTEX (benzene, toluene, ethylbenzene, and xylene), which are harmful to marine and human life [34]. Gas venting and flaring also contribute to global warming and air pollution [33]. These are just some examples of a big list of open cycle manufacturing and their adverse consequences on the environment and society.

Today, the way forward is nothing but an integrated approach to close product life cycles by recycling waste to the production system. This is the only approach, which can curb the current environmental crises. Various mechanisms have been introduced, such as the 3R (Reduce, Reuse, Recycle) and 6R (Reduce, Reuse, Recover, Redesign, Remanufacture, and Recycle) programs (Fig. 12) [35], which are generally founded on the three main pillars of reducing consumption, reusing products, and recycling them to the system instead of further utilizing nature's resources. For instance, one of the options to tackle the climate change crisis is to recycle greenhouse gases in order to reduce their accumulation in the atmosphere. There are currently an increasing rate of academic research for finding pathways to reuse CO_2 in chemical and fuel production. Detailed discussion about CO_2 capture and utilization is given in Chapter 8.

Reuse is one of the approaches that require more attention and support from governments. In a world with 7.6 billion population, product refurbishment and the second-hand market can significantly reduce damage to the environment. There are some successful historical stories of product remanufacturing [36]. The internet world and online shops such as eBay and Amazon have facilitated an easy second-hand market, which also promotes refurbished products. There is also space for governments to promote the reuse culture and change the social attitude toward second-hand or remanufactured products.

In summary, governments and industries are responsible for recycling bulk wastes to avoid their accumulation in the nature. However, waste management is a task which requires the cooperation of all entities in the commodity supply chains from supplier to consumer. The rise of the prosumer is also expected to contribute to waste reduction through reducing the unnecessary size of supply chain networks.

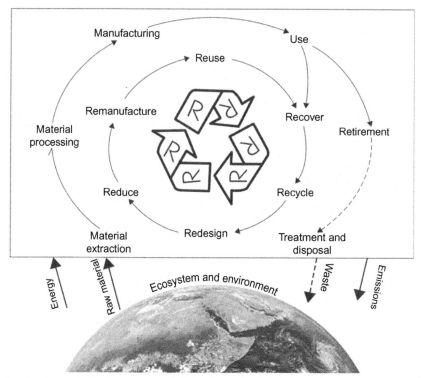

Fig. 12 The 6R (Reduce, Reuse, Recover, Redesign, Remanufacture, and Recycle) concept for a closed-loop product life-cycle system. *(Image reproduced with permission from Jaafar IH, et al. Product design for sustainability: a new assessment methodology and case studies. Environmentally conscious mechanical design; 2007.)*

4 CONCLUSION

Before the Industrial Revolution, while humans had trivial dominance over nature, the biosphere and atmospheric systems were operating in full integration achieved over billions of years. Humans and anything related to them were scattered around the world and disconnected. This disconnection and the demands of small-size communities necessitated living a prosumer life. The Industrial Revolution empowered humans to take mastery and control over nature, leading to today's disintegrated Earth with the damaged biosphere, vast deforestation, polluted and acidified oceans, and energy-food-water struggles. Nevertheless, human ingenuity has succeeded in developing a sociotechnically connected world. Complex, centralized supply chain networks have also separated supplier from the consumer and have converted the old prosumers to consumers.

The way forward for having a sustainable Earth and social development is the decentralization of inefficient supply chain systems. Along with this, consumers should take proactive responsibility in the supply chain networks by becoming prosumers. Additionally, we discussed the sustainable utilization of carbonaceous fuels, various energy vectors such as hydrogen, and the critical role of waste recycling.

REFERENCES

[1] Hawkley LC, Cacioppo JT. Loneliness matters: a theoretical and empirical review of consequences and mechanisms. Ann Behav Med 2010;40(2):218–27.

[2] Lieberman P. The evolution of human speech: its anatomical and neural bases. Curr Anthropol 2007;48(1):39–66.

[3] Daniels PT, Bright W. The world's writing systems. New York: Oxford University Press; 1996.

[4] Schmidt VA. Democratizing France: the political and administrative history of decentralization. Cambridge University Press; 2007.

[5] Kohr L. The breakdown of nations. Routledge & Kegan Paul; 1986.

[6] Schumacher EF. Small is beautiful: economics as if people mattered: 25 years later... with commentaries. Hartley & Marks Publishers; 1973.

[7] Toffler A. Future shock.

[8] Toffler A. The third wave. Morrow; 1980.

[9] Naisbitt J. Megatrends: ten new directions transforming our lives. Warner Books; 1984.

[10] Rockström J, et al. A safe operating space for humanity. Nature 2009;461:472.

[11] UNEP. UNEP year book 2010: new science and developments in our changing environment. United Nations Environment Programme; 2010.

[12] Hansen J, et al. Climate impact of increasing atmospheric carbon dioxide. Science 1981;213(4511):957–66.

[13] Maslin M. Global warming: a very short introduction. OUP Oxford; 2008.

[14] Fleming J. The callendar effect: the life and work of guy stewart callendar (1898-1964). American Meteorological Society; 2013.

[15] Shapiro J. Mao's war against nature: politics and the environment in revolutionary China. Cambridge University Press; 2001.

[16] Summers-Smith D. In search of sparrows. Bloomsbury Publishing; 2010.

[17] Peng X. Demographic consequences of the great leap forward in China's provinces. Popul Dev Rev 1987;13(4):639–70.

[18] Goudie AS. Human impact on the natural environment. Wiley; 2018.

[19] Rolls EC. They all ran wild: the story of pests on the land in Australia. Angus and Robertson; 1969.

[20] ACCC. Restoring electricity affordability and Australia's competitive advantage: retail electricity pricing inquiry—final report. Canberra: Australian Competition and Consumer Commission; 2018.

[21] Khalilpour R, Vassallo A. Community energy networks with storage: modeling frameworks for distributed generation. Energy policy, economics, management & transport. Singapore: Springer; 2016.

[22] Khalilpour KR. Polygeneration with polystorage: for chemical and energy hubs. Elsevier Science; 2018.

[23] Khalilpour R. Flexible operation scheduling of a power plant integrated with PCC processes under market dynamics. Ind Eng Chem Res 2014;53(19):8132–46.

[24] Khalilpour R, Karimi IA. Investment portfolios under uncertainty for utilizing natural gas resources. Comput Chem Eng 2011;35(9):1827–37.

[25] Shipman J, et al. An introduction to physical science. Cengage Learning; 2015.

[26] McCarthy JJ, et al. Climate change 2001: impacts, adaptation, and vulnerability: contribution of working group II to the third assessment report of the intergovernmental panel on climate change. Cambridge University Press; 2001.

[27] Reviews CT. e-Study guide for: energy, environment, and climate by Richard Wolfson. Cram101; 2013. ISBN 9780393912746.

[28] Orecchini F. The era of energy vectors. Int J Hydrogen Energy 2006;31(14):1951–4.

[29] Yergin D. The prize: the epic quest for oil, money & power. Free Press; 2011.

[30] Watts DJ. A twenty-first century science. Nature 2007;445:489.

[31] Gao Y, et al. Assessing deforestation from biofuels: methodological challenges. Appl Geogr 2011;31(2):508–18.

[32] Sachs I. In: The food-energy nexus: seeking local solutions to global problems. Energy and agriculture. Proc. UN University symposium, Paris, 1982; 1984. p. 25–40.

[33] Khalilpour R, Karimi IA. In: Evaluation of LNG, CNG, GTL and NGH for monetization of stranded associated gas with incentive of carbon credit. International petroleum technology conference (IPTC), Doha, Qatar; 2009.

[34] Khalilpour R. Produced water management: an example of a regulatory gap. Society of Petroleum Engineers; 2014.

[35] Jaafar IH, et al. In: Product design for sustainability: a new assessment methodology and case studies. Environmentally conscious mechanical design; 2007.

[36] Haynsworth H, Lyons RT. Remanufacturing by design, the missing link. Prod Invent Manage 1987;28(2):24–9.

FURTHER READING

[37] Francesco C, et al. Toward a common classification approach for biorefinery systems. Biofuels Bioprod Biorefin 2009;3(5):534–46.

CHAPTER 3

Energy Hubs and Polygeneration Systems: A Social Network Analysis

Faezeh Karimi*, Kaveh Rajab Khalilpour†
*Faculty of Engineering and Information Technology, The University of Sydney, Sydney, NSW, Australia
†Faculty of Engineering and Information Technology, Monash University, Melbourne, VIC, Australia

Abstract

Enforced by sustainability concerns, the energy industry has been through a transformation from a conventional centralized and inflexible fossil-fuel-based system to decentralized, cleaner, and flexible systems. The concept of an energy hub and all relevant terms such as "polygeneration," "microgrid," "mesogrid," "nanogrid," "energy internet," "community energy network," "social energy network," and "virtual power plant" are introduced to address decentralized energy systems, which can utilize diverse resources and technologies in the vicinity for energy generation and storage to supply sustainable, reliable, and affordable sources of energy. This study investigates the evolutionary trends of major research topics in this field by exploring the co-occurrence network of keywords associated with publications. The analysis is based on over 3000 related publications from 1962 to 2017. The term "polystorage" is also introduced as an identical to "polygeneration" in the context of energy storage diversification.

Keywords: Energy hub, Chemical hub, Polyenergy, Multienergy, Polygeneration, Multigeneration, Microgrid, Mesogrid, Nanogrid, Energy internet, Community energy network, Social energy networks, Social network analysis, Polystorage

1 INTRODUCTION

1.1 Energy and Sustainability Challenges

Until the Industrial Revolution, life around the world was decentralized and relied mainly on the resources in the vicinity. The producers were consumers of their products (e.g., farms and agriculture), and redundancies were supplied to neighborhood community markets or shortages were remedied from the neighborhood. Energy resources were mostly renewable (e.g., wood, solar thermal, wind turbines, water mills) and humans and other creatures had learned to adapt to the variability of renewable resource availability

Polygeneration with Polystorage
https://doi.org/10.1016/B978-0-12-813306-4.00003-3

with, for example, maximum utilization of daylight for work (Chapter 1). The development of steam engines and the subsequent Industrial Revolution transformed the lifestyle by centralizing energy and commodity production systems in order to benefit from the economy of scale. The newborn giant industries driven by steam turbines needed reliable and economic sources of energy when renewable energies could not satisfy. Though renewable resources have several critical advantages, including abundance and relatively scattered geographic distribution, a combination of reasons, including their intermittency and limited availability, have made them among the most expensive energy sources. These constraints first result in a low capacity utilization factor and thus high investment costs (though negligible subsequent operation costs). Secondly, because of the unavailability of the energy source (solar radiation, wind, biomass, etc.) at particular times (day, week, season, etc.), either an auxiliary energy source (such as other types of generation or connection to a grid) or energy storage is required. Without this consideration, energy security and autonomy with renewables is almost impossible at both macro- and microlevel.

As such, the need for continuous sources of energy led to the utilization of fossil fuels, and gradually the global energy supply was dominated by cheap and reliable fossil fuels which beat most renewable energy technologies out of the game. Fig. 1 shows this historical transition. Although it is focused on the United States, it can be extrapolated to the global energy history. According to Fig. 1, until around one and a half centuries ago, wood was the only energy source. Coal was explored and gradually became the

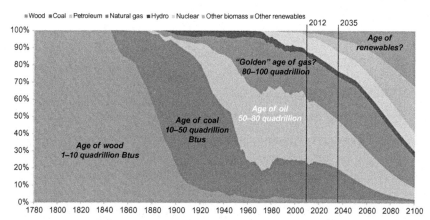

Fig. 1 Evolution of the US primary energy mix from 1780 to the present, and the business as a usual projection out to 2100 [2].

dominant source of energy in the first half of the 1900s. The age of coal continued until oil and gas joined the mix. Gradually, along with these energy sources, hydro and nuclear sources diversified the energy portfolio. By the second half of the 20th century, however, concern about the environmental consequences of fossil fuels began to be expressed globally. The energy crisis of the 1970s brought renewable energies to the front of discussions in energy-dependent countries. As the motivation during this period was mostly on energy security and least on sustainability, the fall of oil prices in the 1980s further decreased the competitiveness of renewable energy technologies. However, the severe climate change and ozone layer depletion caused by greenhouse gases (GHG) emissions led to the two frameworks of the Montreal Protocol (1987, for ozone-depleting substances) and the Kyoto Protocol (1997, mainly for CO_2) and also establishment of the United Nations Framework Convention on Climate Change (UNFCCC) in 1992 [1].

After a few decades of quests for practical solutions, the sustainable energy policy for mitigating the environmental crises is considered to have three pillars: (1) efficiency improvement, (2) replacing fossil fuels with clean energy sources, and (3) carbon capture, storage, and utilization [3, 4]. Efficiency improvement is an important but limited step toward a clean and sustainable environment. The advantage of this approach lies in its contribution to both energy demand and climate change mitigation (CCM) objectives. Renewable energies (undoubtedly), and nuclear energy (arguably) are superior energy sources compared to fossil fuels, due to their minimum associated CO_2 emissions and other pollutants. Although these energy sources are better methods for solving carbon emission problems, most of them have not been technoeconomically competitive with fossil fuels in various geographies. As such, for the time being, the development of most renewable energy technologies requires government support. This limitation brings about the third alternative solution, which is carbon capture, storage, and utilization (CCSU). This is an intermediate and bridging option while zero-emission technologies are being developed. The energy industry must rely on fossil fuels for some more time.

Therefore, current energy industries are moving away from traditional fossil-fuel-dominant systems to more complex and integrated systems in which conventional energy systems are hybridized with renewable energy technologies as well as GHG abatement schemes such as CCSU. These have led to the development of integrated energy systems such as "polygeneration" and "energy hubs." For instance, the term "polygeneration" appears to have been first coined by the NASA (National Aeronautics and Space Administration) in attempts to develop a more sustainable coal-based power generation system, which converts coal to syngas, to subsequently produce power as well as hydrogen and several other synthetic fuels [5]. This polygeneration process is known as an integrated gasification combined cycle (IGCC) [6] (Chapter 7). The terms "microgrid," "mesogrid," "nanogrid," "energy internet," "community energy network," "social energy network," and "virtual power plant" are similar concepts, which have been introduced over the last decade with more focus on electricity grids, especially with renewable energy integration [7]. Here, we use keyword analysis to investigate the history of this field and identify the research trends.

1.2 History of Hub

The core meaning of the term "hub" is centrality. Its origin is unknown, but according to the Oxford dictionary, in the 16th century, it referred to the shelf at the side of a fireplace used for heating pans (OED 2017). Later, it was referred to as the "central part of a wheel, rotating on or with the axle, and from which the spokes radiate." Today this word is metaphorically used for "*the effective centre of an activity, region, or network*" where critical actions occur or decisions are made. It is also synonymous with the center of activity, core, kernel, heart, focus, focal point, middle, and nucleus. Today, the hub is not only used for the physical center of activities but also refers to virtual centers such as "social hubs" in social media [8].

In any network and supply chain design, identification of the location of a hub is a complex problem. It is claimed that hub location was born as a new academic field in 1986 by the two works of Morton E. O'Kelly [9, 10]. The initial motivation behind hub location was the transportation (airlines, trucking, etc.) deregulation of the United States in the 1970s, which reduced entry barriers and led to the development of new large-scale transportation firms. These firms were interested in optimizing their terminal locations to maximize the use of their facilities and carriers for economic benefits [11]. Today, we can define the hub location problem as locating consolidating facilities when flows (passengers, cargoes, mail, bits (electronic data), electrons, fluids, etc.) must be sent from origin to destination nodes through network paths (e.g., roads, railways, airlines, data cables, power lines, telecommunications, and pipelines) with the most technoeconomically efficient approach. This generally leads to the identification of one or more central nodes. With hubs, rather than developing paths between all origin-destination (OD) nodes, OD nodes with weak connections can be indirectly linked through a hub to avoid the increased cost of building underutilized paths [12]. Therefore, hub location problems are indeed network design problems [13], with the key elements being OD nodes for which network paths should be identified with four key features: (1) identification of hub nodes, (2) connection paths between hubs, (3) indirect connection paths between nonhub OD nodes through hub nodes, and (4) direct connection paths between nonhub OD nodes [11].

The application of hub location theory in the energy and chemical sectors is fairly new. According to Scopus records, the first incidence of "energy hubs" occurred in 1995 (see Table 1) [14]. Most of the initial uses referred to regional energy hub developments around the world, such as the Caspian Sea energy hub [14] or Singapore's Jurong Island as a chemical hub [15, 16].

In recent years, the terms "multienergy" and "polyenergy" have received increasing attention from renewable energy and microgrid communities for addressing energy systems with multiple generation technologies (see Fig. 2). However, it seems that these terms were originally used in the discipline of plasma physics with reference to laser-matter interaction for multienergy ion implantation [17]. Thus, they have a longer history of use, dating back to the early 1960s (see Table 1).

Another common term in the energy sector is "multigeneration." However, this term also seems to have originated from the field of biology and social science, where it refers to behavioral or biological interactivities across multiple generations of living things including humans [18, 19, 24]. In the next section, we carry out keyword analysis to identify the historical development of these keywords and their interactions.

This chapter focuses on energy and chemical hubs as well as polygeneration systems and investigates how the associated research has evolved over the past few decades based on keyword co-occurrences.

2 DATA ANALYSIS APPROACH

The publication dataset used in this study is built based on the publication records extracted from Scopus.com. This bibliometric website was searched for publications having "energy hub," "energy-hub," "chemical hub," "chemical-hub," polyenergy, poly-energy, multi-energy, multienergy, "multi energy," polygeneration, "poly-generation," "multigeneration," and "multigeneration" in their titles, abstracts, or keywords. The terms

Table 1 Keyword birth for energy hub and polygeneration field

Keyword	Year first appeared	Introduced by	Article(s) title	Reference
Energy hub or energy-hub	1995	Barylski	Russia, the West and the Caspian Energy Hub	[14]
Chemical hub or chemicals hub or chemical-hub	2000	- Anon - Chemical market reporter	- Storage activity heats up - An industrial hub - The Jurong jewel	[15,16,20]
Polyenergy	1987	Tomashov et al.	The corrosion behavior of the surface layers of a titanium-palladium alloy obtained by polyenergy implantation of Pd ions in titanium	[21]
Poly-energy or poly energy	1986	Van Gool	Poly-energy Handbook (Poly-energyie Zakboekje)	[22]
Multi energy or multi-energy	1962	Holland and Paul	Effect of pressure on the energy levels of impurities in semiconductors. III. Gold in germanium	[23]
Multienergy	1964	Giesing	Vorticity and kutta condition for unsteady multienergy flows	[24]
Polygeneration	1981	Burns	- Integrated gasifier combined cycle polygeneration system to produce liquid hydrogen - Polygeneration at the John F. Kennedy Space Center, project overview	[5,25]
Poly-generation	1995	Peng	Comparison of growth dynamics between first generation and regeneration of *Pinus massoniana*	[26]
Multi-generation	1958	Mendell and Fisher	A multi-generation approach to treatment of psychopathology	[27]
Multigeneration	1964	Hutchinson et al.	- Flour treated with chlorine dioxide: Multigeneration tests with rats - Progress report on multigeneration reproduction studies in rats fed butylated hydroxytoluene (BHT)	[18,19]

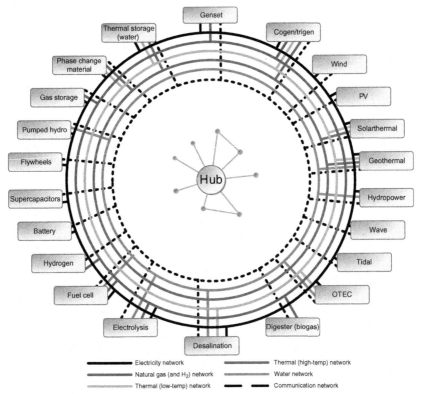

Fig. 2 Illustration of an energy hub, which is a true internet of things (IoT).

polygeneration and multigeneration are also used for referring to multiple generations in human, animal, or plant species. To avoid this confusion, we filtered out references related to the fields of medicine and agriculture.

Fig. 3 shows the publication trend over time. As evident, publications were issued from 1962 to 2017. The search was conducted in late December 2017 and resulted in 3338 publication records in English. This study explores the changes in the publications over different periods within 55 years.

The keywords, carefully assigned to scientific papers by authors, describe the main topics and research foci of the scientific papers. Exploring these keywords and identifying the most common terms and their associations in a discipline can unfold its knowledge structure [28, 29]. Coword analysis is a technique, which uses keywords to build a network establishing relations between concepts and ideas in a corpus. Two keywords are linked in the network if they appear together in a scientific paper [28, 30]. Researchers

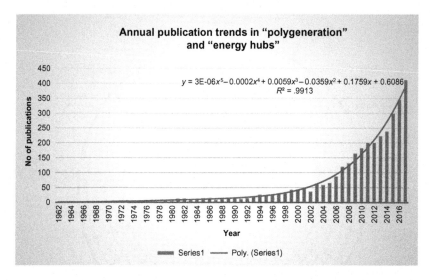

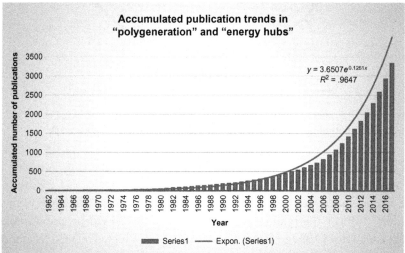

Fig. 3 Publication frequency of energy hubs and polygeneration systems ("energy hub," "energy-hub," "chemical hub," "chemical-hub," polyenergy, poly-energy, multi-energy, multienergy, "multi energy," polygeneration, "poly-generation," "multigeneration," and "multigeneration"). *Top*: Annual profile; *lower*: accumulated profile.

have used this approach in different fields including renewable energy [28], service innovation [31], hydrogen energy and fuel cells [32], ecology [33], climate change [1], polymer chemistry [34], zoonotic disease propagation [35], simulation optimization of supply chains [36], and information retrieval [37].

This study examines keyword co-occurrence networks (keywords networks for short). Each network consists of a set of nodes and ties connecting certain nodes together. In this study, nodes are keywords, and a tie between two keyword nodes is an indication of the two keywords appearing in the list of keywords in a paper.

The MATLAB 2017a software package was used to write programs translating information regarding co-occurrences of keywords into adjacency matrices representing the keywords networks. VOSviewer 1.6.6 is used for mapping countries and keywords networks in different time intervals. VOSviewer maps are distance-based rather than graph-based. The two types of maps differ in the interpretation of the distance allocated between the nodes. In distance-based maps, this distance represents the strength of the relationship between the nodes. In graph-based maps, the distance does not necessarily indicate relationship strength. In more detail, in VOSviewer maps, nodes are placed in such a way that the distance between each pair of nodes (e.g., i and j) represents their similarity (s_{ij}) as accurately as possible. The similarity measure used by VOSviewer (also known as the association strength) is calculated as $s_{ij} = c_{ij}/(w_i\,w_j)$ where c_{ij} is the number of nodes i and j's co-occurrences, and where w_i and w_j are the total number of occurrences/co-occurrences for nodes i and j, respectively.

VOSviewer provides different views of network maps. In the label view, each node is represented by a circle and its corresponding label. The sizes of the circles and labels vary based on the weight of the nodes. The weight of a node is assigned as the total strength of all the links of the node. To avoid overlapping, the set of all labels is shown partially.

In the density view, labels are represented in a similar way to the label view. The colors of the labels are by default between red to blue and are determined based on the density of the nodes. Colors close to red represent higher density, while colors close to blue demonstrate lower density. Density of a node is defined based on the number of its neighboring nodes and their weights. This view can be used to identify important areas of the map [38, 39].

3 PUBLICATION TREND OF ENERGY HUBS AND POLYGENERATION SYSTEMS

3.1 Publications by 1990

From the annual publication profile (Fig. 3), it can be seen that there was an insignificant number of publications under the topic of study from the first publication incidence (i.e., 1962) until the 1980s. There are only 166 publications by the year 1990 (5% of total records). An increasing trend of publication started from this period. Therefore, it would not be anticipated to notice a meaningful developed field of research by studying the literature up to 1990. Fig. 4 proves this expectation by revealing that the field's keyword network is very limited and we cannot identify any of the keywords of interest from this network, which contains keywords mostly relevant to the field of plasma physics.

3.2 Publications by 2000

The topic of the study showed a linear increasing pace from the 1980s and, according to Fig. 3, this continued during the 1990s. By the year 2000, the total number of publication records reached 406 records (~12% of total records). The density map of the keywords network until 2000 (Fig. 5) clearly shows the emergence of new fields "multigeneration" (Fig. 5). However, as evident from the other keywords in this network, this term was also used in medical fields (in connection with keywords "rat," "reproduction," and "open field [test]").

3.3 Publications by 2010

As evident from the annual publication profile (Fig. 3), from the early 2000s, this field observed exponential growth. Whereas for almost four decades

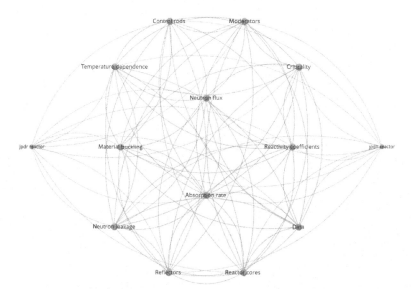

Fig. 4 Keywords network map for all the records until 1990.

Fig. 5 Keywords network map (heat map) for all records until 2000.

(from the first record in 1962 until the year 2000) only 406 publications had appeared, within the decade to 2010 the publication records tripled to 1231. The density map of the keywords network until 2010 (Fig. 6) clearly shows the development of a distinct cluster with "polygeneration" as the overarching keyword (see the centre of Fig. 6). Fig. 7 shows the connection map for a multilayer network with other keywords.

As evident from Fig. 7, the other strong keywords in this cluster include "multigeneration," "polygeneration system," "optimization," "ion implantation," "cogeneration," "simulation," "simulation," "trigeneration," "modeling," "energy hub," and "distributed generation." These prove that polygeneration emerges to dominantly represent the energy field.

3.4 Publications by 2017

In less than a decade, from 2010 to 2017, the number of publication records almost tripled and reached 3337 (Fig. 3). The density and network maps of the keywords until 2017 (Figs. 8 and 9) clearly show the development of a few intensely integrated research fields. Further to "polygeneration," which appeared in the previous period, two new clusters, "energy hub" and "exergy and efficiency," have emerged. The energy hub cluster is strongly tied with the polygeneration cluster and, in fact, they together form a cluster. The other cluster is "exergy and efficiency," which has an obvious and interesting implication. We already discussed that energy efficiency is one of the three pillars of climate change mitigation. Polygeneration systems and energy hubs are solutions for transforming the energy industry into sustainable systems. As such, energy hubs and energy efficiency are two sides of a coin and appear as two strong clusters.

In Table 2 we list the top 50 keywords, which had emerged by the end of 2017. Further to our earlier discussions, it is evident that "multigeneration" and "multienergy" keywords have also been popular. Optimization and exergy appear to be two of the top keywords with inherent implications as key approaches for designing and operating energy hubs with the highest efficiency and exergy. The terms "microgrid" and "smart grid," being close to energy hub in definition, are also among the top ten keywords. Gasification has a central role in polygeneration processes and, as mentioned in the introduction section, the term polygeneration first appeared with integrated gasification combined cycle (IGCC) systems. This could justify its appearance within the top ten keywords. The keywords related to renewable energy are the most dominant terms within the top 50 keywords.

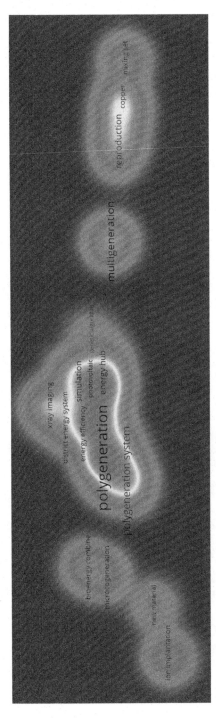

Fig. 6 Keywords network map (heat map) for all the records until 2010.

(Continued)

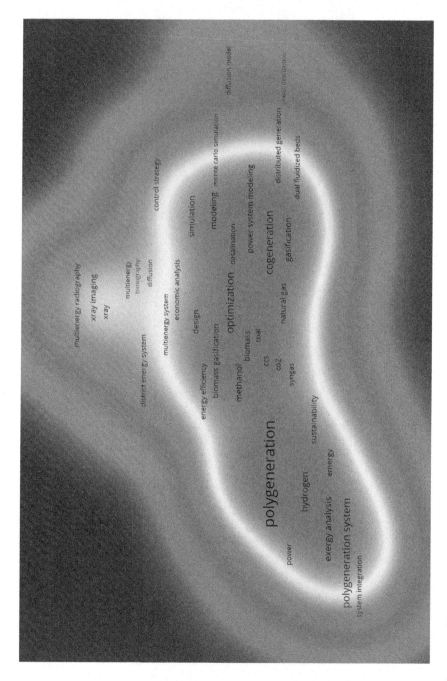

Fig. 6, cont'd

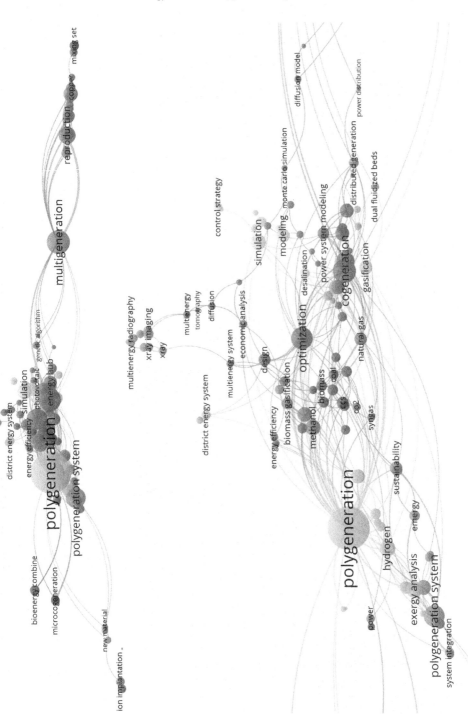

Fig. 7 Keywords network map till 2010: focus on strong clusters.

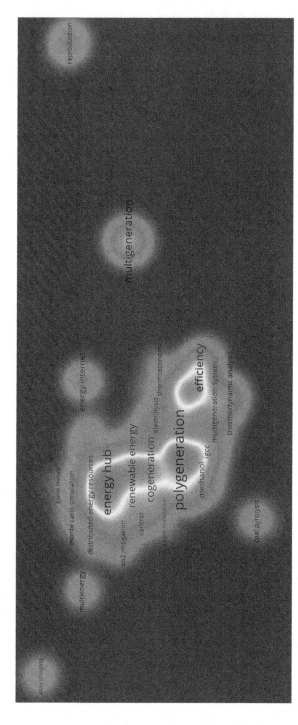

Fig. 8 Keywords network map for all records until 2017.

(Continued)

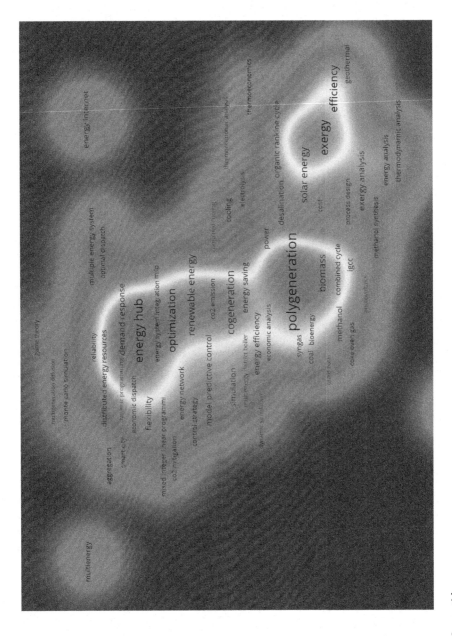

Fig. 8, cont'd

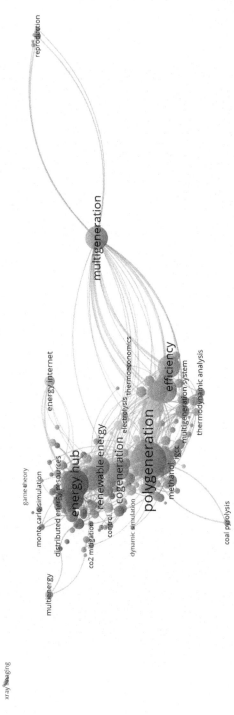

Fig. 9 Keywords network map till 2017: focus on strong clusters.

(Continued)

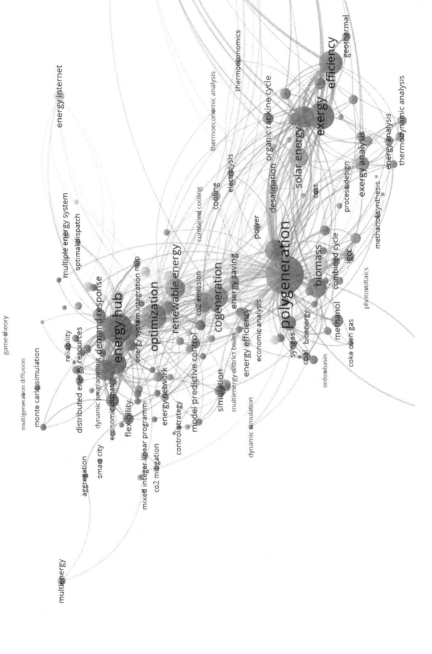

Fig. 9, cont'd

This implies that the increased uptake of renewable energy resources is the key driver behind the development of energy hubs and the polygeneration field. A complement to renewable energy technologies is energy storage (the 15th top keyword in Table 2). With the growing share of renewable technologies, energy storage will also diversify, and various types of storage systems will emerge (Chapter 4). As such, Khalilpour (*Energy Hubs and*

Table 2 List of most frequently occurring keywords over time

Keyword	No. occurrence	Keyword	No. occurrence
Polygeneration	306	CO_2 capture	29
Energy hub	180	Simulation	29
Multigeneration	108	Combined heat and power	27
Optimization	102	Bond graph	26
Multienergy system	73	Multienergy	25
Microgrid	73	Demand response	25
Smart grid	67	Energy efficiency	24
Exergy	66	Coal gasification	24
Polygeneration system	60	Modeling	23
Gasification	58	IGCC	22
Renewable energy	53	Energy management system	21
Biomass	52	Sustainability	20
Cogeneration	51	Hydrogen production	19
Efficiency	49	Reproduction	19
Solar energy	49	Multiple energy carriers	19
Energy storage	46	Methanol	19
Energy	43	Flexibility	18
Hydrogen	41	Model predictive control	18
Trigeneration	37	Risk assessment	18
Energy internet	34	Desalination	17
Integrated energy system	34	Optimal power flow	17
Energy management	34	Natural gas	17
Distributed generation	33	System integration	17
Exergy analysis	32	Geothermal energy	16
Ion implantation	30	Uncertainty	16

Polygeneration Systems: A Social Network Analysis) has also introduced a "polystorage" keyword as complementary to the polygeneration which will emerge over time.

4 CONCLUSION

We investigated the keywords of over 3000 publications relevant to energy hub, chemical hub, polyenergy, multienergy, polygeneration and multigeneration. The keywords clearly show the emergence of a new field overarching "polygeneration" and "energy hub," within which renewable technologies and systems engineering (e.g., modeling, simulation, optimization, planning, scheduling, efficiency, and exergy) play key roles. While several keywords such as microgrid, mesogrid, nanogrid, energy internet, community energy network, and social energy networks have been introduced, we recommend "energy hub" as the most suitable overarching keyword across various research and commercial communities.

REFERENCES

[1] Karimi F, Khalilpour R. Evolution of carbon capture and storage research: trends of international collaborations and knowledge maps. Int J Greenh Gas Control 2015;37:362–76.

[2] Channell J, Jansen HR, Syme AR, Savvantidou S, Morse EL, Yuen A. Energy Darwinism: the evolution of the energy industry. In: Citi GPS: global perspectives & solutions. Citigroup; 2013.

[3] Damm DL, Fedorov AG. Conceptual study of distributed CO_2 capture and the sustainable carbon economy. Energ Convers Manage 2008;49:1674–83.

[4] Khalilpour R. Multi-level investment planning and scheduling under electricity and carbon market dynamics: retrofit of a power plant with PCC (post-combustion carbon capture) processes. Energy 2014;64:172–86.

[5] Burns RK, Staiger PM, Donovan R. Integrated gasifier combined cycle polygeneration system to produce liquid hydrogen.

[6] Anon. Coal gasification-combined-cycle projects involving fuel for industrial use. Ind Heat 1981;48:13–4.

[7] Khalilpour KR, Vassallo A. Community energy networks with storage: modeling frameworks for distributed generation. Singapore: Springer; 2016.

[8] Salvi R, Turnbull J. The discursive construal of trust in the dynamics of knowledge diffusion. Newcastle upon Tyne, UK: Cambridge Scholars Publishing; 2017.

[9] O'Kelly M. The location of interacting hub facilities. Trasnport Sci 1986;20:92–106.

[10] O'Kelly ME. Activity levels at hub facilities in interacting networks. Geogr Anal 1986;18:343–56.

[11] Campbell JF, O'Kelly ME. Twenty-five years of hub location research. Transplant Sci 2012;46:153–69.

[12] Etemadnia H, Goetz SJ, Canning P, Tavallali MS. Optimal wholesale facilities location within the fruit and vegetables supply chain with bimodal transportation options: an LP-MIP heuristic approach. Eur J Oper Res 2015;244:648–61.

[13] Contreras I, Fernández E. General network design: a unified view of combined location and network design problems. Eur J Oper Res 2012;219:680–97.

[14] Barylski RV. Russia, the west, and the Caspian energy hub. Middle East J 1995;49:217–32.

[15] Chemical-Market-Reporter. An industrial hub. Chem Market Rep 2003;264:10.

[16] Chemical-Market-Reporter. The jurong jewel. Chem Market Rep 2004;265:27.

[17] Dening W, Weiyuan W. A study of multi-energy ion implantation. J Electron 1987;4:39–45.

[18] Hutchinson JB, Moran T, Pace J. Flour treated with chlorine dioxide: multigeneration tests with rats. J Sci Food Agr 1964;15:725–32.

[19] Frawley JP, Kohn FE, Kay JH, Calandra JC. Progress report on multigeneration reproduction studies in rats fed butylated hydroxytoluene (BHT). Food Cosmet Toxicol 1965;3:377–86.

[20] Anon. Storage activity heats up. Oil Gas J 2000;98:74–5.

[21] Tomashov ND, Tashlykov IS, Zhil'tsova OA, Chernova GP, Guseva MI, Vladimirov BG. Corrosion behavior of the surface layers of a Ti-Pd alloy obtained by polyenergetic implantation of Pd+ ions in titanium. Prot Met 1988;23:579–82.

[22] Gool WV. Poly-energy handbook (poly-energyie zakboekje). The Hague: PBNA; 1986.

[23] Holland MG, Paul W. Effect of pressure on the energy levels of impurities in semiconductors, III Gold in Germanium. Phys Rev 1962;128:43–55.

[24] Giesing JP. Vorticity and kutta condition for unsteady multienergy flows. J Appl Mech Trans ASME 1964;36:608–14.

[25] Manfredi L, Gutkowski G. Polygeneration at the john f. kennedy space center, project overview. In: Veziroglu TN, Taylor JB, editors. Hydrogen energy progress V, proceedings of the 5th world hydrogen energy conference. New York, NY, USA Toronto, ON, Canada: Pergamon Press; 1984. p. 183–91.

[26] Peng S. Comparison of growth dynamics between first generation and regeneration of Pinus massoniana. Chin J Appl Ecol 1995;6:11–3.

[27] Mendell D, Fisher S. A multi-generation approach to treatment of psychopathology. J Nerv Ment Dis 1958;126:523–9.

[28] Romo-Fernandez LM, Guerrero-Bote VP, Moya-Anegon F. Co-word based thematic analysis of renewable energy (1990-2010). Scientometrics 2013;97:743–65.

[29] Yi S, Choi J. The organization of scientific knowledge: the structural characteristics of keyword networks. Scientometrics 2012;90:1015–26.

[30] Ravikumar S, Agrahari A, Singh SN. Mapping the intellectual structure of scientometrics: a co-word analysis of the journal Scientometrics (2005–2010). Scientometrics 2015;102:929–55.

[31] Zhu WJ, Guan JC. A bibliometric study of service innovation research: based on complex network analysis. Scientometrics 2013;94:1195–216.

[32] Chen YH, Chen CY, Lee SC. Technology forecasting of new clean energy: the example of hydrogen energy and fuel cell. Afr J Bus Manage 2010;4:1372–80.

[33] Budilova EV, Drogalina JA, Teriokhin AT. Principal trends in modern ecology and its mathematical tools: an analysis of publications. Scientometrics 1997;39:147–57.

[34] Callon M, Courtial JP, Laville F. Co-word analysis as a tool for describing the network of interactions between basic and technological research – the case of polymer chemistry. Scientometrics 1991;22:155–205.

[35] Hossain L, Karimi F, Wigand RT, Crawford JW. Evolutionary longitudinal network dynamics of global zoonotic research. Scientometrics 2015;103:337–53.

[36] Huerta-Barrientos A, Elizondo-Cortes M, de la Mota IF. Analysis of scientific collaboration patterns in the co-authorship network of simulation-optimization of supply chains. Simul Model Pract Th 2014;46:135–48.

[37] Ding Y, Chowdhury GG, Foo S. Bibliometric cartography of information retrieval research by using co-word analysis. Inform Process Manag 2001;37:817–42.

[38] van Eck NJ, Waltman L. VOSviewer: a computer program for bibliometric mapping. Pro Int Conf Sci Inf 2009;2:886–97.

[39] van Eck NJ, Waltman L. Software survey: VOSviewer, a computer program for bibliometric mapping. Scientometrics 2010;84:523–38.

CHAPTER 4

Single and Polystorage Technologies for Renewable-Based Hybrid Energy Systems

Zainul Abdin*, Kaveh Rajab Khalilpour†
*Faculty of Information Technology, Monash University, Melbourne, VIC, Australia
†Faculty of Engineering and Information Technology, Monash University, Melbourne, VIC, Australia

Abstract

The uptake of renewable energy technologies is accelerating. This has led to the emergence of generation intermittency and fluctuation along with the already existing load variability. Consequently, there are critical concerns about the stability and reliability of future energy networks. Energy storage systems provide a wide array of technological approaches to manage power supplies in order to create a more resilient energy infrastructure and bring cost savings to utilities and consumers. This chapter provides an in-depth analysis of different electrical energy storage technologies currently deployed around the world. It also discusses the need for polystorage systems to hybridize the advantages of multiple storage technology while avoiding the shortcomings when used alone.

Keywords: EES, DSM, Pumped hydroelectric storage, Compressed air energy storage, Flywheel, Electrochemical storage, Electrical storage, Thermal storage, Chemical storage, Hybrid energy storage

1 INTRODUCTION

More than 80% of primary energy supply in the world comes from fossil fuels, and the demand has been projected to grow at about 2.3% per year from 2015 to 2040 [1]. This is threatening to further increase carbon dioxide levels and the average global temperature. At the United Nations Climate Change Conference in December 2015, world leaders agreed to limit the global average temperature rise to 2°C and to work toward an even lower rise of 1.5°C [2]. To minimize the adverse effects of global warming and consequent climate change, the integration of sustainable and renewable energy technologies is vital for the environment. The key challenge with renewable energy technologies is their variability—adding generation

Polygeneration with Polystorage
https://doi.org/10.1016/B978-0-12-813306-4.00004-5
77

intermittency and fluctuation to the existing load variability leads to critical concerns about the stability and reliability of future energy networks.

Different alternatives have been proposed to mitigate this problem including electrical energy storage (EES); demand side management (DSM) with load shifting; and interconnection with external grids [3]. All options have their niche potential for addressing the challenge; however, as renewable energy penetration into the grid increases, EES becomes the most critical option—not only for power quality, but also for energy management. EES can maximize the utilization of existing generation and transmission infrastructures while also preventing costly network upgrades. In relation to conventional power production, EES can improve overall power quality and reliability, which is becoming increasingly important for modern commercial applications.

EES is not a new technology or concept; it has been practiced for over a century. Twenty years after the invention of rechargeable lead acid batteries in 1859 [4], Thomas Edison invented the light bulb in 1879 and later developed the first centralized power plant in 1882 in New York City's financial district to light the shops and attract customers [5]. Demand soon increased, and lead acid batteries were found as a solution for storing electricity at low demand times and selling it to the shops at peak evening times. In 1896, a 300-ton, 400-kWh lead acid battery was used at a hydropower station to avoid outages from equipment breakdowns [6].

Over the last one and half centuries, battery storage has been developing along with other energy storage types, each with a certain learning rate. The objectives of electricity storage have also far exceeded the initial intention of peak shaving or short-term outage prevention [7]. Today, EES is used for many other reasons, such as delaying capacity/network expansion, regulating frequency and balancing voltage (preventing brownouts) [8]. As such, each energy storage technology is suitable for a given objective. Fig. 1 shows the role of EES at various locations across an electricity network.

At present, an electric power infrastructure functions largely as a just-in-time inventory system, in which a majority of energy is generated and then transmitted to the user as it is consumed. Without the ability to store energy, there must be sufficient generation capacity on the grid to handle peak demand requirements, despite the likelihood that much of that capacity sits idle daily as well as for large portions of the year. Correspondingly, the transmission and distribution systems must also be sized to handle peak power transfer requirements, even if only a fraction of that capacity is used during

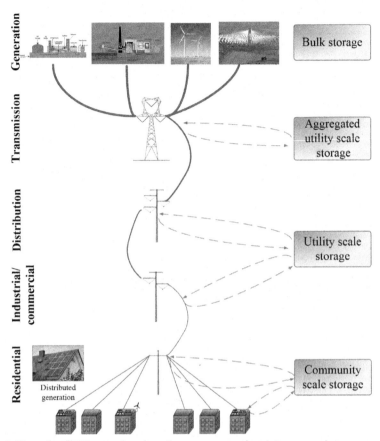

Fig. 1 The role of EES at various locations across an electricity network.

most of the year. Operationally, electrical power generation must be continuously ramped up and down to ensure that the delicate balance between supply and demand is maintained. By decoupling generation and load, grid energy storage simplifies the balancing act between electricity supply and demand and overall grid power flow. EES systems have potential applications throughout the grid, from bulk energy storage to distributed energy functions. Table 1 shows the functional uses of EES systems and their associated value metrics.

EES technologies are usually categorized based on the application time scale: instantaneous (less than a few seconds); short term (less than a few minutes); midterm (less than a few hours); and long term (days) [9]. A detailed background of the historical development of various energy

Table 1 Functional uses of EES systems and their associated value metrics

	Functional use	Value metric
Bulk energy	Electric energy time shift	The price differential between energy prices during charge and discharge, which includes the following: • Arbitrage • Renewable energy firming and integration • Electric supply capacity: the avoided cost of new generation capacity (procurement or build capital cost) to meet requirements.
Transmission & distribution	Transmission upgrade deferral	The avoided cost of deferred infrastructure
	Distribution upgrade deferral	The avoided cost of deferred infrastructure
	Transmission voltage support	The avoided cost of procuring voltage support services through other means
	Distribution voltage support	The avoided cost of procuring voltage support services through other means
Reserve	Synchronous	Regulated env[a]: the avoided cost of procuring reserve service through other means Market env: the market price for synchronous reserve
	Nonsynchronous	Regulated env: the avoided cost of procuring reserve service through other means Market env: the market price for nonsynchronous reserve
	Frequency regulation	Regulated env: the avoided cost of procuring reserve service through other means Market env: the market price for frequency regulation service
Customer	Power reliability	The avoided cost of new resources to meet reliability requirements
	Power quality	The avoided cost of new resources to meet power quality requirements, or avoided penalties if requirements are not being met

[a] *Env*, environment.

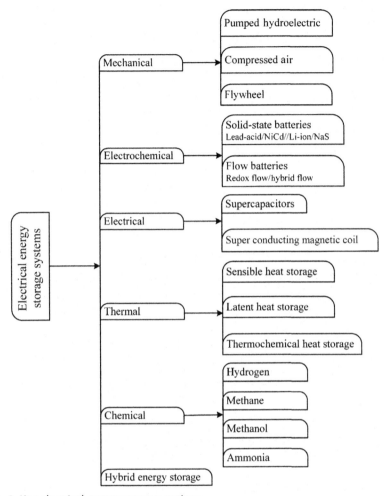

Fig. 2 Key electrical energy storage options.

storage options can be found in the *Electricity Storage Handbook* published by Sandia National Laboratories [10].

There are numerous potential energy storage options for the electricity sector, each with unique operational, performance, cycling, and durability characteristics. A widely used approach for classifying EES is the determination according to the form of energy used (Fig. 2). A comprehensive picture of the state-of-the-art technologies available for EES is addressed in this chapter.

2 MECHANICAL STORAGE

2.1 Pumped Hydroelectric Storage

Dams, built on the natural pathways of rainwater to create reservoirs, have been used as energy harvesting and storage systems for a long time in human civilization history. Thus, water can be collected during high-rain seasons to be used in a planned way, not only for water supply, but also for power generation. However, a pumped hydroelectric storage (PHS) system is a relatively new system designed for load management. With the development of inflexible, large-scale power plants, such as nuclear in the 20th century, there was a necessity for energy supply demand misbalance management. Technologies were required to respond quickly to market signals by absorbing the surplus electricity from the grid and supplying it during demand periods. Given the high costs of large-scale battery storage in those days, PHS became a successful alternative. A PHS system requires development of a second reservoir, called a lower reservoir, in a hydro power plant (Fig. 3). During electricity oversupply periods, the energy is used to pump the water from the lower reservoir to the upper one. Therefore, it stores energy in the form of water; that is, gravitational potential energy, which is later converted into electrical energy by allowing the water to flow back from the upper to the lower reservoir through a turbine generator. In this manner, energy is converted from electrical to kinetic to gravitational potential, then back to kinetic, and finally back to electrical again. PHS is now the current dominant energy storage technology. As of 2017, it accounted for >96% of worldwide bulk storage capacity and contributed to about 3% of a global generation with a total installed nameplate capacity of over 168 GW [11].

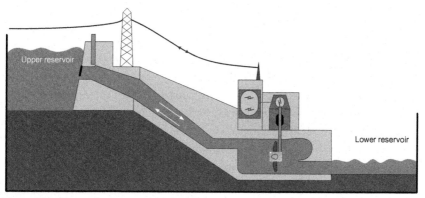

Fig. 3 Schematic of a pumped hydroelectric storage system.

The roundtrip efficiency of this process depends on the efficiencies of pump, motor, turbine, and generator as well as evaporation rates. Various PHS plants exist with power ratings ranging from 1 MW to 3003 MW, with approximately 70%–85% cycle efficiency and a lifetime of >40 years [12]. Due to the low energy density of pumped storage schemes, they are only appropriate for large-scale grid applications. PHS can be used to provide substantial benefits to the energy system including frequency control, ramping/load leveling and peak shaving, load following, and provision of standby reserve capacity.

The main downside of PHS is the need for favorable geography; both a lower and a higher reservoir are required as well as favorable landscape between them. The storage capacity depends on the sizes of the reservoirs, the power of the flow and the head of the water. The energy density is proportional to the height (head) between the two reservoirs; as such, it is only viable to construct a PHS facility if a certain head is available between the two reservoirs.

PHS projects have been providing energy storage capacity and transmission grid ancillary benefits in the United States (US) and Europe since the 1920s [13]. In late 2014, there were 51 PHS projects operating in the US with a total capacity of 39 GW [14]. Recently, with the advance of technology, investigations into underground PHS (U-PHS) and seawater PHS schemes are underway, such as a 300-MW seawater PHS scheme in Japan [12]. There are several large-scale pumped storage activities around the world, including Germany (1400-MW Atdorf plant) [15], Switzerland (1000-MW Linthal 2015 project) [16], and Portugal (136-MW Alvito project) [17]. In addition, wind or solar power generation coupled with PHS is now being developed, such as the Ikaria Island power station (Greece) [12].

2.2 Compressed Air Energy Storage

Compressed air energy storage (CAES) is largely equivalent to PHS in terms of applications, output, and storage capacity. Unlike a PHS, where energy is storage in elevation, CAES energy is stored in ambient air by compressing it and then storing it in underground caverns or aboveground facilities. When electricity is required, the pressurized air is heated and expanded in a turbine for power production.

CAES comprises a number of key subsystems such as compression, air storage, heat regeneration, and electric power generation (Fig. 4). The compression unit uses surplus electricity (from renewable or nuclear generators)

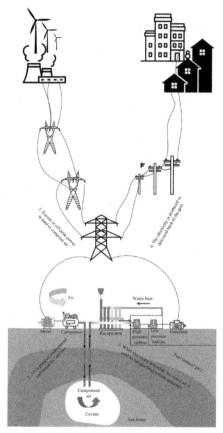

Fig. 4 Schematic of a compressed air energy storage system [4].

to drive compressors to produce high-pressure air, along with high-temperature compression thermal energy. The air storage subsystem is used to store high-pressure air (molecular potential energy) generated by the compressor. During the expansion process, the heat regeneration subsystem heats the high-pressure air, improving the entrance enthalpy of air expansion. The electric power generation subsystem is employed to drive the turbine to generate electricity with high-temperature air and, therefore, to achieve the ultimate conversion of molecular potential energy and thermal energy to electricity.

A CAES system stores air by three possible ways: diabatic, adiabatic, and isothermal. *Diabatic CAES* is basically the same as a conventional gas turbine except that the compression and expansion stages occur at different time periods. For example, when electricity is in excess, air is compressed and stored in a reservoir, and when electricity is needed, air is heated with natural gas and expanded through a turbine. Worldwide, there are two diabatic CAES plants in operation: the Huntorf plant (290 MW) in Germany and the McIntosh plant (110 MW) in Alabama, USA [18]. The Huntorf plant was commissioned in 1978 to become the world's first CAES plant. The McIntosh plant incorporates a recuperator to reuse the exhaust heat energy [9]. When compared with the Huntorf plant, this recuperator reduces fuel consumption by 22%–25% and improves cycle efficiency by 42%–54% [12]. These two CAES plants have consistently shown good performances with 91.2%–99.5% starting and running reliabilities [12].

Adiabatic CAES is a system in which the heat produced due to the compressing of air is captured via a thermal energy storage system. When the electricity is needed, this stored heat is returned to the air before expansion through the turbine. This method does not require the use of premium fuels to heat the compressed air before expansion as the diabatic method requires. The world's first adiabatic CAES system is ADELE at Saxony-Anhalt in Germany, with a storage capacity up to 360 MWh and an electric output of 90 MW, aiming for around 70% cycle efficiency [9].

Isothermal CAES is an evolving technology that attempts to overcome some of the limitations of conventional CAES (diabatic or adiabatic). For example, current CAES systems use turbomachinery to compress air to around 70 bars before storage and, in the absence of intercooling, the air heats up to around 900 K, making it impossible to process and store. Isothermal CAES is technologically challenging, since it requires heat to be removed continuously from the air during the compression cycle and added continuously during expansion to maintain an isothermal process. There are currently no isothermal CAES plants available around the world, but several possible solutions have been proposed based on reciprocating machinery with a cycle efficiency of 70%–80% [9].

2.3 Flywheel

With the increasing share of renewable energy technologies in electricity networks, one of the emerging challenges of future smart grids is *inertia reduction*. Inertia is literally defined as "the resistance of any physical object to any change in its state of motion." In power systems, the large rotating masses of the synchronous generators protect the grid by overcoming the immediate imbalance between power supply and demand. There is an ongoing debate that, with such a high penetration of renewable energies (without rotating mass) to the grid, the inertia may decline to a level that puts the entire grid at risk. A counterargument is that although inertia will decline, there are several power electronic components, which can be used to protect the grid. One such option is the flywheel.

Flywheel energy storage (FES) follows the rotating mass principle to store energy in the form of rotational kinetic energy. An FES comprises a spinning rotor, motor generator, bearings, a power electronics interface, and a vacuum chamber or housing (Fig. 5). It is placed in a high-vacuum environment for reducing wind shear and energy loss from air resistance [12]. The FES is coaxially connected to a motor generator since it can interact with the utility grid through advanced power electronics. The amount of energy stored is dependent on the rotating speed of the flywheel and its inertia [19]; the flywheel is accelerated during energy storage and deaccelerated when discharging. FES systems can be classified by speed: two low-speed groups (steel material, rotation speed $<6 \times 10^3$ rpm) and one high-speed group (advanced composite materials—for example, carbon-fiber—speeds up to $\approx 10^5$ rpm) [12].

Some of the key advantages of FES are a high cycle life (hundreds of thousands); long calendar life (>20 years); fast response; high cycle efficiency (up to $\approx 95\%$ at rated power); high charge and discharge rates; high power

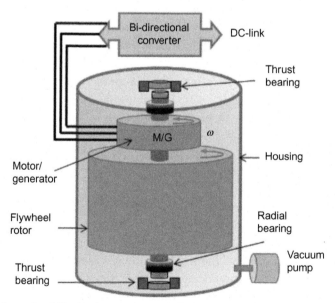

Fig. 5 Schematic of flywheel energy storage [19].

density; and low environmental impacts [19]. Additionally, FES can bridge the gap between short-term, ride-through power and long-term energy storage with excellent cyclic and load following characteristics.

FES systems can be used as substitutes for batteries, or in combination with batteries in uninterruptible power source (UPS) systems, because they can deal with shorter interruptions (subseconds), while batteries can be used for longer brownouts (seconds and minutes). This can protect batteries from frequent charges-discharges [19]. Flywheels are used in hybrid and electric vehicles for severe conditions such as harsh acceleration and uphill climbs. In 2014, VYCON Inc. installed an FES for the Los Angeles Red Line trains to recover their braking energy [12]. FES can also assist in the penetration of wind- and solar-based energy systems to improve their system stability. There has been a wide range of flywheel systems developed for integration with renewable energy systems; for example, ABB's PowerStore; Urenco Power; Beacon Power; and VYCON technology. These have all provided FES-based systems for wind- and solar-based energy [19].

3 ELECTROCHEMICAL STORAGE

Electrochemical storage can be categorized into two battery types: solid state and flow. Fig. 6 illustrates the various electrochemical storage options; Table 2 summarizes the characteristics of different battery technologies.

3.1 Solid-State Batteries

A solid-state battery is a rechargeable, portable voltaic cell. It comprises one or more electrochemical cells that accumulate and store energy through a reversible electrochemical reaction. Rechargeable batteries are produced in many different shapes and sizes, ranging from button cells to megawatt systems connected to stabilize an electrical distribution network.

3.1.1 Lead-Acid Battery

Lead-acid batteries have been used for >130 years [5] in many different applications, and they are still the most widely used rechargeable electro-chemical devices for small- and medium-scale storage applications, currently occupying >60% of the total battery market, which has not been reduced by the rapid development of Li-ion batteries and other technologies [20]. Lead-acid battery applications include vehicles (70%), communications (21%), grid and off-grid energy storage (5%), and others (around 4%) [21].

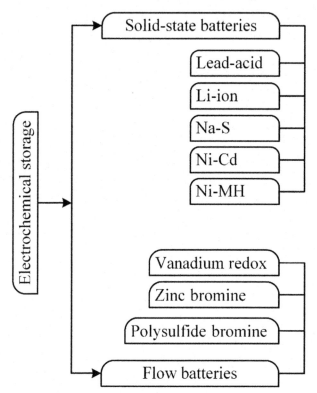

Fig. 6 Different electrochemical storage options.

Lead-acid batteries have high reliability within their lifetime, strong surge capabilities, and medium-to-high efficiency. They are usually good for uninterruptible power supplies, power quality, and spinning reserve applications [4]. However, they are poor for energy management purposes because they have a short life, require regular maintenance, have low energy density (Table 2), emit explosive gas and acid fumes, and have poor performance in cold conditions, which may require a thermal management system [4]. Although valve-regulated lead-acid batteries require less maintenance, and create fewer gas emissions and self-discharges than their nonvalve-regulated counterparts, they are primarily designed for backup power supply and telecommunication applications due to their decreased cycle life [21]. While cheap to install, the short lifetime of lead-acid batteries translates to a higher LCOE (levelized cost of energy).

Table 2 Characteristics of different battery technologies [12, 22]

Battery	Energy density (Wh/L)	Power density (W/L)	Specific energy (Wh/kg)	Specific power (W/kg)	Power rating (MW)	Rated energy capacity (MWh)	Daily self-discharge (%)	Lifetime (years)	Cycling times (cycles)	Cycle efficiency (%)	Discharge efficiency (%)
Lead-acid	50–90	10–400	25–50	75–300	0–40	0.001–40	0.1–0.3	5–15	500–1800	70–90	85
Li-ion	200–500	1500–10,000	75–200	150–2000	0.1–100	0.004–10	0.1–5	5–16	1000–20,000	75–90	85
Na-S	150–300	~140–180	100–240	90–230	8–34	0.4–244.8	≈0	10–20	2500–4500	75–90	85
Ni-Cd	15–150	80–600	45–80	150–300	0–40	6.75	0.03–0.6	10–20	2000–3500	60–83	85
Ni-MH	190–490	–	60–120	200–2000	–	–	0.5–4	2–5	180–2000	70–90	66–92
VRB	16–33	~<2	10–30	166	~0.03–50	2–60	Very low	5–20	12,000–14,000	75–85	75–82
ZnBr	20–65	~<25	30–80	45–100	0.05–10	0.05–4	Small	5–10	1500–2000[+]	~65–75	~60–70
PSB	~20–30	~<2	~15–30	–	0.004–15	0.06–120	Small	10–15	–	60–75	–

3.1.2 Li-Ion Battery

Lithium-ion batteries (Li-ion) have many desirable characteristics such as high efficiencies, a long cycle life, high energy density, and high power density (Table 2) [4]. These characteristics, along with their capability for fast discharge, have made them nearly ideal for portable electronics applications. The main downsides are that the DOD (Depth of Discharge) cycle can affect the Li-ion battery's lifetime, and the battery pack usually requires an onboard computer to manage its operation, thereby increasing overall costs.

Moreover, protection circuits are required due to Li-ion battery fragility and the use of flammable organic electrolytes raises issues about security and greenness. Nevertheless, Li-ion batteries have been successfully installed in both grid-connected and offgrid systems. Some large-scale energy storage projects have been installed around the world. For example, Japan's Sendai Substation installed 40 MW/20 MWh of a Li-ion battery pilot project for frequency regulation and voltage support. In addition, Japan's Tohoku Minami-Soma Substation installed 40 MW/40 MWh of Li-ion battery storage, with the aim of improving the balance between the renewable energy supply and the power demand [23]. The Zhangbei National Wind and Solar Energy Storage and Transmission Demonstration Project, Northern China, installed 14 MW/63 MWh of Li-ion batteries to provide electric energy time shift, renewable capacity firming, and ramping and frequency regulation in combination with a wind and a solar power plant [23]. A123 Systems has installed 36 MW/9 MWh of grid-connected lithium-ion battery storage in various locations, serving needs that include renewable integration and grid stability [24]. Tesla has installed the world's largest Li-ion battery storage 100 MW/129MWh paired with Neoen's Hornsdale wind farm in Jamestown, Australia [25].

3.1.3 Na-S Battery

Sodium-sulfur (Na-S) batteries were originally developed by the Ford Motor Company in the 1960s, and subsequently the technology was sold to the Japanese company NGK. They have high power and energy density, high efficiency of charge/discharge, and a long cycle life (Table 2). These batteries also have pulse power capability over six times their continuous rating (for 30 s) [4, 26]. This feature makes Na-S batteries a candidate for applications in combined power quality and peak shaving.

Na-S batteries are currently being used by 190 locations in Japan, North America, Middle East, and Europe, providing an overall capacity of 530 MW and 3700 MWh for load leveling, renewable energy stabilization, transmission and distribution network management, in microgrids, and for ancillary services [27]. For example, 108 MW of Na-S battery systems are being used by Abu Dhabi's main utility for grid-scale demand management to operate thermal generation efficiently. In Italy, 35 MW Na-S facilities, operated by the transmission operator Terna, store the surging supply of wind energy generated in the south of that country for transmission across the grid to the large power users in the north, thereby reducing transmission congestion and the curtailment of wind generation in the Italian grid. Moreover, in Japan, a 50 MW/300 MWh Na-S battery system was delivered to optimize the balance of supply and demand of power by absorbing excess solar PV generation to avoid limiting the output from solar facilities [27].

3.1.4 Ni-Cd Battery

Nickel-cadmium (Ni-Cd) batteries have high power and energy density, high efficiency of charge/discharge, and a low cycle life (Table 2). The primary demerit of Ni-Cd batteries is a relatively high cost because the manufacturing process is expensive. In contrast, Cadmium is a toxic-heavy metal, hence posing issues associated with the disposal of Ni-Cd batteries. Ni-Cd batteries also suffer from "memory effect," where the batteries only take a full charge after a series of full discharges [4].

To date, there have been very few commercial successes using Ni-Cd batteries for utility-scale EES applications. For example, in 2003, Golden Valley Electric Association BESS used 27 MW of a Ni-Cd battery bank for energy-storage applications. Ni-Cd has also been used for stabilizing wind-energy systems, with a 3-MW system on the island of Bonaire commissioned in 2010 as part of a project for the island with 100% of its power derived from sustainable sources [28].

3.1.5 Ni-MH Battery

Nickel-metal hydride (Ni-MH) batteries have a 1.5 to 2 times higher energy density than Ni-Cd [29, 30]. They exhibit high power capability, tolerance to overcharge/discharge, and environmental compatibility and safety, which make them appropriate for portable power tools and HEVs, although their energy density is relatively low compared to Li-ion batteries (Table 2) [22].

The major use of Ni-MH batteries is in the production of hybrid cars (hybrid electric vehicles, HEVs). In the year 2000, the total production of small Ni-MH batteries was up to 1 billion in Japan. China was one of the few countries to participate in the early development of Ni-MH batteries. Since 1995, China has built a number of manufacturing bases for the large-scale production of Ni-MH batteries, such as the Tianjin Peace Bay Company, the Shenyang Sanpu Company, etc. China is now ranked the world's first in the production of Ni-MH batteries [22]. As of 2017, 85% of listed HEVs were based on Ni-MH batteries; however, with the rapid market development of HEVs, extensive research is ongoing to improve the energy density of Ni-MH batteries and their cycle lives.

3.2 Flow Battery

A recent commercial entrant to the electricity storage field is the flow battery, or redox flow battery (RFB). This is a type of rechargeable battery where rechargeability is provided by two chemical components dissolved in liquids contained within the system and separated by a polymeric membrane (Fig. 7). This technology is similar to a polymer electrolyte membrane fuel cell. The principle behind an RFB cell is two electrochemical reduction and oxidation reactions occurring in two liquid electrolytes. The reduction half-reaction at one electrode extracts electrons and ions from one electrolyte, while the oxidation half-reaction at the other electrode recombines them into the other electrolyte (Fig. 7). Ions migrate from one electrode to the other (from anode to cathode) through an electrolyte impermeable to electrons, which are thus forced through an external circuit and produce electricity. In order to keep the solutions in the liquid phase, the cell must operate at near room temperature. Both half-cells are connected to external

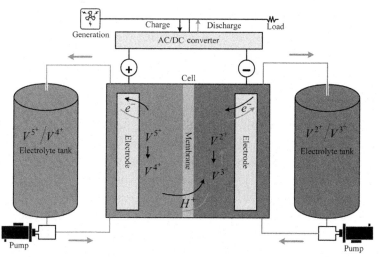

Fig. 7 Schematic of redox flow battery.

storage tanks to provide the required volume of electrolyte solutions circulated by pumps.

The potential applications of RFBs are numerous, including load leveling and peak shaving, continuous power supplies, emergency backup, and facilitation of wind and photovoltaic energy delivery. With regard to both economic and safety considerations, redox flow batteries are recognized as one of the most realistic candidates among electrochemical technologies for energy storage in the range of several kW/kWh up to tens of MW/MWh.

3.2.1 Vanadium Redox Flow Battery

Vanadium redox flow battery (VRFB) systems are the most developed among flow batteries because of their active species remaining in solution at all times during charge/discharge cycling, their high reversibility, and their relatively large power output (Table 2). However, the capital cost of these systems remains far too high for deep market penetration. In order to meet the proposed cost targets, recent investigations have highlighted the use of organic active materials in solid-state organic batteries, in which energy is stored within the cell, mainly in the form of a radical polymer [31].

Recently, it has been installed for several purposes. Examples of this technology in use are a 275-kW output balancer on a wind power project in the Tomari Wind Hills, Japan; a 200-kW output leveler on King Island, Australia; a 250-kW, 2-MWh load leveler in Utah, USA; and two wind and solar power projects (5 kW each) installed in Kenya [32, 33]. Additionally, Hokkaido Electric Power Co. Inc. (HEPCO) and Sumitomo Electric Industries (SEI) Ltd. installed 15 MW/60MWh of VRFB at the Minamihayakita Transformer Station in Abira-chou, Hokkaido, Japan—this is one of the world's largest redox flow battery operations in use [34].

3.2.2 Zinc-Bromine (ZnBr) Flow Battery

Zinc-bromine (ZnBr) flow batteries exhibit relatively high energy density, deep discharge capability, and good reversibility (Table 2). The disadvantages include material corrosion, dendrite formation, and relatively low cycle efficiencies compared to traditional batteries, which can limit its applications [12, 35].

Integrated ZBB (ZBB Energy Corporation) energy storage systems have been tested on transportable trailers (up to 1 MW/3 MWh) for utility-scale applications. Multiple systems of this size could be connected in parallel in much larger applications. ZBB systems are also being supplied at the

5 kW/20 kWh community energy storage (CES) scale, and are now being tested by utility companies—mostly in Australia [36].

3.2.3 Polysulfide-Bromine (PSB) Flow Battery

Polysulfide-bromine batteries (also called regenerative fuel cells or Regenesys) have a very fast response time; they can react within 20 milliseconds. Under normal conditions, PSB batteries can start charging or discharging within 0.1 s [37]; therefore, PSB batteries are particularly useful for frequency response and voltage control.

PSB systems have been tested in the laboratory and demonstrated at the multi-kW scale in the UK. In 2002, a 15-MW 120-MWh Regenesys PSB flow battery (24,000 cell) was built at Innogy's Little Barford Power Station in the UK to support a 680-MW combined cycle gas turbine plant. The Tennessee Valley Authority (TVA) in Columbus wanted a 12-MW, 120-MWh battery to avoid upgrading the network: the exact details of the project are not available as the project was never fully commissioned [38].

4 ELECTRICAL STORAGE

4.1 Supercapacitors

Supercapacitors store electrical energy without any conversion. They have low energy density but high power densities (5–15 kW/kg [39]) with lifetimes of about one million charge/discharge cycles. Depending on the charge storage mechanism, supercapacitors can be classified into three categories: electric double layer capacitors (EDLCs); pseudocapacitors; and hybrid supercapacitors formed by a combination of EDLCs and pseudocapacitors (Fig. 8).

In EDLCs, the capacitance is produced by the electrostatic charge separation at the interface between the electrode and the electrolyte. To maximize the charge storage capacity, the electrode materials are usually made from highly porous carbon materials [40]. In contrast, pseudocapacitors are based on faradaic redox reactions involving high energy electrode materials based on metal oxides, metal-doped carbons or conductive polymers. These electrode materials allow supercapacitors with higher energy density. The hybrid supercapacitors incorporate mechanisms from both EDLCs and pseudocapacitors [41].

The characteristics of supercapacitors enable them to meet instantaneous variations in electricity demand in combination with one or more

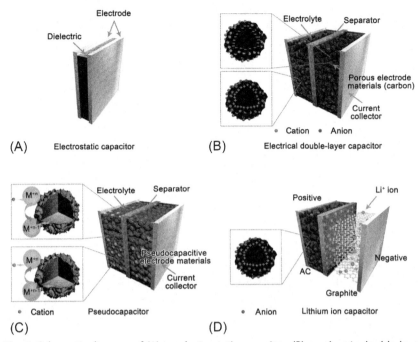

Fig. 8 Schematic diagram of (A) an electrostatic capacitor, (B) an electric double-layer capacitor, (C) pseudocapacitors, and (D) hybrid supercapacitors [40].

technologies that have superior energy density but lower effective power density, such as hydrogen stored as a metal hydride.

4.2 Superconducting Magnetic Energy Storage

In 1969, Ferrier [42] originally introduced the superconducting magnetic energy storage (SMES) system as a source of energy to accommodate the diurnal variations of power demands [43]. An SMES system contains three main components: a superconducting coil (SC); a power conditioning system (PCS); and a refrigeration unit (Fig. 9). It stores energy in a superconducting coil in the form of a magnetic field generated by a circulating current. The maximum stored energy is determined by two factors. The first is the size and geometry of the coil, which determines the inductance of the coil. Obviously, the larger the coil, the greater the stored energy. The second factor is the conductor characteristics, which regulate the maximum current. Superconductors can carry substantial currents in high magnetic fields [44].

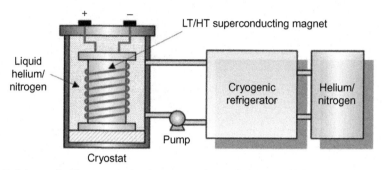

Fig. 9 Schematic of superconducting magnetic energy storage [4].

The magnetic field is created with the flow of a direct current (DC) through the superconducting coil. In SMESs, the superconducting coils are usually made of niobium–titanium (NbTi) filaments with a critical temperature of about 9.2 K [4]. To maintain the system charge, the coil must be cooled adequately. This has been achieved by cryogenically cooling to a temperature below its superconducting critical temperature, thereby enabling the current to circulate indefinitely with almost zero loss. Most importantly, the only conversion process in the SMES system is from AC to DC; thus, there are no inherent thermodynamic losses associated with the conversion. This leads to higher cycle efficiency, instantaneous charge and discharge (within a few milliseconds), and high storage efficiency [4].

Due to its characteristics, SMESs are more suitable for high power and short duration applications, since they are cheap on the output power basis—with a high-power density—but expensive in terms of energy storage capacity. As a result, SMESs have attracted attention for applications by solving voltage stability and power quality problems for large industrial customers, electric utilities, and the military (for example, microchip manufacture) [45]. Onsite SMES is suitable to mitigate the negative impacts of renewable energy in power quality related issues, especially with power converters (needed for solar photovoltaic and some wind farms), wind power oscillations, and flicker [46].

The main drawbacks of an SMES system are the need for significant power to keep the coil at low temperatures combined with the high overall cost of its employment [47]. Additionally, this technology is economically suitable for short cyclic periods only, with a maximum of hours of duration in storage; this is due to a high self-discharge ratio for longer periods (10%–15% per day) [4] and mechanical stability problems [44].

Since 2011, the U.S. Department of Energy Advanced Research Projects Agency for Energy (ARPA-E) has awarded a $4.2 million grant to Swiss-based engineering firm ABB to create a 3.3-kW-hour, proof-of-concept SMES prototype. ABB is collaborating with superconducting wire manufacturer SuperPower, Brookhaven National Laboratory, and the University of Houston. The group's ultimate goal is to develop a 1–2 MWh commercial-scale device that is cost competitive with lead-acid batteries [48].

5 THERMAL ENERGY STORAGE

Thermal energy storage (TES) is generally considered as energy management options attached to renewable energy generation units; however, its standalone application at the demand side is also possible. In recent decades, TES systems have demonstrated a capability to shift electrical loads from high-peak to off-peak hours, so they have the potential to become powerful instruments in demand-side management programs [49]. TES can also be defined as the temporary storage of thermal energy at high or low temperatures; it has the potential to increase the effective use of thermal energy equipment and to facilitate large-scale switching. TES is normally useful for correcting the mismatch between supply and demand energy [50]. Certainly, TES is of particular interest and significance for solar thermal applications such as heating, hot water, cooling, air conditioning, etc. because of its intermittent nature. In these applications, a TES system must be able to retain the energy absorbed for at least a few days in order to supply the energy needed on cloudy days when the energy input is low [49].

TES comprises three parts: a storage medium, a heat transfer mechanism, and a containment system. TES stores the thermal energy in one of three forms such as sensible heat, latent heat or vaporization, and thermochemical reactions. At present, synthetic oil and molten salt are the most widely used sensible heat storage materials in large-scale concentrating solar power (CSP) systems. Other systems that utilize latent heat, thermochemical and novel sensible heat materials are still under development [51]. The characteristic data of each thermal energy storage are given in Table 3. Fig. 10 shows the different materials and systems of thermal energy storage.

5.1 Sensible Heat Energy Storage

Sensible heat storage means shifting the temperature of a storage medium without phase change. It is the most common simple, low-cost, and long-standing method. This storage system exchanges the solar energy into

Table 3 Characteristics and comparisons of the thermal energy storage systems [52]

Particulars		Sensible heat storage	Latent heat storage	Thermochemical storage
Energy density	Volumetric	Small: ~50 kWh m^{-3} of material	Medium: ~100 kWh m^{-3} of material	High: ~500 kWh m^{-3} of reactant
	Gravimetric	Small: ~0.02–0.03 kWh kg^{-1} of material	Medium: ~0.05–0.1 kWh kg^{-1} of material	High: ~0.5–1 kWh kg^{-1} of reactant
Storage temperature		Charging step temperature	Charging step temperature	Ambient temperature
Storage period		Limited (thermal losses)	Limited (thermal losses)	Theoretically unlimited
Transport		Small distance	Small distance	Distance theoretically unlimited
Maturity		Industrial scale	Pilot scale	Laboratory scale
Technology		Simple	Medium	Complex

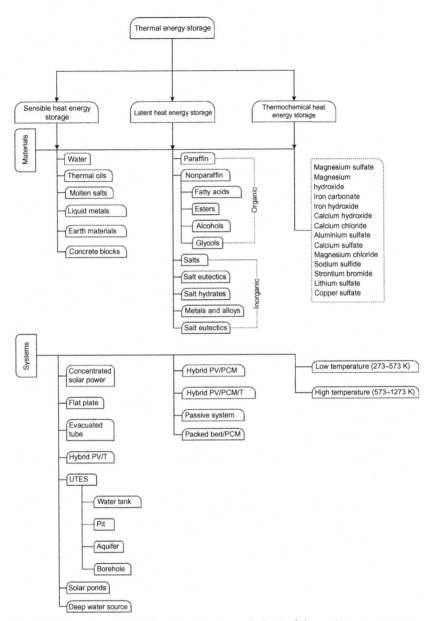

Fig. 10 Different materials (above) and systems (below) of thermal energy storage.

sensible heat in a storage medium (usually solid or liquid) and releases it when necessary. The amount of stored sensible heat in a material depends on its heat capacity (energy density) and the thermal diffusivity (rate at which the heat can be released and extracted) [51]:

$$Q = \int_{T_i}^{T_f} mC_p dT = mC_p \left(T_f - T_i \right) \tag{1}$$

where Q is the amount of heat stored, T_i is the initial temperature, T_f is the final temperature, m is the mass of heat storage medium, and C_p is the specific heat.

The storage materials absorb heat by the conventional heat transfer mechanisms of radiation, conduction, and convection. As the materials cool at night or on cloudy days, the stored heat is released by the same modes.

In terms of containment, sensible heat storage materials can be stored above ground or underground. The major methods employed for underground thermal energy storage (UTES) are aquifer storage and underground soil storage. Sensible heat storage is mainly demonstrated in space heating and domestic hot water supply for which the required temperature ranges from 40°C to 80°C [53]. Hence, water, rock-sort materials (for example, gravel, pebbles, and bricks) and ground or soil have been widely used as storage media in large-scale demonstration projects around the world [53]. Furthermore, all of the currently installed TES systems in utility-scale solar thermal electric plants store energy using molten salts, synthetic oil, liquid metals, or powders [51].

Water-based sensible heat storage systems use water as the storage medium or heat carrier fluid for storing/transferring heat. These systems can be classified as water tank and aquifer storage systems. Water tank or pit storage systems store water in an artificial structure made of stainless steel or reinforced concrete surrounded by thick insulation, and are usually buried underground (also called water pits) or placed on the roof or outside of a building [54]. Aquifer storage, however, uses natural water directly from the underground layer. An aquifer needs to drill at least two thermal wells (hot and cold). The aquifer geologic formation is employed as the storage medium and the groundwater is used as the heat carrier fluid [53]. One of the largest water-based sensible heat storages is in Pimlico, London, UK, consisting of three 8-MW$_{th}$ boilers; two 2-MW$_{th}$ combined heat and power (CHP) engines; and an accumulator that can store $2500\,\text{m}^3$ of water at just $<100°C$. It provides heating and hot water to 3256 homes, 50 commercial properties, and three schools [55].

In rock-bed, sensible heat storage, the rock (for example, pebble, gravel or bricks) bed is usually disseminated with heat transfer fluid (water or air) to exchange heat (gained in summer and released in winter). Rock-based systems can endure higher temperatures in comparison with water-based systems; however, the energy density is low. Hence, they need larger volumes to store the same amount of heat storage, which is approximately three times higher than water-based storage systems [53]. In Qinhuang Island, China, a 300-m^3 pebble bed was constructed to store surplus heat from the 473.2-m^2 solar collector during the day and provide heating and hot water during the night [56].

Ground or soil sensible heat storage is another application of UTES aside from aquifer systems. In this system, ground or soil itself is used directly as a storage medium. The underground structure can store a large amount of solar heat, which is collected in the summer for later use in winter. In this storage system, the ground is excavated and drilled to insert vertical or horizontal tubes, so it is also called borehole thermal energy storage (BTES) or duct heat storage [53]. The Drake Landing Solar Community, Alberta, Canada, provides heating and hot water to 52 homes (around 97% of their year-round heat). The heat energy is captured from 800 solar thermal collectors located on the roofs of all 52 houses' garages. This is enabled by interseasonal heat storage in a large mass of underground rock. The thermal exchange occurs via 144 boreholes, drilled 37 m into the earth [57].

Before installing sensible heat energy storage systems, a few aspects need to be considered, such as local geological conditions, available site size, temperature levels of the reservoir, and legal issues about drilling and investment costs. Besides, sensible heat storage system suffers from heat loss problems. If thick insulation is installed, the temperature of the heat source may not allow the extracted heat to be used directly in the heating season. Therefore, the storage unit needs additional equipment, such as a heat pump, to upgrade the temperature level to satisfy the required heat load, which also incurs higher investment cost.

5.2 Latent Heat Energy Storage

Latent heat energy storage is a near-isothermal process that can provide significantly high storage density with smaller temperature swings in comparison with sensible storage systems. In addition, latent heat storage has the capacity to store heat of fusion at a constant or near-constant temperature that corresponds to the phase transition temperature of the phase change material (PCM). Latent heat storage is based on the heat absorption or release when a storage material undergoes a phase transformation from solid to solid, liquid to gas, and solid to liquid, or vice versa. The storage capacity of a latent heat storage system with PCM is given by:

$$
\begin{aligned}
Q &= \int_{T_i}^{T_m} mC_p dT + ma_m \Delta H_m + \int_{T_m}^{T_f} mC_p dT \\
&= m\left[a_m \Delta H_m + C_p(T_m - T_i) + C_p(T_f - T_m)\right]
\end{aligned}
\tag{2}
$$

where a_m is the melted fraction and ΔH_m is the heat of fusion.

The most practicable routes of latent heat storage are solid to liquid and solid to solid. Solid-to-gas and liquid-to-gas transitions have a higher latent heat of fusion, but their large volume expansion, associated with containment problems, rule out their potential use in thermal storage systems [58]. Large volume changes render such systems complex and impractical. In solid-to-solid transitions, heat is stored when the material is transformed from one crystalline structure to another. This transition generally has less latent heat and slighter volume changes than a solid-to-liquid transition. Solid-to-solid PCMs offer the advantages of less rigorous container requirements and better design flexibility. Solid-to-liquid transformations have comparatively smaller latent heat than liquid to gas; however, these transformations involve only small changes in volume (<10%). Solid-to-liquid transition has proved its economic viability for use in thermal energy storage systems [58, 59].

PCMs themselves cannot serve as heat transfer mediums. Heat transfer mediums with a heat exchanger are required to transfer energy from source to PCM and from PCM to load. Therefore, heat exchanger design is an important part in improving heat transfer in latent heat storage systems. PCMs also have positive volumetric expansion on melting, and so volume design of the containers is also necessary, which should be compatible with the PCM used. Any latent heat storage system must possess at least these three properties: a suitable PCM with its melting point in the desired temperature range, a suitable heat exchange surface, and a suitable container compatible with the PCM [59].

During the phase change, the materials remain, theoretically, at constant temperature (real systems show a temperature stabilization around the melting temperature). PCMs undergo phase-changing processes by absorbing (endothermic process) and releasing (exothermic process) heat in the form of latent heat of fusion without the temperature changing in each period. The phase-changing temperatures of PCMs differ across a wide range, making them applicable for various situations. The energy density of latent heat storage is typically 5 to 10 times higher than sensible heat storage. For example, Morrison and Abdel-Khalik [60] and Ghoneim [61] showed that to store the same amount of energy from a unit collector area, rock requires more than seven times the storage mass of Paraffin 116 Wax (P116-Wax), five times the storage mass of medicinal paraffin, and more than eight times the storage mass of $Na_2SO_4 \cdot 10H_2O$.

A wide range of PCM is available for use in latent heat energy storage systems, thus providing application potential in a range of different temperatures from freezing to high-temperature storage applications. The different materials proposed for use as phase change materials include aqueous salt solutions, water, gas hydrates, paraffins, fatty acids, salt hydrates and eutectic mixtures, sugar alcohols, nitrates, hydroxides, chlorides, carbonates, and fluorides. The temperature range covered by these materials is from below $-20°C$ to above $700°C$.

Many of the phase change materials have low thermal conductivity, which can dwindle the rates of heat charging and discharging. To address this limitation, different approaches have been introduced. For example, Fukai, Hamada [62] introduced carbon-fiber brushes on the shell side of a heat exchanger to enhance the conductive heat transfer rates in a PCM. In the charging process, the brushes prevent the natural convection; however, the charge rate with the brushes is 10%–20% higher compared to the case without fibers. The brushes also improve the discharge process. The discharge rate using the brushes with one volume percent is about 30% higher than that without fibers. Zhang, Zhang [63] used expanded graphite (EG) to absorb liquid paraffin. The new composite resulted in a reduction in thermal energy storage charging time compared with using paraffin only. The EG exhibited the maximum sorption capacity of 92 wt% for paraffin at 800 W (microwave irradiation) for 10 s.

Previously, latent heat storage was mainly employed with construction as a passive method in building envelopes (walls, windows, ceilings, or floors). Here, "passive" means that the phase-change processes occur without resorting to mechanical equipment. PCMs are often mixed with other construction materials, such as concrete, or used alone to play a role in building envelopes. In greenhouse applications, north wall storage incorporated with PCMs is the most common technology for increasing the indoor temperature. However, most passive applications are not highly controllable and have low solar fractions, so they are limited to short-term storage [53]. However, for long-term storage, PCMs would provide more potential for inactive storage such as when integrated with solar collector systems (for example, agriculture greenhouses and heat pumps) [53, 64]. In order to increase the storage capacity of a CSP system, it has been offered that latent heat storage could be used in conjunction with a steam accumulator by either inserting PCMs directly inside the steam accumulator or from the outside into the steam accumulator [51].

PCMs are beneficial with respect to reasonable cost, higher energy density, simplicity in system design, and the delivery of heat at constant temperature. However, for applications in heating and cooling systems, the technology should be improved in a few aspects (but not limited to), such as the need to improve the storage density to enable the integration of PCMs into buildings and HVAC systems, and to enhance the rate of heat discharge from PCMs that can be used for domestic hot water production.

An alternative option, in this category, is energy storage in molten metals. For instance, the Perryman Thermal Battery^{TM} is a technology that stores energy in molten metals (e.g., nickel), with ceramic embodiment, at temperatures up to 2000°C [65]. The technology can be charged using a standard induction unit directly from AC or DC power sources. The efficiency loss from converting electricity to thermal energy using induction is only around 1.68% for AC electricity and 2.74% for DC electricity. Unlike the electrochemical cells, this technology has a negligible loss ($\sim$ 2% per month) without an aging problem. Depending on the conditions of use, it can last for many decades. Particularly, the ceramic embodiment has a long life with some working examples being in service for over 80 years. Furthermore, the technology does not involve materials with environmental impacts, and it faces minimal concerns for decommissioning and recycling. Perryman Thermal Batteries^{TM} can be varied in scale from the size of a large basketball to very large industrial units holding over 30 MWh of thermal energy. Given its very high operating temperature, the technology has a high energy density (J/m^3) with a standard 29MWh industrial unit being approximately 6m tall, 2.4m wide and 26m^3 in volume (Fig. 11).

5.3 Thermochemical Heat Energy Storage

In a thermochemical heat storage system, reactions are reversible:

$$A \overset{\Delta H}{\Leftrightarrow} B + C \tag{3}$$

In this storage system, the thermochemical heat reserve is associated with the reaction enthalpy, ΔH. During the charging step, thermal energy is used

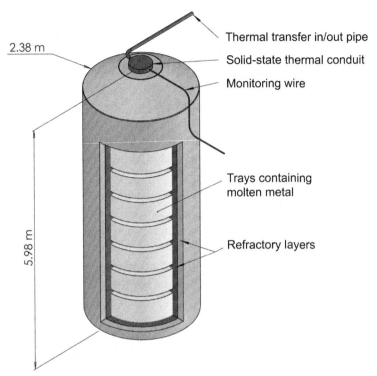

2.38 m

5.98 m

Thermal transfer in/out pipe

Solid-state thermal conduit

Monitoring wire

Trays containing
molten metal

Refractory layers

Fig. 11 Schematic of a 29 MWh Perryman Thermal Battery™ for energy storage in molten metals. *(Image Courtesy of Virgil Perryman)*

to dissociate a chemical reactant (A) into products (B and C). This reaction is endothermic. During the releasing step, the products of the endothermic reaction (B and C) are mixed together and react to form the initial reactant (A). This reaction is exothermic and releases heat. The products of both reactions can be stored either at ambient temperature or at working temperature. In this process, substance B can be a hydroxide, hydrate, carbonate, ammoniate, etc., and C can be water, CO, ammonia, hydrogen, etc. There is no restriction on phases, but usually A is a solid or a liquid and B and C can be any phase. The thermal energy stored in thermochemical material can be expressed as:

$$Q = n_A \Delta H \tag{4}$$

where n_A is the mol number of reactant A.

Thermochemical storage systems have several advantages. Their energy densities are 5 to 10 times higher than latent heat storage systems and sensible heat storage systems, respectively. Both storage period and transport are theoretically unlimited because there is no thermal loss during storage, as products can be stored at ambient temperature for long periods with little or no degradation in stored energy content. However, the thermochemical energy storage is the least developed storage technology, requiring complex reactor design to achieve the desired operational performance [51]. Nevertheless, these systems suffer from the impediments commonly exhibited by other systems, including limitations in heat transfer, cycling stability, operating condition (pressure and temperature), reversibility, and costs. In addition, thermochemical systems can be restricted by reaction kinetics [66].

If successful reactors could be designed with performances close to predicted theoretical levels with no materials degradation on repeated cycling, then this may allow the realization of long-term, effective heat storage with compact energy storage systems. However, far more research is required on materials and reactor design—including humidification and regeneration processes—before this to be achieved [67]. Two review articles have investigated low-temperature (273–573 K), thermochemical TES [68, 69]. One is a state-of-the-art review on sorption and chemical reaction processes for thermochemical TES applications by Cot-Gores, Castell [70]. Recently, Pardo, Deydier [52] reviewed high-temperature (573–1273 K) thermochemical TES systems, which have the potential to become an important part of the sustainable handling of energy in the near future.

6 CHEMICAL ENERGY STORAGE

Chemical energy storage includes sorption and thermochemical reactions. *Sorption* is a physical and chemical process by which one substance becomes attached to another. Sorption is the common term used for both absorption and adsorption. *Absorption* is the incorporation of a substance in one state into another of a different state (for example, liquids being absorbed by a solid or gases being absorbed by water). *Adsorption* is the physical adherence or bonding of ions and molecules onto the surface of another molecule. In thermochemical reactions, energy is stored after a dissociation reaction and then recovered in a chemically reverse reaction (discussed in Section 5.3).

6.1 Hydrogen Storage

Batteries might be argued as suitable candidates for short-term energy storage and power quality management. However, their feasibility for energy management and long-term storage (that is, days, months and years) is questionable. In contrast, though hydrogen might not be a competitive solution

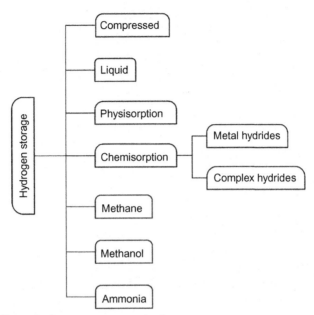

Fig. 12 Different hydrogen storage methods.

for short-term storage, it has the capability of energy storage at massive volumes; for example, terawatt hours of volume for a long period in various forms. Furthermore, any large-scale hydrogen distribution system must deal with storage to provide a buffer between production facilities and fluctuations in demand. Economically viable, environmentally benign, and sustainable mass storage techniques are therefore major research goals. Fig. 12 shows the available hydrogen storage methods.

The lower heating value (LHV) of hydrogen is 120 MJ/kg, compared to about 50 MJ/kg for methane and around 40 MJ/kg for petroleum products. The basic problem of storing hydrogen is its low density at ambient conditions ($0.0823 \, kg/m^3$ at 298 K and atmospheric pressure [71]). Storing 1 kg of hydrogen at atmospheric pressure and 25°C, therefore, requires a tank with internal volume $12.3 \, m^3$, which is impractical in most situations (such as even a small fuel-cell vehicle requiring at least 5 kg of H_2, or an offgrid energy system requiring many kg of storage capacity). Therefore, there is a need for improving the volume intensity of hydrogen.

While liquefaction or cryocooling of hydrogen gas is possible [72], current storage systems (such as fuel-cell vehicles, onsite storage, or stand-alone applications) are mostly limited to compressed gas storage in cylinder/canisters. The drawback of compressed hydrogen is its lower volume intensity, which makes the transportation costs higher—especially for delivery to distant locations. Solid-state hydrogen storage in metal hydride or complex hydride could be a potential option for standalone or stationary use due to favorable characteristics such as decoupling of power and energy ratings, low operating pressure, safety, and high volumetric energy intensity. However, various challenges need to be addressed, depending on the nature of the storage material, including the overall weight of the storage system, the limited storage capacity of the material used, degeneration with cycles, and cost.

6.1.1 Compressed

The simplest storage system is compressed H_2 gas. The advantages of this method include the ease of operation at ambient temperature and simplicity of storage and retrieval. Compressing hydrogen increases H_2 density to $23.32 \, kg/m^3$ at 350 bar (common in fuel-cell buses) and to $39.22 \, kg/m^3$ at 700 bar (common in commercial fuel-cell sedans) [71]. Cryocompression to 200 bar at 100 K achieves a density of $39.52 \, kg/m^3$ [71], matching that of 700-bar compression at 25°C and partially trading one technical difficulty (high pressure) for another (cryogenic temperature).

Compressed hydrogen storage is now a commercial reality in fuel-cell vehicles and refueling stations. The so-called type IV hydrogen storage tanks used in vehicles have a cylindrical composite structure with wound carbon fiber over a hydrogen-impermeable liner [73, 74]. From an electrical energy storage perspective, compressed hydrogen storage is technically viable, but (i) it necessitates expensive compression technology, which constitutes a parasitic load on the energy system, and (ii) it operates at a much higher pressure than common electrolyzers and fuel cells.

6.1.2 Liquid

Traditionally, liquid hydrogen storage is technically viable at small scales and has been trialed in vehicles but has been overtaken by compressed hydrogen storage. Its potential role in energy systems is not established; nonetheless, cryogenic storage at the scale of many m^3 of liquid is a well-established technology in the space industry. Liquefied hydrogen becomes mandatory for large-scale export of pure hydrogen. Kawasaki Heavy Industry (Japan) is moving forward with the construction of small liquefied hydrogen carriers, initially at the 200-t scale [75].

Liquid hydrogen suffers from unavoidable losses from boil-off owing to heat flow into the reservoir from the exterior. Furthermore, hydrogen liquefaction constitutes a parasitic load consuming around 35% of the LHV energy content of the liquefied hydrogen. It is best suited to centralized liquefaction plants with their attendant economies of scale. Remaining challenges include the total system volume and weight, the high cost of the tank, and the ortho-para conversion [76].

6.1.3 Physisorption

Hydrogen can be stored in its molecular form by physical adsorption onto the surface of a porous solid material. Dalebrook, Gan [77] reviewed different new techniques of hydrogen absorption and desorption and categorized them as physisorption storage (for example, zeolites, metal-organic frameworks (MOFs)) and chemical storage (for example, amines, formic acid). In physisorption, hydrogen remains in its molecular form and is absorbed and desorbed reversibly. In this process, the surface densities are higher than the bulk gas concentration due to gas-solid interactions [60], and storage materials may express gravimetric loadings as either total capacity (the entire quantity of deliverable hydrogen) or excess capacity (the gain in uptake promoted by an adsorbent).

Carbon-based materials (such as activated carbon, graphite, carbon nanotubes, and carbon foams) have received significant attention due to characteristics such as low weight, high surface area, and chemical stability [78, 79]. Graphene is a single layer of graphite and has a theoretical surface area of $2630 \, m^2 \, g^{-1}$ [80, 81], making it a suitable candidate for physisorption storage. Burress, Gadipelli [82] showed that using different activation techniques rather than heat treatment could reduce the oxygen-to-carbon (O/C) ratio, which helps to improve the surface area and hydrogen uptake of graphene oxide frameworks (GOFs).

Microporous organic polymers (MOPs) are desirable for energy storage due to their high specific surface areas and tailored porosity. Wood, Tan [83] introduced a series of synthetic routes to enhance the interaction with hydrogen molecules. Recently, polymers of intrinsic microporosity (PIMs) [84] and hypercrosslinked polymers (HCPs) [85] have been investigated for physisorption-based hydrogen storage at low temperature. Covalent organic frameworks (COFs) show much finer control over the crystallinity and porosity properties in comparison to HCPs and PIMs [86]. Ding and Wang [86] critically reviewed COFs, covering materials design to applications. Zeolites contain well-defined open-pore structures, often with tuneable pore sizes, and demonstrate notable guest-host chemistry with important applications in catalysis, gas adsorption, purification, and separation [87]. Additionally, these materials are inexpensive and have been widely used in industrial processes for many decades.

An extensive experimental survey has shown that the hydrogen storage capacity of zeolites to be $<2\,wt\%$ at cryogenic temperatures and $<0.3\,wt\%$ at room temperatures and above [88]. Recently, novel nanoporous materials (such as MOFs) have been applied to hydrogen storage [89–91]. MOFs are porous materials constructed by coordinate bonds between multidentate ligands and metal atoms or small metal-containing clusters, and are highly crystalline, inorganic-organic hybrid structures that contain metal clusters or ions (secondary building units) as nodes and organic ligands as linkers [92]. MOFs can be synthesized via self-assembly from different organic linkers and metal nodules. Due to the variable building blocks, MOFs have very large surface areas, high porosities, uniform and adjustable pore sizes and well-defined hydrogen occupation sites. These features make MOFs promising candidates for hydrogen storage based on physisorption.

6.1.4 Chemisorption
Storing hydrogen as a chemical metal hydride or complex hydride can provide high volumetric densities and low absorption pressures during hydrogen uptake.

Metal Hydrides
The class of materials collectively known as metal hydrides (MHs) contains a very wide variety of materials, including elemental metals, alloys, and stoichiometric nonmetallic compounds, with the common ability to dissociate hydrogen molecules at the surface of the material and absorb hydrogen atoms into the interior crystal structure. Absorption and desorption take place in a very wide range of pressures and temperatures, as illustrated in Fig. 13, where the most practicable materials are included from stationary hydrogen storage, namely, relatively heavy metal hydrides with high volumetric hydrogen density.

The advantages of metal-hydride storage in these situations come from (i) the decoupling of power and energy ratings, which makes it advantageous for long-term storage compared to batteries [93]; (ii) its ability to be tuned for low operating pressure suitable for direct coupling to an electrolyzer and near-ambient operating pressure; (iii) excellent safety coming from the low pressure and relatively slow kinetics of hydrogen release; and (iv) high volumetric energy density. For example, the classic intermetallic hydride $LaNi_5H_6$ contains only 1.4 mass% hydrogen but has a 100%-dense

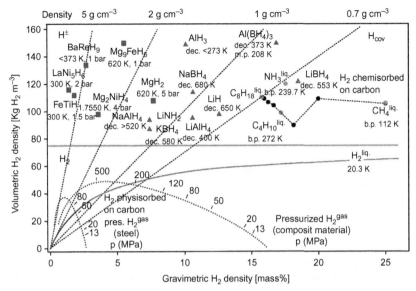

Fig. 13 Comparison of metal-hydride and other hydrogen storage materials [101].

volumetric capacity of approximately $115 \, \mathrm{kg \, m^{-3}}$ at room temperature and $<1 \, \mathrm{MPa}$ pressure [94] compared to $70.8 \, \mathrm{kg \, m^{-3}}$ at $20.3 \, \mathrm{K}$ and $0.1 \, \mathrm{MPa}$ for liquid hydrogen. A more detailed review of solid-state hydrogen storage is available in several review papers and books published within the recent decade [95–97].

Light-weight metals—such as Li, Na, and Mg—all form hydrides with high gravimetric hydrogen capacities. The release of the hydrogen, however, requires high temperatures ($>650°C$ for Li, for example) due to the high enthalpy of formation. Magnesium hydride offers the greatest potential with 7.6 wt% and good reversibility, but desorption is kinetically limited. Webb [96] studied MgH_2 to improve the hydrogen sorption through mechanical modification of the material and through the addition of catalytic materials [96].

Alloying of different metals can modify the enthalpy of the hydride, and this has been investigated for hydrogen storage applications. Most useable alloys are AB_5 intermetallic compounds (for example, $LaNi_5$) for hydrogen storage with 1.4 wt% and equilibrium pressure $<2 \, \mathrm{bar}$ at room temperature [95]. AB_2 compounds are derived from the Laves phases with a hydrogen storage capacity of up to 2 wt% [77]. Body-centered cubic (BCC) alloys have maximum hydrogen capacities up to 4 wt% with reversible capacities of $>2 \, \mathrm{wt\%}$ [98–100].

Mechanical modification can provide fresh surfaces and defects for fast hydrogen diffusion. Ball milling is a mechanical process, which is widely used to enhance the surface area of metal hydrides so as to create micro/ nanostructures and annealing treatments of the material [102]. In addition, ball milling can offer a synthesis path for new alloys. For example, the milling of magnesium and titanium with different ratios of hydrogen produces a ternary hydride with 3.7 wt% hydrogen content and improves the desorption kinetics promoted by the titanium component [103]. The crystalline Mg_2Ni alloy obtained by ball milling has excellent surface properties compared with those prepared by a conventional metallurgical method [104]. Nanostructuring techniques—such as ball milling, cold rolling, thin films, and nanoconfinement—all positively influence the reaction kinetics [105, 106].

Complex hydrides

Complex hydrides are metal salts, typically containing more than one metal or metalloid, where the anion contains the hydride. These complex metal hydrides typically have higher hydrogen gravimetric storage capacities and volumetric densities than simple hydrides. For example, the hydrogen content of $LiBH_4$ is 18 wt% [95]. Low-weight complex hydrides include the alanates $[AlH_4]^-$, amides $[NH_2]^-$, imides $[NH_4]^-$, and borohydrides $[BH_4]^-$. Alanates and borohydrides received attention due to their light weight and hydrogen capacity per metal atom. However, both have issues with poor reversibility and high stability, decomposing only at elevated temperatures.

Sodium alanate is a complex hydride of aluminum and sodium ($NaAlH_4$). Theoretically, $NaAlH_4$ contains 7.4 wt% of hydrogen [107]; however, the decomposition to release hydrogen is a multistep reaction with details in Refs. [108, 109]. The kinetics of the sodium alanate sorption can be improved with transition metal additives such as Ti [110].

Lithium alanates are very attractive due to the higher hydrogen content of 10.5 wt% and 11.2 wt% for $LiAlH_4$ and Li_3AlH_6, respectively [111]. It has been theoretically predicted that lithium nitride (Li_3N) can store 10.4 wt% hydrogen, but the reversible hydrogen capacity is only 6 wt% [112].

Schlesinger and Brown [113] developed lithium borohydride ($LiBH_4$) from ethyl lithium and diborane, yielding a material containing 18.3 wt% hydrogen [114, 115]. Many attempts to destabilize the highly stable complex hydrides have been made by doping, catalyst addition, $H^{\delta+}/H^{\delta-}$ coupling, size effects, partial cation substitution, or direct reaction with other hydride materials [116]. A quaternary hydride system (that is, $Li_3BN_2H_8$) with a theoretical hydrogen capacity of 11.9 wt% was prepared by ball milling [117, 118]. Solid sodium borohydride exhibits adverse thermodynamic effects for use as a recyclable hydrogen storage material [119]. Other borohydrides of potassium, magnesium, calcium, and zinc also decompose to hydrogen (details in Refs. [120–124]). Beryllium borohydride claims the highest hydrogen content at ≈ 20.7 wt% hydrogen, but the high toxicity and cost of the metal discourages its use in hydrogen storage applications [125]. A mixture of lithium borohydride, calcium hydride, and magnesium hydride has also been studied by Zhou, Liu [126].

Recent review papers provide far more detail on complex hydrides. Paskevicius, Jepsen [127] report that there are now well over 200 known compounds in the borohydride class of complex metal hydrides. Other articles highlighting the wide range of metal hydrides include a review of B-N and Al-N compounds [128], a review of metal boranes [129], and reviews of complex transition metal hydrides [130–133]. This is a nonexhaustive list that does not include amide/imide compounds.

6.2 Methane

The currently most developed hydrogen conversion process is methanation. Hydrogen reacts with CO_2 to yield methane according to the Sabatier reaction or biological methanation resulting in an extra energy conversion loss of 8% [134]. Müller, Müller [135] reported that 95% of the CO_2 is converted to methane in a demonstration plant. The reaction is highly exothermic, which leads to high conversion losses when the heat is not completely used.

The key advantage of power conversion to methane is the availability of a natural gas distribution network as well as the market. The first industry-scale, power-to-methane plant was developed by ETOGAS for Audi AG in Werlte, Germany [136]. The plant is using CO_2 from a waste-biogas plant and intermittent renewable power to produce synthetic natural gas (SNG), which is directly fed into the local gas grid [136]. In April 2014, KIT started a research project named HELMETH (Integrated High-Temperature ELectrolysis and METHanation for Effective Power to Gas Conversion, financed by the European Union). The objective of this project is the proof of concept of a highly efficient power-to-gas technology by integrating high-temperature electrolyzers such as SOEC with CO_2 methanation [136].

6.3 Methanol

Methanol is usually considered as a hydrogen carrier. It is a versatile fuel with many advantages over pure hydrogen. The key benefit is that its liquid phase at ambient conditions, thereby making it possible to be handled and distributed with an identical type of infrastructure by which liquid gasoline is distributed today. Methanol is easily turned into hydrogen through a catalytic process, using a fuel reformer. It can be done at a temperature of 200–300°C. Many other liquid fuels can be reformed into hydrogen, but the process requires a considerably higher temperature and greater energy consumption than methanol.

Usually, the electricity consumption for methanol production is higher than that of methanation because additional compressor power is required for the recycle stream. However, methanol synthesis has fewer conversion losses because it is less exothermic. Methanol seems to be a promising storage option because it has a higher volumetric energy density than hydrogen and methane, and it can be combusted in gasoline engines [134].

In 2011, nearly 17 Mtonnes of methanol were used in fuel and energy applications. From 2012 to 2016, overall methanol consumption as a fuel almost doubled (from 20 Mtonnes in 2012 to 38 Mtonnes in 2016), translating to approximately 37% annual growth around the world [137].

Methanol is traditionally produced from syngas ($CO_2 + 3H_2 \rightarrow CH_3OH + H_2$). As the required ratio of CO_2 in the feed gas to the reactor is high, the methanol process is also viewed as a CO_2 utilization process opportunity for recycling CO_2 rather than direct emission. As such, methanol CO_2 utilization has received significant attention over recent years. When H_2 comes from renewable sources, a methanol process can be a double benefit with not only offering an intensified hydrogen carrier, but also reducing CO_2 emissions. Rihko-Struckmann, Peschel [138] investigated CO_2 utilization in membrane processes based on process simulations and assuming equilibrium conversion. They found that nearly 27% of CO_2 is converted in the reactor (250°C and 5 MPa), and the unreacted CO_2 is separated from the raw product gas and recycled to the reactor. As a result, a total conversion of CO_2 is reached ($\approx 96.8\%$).

Jadhav, Vaidya [139] found that the equilibrium conversion efficiency could be obtained with a copper catalyst. Methanol is synthesized from CO_2 at 300°C and around 70 bars in a reactor with a metallic catalyst (copper and zinc oxides on an alumina-based ceramic, $Cu/ZnO/Al_2O_3$). The ceramic is particularly adapted to this highly exothermic reaction.

Since 2009, a pilot plant of 100 tons per year has been built by Mitsui chemicals in Japan. In China, Wei and his coworkers have found a novel catalyst (ZrO_2-doped CuZnO) for methanol synthesis from CO/CO_2 hydrogenation. The ZrO_2-doped CuZnO catalyst showed high activity and high selectivity toward both CO and CO_2 hydrogenation [140].

6.4 Ammonia

As a hydrogen energy carrier, methane and methanol offer outward potentials; methane, because of an existing natural gas distribution infrastructure; and methanol, due to its liquid phase and ease of transportation. However, both of these options contribute to CO_2 emissions. Ammonia is, however, a carbon-free chemical energy carrier. It is not a greenhouse gas (GHG) and has a high hydrogen density, which leads to NH_3 being a favorable alternative to hydrogen. Ammonia is easy to store and deliver in large quantities. It is one of the most transported chemicals worldwide as gas, liquid, or solid form. The ability of ammonia gas to become a liquid at low pressures means that it is a good "carrier" of hydrogen. Liquid ammonia contains more hydrogen by volume than compressed hydrogen or liquid hydrogen. For example, ammonia is over 50% more energy dense per gallon than liquid hydrogen, so ammonia can be stored and distributed easier than elemental hydrogen.

The use of ammonia as an energy carrier has several advantages, such as mature technologies for production and transportation. The synthesis process of ammonia is well studied (the Haber-Bosch process). Although ammonia synthesis is exothermic, in practice the production of ammonia from hydrogen and nitrogen incurs a small energy loss of 1.5 GJ/ton compared to the 28.4 GJ/ton energy stored in the ammonia [141].

The storage of ammonia is more convenient than hydrogen; for example, up to 50,000 tons of ammonia stored in insulated tanks at 1 bar and $-33°C$. The temperature is kept down by slow vaporization, and the ammonia vapor is continually compressed back into a liquid. For small tanks, below 1500 ton, ammonia is stored under pressure in stainless steel spheres [141]. Ammonia can be stored as a liquid; a standard tank with a capacity of $60,000\,m^3$ filled with ammonia contains about 211 GWh of energy, equivalent to the annual production of roughly 30 wind turbines on land [142].

Ammonia has been already used in solid oxide fuel cells (SOFC), and it can also be used in polymer electrolyte membranes (PEMFCs) and alkaline fuel cells (AFCs). However, before feeding to the fuel cell, ammonia needs

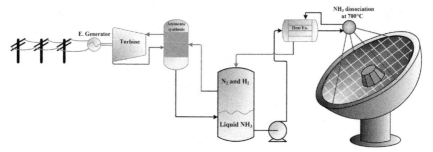

Fig. 14 Schematic of ammonia dissociation and storage [144].

to be split into hydrogen and nitrogen [143]. Dunn, Lovegrove [144] demonstrated that ammonia-based thermochemical storage with concentrating solar power systems are technically achievable (Fig. 14).

7 HYBRID ENERGY STORAGE (POLYSTORAGE)

The penetration of renewable energy sources (for example, wind and solar) into the electrical energy network is accelerating. The problem with these generation systems is that their energy sources are intermittent, and so the generated power is also intermittent, reducing stability, reliability, and power quality in the main electric grid. Energy storage is undoubtedly the best solution to address this challenge, along with demand-side management. However (as discussed in the earlier technology analysis sections), for most of the time, no single energy storage can address all technoeconomic requirements. To overcome these problems, a hybrid energy storage system (HESS) or polystorage system is required [145].

A HESS or polystorage system is differentiated by a favorable combination of two or more EES systems with complementary operating features (such as energy and power density, self-discharge rate, efficiency, lifetime, and so on) to provide an optimal solution not achievable by any single technology. The HESS typically includes storage technologies that separately cover sprinter loads required for fast response, and marathon loads required for peak shaving and load shifting. Therefore, it should have a storage system with both high energy and power densities. However, none of the currently available EES system technologies satisfy these two features. Fig. 15 shows the basic structure of a HESS.

Khalilpour et al. [145, 146] have introduced an optimization algorithm for optimal screening, selection, sizing, and scheduling of multiple energy

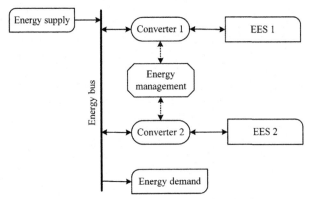

Fig. 15 Basic structure of a HESS.

storage systems along with multiple generator types for a given purpose. Generally, two complementary storage devices, one of high energy density and the other of high power density, form the HESS. The use of a unique EES system exhibiting high energy density but low power density creates problems with power control, since the response of these types of EES system is slow. In addition, a high-power supply may lessen the lifecycle of a storage system. Hence, by adding a short-storage system, the operating conditions of the main storage system are improved, prolonging its life cycle while simultaneously satisfying the power requirements. In addition, the use of a short-storage system in parallel with a long-storage one reduces the size and power losses of the main storage system [147].

Batteries, particularly lithium-ion, play a key role in many HESS-applications. They can be utilized both as "high energy" and as "high power" storage. Compared to batteries, supercapacitors and flywheels exhibit even higher power densities, efficiencies, and cycle lifetimes. For example, valve-regulated lead acid (VRLA) batteries suffer from flexibility in depth of charge and discharge. The mechanism of these batteries is based on the growth of lead sulfate crystals on the anode during discharging, and on dissolution during charging. If the battery is not fully charged and discharged then, over time, some crystals can stick to the electrode and resist dissolution. This mechanism is called *sulfation*, which reduces the life of a battery. On the other hand, a flexible battery is required to operate in various depths of charge and discharge, partial or full. To address this challenge, lead-acid batteries are integrated with supercapacitors to combine the benefits of both technologies, thus enabling a partial state of charge capability with an improved lifetime [148]. An example of such a technology is UltraBattery

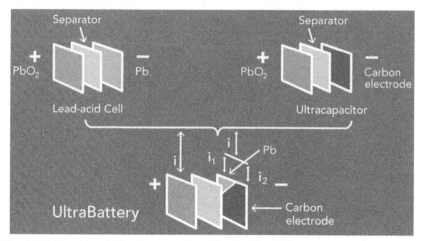

Fig. 16 Schematic of UltraBattery, a hybrid of lead-acid battery and supercapacitor [149].

[149] (see Fig. 16) with its potential use in various applications including hybrid vehicles [150].

Redox-flow batteries are also a promising technology due to their storage immanent decoupling of power and stored energy (similar to the hydrogen and power-to-gas storage path), and due to their good cycle lifetime and recyclability. Renewable hydrogen (H_2) and methane (CH_4) are both very promising options for long-term energy storage. In addition, heat storage and power-to-heat concepts will gain importance in the context of future HESS applications. The storage of heat produced from excessive renewable energy (via electric heating cartridges or heat pumps) and from power-to-gas conversion processes (for example, electrolyzers or fuel cells) will increase the overall utilization rate of renewable energies. Moreover, power-to-heat will enable HESS to perform peak shaving, and thereby significantly reduce the stress for the other storage components and for the public grid. Optimizing design, control, and energy management strategies for HESS at the interface between electricity, heat, and gas sectors will play an important role. It will unfold significant potentials for further improvements of cost, efficiency, and lifetimes of energy systems.

The shifting nature of wind and fluctuating load profiles are problematic for the operation of wind-based power systems, particularly when they operate in standalone mode. The random variation of wind speed leads to fluctuating torque of the wind turbine generator resulting in voltage and frequency excursions in the remote area power supply (RAPS) system. An ideal

ESS should be able to provide both high energy and power capacity to handle situations such as wind gusts or sudden load variations that may exist for a few seconds or even longer [151]. However, a single type of energy storage is not seen to satisfy both the power and energy requirements of the RAPS system; thus, there is a requirement to combine two or more energy storage systems to perform in a hybrid manner. In general, batteries and supercapacitors are perceived to provide high energy and power requirements, respectively. Therefore, the integration of a supercapacitor ensures a healthy operation of the battery storage by preventing it from operating in high depth of discharge (DOD) regions and allowing it to operate in low-frequency power regions (Fig. 17) [151].

Duke Energy recently installed a HESS that combines high-capacity batteries with fast-responding supercapacitors. The pilot project will provide peak demand response, load shifting, and support for a utility-owned 1.2-MW photovoltaic array [152]. Another HESS is ADELE at Saxony-Anhalt in Germany, which uses CAES and TES technologies to enhance efficiency and avoid fossil fuel consumption [9].

Zhao, Dai [153] proposed a HESS, which consists of an adiabatic compressed air energy storage system (A-CAES) and a FESS (Fig. 18). The A-CAES system is the high-power rating, high-energy rating, and slow ramp rate device. The FESS is the low-power rating, low-energy rating, and fast ramp rate device. The power generated from a windfarm is used to meet the system load. If the wind power supply exceeds the load demand, the surplus power will be divided into two branches through an energy dispatch system. One branch is stored in the FESS by accelerating the flywheel rotor via a motor. The electrical energy is converted from the kinetic energy

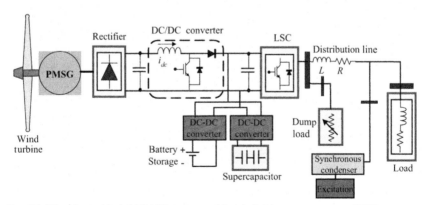

Fig. 17 Wind-based hybrid RAPS system with a hybrid energy storage [151].

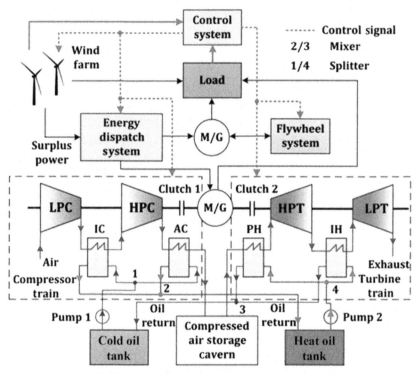

C: Compressor   M: Motor   G: Generator   LP: Low pressure   HP: High pressure
T: Turbine   IC: Intercooler   AC: Aftercooler   PH: Preheater   IH: Interheater
Fig. 18 A hybrid energy storage system with A-CAES and FESS [153].

of the flywheel rotor. The other branch is stored in the A-CAES system. In this situation, clutch 1 meshes whereas clutch 2 separates. The compressor train, driven by electrical energy, compresses the air from ambient pressure to a high-pressure level, and is then stored in the compressed air storage cavern. Meanwhile, the heat released from the compression process is absorbed in the intercooler (IC) and aftercooler (AC) by the thermal oil circulatory system to store in the hot oil tank.

On the other hand, if the wind power cannot match the load demand, the hybrid energy storage system will be discharged to compensate the electricity gap. For the FESS, the flywheel rotor will drive the generator to convert its kinetic energy into electrical energy, and the speed of the flywheel rotor decreases. For the A-CAES system, the compressed air is drawn from the storage cavern, and is heated in the preheater (pH) and interheater (IH) through heat transferring with the thermal oil from the thermal oil circulatory system. Then, the heated air is expanded in the HP turbine and LP turbine successively for converting most of the energy of compressed air into electrical energy.

As obvious from the name, the concept of the CAES technology is formed around the integration of air storage system with a gas-fired power generator. In a different approach, Khalilpour et al. [154] have further integrated the CAES system with inclusion of natural gas storage. Through this hybrid storage arrangement, a gas-power-generating plant installs both air and natural gas storage system to utilize their stored energy as well as the real economic value of natural gas following market dynamics (Fig. 19).

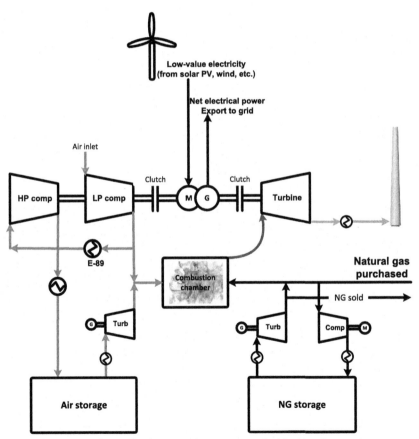

Fig. 19 Schematic of a gas-fired power plant integrated with air and natural gas storage systems [154].

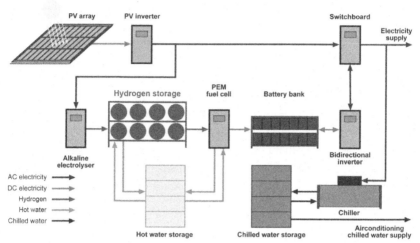

Fig. 20 Installed hydrogen-based energy system: Sir Samuel Griffith Centre, Griffith University, Brisbane, Australia [155].

A renewable and sustainable energy system based on hybrid energy storage (for example, hydrogen and Li-ion batteries) has been installed in the Sir Samuel Griffith Centre [155], Griffith University, Brisbane, Australia (Fig. 20). The goal of the system is to demonstrate an autonomous energy system, with long-term storage of energy in the form of hydrogen ensuring that the building will be operated off-grid as much as possible. The primary energy source is 330 kW of PV modules installed on the building itself. Surplus solar energy is first used to charge a bank of Li-ion batteries, with a capacity of 1.31 MWh electricity, to provide short-term energy storage. Once the batteries are fully charged, the surplus solar energy is used to generate hydrogen using alkaline water electrolyzers. Hydrogen storage is in a proprietary metal-hydride storage system with a total capacity of 120 kg H_2, equivalent to approximately 2.3 MWh electricity. Should the solar generation drop below the building requirements—that is, at night and in dull weather—then the batteries initially supply the building's energy requirements. When the batteries reach a set level of discharge, fuel cells power the building using stored hydrogen.

Recently, Toshiba Corporation installed a solar hydrogen autonomous energy system (Fig. 21) named H_2One at the Henn-na Hotel near Nagasaki, Japan [156]. The installed solar capacity is 62 kW, fuel cell output is up to 54 kW (depending on the number of installed modules), and the hydrogen storage capacity is approx. 1.3 MWh electric equivalent.

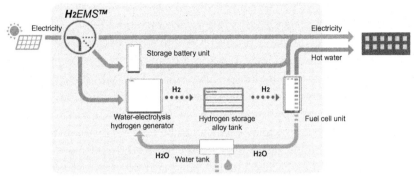

Fig. 21 Installed hydrogen-based energy system: Henn-na Hotel, at the Huis Ten Bosch theme park near Nagasaki, Kyushu, Japan [156].

8 SUMMARY AND CONCLUSIONS

This chapter gives an outline of various current state-of-the-art EES technologies. From this overview, it is apparent that existing EES technologies exhibit a wide range of technological characteristics. Hence, it might be helpful to combine different EES technologies to meet the different requirements of power system and network operations. Fig. 22 [12] illustrates a comparison of power ratings and rated energy capacities of EES technologies. From Fig. 20, EES technologies can be categorized by the nominal discharge time at rated power such as *discharge time* < *1 h* (flywheel, supercapacitors and SMES); *discharge time up to around 10 h* (aboveground, small-scale CAES,

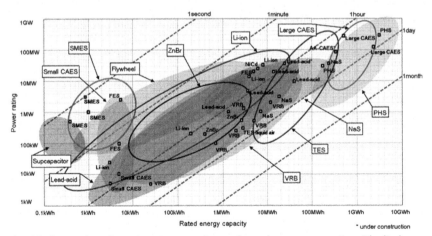

Fig. 22 Comparison between power rating and rated energy capacity with discharge times of EES technologies [12].

lead-acid, Li-ion, NiCd, ZnBr, and PSB); *discharge time longer than 10 h* (PHS, underground large-scale CAES, VRB, hydrogen, and TES).

The level of self-discharge of an EES system is one of the major factors in understanding the exact storage duration. Usually, PHS, CAES, NaS batteries, flow batteries, and hydrogen have small daily self-discharge ratios, which lead to energy that can be stored in long-term duration—possibly up to months. Secondary batteries (except NaS) have daily self-discharge ratios ranging from 0.03% to 5%, which can be used for medium-term storage durations—possibly up to days. Flywheel, SMES and supercapacitors have very high daily self-charge ratios, ranging from 10% to 100%; hence, they can only be used for short-term storage durations—perhaps up to hours. TES includes a variety of technologies, and thus it may be suitable for medium-term and/or long-term storage durations.

A comprehensive economic analysis of EES technologies is necessary to optimize the system output by considering the capital cost, operation cost, maintenance cost, and the impact of the equipment lifetime. Lifetime and cycling times are two factors that affect the overall investment cost. Low lifetime and low cycling times will increase the cost of maintenance and replacement. Besides which, EES plays an important role in energy management for optimizing energy uses and decoupling the timing of generation and consumption of electric energy. Time shifting and peak shaving are typical applications in energy management; hence, dynamic power management systems need to develop power systems and network operations to maximize the availability of power and minimize outages. However, many EES-based projects have been installed globally, but the wide-ranging deployment of such kinds of power system will depend on the advancement of EES technologies and the competitive benefits brought by EES.

REFERENCES

[1] Martinez-Frias J, et al. A coal-fired power plant with zero-atmospheric emissions. J Eng Gas Turb Power 2008;130(2). 023005. https://doi.org/10.1115/1.2771255.

[2] C2ES. Outcomes of the UN climate change conference in Paris., Available from: http://www.c2es.org/docUploads/cop-21-paris-summary-02-2016-final.pdf; 2015. Accessed 27 May 2016.

[3] Barbato A, Capone A. Optimization models and methods for demand-side management of residential users: a survey. Energies 2014;7(9):5787–824.

[4] Chen HS, et al. Progress in electrical energy storage system: a critical review. Prog Nat Sci 2009;19(3):291–312.

[5] Sulzberger C. Pearl street in miniature: models of the electric Generating Station [history]. IEEE Power Energy Mag 2013;11(2):76–85.

[6] Vassallo AM. Chapter 17: Applications of batteries for grid-scale energy storage. In: Lim CMS-KM, editor. Advances in batteries for medium and large-scale energy storage. Cambridge, UK; Waltham, MA, USA: Woodhead Publishing; 2015. p. 587–607.

[7] Baker JN, Collinson A. Electrical energy storage at the turn of the millennium. Power Eng J 1999;13(3):107–12.

[8] Decourt B, Debarre R. Electricity storage. In: Leading the enrgy transition factbook. Gravenhage, The Netherlands: Schlumberger Business Consulting (SBC) Energy Institute; 2013.

[9] Koohi-Kamali S, et al. Emergence of energy storage technologies as the solution for reliable operation of smart power systems: a review. Renew Sustain Energy Rev 2013;25:135–65.

[10] Akhil AA, et al. DOE/EPRI 2013 electricity storage handbook in collaboration with NRECA. US Department of Energy and EPRI: California; 2013.

[11] Global Energy Storage Database, Accessed via http://www.energystorageexchange.org.

[12] Luo X, et al. Overview of current development in electrical energy storage technologies and the application potential in power system operation. Appl Energy 2015;137: 511–36.

[13] Parfomak PW. Energy storage for power grids and electric transportation: a technology assessment. Congr Res Ser 2012;42455.

[14] Hydropower Market Report, 2014. Accessed via https://www.energy.gov/hydropower-market-report.

[15] Atdorf plant, Accessed via https://de.wikipedia.org/wiki/Pumpspeicherkraftwerk_Atdorf.

[16] Linthal 2015 project, Accessed via https://en.wikipedia.org/wiki/Linth–Limmern_Power_Stations.

[17] Portugal utility wins auctions for 256-MW Fridao, 136-MW Alvito, Accessed via http://www.hydroworld.com.

[18] Compressed air energy storage, Accessed via https://en.wikipedia.org/wiki/Compressed_air_energy_storage

[19] Amiryar ME, Pullen KR. A review of flywheel energy storage system technologies and their applications. Appl Sci 2017;7(3):286.

[20] Liu J. Addressing the grand challenges in energy storage. Adv Funct Mater 2013;23(8): 924–8.

[21] Chang Y, et al. Lead-acid battery use in the development of renewable energy systems in China. J Power Sources 2009;191(1):176–83.

[22] Ouyang L, et al. Progress of hydrogen storage alloys for Ni-MH rechargeable power batteries in electric vehicles: a review. Mater Chem Phys 2017;200(Suppl C): 164–78.

[23] Telaretti E, Dusonchet L. In: Stationary battery systems in the main world markets: Part 2: Main trends and prospects. Environment and Electrical Engineering and 2017 IEEE Industrial and Commercial Power Systems Europe (EEEIC/I&CPS Europe), 2017 IEEE International Conference on IEEE; 2017.

[24] Leadbetter J, Swan LG. Selection of battery technology to support grid-integrated renewable electricity. J Power Sources 2012;216:376–86.

[25] Gray ML, et al. Parametric study of solid amine sorbents for the capture of carbon dioxide. Energy Fuel 2009;23:4840–4.

[26] Schoenung S. Energy Storage Systems Cost Update A Study for the DOE Energy Storage Systems Program (Report SAND2011-2730). New Mexico and California: Sandia National Laboratories; 2011. Retrieved from, http://prod/.sandia.gov/techlib/accesscontrol.cgi/2011/112730.pdf.

[27] Kumar D, et al. Progress and prospects of sodium-sulfur batteries: a review. Solid State Ion 2017;312:8–16.

[28] Vormedal I. The influence of business and industry NGOs in the negotiation of the Kyoto mechanisms: the case of carbon capture and storage in the CDM. Glob Environ Pol 2008;8(4):36–65.

[29] Sakai T, Uehara I, Ishikawa H. R&D on metal hydride materials and Ni–MH batteries in Japan. J Alloys Compd 1999;293–295(Suppl C):762–9.

[30] Liu Y, et al. Advanced hydrogen storage alloys for Ni/MH rechargeable batteries. J Mater Chem 2011;21(13):4743–55.

[31] Leung P, et al. Recent developments in organic redox flow batteries: a critical review. J Power Sources 2017;360(Suppl C):243–83.

[32] Vynnycky M. Analysis of a model for the operation of a vanadium redox battery. Energy 2011;36(4):2242–56.

[33] Al-Fetlawi HA-ZA. Modelling and simulation of all-vanadium redox flow batteries. University of Southampton; 2011.

[34] Mazzoldi A, Hill T, Colls JJ. CO_2 transportation for carbon capture and storage: Sublimation of carbon dioxide from a dry nice bank. Int J Greenh Gas Control 2008;2(2):210–8.

[35] Baxter R. Energy storage: a nontechnical guide. Tulsa, OK, USA: PennWell Books; 2006.

[36] [Anon]Doosan Babcock aims to lead on carbon capture. Prof Eng 2008;21(16):7.

[37] Group, U.S.E.R. Study of Electricity Storage Technologies and Their Potential to Address Wind Energy Intermittency in Ireland. University College Cork; 2004. Final Report RE/HC/103/001.

[38] Blankinship S. Carbon capture and economics dominate COAL-GEN. Power Eng 2008;112(9):20.

[39] Sharma P, Bhatti T. A review on electrochemical double-layer capacitors. Energ Conver Manage 2010;51(12):2901–12.

[40] Zhong C, et al. A review of electrolyte materials and compositions for electrochemical supercapacitors. Chem Soc Rev 2015;44(21):7484–539.

[41] Vangari M, Pryor T, Jiang L. Supercapacitors: review of materials and fabrication methods. J Energy Eng 2012;139(2):72–9.

[42] Ferrier M. Stockage d'energie dans un enroulement supraconducteur. Low Temp Electr Power 1970;425–32.

[43] Buckles W, Hassenzahl WV. Superconducting magnetic energy storage. IEEE Power Eng Rev 2000;20(5):16–20.

[44] Yuan W. Second-generation high-temperature superconducting coils and their applications for energy storage. London: Springer Science & Business Media; 2011.

[45] Ribeiro PF, et al. Energy storage systems for advanced power applications. Proc IEEE 2001;89(12):1744–56.

[46] Amaro N, et al. Superconducting magnetic energy storage.

[47] Kaldellis JK. Stand-alone and hybrid wind energy systems: technology, energy storage and applications. Cambridge, UK: Woodhead Publishing; 2010.

[48] Mckenna P. Superconducting magnets for grid-scale storage. MIT Technology Review Energy; 2011.

[49] Kousksou T, et al. Energy storage: applications and challenges. Sol Energy Mater Sol Cells 2014;120:59–80.

[50] Dincer I, Rosen M. Thermal energy storage: systems and applications. Chichester, UK: John Wiley & Sons; 2002.

[51] Kuravi S, et al. Thermal energy storage technologies and systems for concentrating solar power plants. Prog Energy Combust Sci 2013;39(4):285–319.

[52] Pardo P, et al. A review on high temperature thermochemical heat energy storage. Renew Sust Energy Rev 2014;32(Suppl C):591–610.

[53] Xu J, Wang RZ, Li Y. A review of available technologies for seasonal thermal energy storage. Sol Energy 2014;103(Suppl C):610–38.

[54] Bauer D, et al. German central solar heating plants with seasonal heat storage. Sol Energy 2010;84(4):612–23.

[55] Charles D. Energy research. Stimulus gives DOE billions for carbon-capture projects. Science 2009;323(5918):1158.

[56] Zhao D, et al. Optimal study of a solar air heating system with pebble bed energy storage. Energ Conver Manage 2011;52(6):2392–400.

[57] ElKady AM, et al. Application of exhaust gas recirculation in a DLN F-class combustion system for postcombustion carbon capture. J Eng Gas Turbines Power 2009;131 (3)034505. https://doi.org/10.1115/1.2982158.

[58] Wang X, et al. Heat storage performance of the binary systems neopentyl glycol/pentaerythritol and neopentyl glycol/trihydroxy methyl-aminomethane as solid–solid phase change materials. Energ Conver Manage 2000;41(2):129–34.

[59] Sharma SD, Sagara K. Latent heat storage materials and systems: a review. Int J Green Energy 2005;2(1):1–56.

[60] Morrison D, Abdel-Khalik S. Effects of phase-change energy storage on the performance of air-based and liquid-based solar heating systems. Sol Energy 1978;20(1):57–67.

[61] Ghoneim A. Comparison of theoretical models of phase-change and sensible heat storage for air and water-based solar heating systems. Sol Energy 1989;42(3):209–20.

[62] Fukai J, et al. Improvement of thermal characteristics of latent heat thermal energy storage units using carbon-fiber brushes: experiments and modeling. Int J Heat Mass Transfer 2003;46(23):4513–25.

[63] Zhang Z, et al. Preparation and thermal energy storage properties of paraffin/expanded graphite composite phase change material. Appl Energy 2012;91(1):426–31.

[64] Benli H, Durmuş A. Performance analysis of a latent heat storage system with phase change material for new designed solar collectors in greenhouse heating. Sol Energy 2009;83(12):2109–19.

[65] Perryman, V.D., Thermal energy storage and delivery system. 2015, United States. https://patents.google.com/patent/US9115937B2/en.

[66] Schaube F, Antje W, Tamme R. High temperature thermochemical heat storage for concentrated solar power using gas–solid reactions. J Sol Energy Eng 2011;133(3) 031006.

[67] Aydin D, Casey SP, Riffat S. The latest advancements on thermochemical heat storage systems. Renew Sustain Energy Rev 2015;41:356–67.

[68] N'Tsoukpoe KE, et al. A review on long-term sorption solar energy storage. Renew Sustain Energy Rev 2009;13(9):2385–96.

[69] Wongsuwan W, et al. A review of chemical heat pump technology and applications. Appl Therm Eng 2001;21(15):1489–519.

[70] Cot-Gores J, Castell A, Cabeza LF. Thermochemical energy storage and conversion: a-state-of-the-art review of the experimental research under practical conditions. Renew Sustain Energy Rev 2012;16(7):5207–24.

[71] Smid K. Carbon dioxide capture and storage: a mirage. GAIA Ecol Perspect Sci Soc 2009;18(3):205–7.

[72] Stolzenburg K, Mubbala R. Hydrogen liquefaction report. Integrated Design for Demonstration of Efficient Liquefaction of Hydrogen (IDEALHY); 2013.

[73] Klell M, Kindermann H, Jogl C. In: Thermodynamics of gaseous and liquid hydrogen storage. Proceedings of international hydrogen energy congress and exhibition IHEC; 2007.

[74] Hirose K, Hirscher M. Handbook of hydrogen storage: new materials for future energy storage. New York: John Wiley & Sons; 2010.

[75] Schumann D. Public acceptance of carbon dioxide capture and storage research approaches for investigating the impact of communication. GAIA Ecol Perspect Sci Soc 2009;18(3):261–3.

[76] Sadaghiani MS, Mehrpooya M. Introducing and energy analysis of a novel cryogenic hydrogen liquefaction process configuration. Int J Hydrogen Energy 2017;42(9): 6033–50.

[77] Dalebrook AF, et al. Hydrogen storage: beyond conventional methods. Chem Commun 2013;49(78):8735–51.

[78] Durbin DJ, Malardier-Jugroot C. Review of hydrogen storage techniques for on board vehicle applications. Int J Hydrogen Energy 2013;38(34):14595–617.

[79] Schlapbach L, Züttel A. Hydrogen-storage materials for mobile applications. Nature 2001;414(6861):353–8.

[80] Peigney A, et al. Specific surface area of carbon nanotubes and bundles of carbon nanotubes. Carbon 2001;39(4):507–14.

[81] McAllister MJ, et al. Single sheet functionalized graphene by oxidation and thermal expansion of graphite. Chem Mater 2007;19(18):4396–404.

[82] Burress JW, et al. Graphene oxide framework materials: theoretical predictions and experimental results. Angew Chem Int Ed 2010;49(47):8902–4.

[83] Wood CD, et al. Hydrogen storage in microporous hypercrosslinked organic polymer networks. Chem Mater 2007;19(8):2034–48.

[84] Ramimoghadam D, Gray EM, Webb C. Review of polymers of intrinsic microporosity for hydrogen storage applications. Int J Hydrogen Energy 2016;41(38):16944–65.

[85] Germain J, Fréchet JM, Svec F. Hypercrosslinked polyanilines with nanoporous structure and high surface area: potential adsorbents for hydrogen storage. J Mater Chem 2007;17(47):4989–97.

[86] Ding S-Y, Wang W. Covalent organic frameworks (COFs): from design to applications. Chem Soc Rev 2013;42(2):548–68.

[87] Liu S, Yang X. Gibbs ensemble Monte Carlo simulation of supercritical CO_2 adsorption on NaA and NaX zeolites. J Chem Phys 2006;124(24):244705.

[88] Weitkamp J, Fritz M, Ernst S. Zeolites as media for hydrogen storage. Int J Hydrogen Energy 1995;20(12):967–70.

[89] Li Y, Yang RT. Significantly enhanced hydrogen storage in metal-organic frameworks via spillover. J Am Chem Soc 2006;128(3):726–7.

[90] Graetz J. New approaches to hydrogen storage. Chem Soc Rev 2009;38(1):73–82.

[91] Cote AP, et al. Porous, crystalline, covalent organic frameworks. Science 2005;310 (5751):1166–70.

[92] O'Keeffe M, et al. Frameworks for extended solids: geometrical design principles. J Solid State Chem 2000;152(1):3–20.

[93] Gray EM, et al. Hydrogen storage for off-grid power supply. Int J Hydrogen Energy 2011;36(1):654–63.

[94] Buschow KH, Van Mal HH. Phase relations and hydrogen absorption in the lanthanum-nickel system. J Less Common Met 1972;29(2):203–10.

[95] Sakintuna B, Lamari-Darkrim F, Hirscher M. Metal hydride materials for solid hydrogen storage: a review. Int J Hydrogen Energy 2007;32(9):1121–40.

[96] Webb C. A review of catalyst-enhanced magnesium hydride as a hydrogen storage material. J Phys Chem Solid 2015;84:96–106.

[97] Ren J, et al. Current research trends and perspectives on materials-based hydrogen storage solutions: a critical review. Int J Hydrogen Energy 2016;42:289–311.

[98] Tamura T, et al. Protium absorption properties and protide formations of Ti–Cr–V alloys. J Alloys Compd 2003;356–357:505–9.

[99] Challet S, Latroche M, Heurtaux F. Hydrogenation properties and crystal structure of the single BCC $(Ti_{0.355}V_{0.645})_{100-x}M_x$ alloys with M=Mn, Fe, Co, Ni (x=7, 14 and 21). J Alloys Compd 2007;439(1–2):294–301.

[100] Cho S-W, et al. Hydrogen absorption–desorption properties of $Ti_{0.32}Cr_{0.43}V_{0.25}$ alloy. J Alloys Compd 2007;430(1–2):136–41.

[101] Karkamkar A, Aardahl C, Autrey T. Recent developments on hydrogen release from ammonia borane. Mater Matters 2007;2(2):6–9.

[102] Ares JR, Cuevas F, Percheron-Guégan A. Influence of thermal annealing on the hydrogenation properties of mechanically milled AB5-type alloys. Mater Sci Eng B 2004;108(1–2):76–80.

[103] Cuevas F, Korablov D, Latroche M. Synthesis, structural and hydrogenation properties of Mg-rich MgH_2–TiH_2 nanocomposites prepared by reactive ball milling under hydrogen gas. Phys Chem Chem Phys 2012;14(3):1200–11.

[104] Chen J, Dou SX, Liu HK. Crystalline Mg_2Ni obtained by mechanical alloying. J Alloys Compd 1996;244(1–2):184–9.

[105] Zhong Y, et al. Ab initio computational studies of mg vacancy diffusion in doped MgB_2 aimed at Hydriding kinetics enhancement of the $LiBH_4$+ MgH_2 system. J Phys Chem C 2010;114(49):21801–7.

[106] Berube V, et al. Size effects on the hydrogen storage properties of nanostructured metal hydrides: a review. Int J Energy Res 2007;31(6–7):637–63.

[107] Lozano Martinez GA. Development of hydrogen storage systems using sodium alanate.

[108] Nielsen TK, et al. Improved hydrogen storage kinetics of nanoconfined $NaAlH_4$ catalyzed with $TiCl_3$ nanoparticles. ACS Nano 2011;5(5):4056–64.

[109] Bogdanović B, et al. Metal-doped sodium aluminium hydrides as potential new hydrogen storage materials. J Alloys Compd 2000;302(1):36–58.

[110] Pitt M, et al. Hydrogen absorption kinetics and structural features of $NaAlH_4$ enhanced with transition-metal and Ti-based nanoparticles. Int J Hydrogen Energy 2012;37(20):15175–86.

[111] Zaluska A, Zaluski L, Ström-Olsen JO. Method of fabrication of complex alkali metal hydrides; 2001. Google Patents.

[112] Hu YH, Ruckenstein E. H_2 storage in Li_3N. Temperature-programmed hydrogenation and dehydrogenation. Ind Eng Chem Res 2003;42(21):5135–9.

[113] Schlesinger H, Brown HC. Metallo borohydrides. III. Lithium borohydride. J Am Chem Soc 1940;62(12):3429–35.

[114] Züttel A, et al. Hydrogen storage properties of $LiBH_4$. J Alloys Compd 2003;356–357: 515–20.

[115] Züttel A, Borgschulte A, Orimo S-I. Tetrahydroborates as new hydrogen storage materials. Scr Mater 2007;56(10):823–8.

[116] Christian M, Aguey-Zinsou K-F. Destabilisation of complex hydrides through size effects. Nanoscale 2010;2(12):2587–90.

[117] Pinkerton FE, et al. Hydrogen desorption exceeding ten weight percent from the new quaternary hydride $Li_3BN_2H_8$. J Phys Chem B 2005;109(1):6–8.

[118] Vajo JJ, et al. Hydrogen-generating solid-state hydride/hydroxide reactions. J Alloys Compd 2005;390(1–2):55–61.

[119] Urgnani J, et al. Hydrogen release from solid state $NaBH_4$. Int J Hydrogen Energy 2008;33(12):3111–5.

[120] Orimo S, Nakamori Y, Züttel A. Material properties of MBH_4 (M=Li, Na and K). Mater Sci Eng B 2004;108(1–2):51–3.

[121] Chłopek K, et al. Synthesis and properties of magnesium tetrahydroborate, $Mg(BH_4)_2$. J Mater Chem 2007;17(33):3496–503.

[122] Rönnebro E, Majzoub EH. Calcium borohydride for hydrogen storage: catalysis and reversibility. J Phys Chem B 2007;111(42):12045–7.

[123] Miwa K, et al. First-principles study on thermodynamical stability of metal borohydrides: aluminum borohydride $Al(BH_4)_3$. J Alloys Compd 2007;446–447:310–4.

[124] Jeon E, Cho Y. Mechanochemical synthesis and thermal decomposition of zinc borohydride. J Alloys Compd 2006;422(1–2):273–5.

[125] van Setten MJ, de Wijs GA, Brocks G. First-principles calculations of the crystal structure, electronic structure, and thermodynamic stability of Be(BH$_4$)$_2$. Phys Rev B 2008;77(16):165115.

[126] Zhou Y, et al. Functions of MgH$_2$ in hydrogen storage reactions of the 6LiBH$_4$–CaH$_2$ reactive hydride composite. Dalton Trans 2012;41(36):10980–7.

[127] Paskevicius M, et al. Metal borohydrides and derivatives–synthesis, structure and properties. Chem Soc Rev 2017;46(5):1565–634.

[128] Dovgaliuk I, Filinchuk Y. Aluminium complexes of B- and N-based hydrides: synthesis, structures and hydrogen storage properties. Int J Hydrogen Energy 2016;41(34):15489–504.

[129] Hansen BR, et al. Metal boranes: progress and applications. Coord Chem Rev 2016;323:60–70.

[130] Yvon K, Renaudin G. Hydrides: solid state transition metal complexes. Chichester, UK: John Wiley & Sons; 2005.

[131] Yvon K, Bertheville B. Magnesium based ternary metal hydrides containing alkali and alkaline-earth elements. J Alloys Compd 2006;425(1):101–8.

[132] Sheppard D, Humphries T, Buckley C. Sodium-based hydrides for thermal energy applications. Appl Phys A 2016;122(4):406.

[133] Humphries TD, Sheppard DA, Buckley CE. Recent advances in the 18-electron complex transition metal hydrides of Ni, Fe, Co and Ru. Coord Chem Rev 2017; https://doi.org/10.1016/j.ccr.2017.04.001.

[134] Sternberg A, Bardow A. Power-to-what? – environmental assessment of energy storage systems. Energ Environ Sci 2015;8(2):389–400.

[135] Müller B, et al. Energiespeicherung mittels Methan und energietragenden Stoffen–ein thermodynamischer Vergleich. Chemie Ingenieur Technik 2011;83(11):2002–13.

[136] Pielke RA. An idealized assessment of the economics of air capture of carbon dioxide in mitigation policy. Environ Sci Policy 2009;12(3):216–25.

[137] Räuchle K, et al. Methanol for renewable energy storage and utilization. Energ Technol 2016;4(1):193–200.

[138] Rihko-Struckmann LK, et al. Assessment of methanol synthesis utilizing exhaust CO$_2$ for chemical storage of electrical energy. Ind Eng Chem Res 2010;49(21):11073–8.

[139] Jadhav SG, et al. Catalytic carbon dioxide hydrogenation to methanol: a review of recent studies. Chem Eng Res Des 2014;92(11):2557–67.

[140] Amouroux J, et al. Carbon dioxide: a new material for energy storage. Prog Nat Sci: Mater Int 2014;24(4):295–304.

[141] Klerke A, et al. Ammonia for hydrogen storage: challenges and opportunities. J Mater Chem 2008;18(20):2304–10.

[142] Anon. Renewed interest for US IGCC, carbon-capture coal power plant. Hydrocarb Process 2009;88(5):17.

[143] Lipman T, Shah N. Ammonia as an alternative energy storage medium for hydrogen fuel cells: Scientific and technical review for near-term stationary power demonstration projects. Final Report, UC Berkeley Transportation Sustainability Research Center; 2007.

[144] Dunn R, Lovegrove K, Burgess G. A review of ammonia-based thermochemical energy storage for concentrating solar power. Proc IEEE 2012;100(2):391–400.

[145] Khalilpour KR, Vassallo A. A generic framework for distributed multi-generation and multi-storage energy systems. Energy 2016;114:798–813.

[146] Khalilpour R, Vassallo A. Community Energy Networks With Storage: Modeling Frameworks for Distributed Generation. In: Energy policy, economics, Management & Transport. Singapore: Springer; 2016.

[147] Li W, Joos G. In: A power electronic interface for a battery supercapacitor hybrid energy storage system for wind applications. Power Electronics Specialists Conference, 2008. PESC. IEEE; 2008.

[148] Lam LT, et al. VRLA Ultrabattery for high-rate partial-state-of-charge operation. J Power Sources 2007;174(1):16–29.

[149] Cooper A, et al. The UltraBattery—a new battery design for a new beginning in hybrid electric vehicle energy storage. J Power Sources 2009;188(2):642–9.

[150] Lam LT, Louey R. Development of ultra-battery for hybrid-electric vehicle applications. J Power Sources 2006;158(2):1140–8.

[151] Mendis N, Muttaqi KM, Perera S. Management of battery-supercapacitor hybrid energy storage and synchronous condenser for isolated operation of PMSG based variable-speed wind turbine generating systems. IEEE Trans Smart Grid 2014;5(2):944–53.

[152] Yang WC, Hoffman J. Exploratory design study on reactor configurations for carbon dioxide capture from conventional power plants employing regenerable solid sorbents. Ind Eng Chem Res 2009;48(1):341–51.

[153] Zhao P, Dai Y, Wang J. Design and thermodynamic analysis of a hybrid energy storage system based on A-CAES (adiabatic compressed air energy storage) and FESS (flywheel energy storage system) for wind power application. Energy 2014;70:674–84.

[154] Khalilpour KR, Grossmann I, Vassallo A. Integrated power-to-gas and gas-to-power with air and natural gas storage; 2017. Sydney.

[155] [Anon]Carbon capture could earn UK 5bn pound a year. Prof Eng 2009;22(16):13.

[156] Jockenhovel T. CCS-project on the rise. Bwk 2009;61(6):28–9.

CHAPTER 5

Interconnected Electricity and Natural Gas Supply Chains: The Roles of Power to Gas and Gas to Power

Kaveh Rajab Khalilpour
Faculty of Engineering and Information Technology, Monash University, Melbourne, VIC, Australia

Abstract

Historically, the interaction between the gas and electricity networks has been one directional: the electricity network has been one of the consumers of the natural gas network for running gas power plants (i.e., gas to power, GtP). With the increased penetration of renewable energies into the electricity grid, a new challenge of electron oversupply and undersupply arises due to variable renewable resource availability. This challenge has created abundant interest in the "power-to-gas (PtG)" concept, where surplus electrons from the electricity network can be utilized for hydrogen or methane production to feed into the natural gas network. In this chapter, we investigate the PtG and GtP concepts and analyze the challenges and opportunities of available options.

Keywords: Energy storage, Electricity supply chains, Natural gas polyutilization, Power to gas, Gas to power, PtG, P2G, GtP, G2P, Hydrogen economy, Water electrolysis

1 NATURAL GAS SUPPLY CHAIN

1.1 Natural Gas Polyutilization and the Challenges

Until a few centuries ago, wood was the only energy source before coal was explored and gradually became the dominant source of global energy. The age of coal continued until oil and gas joined the mix. Gradually, along with these energy sources, hydro, nuclear, and renewables diversified the energy portfolio. Unlike renewables, fossil fuel sources are not distributed evenly around the world, and this has been the original cause of numerous economic and political crises within the last century. The so-called value chain of fossil fuel energy sources has several components: exploration/

Polygeneration with Polystorage
https://doi.org/10.1016/B978-0-12-813306-4.00005-7

133

production, storage (at production), shipping, storage (consumer side), and consumption.

In many cases, producers and consumers are distant from each other. Shipping accounts for a major proportion of the delivered energy costs over a value chain [1], a factor that indicates the importance of the (volumetric) energy intensity of fuels. For instance, coal as a solid fuel has very high energy intensity but its loading/offloading is difficult and pipeline transportation is not a viable option. Crude oil is an excellent fuel source with high energy intensity; furthermore, its liquid phase makes transportation easy using pipeline or giant tankers (easy loading/offloading). Due to these features, regions with major crude oil sources have been the focus of political sensitivities since their discovery in nineteenth century [2]. On the other hand, natural gas (NG) is the poorest fossil fuel in terms of energy intensity but the best for environmental impact. Table 1 summarizes the typical emission data for fossil fuels. Though coal is the most abundant fuel worldwide, it is of great concern due to its high carbon intensity (CI). The typical CIs of coal, oil, and NG are 24.5, 20.3, and 13.8 kg-C/MJ, respectively [3]. This means that NG, on average, emits close to half the CO_2 emitted by coal (and around 70% of the emissions from crude oil).

A lower C/H ratio and thus lower carbon emissions compared to oil and coal, along with reduced emission of oxides (nitrogen and sulfur) and particulates, make NG a very attractive environmental option. More importantly, the costs of processes based on NG, such as power generation, are much lower than for coal or oil [5]. As policies and/or mechanisms for carbon taxes/credits/penalties are considered, the economic advantages of NG increase. All these features contribute to the rapid growth of NG exploration, processing, and consumption. The current growth in energy demand, along with global concerns and treaties relating to climate protection, will continue to foster NG demand. It is projected that NG demand will dominate coal by 2035 [6].

Table 1 Fossil fuel emission levels lb./10^8Btu of energy input [4]

Pollutant	Natural gas	Oil	Coal
Carbon dioxide	117,000	164,000	208,000
Carbon monoxide	40	33	208
Nitrogen oxides	92	448	457
Sulfur oxides	1	1122	2591
Particulates	7	84	2744
Mercury	0.000	0.007	0.016

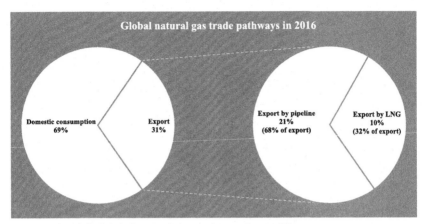

Fig. 1 Status of the global natural gas trade as of 2016 [8].

Although NG is the most favored fossil fuel resource, its transportation to demand sites poses a challenge due to its gaseous state. In other words, the disconnect between remote and offshore gas reservoirs on the one hand and markets on the other has obstructed a fully developed market and globally traded commodity status for NG [1, 7]. As a result, NG trade has largely been through pipelines between limited supply countries and their neighbors. This has prevented the NG market from becoming fully developed, like crude oil, and NG is not yet a globally traded commodity. Fig. 1 shows the global NG market in 2016. More than two-thirds of the 125.4 trillion cubic feed (tcf) NG produced in that year was consumed domestically; only 31% (i.e., 38.3 tcf) was exported. Of the export, more than two-thirds was carried out by pipeline, with liquefied natural gas (LNG) forming the other third [8].

For these reasons, extensive research has been underway over the last six decades to develop alternative gas utilization options for NG: LNG, compressed natural gas (CNG), gas to liquids (GTL), gas to chemicals (GTC), gas to solids (GTS), and gas to wire (GTW) [9] (Fig. 2). All these aim to reduce NG volume, to make its transport practical for export over long distances and reduce transport costs. For instance, NG liquefaction reduces the volume about 600-fold and notably reduces transportation costs. These utilization options and their "sweet spots" have been discussed elsewhere [10].

1.2 NG Demand-Side Management Issues

The volume intensity issue for national gas companies does not just apply to procurement and delivery from overseas suppliers to local consumers. It even causes problems for local users. Given the difficulty of storage, end

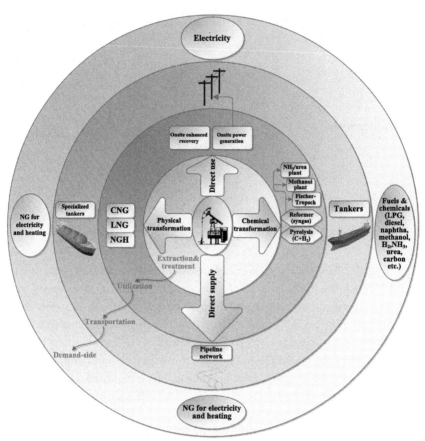

Fig. 2 NG supply chain and various utilization and transportation routes for its ultimate consumption.

users become vulnerable to any interruption in NG supply. Furthermore, in liberalized local markets, electricity and gas prices fluctuate dynamically, based on supply-demand balances. For local consumers (e.g., a gas-powered electricity generator), therefore, it becomes critical to plan energy demand and make future contracts, which minimize the NG purchase costs from the (generally high-priced) spot market. Again, the main obstacle in this business, for the market players, is the low-volume intensity of NG, which makes the storage a challenge. This issue was addressed a few decades ago by storing NG in depleted reservoirs or by building small LNG processing plants, called peak shavers [11,12]. The first successful LNG peak-shaving unit was built in Racine, Wisconsin, in 1965, with a storage size of ~25 million scf. This was followed by a larger unit (~600 million scf) in Portland, Oregon, in 1967 [13].

Take-or-pay contracts are common traditional mechanisms, which still exist in the NG market. The advantage of such contracts is the lower contracted price of the commodity; the disadvantage is lack of compensation for surplus quantity. However, peak shavers allow a gas power generating company to store the surplus contracted NG during low-demand electricity periods for consumption during high-demand periods. Moreover, such storage arrangements can provide companies the flexibility of buying and storing NG in low-price periods for later use when NG and electricity prices are high.

In some cases, gas-fired power companies have used gas pipelines to procure and store NG at elevated pressures (and thus in greater quantity) within pipes for later use. An example is Coloongra, a 667-MW peak-following power plant in Australia. Because the existing gas pipeline could not supply the peak demand of the power plant, the company installed nine kilometers of looped 42-in. storage pipeline along with a compressor station that increased gas pressure from 34 bar to 130 bar and a let-down station [14]. We continue this discussion further after introducing the electricity supply chain and its contemporary management issues.

2 ELECTRICAL ENERGY SUPPLY CHAIN

2.1 Electricity Generation and Transmission Issues

The supply chain management of electron energy is very different from that of fuel energies (gas, liquid, or solid form). Fig. 3 presents a general schematic of the conventional electricity transmission and distribution network, which comprises a generator, transmission and distribution network (analogous to the pipeline for NG), and end user. On the one hand, electricity can be transmitted instantaneously from one location to another, where it would be very costly and time consuming for fuel energies. On the other hand, management of electron during oversupply periods is a challenge compared with that of fuels. This limitation immediately provoked the need for energy storage from the very early days of electricity market development.

2.2 Storage Options for Electricity

It was 20 years after the invention of rechargeable lead-acid batteries, in 1859 [15], when Thomas Edison invented the light bulb, subsequently developing the first commercial power plant in 1882 in New York City's financial district for lighting the shops and attracting customers [16]. Soon the demand increased, and lead-acid batteries were found as a solution, storing

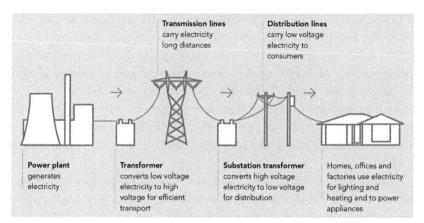

Fig. 3 Schematic of electricity transmission and distribution network. *(Image: courtesy of Australian Energy Market Operator, AEMO.)*

electricity during low-demand periods and selling it to the shops at peak evening times. Since then, battery storage and other energy storage types have been developed. Fig. 4 shows various types of energy storage options. A comprehensive review of energy storage technologies can be found in the plentiful literature, especially in [17,18].

The easiest and least-cost form of storage might be thermal storage, in which energy is used to heat (sensible or latent) or cool a solid/liquid material. The stored energy can then be used later for heating, cooling, or electricity generation. Phase change materials (PCM) play a key role in the advancement of thermal storage with special attention to solar energy storage [19,20].

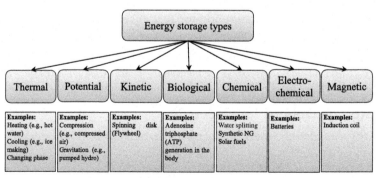

Fig. 4 Various types of electrical energy storage.

Potential energy is another form of storage, achieved by introducing pressure or tension to a medium. For instance, the compression of NG not only reduces its volume and facilitates transportation, but also means that the pressure can be recovered (subject to some loss) at a later time or another location to generate work. In the electricity industry, pumped hydro has a long history of electricity generation and storage. Compressed air has also been introduced recently as an economical approach to electricity storage [18]. The advantage of potential energy storage is its resilience in terms of duration of storage.

In contrast to potential energy storage, kinetic energy storage is used for temporary storage (in seconds or minutes). An example is a flywheel, which is still in development phase. The most complex form of energy storage might be chemical and biological storage, which are the easiest to observe, in our own bodies! Batteries are categorized as electrochemical energy storage. Magnetic energy storage is the form of storing electrons under an electromagnetic field.

The objective of electricity storage has also become far more than the initial intention of peak-shaving or short-term outage prevention [21]. Today, EES is used for many other reasons, some of which are delay of capacity/network expansion, frequency regulation, and voltage balancing (prevent brownouts) [22]. As such, each of the energy storage technologies is suitable for a given objective. They are usually categorized based on a time scale of applications, such as instantaneous (less than a few seconds), short term (less than a few minutes), midterm (less than a few hours), and long term (days) [23]. A detailed background of the historical development of various energy storage options can be found in the *Electricity Storage Handbook* published by Sandia National Laboratories [24].

Attention to electricity storage was triggered when a number of intermittent renewable power sources, especially PV and wind, emerged in various sizes from a few kilowatts to hundreds of megawatts. These power sources, whether grid-connected or offgrid (standalone), require storage (for load balancing) due to their output intermittency as a result of weather/seasonal fluctuations. Historically, pumped hydro has been the dominant option for electricity storage at large centralized power stations, due to its notably lower cost compared with others [25]. However, this popular storage option is geographically limited and is not available/feasible for all levels of grid use including distribution and community level (e.g., residential and commercial), for obvious reasons. As such, a portfolio of storage options is investigated for various applications based on their typical power ratings and discharge times (Chapter 4). Those with lower discharge time are suitable for power quality management and uninterruptible power supply. Those of large size and very high discharge times are suitable for bulk power management. In the middle are storage systems suitable for transmission and distribution (T&D) support and load shifting.

3 COOPERATIVE GAS AND ELECTRICITY GRID

3.1 Bidirectional Interaction

We have already separately discussed NG and electricity supply chains. Traditionally, the linkage between these two has been one-way, in that the electricity supply chain has been a consumer of NG (see Fig. 5A). In other words, gas has been converted to power (gas to power, GtP) in gas-fired power plants. In recent years, however, with the increase in renewable power generation technology installations, the reverse direction has also received attention, i.e., the conversion of surplus power to gas (PtG) [26]. Thus, the traditional one-directional relationship of NG and electricity supply chains is being transformed into a bidirectional crosslinked form (see Fig. 5B). This bidirectional interaction could improve the flexibility of both markets, reduce operational risks, and result in a lower cost of delivered energy for end users.

The core motivation of PtG is to utilize surplus electrical power from the electricity network for application in the NG network. We have already discussed various options for electrical energy storage during oversupply

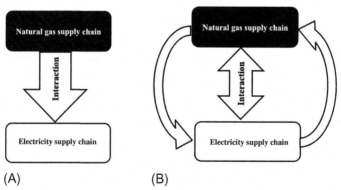

(A)   (B)

Fig. 5 (A) Conventional one-directional electricity and NG market interaction where electricity market is a consumer of NG fuel grid versus (B) the future bidirectional market.

(Fig. 4). From among these, PtG could occur in two ways: (1) a chemical approach, involving use of power for the electrochemical production of gases, especially hydrogen and methane; (2) a physical approach: use of power for pressurization or liquefaction of a gas (air, NG, etc.). We explain each route next. Also, a schematic of bidirectional electricity and the NG network is illustrated in Fig. 6, which we refer to in several sections of this chapter.

3.1.1 PtG: An Electrochemical Approach

Usage of electricity in an electrochemical process is the most important PtG option and has a long history. Interestingly, the science of electrochemistry was born with PtG experiments, which will be soon discussed. Electrochemical PtG begins with hydrogen generation and can have various subsequent processes for the production of complex chemicals. Here, we divide electrochemical PtG discussion into electrolysis and chemical steps (Fig. 7) and explain each separately.

3.2 Electrolysis

Electrolysis is not a new technology. It was invented in 1800 by William Nicholson and Anthony Carlisle, using voltaic current. The invention has an interesting story: It was a few weeks after Alessandro Volta revealed his invention of the voltaic pile that William and Anthony decided to replicate Volta's experiment. In brief, during the experiment they accidentally contacted wires with water and observed some gases, which were found to be hydrogen and oxygen. This led to the birth of the new science, "electrochemistry" [27].

The main technical route in the PtG approach is the use of electrical energy for electrolysis of water to hydrogen ($H_2O(l) \rightarrow \frac{1}{2} O_2(g) + H_2(g)$ $\Delta H_r = +285.8 \, kJ/mol\text{-}H_2O$). The two products of this process are hydrogen and oxygen, with hydrogen being the most favored clean energy source. It can then be used in numerous hydrogen-demanding applications as a sustainable feed. It is noteworthy that majority of the ~50 million tonnes annual hydrogen demand [28] is currently supplied by fossil fuels through syngas generation, which is a sustainability concern.

An alternative application for the generated hydrogen is to convert it back to electrical power in fuel cells (see Fig. 6) with the combustion exhaust being pure water. Bergen [29], in his PhD thesis, and Gahleitner [28] carried out comprehensive studies of PtG pilot plants for renewable-based hydrogen production. The earliest plant in the lists of the two studies is

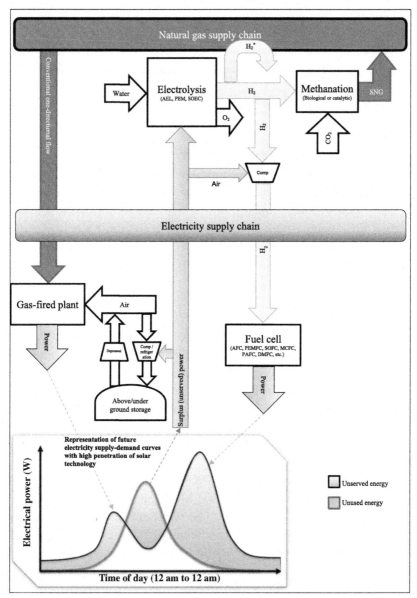

Fig. 6 Schematic of two grids (natural gas and electricity) with bidirectional interactions (H_2* refers to hydrogen injected into NG pipeline for transport to demand location and then separate from NG).

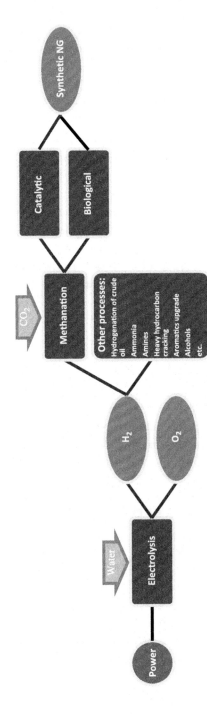

Fig. 7 Electrochemical PtG routes for SNG production.

Solar-Wasserstoff-Bayern, built in Germany in 1986. The system comprised a 370-kW$_p$ PV system, two low-pressure and high-pressure alkaline electrolysis technologies (211 kW$_{el}$ at 1 bar and 100 kW$_{el}$ at 31 bar), and three types of fuel cell technology [30].

Over the years, three major electrolysis systems have been developed: alkaline electrolysis (AEL), polymer electrolyte membrane (PEM), and solid oxide electrolysis (SOEC). As discussed by Kreuter and Hofmann [31], the technologies can be assessed based on the four key factors, efficiency, operability, safety, and costs. Given that several review studies of electrolysis exist (e.g., [32,33]), we briefly list the key features of the three electrolysis technologies in Fig. 8. The most mature technology is AEL, which is cheaper than the others in installation costs, though expensive in maintenance due to corrosiveness. AEL, however, suffers from operational inflexibility when integration with renewable-based fluctuating power sources is desired. It can take up to 1 h from a cold start. PEM, due to its membrane solid nature, is very flexible in operation. However, as is general for all membranes, its life span is short and thus it is expensive in comparison with AEL. Both AEL and PEM are mature technologies, whereas SOEC is in the laboratory development stage. SOEC operates at high temperatures to minimize power consumption by reducing cell voltage. Due to its superior efficiency compared with the other two technologies, SOEC is a promising technology, although there are numerous technical issues (especially heat integration) to be resolved [34]. In terms of pressure, all three technologies favor high pressures for higher hydrogen purity and overall efficiency. Note that other renewable-based H$_2$ generation technologies exist, such as photolysis [35], which are beyond the scope of this study of electrical power utilization.

3.3 Chemical Conversion of Hydrogen

The two product gases of electrolysis (hydrogen and oxygen), after separation, can be used in various industrial applications. Of the two gases, hydrogen is the most preferred, due to its high heat of combustion (HHV: 141.80 MJ/kg). Furthermore, it is not naturally available, whereas oxygen can be obtained from air using, e.g., a membrane air separation unit. Purified hydrogen from electrolysis can be supplied directly for use in fuel cells for electricity generation. It can also be used as a feed in various chemical industries (see Fig. 7) especially hydrogenation units of crude oil refineries for cracking components with long carbon chains or treating and upgrading fuel quality [36]. Among chemical applications, ammonia has traditionally been

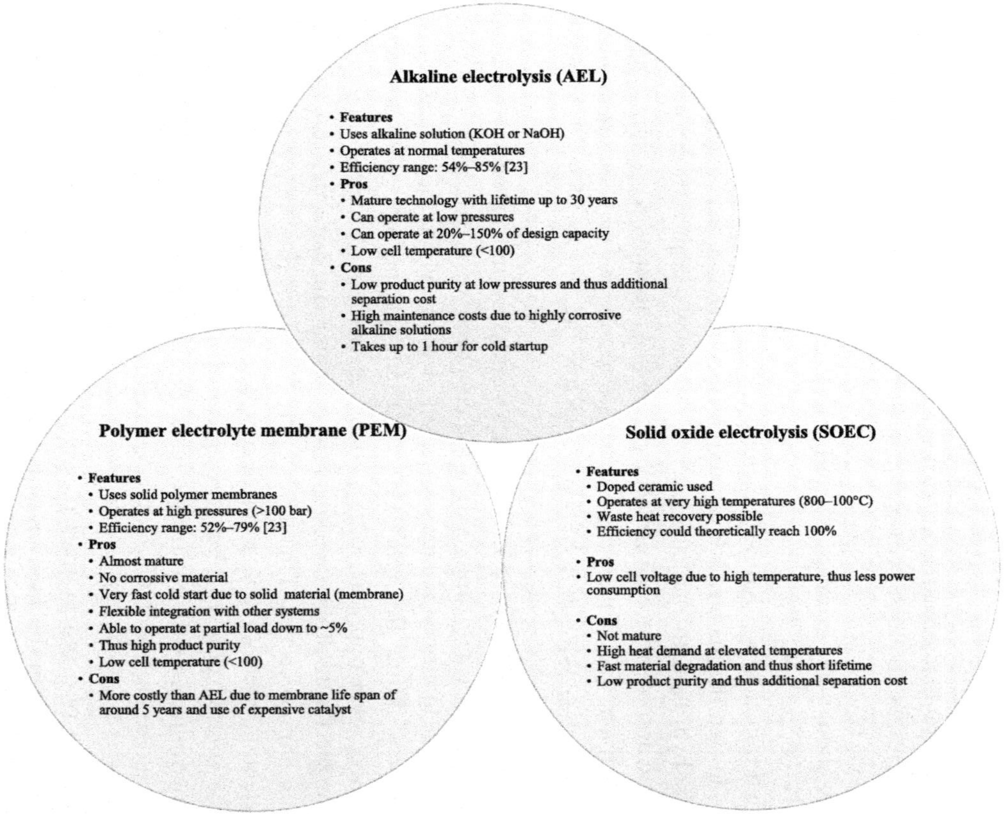

Alkaline electrolysis (AEL)

- **Features**
  - Uses alkaline solution (KOH or NaOH)
  - Operates at normal temperatures
  - Efficiency range: 54%–85% [23]
- **Pros**
  - Mature technology with lifetime up to 30 years
  - Can operate at low pressures
  - Can operate at 20%–150% of design capacity
  - Low cell temperature (<100)
- **Cons**
  - Low product purity at low pressures and thus additional separation cost
  - High maintenance costs due to highly corrosive alkaline solutions
  - Takes up to 1 hour for cold startup

Solid oxide electrolysis (SOEC)

- **Features**
  - Doped ceramic used
  - Operates at very high temperatures (800–100°C)
  - Waste heat recovery possible
  - Efficiency could theoretically reach 100%
- **Pros**
  - Low cell voltage due to high temperature, thus less power consumption
- **Cons**
  - Not mature
  - High heat demand at elevated temperatures
  - Fast material degradation and thus short lifetime
  - Low product purity and thus additional separation cost

Polymer electrolyte membrane (PEM)

- **Features**
  - Uses solid polymer membranes
  - Operates at high pressures (>100 bar)
  - Efficiency range: 52%–79% [23]
- **Pros**
  - Almost mature
  - No corrosive material
  - Very fast cold start due to solid material (membrane)
  - Flexible integration with other systems
  - Able to operate at partial load down to ~5%
  - Thus high product purity
  - Low cell temperature (<100)
- **Cons**
  - More costly than AEL due to membrane life span of around 5 years and use of expensive catalyst

Fig. 8 Three major electrolysis technology types.

one of the most intensive hydrogen consumers, in which one atom of nitrogen combines with three atoms of hydrogen to build one molecule of ammonia ($3H_2 + N_2 \rightarrow 2NH_3$). Another approach, which historically has been less attractive, is synthesis of methane. Hydrogen can react with a carbon source (e.g., CO_2) to generate synthesized natural gas (SNG) for subsequent injection into NG network ($CO_2(g) + 4\,H_2(g) \leftrightarrow CH_4\,(g) + 2\,H_2O$ (g) $\Delta H_r = -165.1\,kJ/mol$). This process, called the Sabatier reaction, was invented by Paul Sabatier in 1913, 1 year after he received the Nobel Prize in Chemistry for his discovery of nickel as a hydrogenation catalyst [37]. A quick glance at the Sabatier reaction may raise a legitimate question: Why should one convert H_2 with HHV of 141.80 MJ/kg to CH_4 with HHV of 55.50 MJ/kg?

Historically, NG has been a cheap fuel compared with hydrogen. This was, in fact, the main reason behind the technoeconomical unattractiveness of the hydrogen-to-methane reaction, and why this process did not attract interest until very recent years. Why, then, has this process recently emerged as an attractive topic in energy systems? The answer should be sought in the contemporary revolution in renewables and the climate change mitigation context:

- Formerly, the business structure was formed around a process with a certain product (say, ammonia) as the economic revenue objective. In this context, hydrogen is a necessary feed, without which the process cannot operate. From the renewable electricity perspective, with projected power oversupply at some periods, electrolysis (or other method) is sought for hydrogen production as an intermediate step for utilizing the unused power. Therefore, hydrogen is not the immediate requirement. Rather, it is an intermediate step or an "energy carrier" in utilizing the unused power. Therefore, given the commercial and geographic environment of the electrolysis plant, the cheapest monetization approach would be preferred. Undoubtedly, the highest revenue could be achieved when a hydrogen fleet or an H_2-consuming industry is nearby. In the absence of a nearby market, however, the fuel cell might be the best option, given the high investment costs of a new hydrogen fleet infrastructure. Alternatively, when a natural gas pipeline is nearby, it could be a good option to convert the H_2 to methane and inject it into the available NG network. Although the methanation process is low in efficiency, given elimination of the need for expensive hydrogen fleet, methanation might still be economically feasible at locations with high NG prices and/or low supply security.

- The world is striving to find routes for curbing greenhouse gas (GHG) emissions. Among the options, carbon capture and storage (CCS) has been considered important. However, both capture and storage of CO_2 are costly, and there are extensive debates about the safety of storing CO_2 underground for hundreds of years [38]. The Sabatier SNG process (also called the CO_2 hydrogenation process) uses CO_2 and converts it to methane. Therefore, this process can be considered carbon-negative and is placed economically as a carbon credit. Interestingly, around two decades ago, a Japanese team led by Koji Hashimoto [39] proposed a large-scale value chain for installation of a huge PV system (13.4 million ha) in the Arabian Peninsula and using the generated electricity for sea water electrolysis (with amorphous alloy electrode) at nearby coasts. The hydrogen then could go under catalytic methanation (with amorphous alloy catalysts) with CO_2. According to the authors, the methanation process could consume all the global CO_2 emissions of the year 1990 (i.e., 6 Gtonnes). The methane could then be exported to the demand market (including Japan) in the form of LNG.

In summary, SNG is one of the main routes of PtG, which has found notable attention in recent years due to its potential not only for producing hydrogen and high-value chemicals but also for utilizing unwanted CO_2, which otherwise must be emitted or sequestered at extra costs [38]. SNG is the key factor in making the electricity and NG networks interactive.

Synthetic methane is an attractive alternative for locations with significant investment in NG infrastructure and where the economic benefit is in favor of continuing to use NG. The key challenge of integrated electrolysis and the methanation processes is increasing process efficiency. The Sankey diagram in Fig. 9 shows the typical round-trip efficiency of power flow from a wind turbine to electrolysis (75%), methanation (80%), and NG turbine (60%), resulting in overall efficiency of approximately 36% [40].

As also shown in Fig. 7, there are two methanation processes for SNG production, catalytic and biological. Catalytic methanation is carried out in reactors with high pressures (as high as 100 bar) and high temperatures (as high as 550°C) over selected catalysts (Ni as the best choice). Various types of reactors (packed or structured bed, fluidized bed, and three-phase column) can be used [41]. Biological methanation is in fact catalytic methanation in which the catalysts are biological. The anaerobic reaction is performed by archea methanogenic microorganisms with metabolic pathways

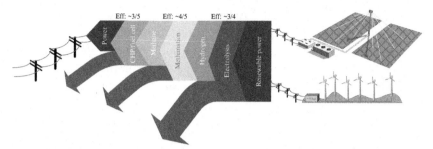

Fig. 9 Illustrative Sankey diagram of the renewable power methane concept.

described elsewhere [42]. Gotz et al. [32] provided a detailed comparative review of these processes and concluded that the biological approach is best for small-scale applications, whereas the catalytic approach is ideal for medium- or large-scale applications. For medium-scale (<100 MW) catalytic processes, reactors of the fluidized bed or three-phase type are best. For larger plants (>100 MW), packed or fixed-bed reactors are considered the most appropriate.

3.3.1 PtG: Physical Approach (Pressurization and Refrigeration)

Referring back to Fig. 6, in addition to the electrochemical PtG approach, there is another way of utilizing surplus electrical energy in gases. This approach of energy storage, unlike the former route, does not involve any chemical reaction. Rather, it changes the potential energy level of gases by pressurization, or in some cases by phase change (e.g., liquefaction or hydrate formation). Compressed air energy storage (CAES) is a promising electrical energy storage option, which has matured through almost four decades of industrial experience since the first plant was built in Huntorf, Germany, 1978 [43]. Fig. 10 shows a schematic diagram of the process. Gas turbines receive NG at high pressures, and the combustion generally occurs at the NG pressure. Therefore, atmospheric-pressure air must be compressed to NG pressure before being supplied to the combustion chamber. The high-temperature and high-pressure exhaust gas is directed to the expander to generate electrical power. However, a notable fraction of the power is used for air compression. On this basis, compression and storage of air can be advantageous in certain scenarios:

- First, when a gas power plant operates under a take-or-pay NG contract, and the demand for electricity is low, the NG can be used for air compression and storage, so that during high electricity prices the stored energy can be utilized for power generation.

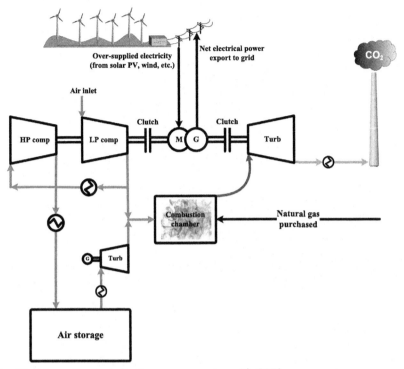

Fig. 10 Schematic of gas-fired power generator with CAES.

- Second, the use of energy for air compression and storage can be used for grid management and power supply reliability by load following in the context of baseload inflexible power plants such as a nuclear power plant [44].
- A third interest is similar to the second in terms of grid management and is mainly motivated by utilizing oversupplied renewable electrical energy. During electricity oversupply periods, the air compressor can be driven by an electrical motor (with energy supplied by the grid) to compress and store air for later use.

The surplus electrical energy can also be used for pressurization in the NG network. For instance, a pipeline route may be overpressurized at the location of unused electrical energy; then the pressured gas is transferred to the demand location where the gas can then be depressurized (by generating electricity) and supplied to the demand market at the desired pressure.

Apart from the CAES, pressurization has received very little attention in the literature as a means of storing unused electrical energy. However, given

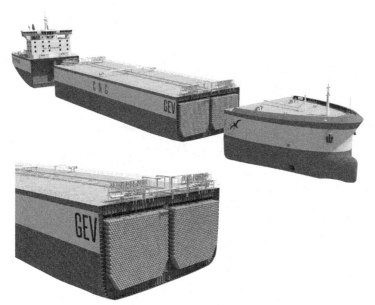

Fig. 11 Conceptual design of compressed natural gas container and carrier. *(Courtesy of Global Energy Ventures Ltd.)*

the lower cost of compression versus liquefaction, it has been proven economical to transport NG in the form of CNG when the demand side is located at a relatively short distance [10]. The "CNG carrier" is a concept achieving this task by pressurizing and storing NG in tubes or vessels. There are a few design concepts for such carriers. For instance, Fig. 11 shows schematics of "CNG Optimum" carrier which is based on a close-packed, high strength pipes that can be accommodated at a ship's storage area (aka cargo hold). The CNG Optimum 200 is composed of multiple Steel pipes with length of 108 m and outside diameter of 400 mm. The carrier can contain a total length of 209 km of such pipes accommodating 200 million scf (~4700 tonne) of NG at pressures up to 248 bar. Another CNG tanker concept is based on modular coiled tubes, called "coiled pipe in a carousel" or Coselle™ [45]. Each module consists of around 21 kilometers of 168-mm-diameter steel pipe, coiled into a reel-like steel support structure, with NG storage capacity of up to 4 million scf (~116,000 scm). CNG carriers (which are not the focus of this research) can accommodate between 16 and 128units of these modules, based on their size.

Such modular concepts can also be used for onshore energy storage. It can be in fact used for storage of any type of gas, including air and NG. Storage of natural gas can have a double economic advantage: the first benefit is pressurization of NG, allowing storage of unused electrical

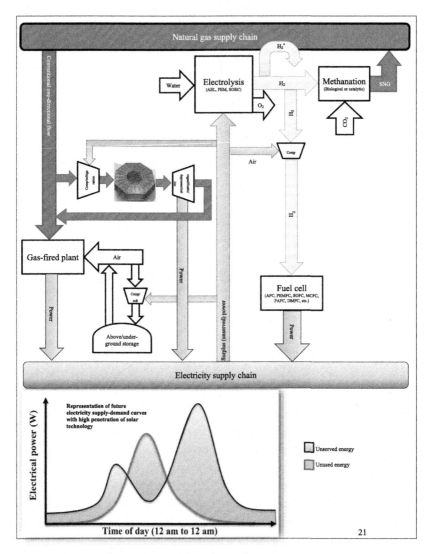

Fig. 12 Schematic of pressurization-based PtG for an interactive gas and electricity network.

energy in the form of potential energy for recovery at a later time. Secondly, NG could be purchased during low-price periods and stored in modules for use or resale at higher priced NG or electricity periods. This double benefit improves the economics of this energy storage option in the face of installation costs of aboveground storage systems. Therefore, our proposed structure of bidirectional NG-electricity supply chains is an improved version of Fig. 6 with the inclusion of potential energy storage shown in Fig. 12.

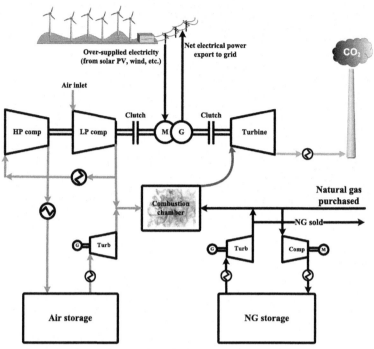

Fig. 13 Schematic of a gas-fired power plant integrated with air and natural gas storage systems.

One such NG pressurization concept has been introduced in connection with CAES. As obvious from its name, the concept of the CAES technology is formed around the integration of a pressurized air storage system and a gas-fired power generator. Khalilpour et al. [46] have further integrated the CAES system with the inclusion of pressurized NG storage (Fig. 13). Through this integrated arrangement, a gas-power-generating plant installs both air and NG storage systems to utilize their stored energy as well as realizing the real economic value of NG following market dynamics and renewable energy availability.

There is some growing interest in air, ammonia, and hydrogen liquefaction. The difference of liquefaction with compression is that the former requires higher energy, but in return needs a much smaller volume of storage due to higher energy intensity.

4 CONCLUSION

This chapter reviewed the ongoing transition in the natural gas and electricity supply chains from the conventional one-directionally connected networks into a more tightly integrated and interconnected networks with bidirectional flow of gas and electron across the networks. This is being mainly facilitated by the emergence of renewable energy resources into the electricity network, which necessitates utilization of the redundant energy generation at occasions of low demand but high renewable resources availability (e.g., mid-day for solar). Such energy could be utilized in natural gas network either physically (for gas pressurization or liquefaction) or electro/chemically for hydrogen and synthetic methane production.

REFERENCES

[1] Khalilpour R, Karimi IA. Selection of liquefied natural gas (LNG) contracts for minimizing procurement cost. Ind Eng Chem Res 2011;50(17):10298–312.

[2] Yergin D. The prize: the epic quest for oil, money & power. New York: Free Press; 2011.

[3] EPRI. Cost of greenhouse gas mitigation. Available from: http://www.treepower.org/globalwarming/CO2-EPRI-EvanHughes.pdf; 2009.

[4] Speight JG. Natural gas: a basic handbook. Houston, TX: Gulf Publishing Company; 2007.

[5] US-DOE. The global liquefied natural gas market: status & outlook. Washington: Energy Information Administration; 2003.

[6] IEA. World energy outlook 2011: are we entering a golden age of gas? Paris: International Energy Agency; 2011.

[7] Thomas S, Dawe RA. Review of ways to transport natural gas energy from countries which do not need the gas for domestic use. Energy 2003;28(14):1461–77.

[8] BP. BP statistical review of world energy 2017. In: BP statistical review of world energy. London: British Petroleum; 2017.

[9] Khalilpour R, Karimi IA. Investment portfolios under uncertainty for utilizing natural gas resources. Comput Chem Eng 2011;35(9):1827–37.

[10] Khalilpour R, Karimi IA. Evaluation of utilization alternatives for stranded natural gas. Energy 2012;40(1):317–28.

[11] Sliepcevich CM. Liquefied natural gas—a new source of energy: Part I. Ship transportation. Am Sci 1965;53(2):260–87.

[12] Sliepcevich CM. Liquefied natural gas—a new source of energy: Part II. Peak load shaving and other uses. Am Sci 1965;53(3):308–16.

[13] Cryogenics-journal. L.N.G. Peak-shaving plant for Oregon. Cryogenics 1967;7(1):54.

[14] Jemena. Australia's largest ever gas storage bottle: the Colongra Lateral Pipeline. In: The Australian pipeliner. Melbourne: Australian Pipeline Industry Association; 2009. p. 6.

[15] Chen HS, et al. Progress in electrical energy storage system: a critical review. Prog Nat Sci 2009;19(3):291–312.

[16] Sulzberger C. Pearl street in miniature: models of the electric generating station [history]. Power Energy Mag IEEE 2013;11(2):76–85.

[17] Hadjipaschalis I, Poullikkas A, Efthimiou V. Overview of current and future energy storage technologies for electric power applications. Renew Sustain Energy Rev 2009;13(6–7):1513–22.

[18] Luo X, et al. Overview of current development in electrical energy storage technologies and the application potential in power system operation. Appl Energy 2015;137:511–36.

[19] Zalba B, et al. Review on thermal energy storage with phase change: materials, heat transfer analysis and applications. Appl Therm Eng 2003;23(3):251–83.

[20] Sharma A, et al. Review on thermal energy storage with phase change materials and applications. Renew Sustain Energy Rev 2009;13(2):318–45.

[21] Baker JN, Collinson A. Electrical energy storage at the turn of the millennium. Power Eng J 1999;13(3):107–12.

[22] Decourt B, Debarre R. Electricity storage. In: Leading the enrgy transition factbook. Gravenhage, The Netherlands: Schlumberger Business Consulting (SBC) Energy Institute; 2013.

[23] Koohi-Kamali S, et al. Emergence of energy storage technologies as the solution for reliable operation of smart power systems: a review. Renew Sustain Energy Rev 2013;25:135–65.

[24] Akhil AA, et al. DOE/EPRI 2013 electricity storage handbook in collaboration with NRECA. US Department of Energy and EPRI: California; 2013.

[25] Vassallo AM. Chapter 17: Applications of batteries for grid-scale energy storage. In: Lim CMS-KM, editor. Advances in batteries for medium and large-scale energy storage. Cambridge, UK; Waltham, MA, USA: Woodhead Publishing; 2015. p. 587–607.

[26] Grond L, et al. Systems analyses power to gas: technology review. In: Energy & sustainability. Groningen, The Netherlands: DNV KEMA; 2013.

[27] Russell C. Enterprise and electrolysis. In: Chemistry world. London, UK: Royal Society of Chemistry; 2003.

[28] Gahleitner G. Hydrogen from renewable electricity: An international review of power-to-gas pilot plants for stationary applications. Int J Hydrogen Energy 2013;38(5):2039–61.

[29] Bergen AP. Integration and dynamics of a renewable regenerative hydrogen fuel cell system. Victoria, BC, Canada: University of Victoria; 2008.

[30] Szyszka A. Demonstration plant, Neunburg vorm Wald, Germany, to investigate and test solar-hydrogen technology. Int J Hydrogen Energy 1992;17(7):485–98.

[31] Kreuter W, Hofmann H. Electrolysis: the important energy transformer in a world of sustainable energy. Int J Hydrogen Energy 1998;23(8):661–6.

[32] Götz M, et al. Renewable power-to-gas: a technological and economic review. Renew Energy 2016;85:1371–90.

[33] Ursua A, Gandia LM, Sanchis P. Hydrogen production from water electrolysis: current status and future trends. Proc IEEE 2012;100(2):410–26.

[34] Buttler A, et al. A detailed techno-economic analysis of heat integration in high temperature electrolysis for efficient hydrogen production. Int J Hydrogen Energy 2015;40(1):38–50.

[35] Abbasi T, Abbasi SA. 'Renewable' hydrogen: prospects and challenges. Renew Sustain Energy Rev 2011;15(6):3034–40.

[36] Hampton GC, Fred H. Hydrogen in oil refinery operations. In: Hydrogen: production and marketing. 67-94. Washington, DC: American Chemical Society; 1980.

[37] Leroy F. A century of nobel prize recipients: chemistry, physics, and medicine. Hoboken, NJ: CRC Press; 2003.

[38] Quadrelli EA, Armstrong K, Styring P. Chapter 16: Potential CO_2 utilisation contributions to a more carbon-sober future: A 2050 vision. In: Armstrong PSAQ, editor. Carbon dioxide utilisation. Amsterdam: Elsevier; 2015. p. 285–302.

[39] Hashimoto K, et al. Global CO_2 recycling—novel materials and prospect for prevention of global warming and abundant energy supply. Mater Sci Eng A 1999;267(2):200–6.

[40] Sterner M. Bioenergy and renewable power methane in integrated 100% renewable energy systems: limiting global warming by transforming energy systems. Kassel: Kassel University Press; 2009.

[41] Kopyscinski J, Schildhauer TJ, Biollaz SMA. Production of synthetic natural gas (SNG) from coal and dry biomass – a technology review from 1950 to 2009. Fuel 2010; 89(8):1763–83.

[42] Thauer RK, et al. Methanogenic archaea: ecologically relevant differences in energy conservation. Nat Rev Microbiol 2008;6(8):579–91.

[43] Giramonti AJ, et al. Conceptual design of compressed air energy storage electric power systems. Appl Energy 1978;4(4):231–49.

[44] Glendenning I. Long-term prospects for compressed air storage. Appl Energy 1976; 2(1):39–56.

[45] Stenning D, Cran JA. In: The Coselle CNG carrier the shipment of natural gas by sea in compressed form. 16th World Petroleum Congress, 11–15 June, Calgary, Canada. World Petroleum Congress; 2000.

[46] Khalilpour KR, Grossmann I, Vassallo A. Integrated power-to-gas and gas-to-power with air and natural gas storage. 2017. Sydney.

CHAPTER 6

Stranded Renewable Energies, Beyond Local Security, Toward Export: A Concept Note on the Design of Future Energy and Chemical Supply Chains

Kaveh Rajab Khalilpour
Faculty of Engineering and Information Technology, Monash University, Melbourne, VIC, Australia

Abstract

Fossil fuels are not spread evenly across the world. This has been a major security concern for several countries. Nevertheless, no country may be found on Earth without a good resource of one or more renewable energies. As such, with the 1973 global energy crisis, the development of renewable energy technologies received considerable attention. Yet in the absence of low-cost renewable energy technologies, attention had been limited to national energy security. However, the fast price decline over the recent decade in some renewable technologies such as photovoltaics (PV) and wind has moved renewable energy development into a new paradigm, where investments are being considered beyond local security, toward revenue generation with renewable energy exports. This chapter discusses the possible export approaches and challenges associated with the required infrastructure. It also proposes a new hybridized supply-chain system for both renewable energies and natural gas.

Keywords: Supergrid, Supersmart grid, Global grid, Stranded energy resources, Hydrogen, Electrolysis, Renewable supply chain, Renewable energy monetization, Polygeneration systems, Energy hubs

1 BACKGROUND

The evolution of primary energy mix over the last two centuries is illustrated in Fig. 1. Until around one and half centuries ago, wood was the major energy source. Following the invention of steam engines, and the consequent Industrial Revolution, coal was explored and gradually became the dominant source by the beginning of the 20th century. While the

Polygeneration with Polystorage
https://doi.org/10.1016/B978-0-12-813306-4.00006-9

157

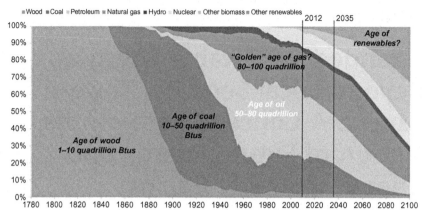

Fig. 1 Evolution of the (US) primary energy mix from 1780 to the present, and business-as-usual projection to 2100 [1].

exploration of coal was climaxing, the Scottish chemist James Young discovered a method of refining crude oil in 1847, which led to the birth of the modern petroleum industry (which earlier was only producing kerosene for lighting) and the development of liquid-fuel-based internal combustion engines. With the gradual exploration of crude oil resources and refining technology development, oil became the dominant primary energy source in the second half of the 20th century (Fig. 1). The energy intensity and ease of transportation compared with coal made crude oil the most attractive energy source in the 20th century, although geographical differences between supply sources and demand regions caused significant geopolitical issues across the world.

Natural gas (NG) is a cleaner fuel than coal or oil. However, widespread application of NG has been hindered by its low volume intensity, which makes transportation cost a significant barrier. Initially NG was used mainly in locations accessible by pipeline; however, with disruptions in the energy market during the 1970s and consequent increases in energy prices, other approaches such as liquefaction became feasible, and today pipeline and liquefied natural gas (LNG) are the two main approaches of NG utilization [2]. Apart from LNG, other NG utilization alternatives such as gas to liquids (GTL), gas to chemicals (GTC), compressed natural gas (CNG), gas to solids (GTS), and gas to wire (GTW) have been used on a smaller scale (e.g., GTL) or are currently under discussion and evaluation [3, 4]. With critical international attention to climate change mitigation since the Kyoto Protocol of 1997 and the gradual loss of hope for carbon capture and storage value propositions [5], interest in NG as an intermediate option has increased to an extent that the International Energy Agency describes our current time as a "golden age of natural gas" [6]. Given the huge energy demand in booming Asian countries such as Japan, Korea, and Singapore, and the reduced role of Indonesia as a traditional Asian LNG exporter, Australia has invested hugely in LNG production over the recent decade.

Needless to say, over the last century several other energy technologies have been developed, including nuclear, large-scale hydropower, wind, geothermal, etc. Especially, since the 1970s, much attention has been paid to renewable technologies. We have had two paradigms in renewable energy development.

2 TWO PARADIGMS IN RENEWABLE ENERGY DEVELOPMENT

2.1 Paradigm 1: Renewables for Energy Security

Renewable energy resources have a few critical advantages, including abundance and relatively scattered geographic distribution. As such, exploration and utilization of local (renewable) energy sources has been an important security matter for energy-importing societies. Furthermore, the possibility of generating energy on the demand side has many advantages in terms of energy efficiency, the cost of delivered energy, and greenhouse gas emissions.

In the early 1970s, the oil embargo of some OPEC exporters increased the price of oil more than six-fold from its per barrel value of $1.8 in 1970 to over $11.5 in 1974. This caused significant energy security concerns in the OECD region, and countries began to look seriously at alternative energy options. A few years later, in the late 1970s, revolution in Iran gave another spike to the oil price and it reached $36.8 in 1980 (more than 20-fold that in 1970). From that time, renewable energies including PV technology received increased attention (still from economic and security perspectives, not sustainability) in developed countries [7]. For instance, the Solar Research Institute (later renamed the National Renewable Energy Laboratory, NREL) of the US Department of Energy was formed in 1977. Although the fall in energy prices in the 1990s to some extent hindered the development of such technologies, the natural technology learning rate, the introduction of carbon penalties following the Kyoto Protocol, and the energy price rise with the US raid on Iraq in the early 2000s helped some technologies such as photovoltaic (PV) cells to evolve and reach market parity in several jurisdictions across the world [8].

2.2 Paradigm 2: Renewables for Export

We are just entering the second paradigm, when renewable energy is sought not only as a resource for local security but also as having potentials worth exploring for export and economic development. This is a completely new realization. The export of renewable energy was formerly analyzed in the academic literature as an "extreme scenario," along with several other logical possibilities. But suddenly, factors including market disruption by PV technology due to a reduction in silicon prices, mass production in China, and the devising of smart policies in countries like Germany and Australia led to widespread PV penetration. This PV uptake began with households, which were paying high electricity tariffs, through some governmental incentives. Today, however, large-scale commercial electricity users, which pay lower electricity tariffs per unit, have found investment in large-scale PV systems feasible even without governmental supports. The price of PV technology has fallen to such an extent that PV installation values have dropped to 1 $/ kW or less [7]. The international large-scale auction values for wind and PV are also in fast declination pace.

This revolution in renewable technologies, initiated by PV, is leading us to a new paradigm where interest in renewable energy utilization shifts from mere "self-security" to "export." For instance, Australia is already the world's first coal and second LNG exporter. However, as the world's sixth largest country, with diverse climate and renewable energy resources, Australia has a legitimate interest in considering the potential of exporting renewable energies to at least the neighboring Asian countries. Otherwise, such resources will be stranded, a situation that is identifiable as resource wastage. Next, potential pathways for stranded renewable resource utilization are discussed.

3 STRANDED RENEWABLE ENERGIES

A resource is called "stranded" when it is either at remote locations (distant from the demand market) or close to customers but with saturated markets [9]. We can imagine that before the Industrial Revolution and the birth of modern supply-chain systems, most resources were either scarce or stranded. But, almost two centuries after industrialization, there are still several examples of stranded resources around the world. The capitalist market operates based on supply-demand balances, and therefore any resource at oversupply periods is subject to underutilization and wastage, unless there are mechanisms such as storage or conversion to other products. Examples include agricultural products and associated NG (which is mostly flared) [10]. In general, a stranded resource can be of any type, including workforce. For instance, there are countries with highly qualified engineering graduates while in other parts of the world there are shortages of professional workers. One emerging stranded resource is renewable energies, which can be stranded in two ways: (1) underutilized, and (2) underdeveloped, discussed next.

3.1 "Underutilized" Renewable Resources

Renewable energy resources generally suffer from two key limitations, intermittency and limited availability. These constraints meanwhile result in low capacity utilization factors and therefore high "here-and-now" investment costs (though negligible "there-and-after" operation costs). The variability of renewable energy generation implies supply-demand misbalance. As a consequence, if no energy storage technology is employed, the surplus energy generation will be stranded, i.e., it should be curtailed. Conversely, the deficit of energy resources (such as solar irradiation rate, wind speed, biomass quantity) at certain times (day, week, season) requires allocation of either an auxiliary power source (such as other types of generation or connection to a grid) or energy storage. In summary, part of the energy generation of most renewable energy technologies has the potential to be stranded. Therefore, smart strategies are required for minimizing such stranded resources. Solutions and strategies can be diverse, given several parameters such as relevant jurisdictions, technology, geography, demand markets, and distribution networks (Chapters 15 and 16).

3.2 "Underdeveloped" Renewable Resources

There is no location on Earth, which does not have one or more source of renewable energies. However, given the high historical installation costs of renewable technologies, almost all parts of the world have renewable resources, which are yet to be developed. These are also stranded resources. For instance, Africa is one of the best places for solar technologies and northern Europe has best wind resources [11]. In countries with high-quality renewable resources, the limitation of investment in renewable technologies for the local demand market implies stranding the remaining energy resources. Unlike fossil fuels and biomass, ambient energies (e.g., solar energy, wind energy, salinity gradients, and kinetic energy) are wasted if they are not recovered spontaneously.

Until recently, renewable energy investments were merely focused on local needs, but with the decline in the cost of generation technologies there is a new prospect for renewable energy harvesting, at least in locations with best-quality resources, in order to export the products to distant markets.

These energies could be harvested and utilized through (1) electricity cables (super grids), (2) renewable fuels and chemicals, or (3) as embodied energy in various products such as those from extractive industries (steel, minerals, etc.). The consequence will be the development of international renewable energy supply chains, with transportation routes through electricity cables (super grids), gas pipelines, or chemical shipments, as discussed next.

4 MONETIZATION OPTIONS FOR STRANDED RENEWABLE ENERGIES

4.1 Direct Utilization

Direct renewable energy utilization includes either power production or conversion of the energy to value-added fuels and chemicals.

4.1.1 Supergrid for Power Export

A "supergrid," "megagrid," or "supersmart grid" [12] is a future grid that interconnects various countries and regions with a high-voltage direct current (HVDC) power grid. It is in fact a wide-area (synchronous) transmission network capable of large-scale transmission of electricity, which makes it possible to trade high volumes of (renewable) electricity across great distances. The extreme case of "supergrid" is "global grid" which connects all continents. The concept of a supergrid dates back to the 1950s. However, interest in such grids has been accelerated recently in the context of monetizing stranded renewable energy resources through transnational export. Fig. 2 illustrates potential options for global supergrids. Some recent supergrid projects include the European supergrid [13–15], the Asia Pacific Super Grid [16], the South-East Asia or ASEAN Power Grid [17], and the North-East Asian Super Grid, which connects the five countries of Japan, South Korea, North Korea, China, and Mongolia. The more extensive version of this proposal is the so-called Asia Super Grid (ASG), which can not only connect the aforementioned five countries, but can also be extended to Taiwan, Thailand, the Philippines, and India (see Fig. 3) [18].

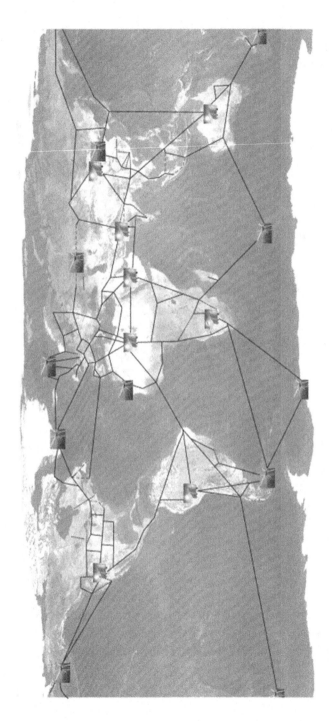

Fig. 2 Potential options for global energy interconnection. *(From Chatzivasileiadis S, Ernst D, Andersson G. The global grid. Renew Energy 57;2013:372-83.)*

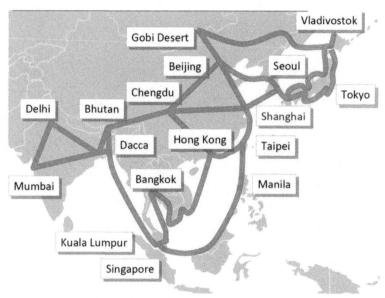

Fig. 3 Asia Super Grid (ASG) concept, proposed by Japan Renewable Energy Foundation, based on utilization of wind and solar resources of the Russian Gobi Desert [19]. *(Used with permission from Renewable Energy Institute.)*

4.1.2 Renewable Fuels and Chemicals

There are several conditions under which supergrid electricity transmission is not feasible. One such condition is a demand market being saturated with electricity supply. Distance and the transmission path can also involve detrimental elements. For instance, the presence of obstacles such as deep ocean between supply and demand markets can significantly affect the economic feasibility of supergrid development. All these factors highlight the need for alternative options for renewable energy export. Water electrolysis is one of the most suitable alternative options (Chapter 5).

Invented in 1800, water electrolysis was not initially considered a commercially attractive technology, for two reasons: its high electricity demand and its low reaction efficiency. However, water electrolysis is now coming increasingly into vogue, particularly because renewable electricity is often in oversupply—at times, it is at low or even negative wholesale prices. This oversupply energy can be utilized in electrolyzers for hydrogen generation. This H_2 can be exported for direct use as fuel in fuel cells or internal combustion engines. It can also be used in the production of value-added fuels/chemicals such as ammonia, methanol, and hydrocarbons (Chapter 4).

Attention to hydrogen as a "future fuel" has a long history. The following inspiring statement is a foresight of John Burdon Sanderson Haldane in 1923 [20], which relates astonishingly well to our time:

"As for the supplies of mechanical power, it is axiomatic that the exhaustion of our coal and oil-fields is a matter of centuries only. As it has often been assumed that their exhaustion would lead to the collapse of industrial civilization, I may perhaps be pardoned if I give some of the reasons which lead me to doubt this proposition. Water-power is not, I think, a probable substitute, on account of its small quantity, seasonal fluctuation, and sporadic distribution. It may perhaps, however, shift the centre of industrial gravity to well-watered mountainous tracts such as the Himalayan foothills, British Columbia, and Armenia. Ultimately we shall have to tap those intermittent but inexhaustible sources of power, the wind and the sunlight. The problem is simply one of storing their energy in a form as convenient as coal or petrol. If a windmill in one's back garden could produce a hundredweight of coal daily (and it can produce its equivalent in energy), our coalmines would be shut down to-morrow. Even to-morrow a cheap, foolproof, and durable storage battery may be invented, which will enable us to transform the intermittent energy of the wind into continuous electric power.

*Personally, I think that four hundred years hence the power question in England may be solved somewhat as follows: The country will be covered with rows of metallic **windmills** working electric motors which in their turn supply current at a very high voltage to great electric mains. At suitable distances, there will be great power stations where during windy weather **the surplus power will be used for the electrolytic decomposition of water into oxygen and hydrogen**. These gasses will be liquefied, and stored in vast vacuum jacketed reservoirs, probably sunk in the ground. If these reservoirs are sufficiently large, the loss of liquid due to leakage inwards of heat will not be great; thus the proportion evaporating daily from a reservoir 100 yards square by 60 feet deep would not be 1/1000 of that lost from a tank measuring two feet each way. In times of calm, the gasses will be recombined in explosion motors working dynamos which produce electrical energy once more, or more probably in oxidation cells. Liquid hydrogen is weight for weight the most efficient known method of storing energy, as it gives about three times as much heat per pound as petrol. On the other hand it is very light, and bulk for bulk has only one third of the efficiency of petrol. This will not, however, detract from its use in aeroplanes, where weight is more important than bulk. These huge reservoirs of liquified gasses will enable wind energy to be stored, so that it can be expended for industry, transportation, heating and lighting, as desired. The initial costs will be very considerable, but the running expenses less than those of our present system. Among its more obvious advantages will be the fact that energy will be as cheap in one part of the country as another, so that **industry will be greatly decentralized**; and that no smoke or ash will be produced.*

It is on some such lines as these, I think, that the problem will be solved. It is essentially a practical problem, and the exhaustion of our coal-fields will furnish the necessary stimulus for its solution. Even now perhaps Italy might achieve economic independence by the expenditure of a few million pounds upon research on the lines indicated. I may add in parenthesis that, on thermodynamical grounds which I can hardly summarize shortly, I do not much believe in the commercial possibility of induced radio-activity."

Interestingly, around two decades ago, a Japanese team led by Koji Hashimoto [21] proposed a large-scale value chain for the installation of a huge PV system (13.4 million ha) in the Arabian Peninsula, using the generated electricity for sea water electrolysis (with amorphous alloy electrode) at nearby coasts. The hydrogen then could go under catalytic methanation (with amorphous alloy catalysts) with CO_2. According to the authors, the methanation process could consume all the global CO_2 emissions of the year 1990 (i.e., 6 Gtonnes). The methane could then be exported to the demand market (including Japan) in the form of LNG.

We earlier discussed the possible limitations of supergrids. One of the critical challenges in any long-term transnational investment is the political risk. Supergrids suffer from this aspect: any international crisis can lead to damage to the supergrid platform. Furthermore, market saturation with one particular product is another key risk. Therefore, investors prefer to minimize risks by diversifying their business portfolios. This is common, for instance, in NG supply chains, when there are interests in diversifying utilization options with gas to power, gas to pipeline, liquefied natural gas (LNG), compressed natural gas (CNG), gas to liquids (GTL), gas to solids (GTS) [9]. In the renewable energy context, a portfolio of renewable power (through a supergrid), along with H_2, and all subsequent products of H_2, including ammonia, urea, and various synthetic hydrocarbons, can reduce investment risks.

5 INDIRECT UTILIZATION: RENEWABLE ENERGY EMBODIED IN OTHER COMMODITIES

Generally, countries have several energy-intensive industries, which heavily rely on energy inputs. Examples include aluminum smelting, concrete manufacturing, iron and steelmaking (Chapter 10), oil and gas (Chapter 11), and various other mineral processing operations. Utilization of renewable energies in the production systems of these industries is another approach for exporting renewable resources in embodied form.

For instance, iron and steel are fundamental components of almost all the infrastructures within which we live. In 2017, around 1.69 billion tons of crude steel were manufactured, with China responsible for approximately half of this production (World Steel Association 2018). This number, divided by the world population of 7.55 billion, translates to an annual 224 kg/capita of crude steel consumption. Add to this the fact that energy accounts for up to 40% of steelmaking costs, which also contributes significantly to greenhouse gas emissions. On this basis, an alternative approach to conventional fossil-fuel-based steelmaking is a renewable hydrogen-based process, which has zero emissions [22]. Fig. 4 shows schematic of conventional and future steelmaking systems. The process on the right is a zero-emission hydrogen-based HYBRIT concept by SSAB. Thereby, countries could utilize their stranded renewable resources in such systems and export the energy in embodied forms in other commodities.

6 THE CHALLENGE OF NEW INFRASTRUCTURE FOR RENEWABLE EXPORT

One of the key supply-chain challenges of renewable energies is the lack of dedicated storage and distribution infrastructure, as well as the fact that this dedicated infrastructure is not viable without significant production. Another issue is that the optimal locations of renewable energies sources (such as wind and solar) are usually scattered and there might not be a single production and export platform. This implies requirements for the development of inland supply chains to bring the product to the borderline for export. A potential approach is integration of the export energy supply chain with the local infrastructures. For instance, hydrogen can be injected into the NG infrastructure at one location and be received at another location. It can also react with carbon dioxide to produce methane and then be injected into the nearest NG infrastructure (Chapter 5). Synthetic methane can be an attractive alternative for locations with significant investment in NG infrastructure and where the economic benefit is in favor of remaining on NG. We project that the optimal integration of export-purpose renewable energy infrastructure with local infrastructures will receive increasing attention over the coming years, as it can notably reduce the supply-chain costs of exported renewable energies.

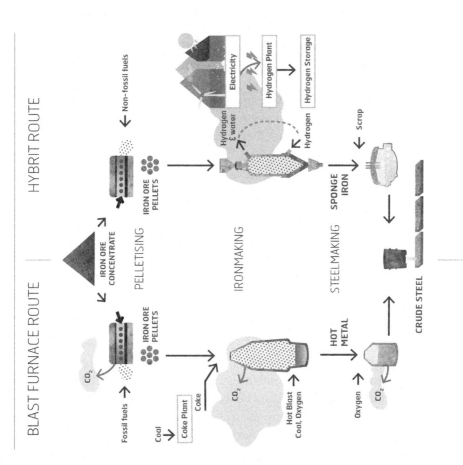

Fig. 4 Steelmaking, past and future directions; *left*: current fossil-fuel-based process, *right*: near-zero-emission hydrogen-based HYBRIT concept by SSAB (Fe₂O₃+H₂ = 2FeO+H₂O and FeO+H₂ = Fe+H₂O) (Image Courtesy of Hybrit Development AB.)

7 A HYBRID CONCEPT FOR A RENEWABLE CHEMICAL AND ENERGY SUPPLY CHAIN

Hydrogen is an energy vector around which substantial infrastructure can be established, especially when it comes to long-distance energy transport and export. Hydrogen is appealing because it is not inherently dependent on the use of carbon, and thus does not directly produce fossil fuels; in fact, hydrogen can be used within the context of an entirely carbon-neutral energy system. However, it is still more expensive to produce than current fossil fuel alternatives; both its storage and transportation are also more expensive, and more difficult. Countries with extensive NG and petrochemical product export supply chains are fortunate to have great renewable energy resources. One example is Australia, the world's first coal and second LNG exporter.

Moreover, Australia is the world's sixth largest country, with one of the best solar energy resources, especially in its north east (see Fig. 5). Similar examples include Middle East petrochemical and gas exporters with great solar resources or Nordic petrochemical and gas exporters with great wind resources.

It is therefore a legitimate interest for such countries to consider the potential for exporting renewable energies to at least the neighboring countries. Given the significant export infrastructure costs of renewable energies, and given that the aforementioned energy exporting countries already possess existing supply-chain platforms, we propose a hybrid concept for renewable energy export through existing petrochemical and NG supply chains.

Our proposed process system is illustrated in Fig. 6. Renewable energy can be sourced from various technologies, such as solar (PV or solarthermal), wind, geothermal, bioenergy, ocean, and hydro. Part of that energy will be used for water electrolysis to split water into hydrogen and oxygen. The separated hydrogen will be compressed and/or liquefied for export or application in other processes. Part of the compressed hydrogen will react with nitrogen (separated from the air) to produce ammonia. The product will be liquefied or converted to other physical forms for export.

We limit our discussion to H_2 and NH_3, as they require more complex supply chain systems due to cryogenic operations.

The key features of this integrated system are:

- Renewable energy will be also used in LNG compression and cooling systems to avoid/reduce natural gas combustion. This can increase the LNG

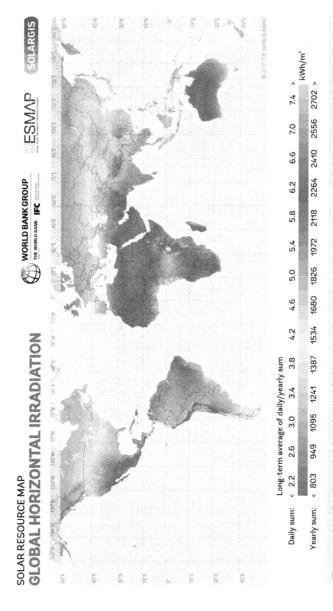

Fig. 5 Map of global horizontal irradiation across the world. (© 2017 The World Bank, Solar resource data: Solargis.)

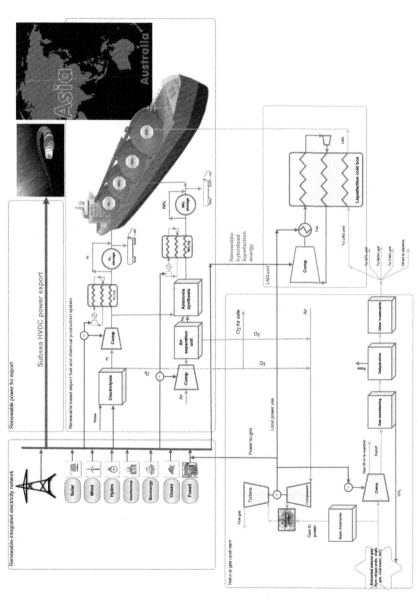

Fig. 6 A concept note for an Australia-Asia renewable energy export supply chain (applicable to all natural gas exporting countries).

production capacity for export and also reduce the CO_2 footprint of LNG production.

- The oxygen from the electrolysis and air separation unit (for ammonia) will be mixed with the air inlet of NG turbines to improve combustion efficiency. This can potentially lead to flue gas with higher CO_2 concentration, thereby reducing carbon capture costs, if of interest.
- Multicompartment cryogenic tankers can be hybridized for multipurpose shipments of NG, ammonia, and hydrogen, and this is expected to offer supply-chain flexibility.
- Electricity will be also exported directly through subsea HVDC cables (supergrid). Identification of the capacity of supergrid and power-to-chemical processes is a complex technoeconomic optimization task to be performed in any given jurisdiction and market structure.
- Onshore stored hydrogen and ammonia will play a short-term energy storage role or serve as a longer-term reserve capacity for the local energy market.
- This will lead to designing next-generation large-scale flexible liquefaction systems that can liquefy NG, hydrogen, and ammonia concurrently.

This new hybrid system will thus not only reduce the export costs of renewable energies, but will also reduce the emission intensity of natural gas and petrochemical supply chains.

8 CONCLUSION

The revolution in renewable technologies, initiated by PV, is leading us to a new paradigm where interest in renewable energy utilization shifts from mere "self-security" to "export." Renewable energy export faces several challenges, including huge infrastructure costs and transnational supply-chain risks. Further to renewable energy embodiment in export commodities, two direct renewable energy monetization options are supergrids and renewable chemicals/fuels. To address supply-chain risks (especially political uncertainties and market demand dynamics), it is conceivable to invest in portfolio of utilization options. To address the huge infrastructure costs, we proposed hybridization of the renewable infrastructure with existing natural gas infrastructures. We propose a hybrid system, which not only facilitates renewable energy integration with natural gas supply-chain systems, but also can reduce the emission intensity of natural gas supply chains.

REFERENCES

[1] Channell J, Jansen HR, Syme AR, Savvantidou S, Morse EL, Yuen A. Energy Darwinism: the evolution of the energy industry. In: Citi GPS: global perspectives & solutions. New York: Citigroup; 2013.

[2] Khalilpour R, Karimi IA. Selection of liquefied natural gas (LNG) contracts for minimizing procurement cost. Ind Eng Chem Res 2011;50:10298–312.

[3] Thomas S, Dawe RA. Review of ways to transport natural gas energy from countries which do not need the gas for domestic use. Energy 2003;28:1461–77.

[4] Khalilpour R, Karimi IA. Evaluation of utilization alternatives for stranded natural gas. Energy 2012;40:317–28.

[5] Karimi F, Khalilpour R. Evolution of carbon capture and storage research: trends of international collaborations and knowledge maps. Int J Greenh Gas Control 2015;37:362–76.

[6] Co-operation OfE, Development, Agency IE. World Energy Outlook: Organisation for Economic Co-operation and. Development, 2011.

[7] Khalilpour R, Vassallo A. Community energy networks with storage: modeling frameworks for distributed generation. Singapore: Springer; 2016.

[8] Khalilpour RK, Vassallo A. Grid revolution with distributed generation and storage. In: Community energy networks with storage: modeling frameworks for distributed generation. Singapore: Springer Singapore; 2016. p. 19–40.

[9] Khalilpour R, Karimi IA. Investment portfolios under uncertainty for utilizing natural gas resources. Comput Chem Eng 2011;35:1827–37.

[10] Khalilpour R, Karimi IA. Evaluation of LNG, CNG, GTL and NGH for monetization of stranded associated gas with incentive of carbon credit. International Petroleum Technology Conference (IPTC). Doha, Qatar; 2009.

[11] MaCilwain C. Energy: supergrid. Nature 2010;468:20.

[12] Blarke MB, Jenkins BM. SuperGrid or SmartGrid: competing strategies for large-scale integration of intermittent renewables? Energ Policy 2013;58:381–90.

[13] Pierri E, Binder O, Hemdan NGA, Kurrat M. Challenges and opportunities for a European HVDC grid. Renew Sustain Energy Rev 2017;70:427–56.

[14] Van Hertem D, Ghandhari M. Multi-terminal VSC HVDC for the European supergrid: obstacles. Renew Sustain Energy Rev 2010;14:3156–63.

[15] Purvins A, Wilkening H, Fulli G, Tzimas E, Celli G, Mocci S, et al. A European supergrid for renewable energy: local impacts and far-reaching challenges. J Clean Prod 2011;19:1909–16.

[16] Andrew B, Joachim L, Anna N. Asia Pacific Super Grid – solar electricity generation, storage and distribution. Green 2012;189. https://doi.org/10.1515/green-2012-0013.

[17] Ahmed T, Mekhilef S, Shah R, Mithulananthan N, Seyedmahmoudian M, Horan B. ASEAN power grid: a secure transmission infrastructure for clean and sustainable energy for South-East Asia. Renew Sustain Energy Rev 2017;67:1420–35.

[18] Bogdanov D, Breyer C. North-East Asian Super Grid for 100% renewable energy supply: optimal mix of energy technologies for electricity, gas and heat supply options. Energ Convers Manage 2016;112:176–90.

[19] Mathews JA. Asia-Pac J 2012;10.

[20] Haldane JBS. Daedalus: or, science and the future; a paper read to the heretics, Cambridge, on February 4th, 1923. New York: E.P. Dutton & Company; 1924.

[21] Hashimoto K, Yamasaki M, Fujimura K, Matsui T, Izumiya K, Komori M, et al. Global CO_2 recycling—novel materials and prospect for prevention of global warming and abundant energy supply. Mater Sci Eng A 1999;267:200–6.

[22] Otto A, Robinius M, Grube T, Schiebahn S, Praktiknjo A, Stolten D. Power-to-steel: reducing CO_2 through the integration of renewable energy and hydrogen into the German steel industry. Energies 2017;10.

CHAPTER 7

Polyfeed and Polyproduct Integrated Gasification Systems

Joel Parraga*, Kaveh Rajab Khalilpour†, Anthony Vassallo*
*School of Chemical and Biomolecular Engineering, University of Sydney, Sydney, NSW, Australia
†Faculty of Engineering and Information Technology, Monash University, Melbourne, VIC, Australia

Abstract

The current dependence on coal-based power generation and the infrastructure developed around it makes it likely that coal will remain a critical energy source for the foreseeable future. There is, however, a great need to utilize this energy source more sustainably. The integrated gasification combined cycle (IGCC) process is an energy conversion system for concurrent power and chemical production. The significant body of research that has been conducted regarding the IGCC process and the well-outlined benefits of this potentially highly flexible polygeneration energy conversion system convey a promising technical and theoretical perspective. This Chapter investigates the polygeneration IGCC system with polyfeed and polyproduct flexibility.

Keywords: Integrated gasification combined cycle, IGCC, Polygeneration, Multienergy, Energy hubs, Flexibility, Waste to energy

1 BACKGROUND

After sun, hydropower and wind, carbonaceous resources have been the main source of energy for humans throughout the history of civilization. Learning to burn wood triggered a revolution to all aspects of early human lifestyle, from diet to comfort and safety. Although there is scattered historical evidence of fossil fuel usage (coal, oil, and gas) across the world, systematic use of coal has a history close to a millennium. Widespread use of oil and gas has a much shorter history starting with James Young (1811–83) who invented crude oil refining in 1840s (Chapter 1 of this book). Given the greater geographic distribution, after wood, coal has traditionally been the primary source of power generation and accounts for approximately two-fifths of global power generation capacity [1]. Fig. 1 classifies various approaches for carbonaceous resource utilization.

Polygeneration with Polystorage
https://doi.org/10.1016/B978-0-12-813306-4.00007-0

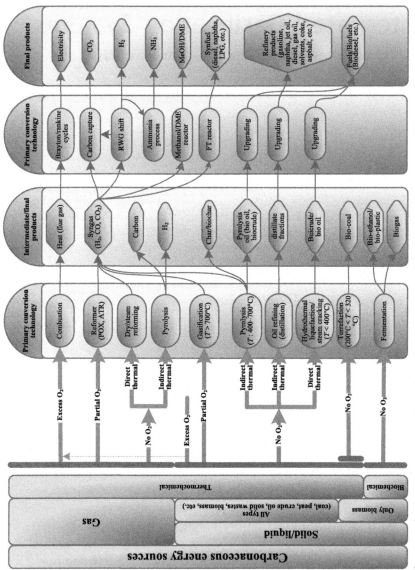

Fig. 1 Classification of various approaches for carbonaceous resource utilization.

Gaseous fuels include natural gas, biogas, and synthetic natural gas, which is mainly composed of methane and some light hydrocarbons such as ethane and propane. The conventional approach of energy utilization from gaseous resources has been direct combustion with excess air to generate thermal energy. This energy could also be recovered in turbines for power generation (e.g., the open loop Brayton cycle). The remaining energy of the effluent gas can be further utilized for steam (and then power) generation using the Rankine cycle. If CO_2 emission removal is required, the exiting flue gas can go through a carbon capture process to remove the CO_2.

An alternative route for gaseous resource utilization is partial oxidation in a process called reformer (POX: partial oxidation reformer) [2]. A third approach for the utilization of gaseous fuels is anaerobic (no oxygen). The initial reactions are endothermic, and the thermal energy is supplied either directly or indirectly. The anaerobic direct-thermal process is still called a reformer as this process also generates syngas through the direct supply of high-temperature steam to the fuel chamber. The anaerobic indirect-thermal process is called pyrolysis, which dissociates gases into their primary components, i.e., largely carbon and hydrogen [3].

Liquid and solid fuels are either fossil-based or renewable biomass. The utilization technologies for these fuels are (1) Thermochemical, and (2) Biochemical. While the first category can process both organic and inorganic fuels, biochemical processes are used only for biomass. Like gaseous fuels, thermochemical processes can be classified into three groups based on the extent of oxygen used. With excess air, they can be directly combusted for heating or power generation (through Rankine cycles). They can also be gasified with partial oxygen levels in high-temperature ($>700°C$) reactors called gasifiers. The product of gasification is also syngas, though it has lower H_2/CO ratios compared with gaseous fuels. Thermochemical conversion of solid/liquid fuels can also be carried out anaerobically through pyrolysis. This also can happen in two modes: direct thermal and indirect thermal. Indirect-thermal pyrolysis happens at lower temperatures compared with gasification ($400–700°C$), and its products contain less gas and liquids (bio-oil) and solids (char/biochar). Biofuels such as biodiesel are produced through this process. In direct-thermal pyrolysis, superheated water is directly supplied to the feed in the reactor. This process is often called "hydrous pyrolysis," "hydrothermal liquefaction," or "steam cracking." There is also another anaerobic indirect-thermal pyrolysis specific for biomass; this process is called "torrefaction" or "mild pyrolysis" due to reactions happening at lower temperatures ($200–320°C$). The advantage of this process is that the product, which is called "torrefied biomass" or "bio-coal," is denser, which improves its energy intensity for efficient transportation or subsequent processing.

Alternatively, organic materials can be converted biochemically through fermentation (digestion) to biogas, bioethanol, bioplastics, etc.

It is no wonder that in the absence of environmental policies combustion-based processes have been the most economical pathways for biomass and fossil fuel energy utilization. Today, pulverized coal combustion (PC) processes are the most established power generation systems. The demand for coal has dramatically increased over the last decade [4]. Coal is abundant in approximately 70 countries throughout the world, and for a number of countries, such as Australia, China, India, and the USA, coal represents a highly reliable and nationally secure energy source, less subject to the conflicts that exemplify the intense competition for oil and natural gas resources. Global proven reserves of coal total 900 billion tonnes, with reserves-to-production ratio indicating over 100 years of continued supply should global production continue at the current rate [6]. Coal is also relatively cheap, the cost of bituminous coal, for example, has historically remained cheaper than oil or gas on a per GJ basis by a significant margin and also offers the benefit of reduced variability [5]. The current dependence on coal-based power generation and the infrastructure developed around it acts to support the industry, making it likely that coal will remain a critical energy source for the foreseeable future in many countries. Even with the fast uptake of renewables, it is projected that coal will account for one-fifth of the world energy supply in 2040 (BP, 2018).

As a consequence, approximately 70% of CO_2 emissions that arise from power generation is attributed to coal consumption ($\sim$40% of global anthropomorphic CO_2 emissions) [4]. The carbon-to-hydrogen ratio of coal is roughly twice that of crude oil and approximately four times that of natural gas. Therefore, direct combustion of coal generates the largest quantity of CO_2 per unit of energy released compared to any other fossil fuel and places this fuel as the least sustainable energy source.

Further to *sustainability challenges,* coal-based power generation suffers from process *flexibility.* Generally, large-capacity PC plants are baseload power sources, operating at maximum output, satisfying the demand security objectives [7]. Baseload power plants are not suitable for "load following." Factors such as the temporal and geographical variability of power demand and supply, the increasing prevalence of variable renewable resources, and the development of the electricity market have resulted in increased demand for more flexible power generating processes [8]. The high variability of many renewable energy sources increases the challenge of the variability of demand. It is probable that future government policy will mandate that an increasing percentage of electricity production be required to come from renewable sources, as evident in several countries [9]. Consequently, power generation processes must be capable of meeting load fluctuations dynamically and able to optimize for changing electricity market conditions rapidly. Achieving reliable operation of power generation processes in the future will, therefore, mandate the continuous balancing between the production and consumption of resources associated with the electricity network and this may prove economically unviable for baseload PC plants. Therefore, there is a significant need to design and establish new energy conversion systems that utilize coal in a more flexible and environmentally sustainable manner in order to allow a safe and affordable transition to the ideal objective of a fully renewable global energy supply.

The gasification process (Fig. 1) involves partial oxidation of a feed material where the exothermic energy results in the autothermal breakdown of the feed to the smallest stable chemical units that can still carry energy [10]. The final mixture of gaseous products is labeled synthesis gas or syngas. The gasification of coal has been practiced since the late 19th century when coal-to-gas plants were common throughout Europe, and coal-derived gas was used for municipal lighting and distributed for lighting, cooking, and heating [11]. In modern applications, the intermediate product of syngas undergoes further downstream processing before conversion to a final product. This includes gas cleaning and a water-gas shift stage (where additional hydrogen is produced). The final product following syngas conversion can be significantly diverse; depending on downstream processing it can vary between gaseous or liquid fuels and chemicals. This system was further advanced through the integration of a gas turbine with the gasification process to increase the diversification of products by generating power. Fig. 2 shows a schematic of a modern polygeneration gasification system with the flexibility of taking multiple feeds and generating multiple products.

The idea of coupling the production of syngas through gasification with a gas turbine for power generation was initially proposed in 1950 [5]. The integrated gasification combined cycle (IGCC) process involves the combustion of syngas in a gas turbine attached to a steam cycle in a combined cycle to accomplish a high-efficiency power generation process. IGCC is being increasingly evaluated as a potential clean coal power generation process with inherent advantages compared to PC with respect to CO_2

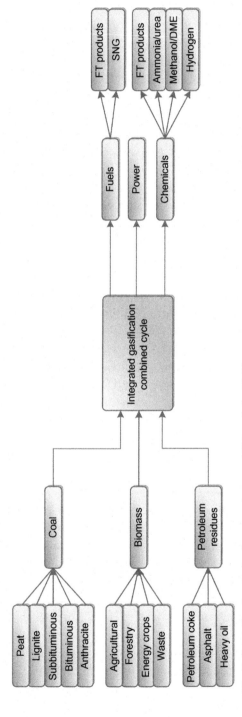

Fig. 2 Cogasification and polygeneration capabilities of the IGCC process.

emissions mitigation. When attempting to capture the CO_2 emissions from the PC process, there is a significant reduction in overall process efficiency, referred to as an energy penalty. This energy penalty is a result of the low CO_2 content (<15% by volume) of the exiting low-pressure flue gas. The consequence of this energy penalty is the need to deploy significant additional capacity solely to enable the capture process, which, in turn, necessitates an increase in coal consumption to achieve the same power output. With IGCC, CO_2 capture and sequestration can occur with a significantly lower energy penalty as the majority of gas handling occurs in the syngas that is contained at high pressure and has not been diluted by combustion air. Therefore, CO_2 removal can be more effective and economical than cleaning up large volumes of low-pressure flue gas as required following PC. Furthermore, overall emissions from an IGCC plant are around one-third to one-tenth of those for a PC plant at equivalent power outputs [5]. There are also additional environmental benefits from IGCC when compared to PC, such as approximately 30% less water consumption and lower leachability of fused slag [12]. All these advantages are making IGCC an attractive option for addressing energy security and sustainability problems concurrently, and the literature shows increasing attention to this technology in recent years [13].

Currently, the gasification of carbonaceous feed material is performed in around 300 plants, utilizing approximately 800 gasifiers across 29 different countries [10]. From 2006 to 2013, the operating global gasification capacity doubled and further growth is expected [10]. Producing chemical products is the dominant application for gasification, e.g., approximately 25% of global ammonia production and 35% of global methanol production are an end product of gasification [10].

2 IGCC WITH MULTIPLE FEED FLEXIBILITY

Generating power from the gasification of coal and other carbonaceous feed material enables a degree of flexibility unattainable for conventional power generation processes. The IGCC process enables polygeneration of power, in addition to fuels and chemicals, based on varying syngas processing (Fig. 2). Interestingly, the term "polygeneration" has been coined in the early 1980s in relation to the production diversity of the IGGC system (Chapter 3). The polygeneration of power, fuels, and chemical allows an aggregated allocation of energy sources in a wide range of power outputs and chemicals for a wide range of applications. This enables flexible modes of operation where the combination of multiple products for different markets can allow tailored income generation by switching between different portfolios of products based on market conditions. The diversification of the product portfolio reduces risk and enables profit maximization. Therefore, processes, such as IGCC, that enable polygeneration have the potential to overcome the issues that arise from the variability of demand in the power sector. Additionally, polygeneration allows the units that are common to multiple processes to operate at a higher capacity and under continuous operating conditions thereby reducing risks associated with thermal load oscillations. Further increasing flexibility and the ability to exploit market conditions for the optimization of profit is the potential for cogasification. As previously stated, any carbonaceous feed material can undergo gasification, this includes different ranks of coal as well as a broad array of biomass and petroleum residues. It should be noted that the heterogeneity of coal, biomass, and petroleum residues will undoubtedly result in an increase to the fluctuation of syngas quantity and composition produced from the gasification process and therefore for the cogasification process to achieve an acceptable degree of consistency coal must act as the buffer. The choice between carbonaceous feed material and selected final product will present a complex technical and economical optimization problem with potentially multiple solutions. Understanding market drivers and influences for feed materials and products in additional process limitations will aid in the development of decision support tools, such as an optimized operational strategies, that improve the commercial attractiveness of the IGGC process [14].

2.1 Biomass

Biomass is a renewable energy resource that results in a negligible net contribution of CO_2 emissions to the atmosphere. Typical biomass sources include wood from plantation forests, residue from agriculture and forestry, and municipal organic waste streams. Collection of forest residue that would otherwise decompose, can also reduce the release of greenhouse gases into the atmosphere, notably CH_4. Hence, biomass has the dual advantage of a carbon offset by acting as an energy substitute for fossil fuels and also as a carbon sink through provided a means of sequestering carbon. It is therefore evident that biomass energy conversion systems have an essential role in the mitigation of CO_2 emissions. Large quantities of biomass are already harvested in well-established systems. For example, the sugar cane industry has considerable experience of harvesting and handling up to 3Mt per annum of biomass at any one facility [15].

Utilizing biomass as the feed material in the gasification process has received notable interest and there has been a significant body of research reviewed elsewhere regarding the current status and perspective of biomass gasification [16–18] and the technoeconomic feasibility [19,20]. With regards to the technical aspects of the biomass gasification process, syngas cleaning is the most prominent field of research, as summarized in [21–24]; however, other areas, such as, cogasification with coal [25–27], cogeneration of products [28], and general process modeling considerations [29,30] have also been extensively explored, as reviewed elsewhere.

There exists significant variation in chemical compositions, energy content, and ash and moisture content, between different sources of biomass and ranks of coal. In comparison with coal, biomass exhibits a notably higher oxygen content, lower carbon content, and lower heating value, as evident in Fig. 3. The high oxygen content in biomass reduces the energy density of the biomass. Research has been conducted [31] to examine the thermodynamic efficiency of biomass and recommended to use a feed material with an oxygen-to carbon ratio <0.4 in order to gasify with high thermodynamic efficiency at 927°C. For gasification, at 1227°C the optimal ratio further decreases to 0.3 or less. Fuels with higher oxygen-to carbon ratios have larger exergy losses due to the high ratio of chemical exergy to heating value.

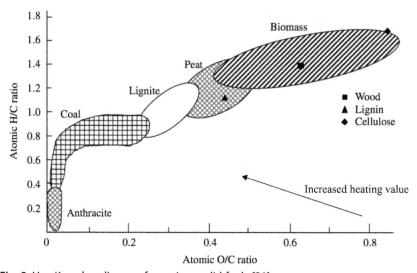

Fig. 3 Van Krevelen diagram for various solid fuels [31].

The study concluded that highly oxygenated biomass, such as biomass with a high cellulose content (see Fig. 3), is not an ideal feed material for gasification from an exergetic viewpoint. However this study, in addition to another [32], both conclude that the combined use of biomass for cogasification with coal proves an effective application for biomass. Cogasification results in an effective method to essentially modify the properties of the solid feed material entering the gasifier through improving the oxygen to carbon ratio and heating value. Further synergy has been identified in a number of studies [33–35] that indicate the higher reactivity of biomass improves the gasification process, acting to catalyze the gasification of coal.

2.2 Petroleum Residues

Gasification is also increasingly viewed as the ideal application for some petroleum-derived products and byproducts. Utilizing petroleum coke for gasification has received a considerable amount of attention, as reviewed in [36–39]. Petroleum coke is a byproduct produced from heavy crude oil residues by thermal cracking processes, most commonly in an oil refinery coker unit. Approximately 31 kg of petroleum coke is produced from the processing of each tonne of crude oil in a refinery [40]. Petroleum coke has often been utilized as an energy source in combustion boilers to generate power. However, this creates issues regarding corrosion and environmental degradation due to the high sulfur, between 5 and 7 wt%, and vanadium, up to 500 ppm, the content of petroleum coke [39]. Furthermore, the increasing consumption of natural gas in combustion boilers has decreased the demand for petroleum coke. With approximately 70 million metric tons of petroleum coke produced in 2015 [39], gasification through the IGCC process may be an effective and economically viable option for utilizing petroleum coke. Comparatively, there is minimal literature produced regarding the gasification of heavy oil and bitumen with a significantly larger body of research focusing on upgrading heavy oil and bitumen for the production of synthetic crude oil. Heavy oil is any petroleum-based energy source that contains the residue obtained following the distillation process of crude oil. Bitumen, also commonly referred to as asphalt, is another byproduct of the distillation process of crude oil, representing the heaviest fraction of crude oil, which is also found in natural deposits. There are some studies [41–45] that highlight the IGCC process as a potential solution for an environmentally responsible option for these energy sources that, similar to petroleum coke, due to the high sulfur and heavy metal content, raise concerns with regards to reuse.

3 IGCC PROCESS OVERVIEW

3.1 Gasification

The process flow diagram of an IGCC process is illustrated in Fig. 4. The first step in the process is the preparation of the feed material and depending on the choice of gasifier type utilized in the process, refer to Table 1, the size of the feed material entering the gasifier may require various methods of preparation in order to enable suitable flow and improve efficiency. The methods of preparation for biomass or petroleum-derived feed material for gasification will further vary depending on the source of the material; however, achieving the appropriate particle size will always be necessary. Additionally, biomass can often have a high moisture content that necessitates drying before gasification. It is generally recommended that biomass moisture content be below 10%–15% before gasification. High moisture content reduces the maximum temperature achievable in the gasification process, potentially causing incomplete gasification. This is an energy-intensive process. However, there are opportunities for integration with waste heat from the gasification process being utilized for this application to minimize the decrease to overall energy efficiency.

When the feed material has been appropriately prepared, it enters the gasifier to undergo autothermal breakdown. The oxygen is supplied by an air separation unit (ASU) process. Cryogenic ASU processes are also increasingly being investigated as a potentially more efficient option [46]. Among various technologies such as membrane, adsorption and ion transport, cryogenic air separation is the most efficient and economical technology for production of large quantities of oxygen. Nevertheless, ASUs represents a substantial fraction of the overall capital expenditure and an ongoing operating expenditure, especially for auxiliary power.

The prepared feed and oxygen react in the gasifier with key occurring reactions summarized in Table 2. There are three main categories of gasifiers, classified by the flow regime inside the reactor, entrained flow, fluidized bed or moving bed. It is evident that entrained flow gasifiers offer preferable operating conditions in terms of capacity, residence time, and steam consumption, however, do require significantly more oxygen.

The composition of syngas can significantly vary and is essentially linked to the gasifier technology in addition to the quality of the feedstock and the operating conditions of the gasification process, most notably temperature and pressure [47]. Each type of gasifier unit is specialized for certain feed materials. Due to this framework, different gasification technologies commercially available form the basis for the competition on the market, as

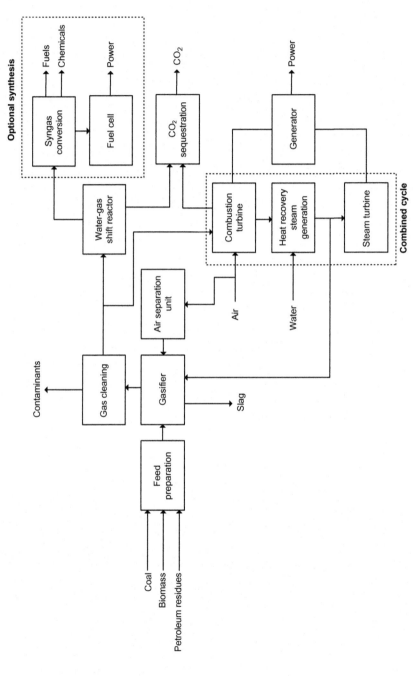

Fig. 4 Process flow diagram for polygeneration through the IGCC process.

Table 1 Gasifier classification by bed type

Characteristic	Moving bed	Fluidized bed	Entrained flow
Flow regime	Concurrent, countercurrent	Stationary, circulating	Upflow, downflow
Feed preparation	Screening, agglomeration	Crushing	Grinding
Feeding system, e.g.	Gravity sluice	Screw feeder	Slurry pump
Tar deterioration	Barely	Predominantly	Completely
Residence time (solids)	Hours	Minutes	Seconds
Specific capacity	Low to moderate	Moderate to high	Very high
Average feed size (mm)	3–60	0–6	<0.25
Oxygen consumption[a]	0.19–0.53	0.4–0.7	0.7–1.0
Steam consumption[b]	0.4–2.0	0.2–0.6	0–0.3
Temperature (exit) (°C)	350–800	800–1000	1300–1700
Pressure (bar)	1–100	1–40	1–86
Carbon conversion (%)	80–90	80–95	>95

[a]m^3(STP)/kg feed material (MAF).
[b]kg/kg feed material (MAF).
MAF, moisture and ash-free basis.
Modified from Gräbner M. Industrial coal gasification technologies covering baseline and high-ash coal. Weinheim: Wiley-VCH; 2014.

Table 2 Key reactions occurring during the gasification of carbonaceous material

Reaction	Equation	Enthalpy change (kJ/mol)
Pyrolysis	$Coal \rightarrow CH_{4(g)} + C_{(s)}$	+ variable
Combustion	$2C_{(s)} + O_{2(g)} \rightarrow 2CO_{(g)}$	−221.3
	$C_{(s)} + O_{2(g)} \rightarrow CO_{2(g)}$	−394.0
	$2CO_{(g)} + O_{2(g)} \rightarrow 2CO_{2(g)}$	−566.7
	$2H_{2(g)} + O_{2(g)} \rightarrow 2H_2O_{(g)}$	−484.2
	$2CH_{4(g)} + O_{2(g)} \rightarrow 2CO_{(g)} + 4H_{2(g)}$	−71.4
Gasification	$C_{(s)} + H_2O_{(g)} \rightarrow CO_{(g)} + H_{2(g)}$	+131.5
	$C_{(s)} + 2H_2O_{(g)} \rightarrow CO_{2(g)} + 2H_{2(g)}$	+90.2
Boudouard	$C(char)_{(s)} + CO_{2(g)} \rightarrow 2CO_{(g)}$	+172.7
Water-gas shift	$CO_{(s)} + H_2O \leftrightarrow CO_{2(g)} + H_{2(g)}$	−41.2
Methanation	$C_{(s)} + 2H_2 \rightarrow CH_{4(g)}$	−74.94
Steam reforming	$CH_{4(g)} + H_2O_{(g)} \rightarrow CO_{(g)} + 3H_{2(g)}$	+206.2
Dry reforming	$CH_{4(g)} + CO_{2(g)} \rightarrow 2CO_{(g)} + 2H_{2(g)}$	+247.4

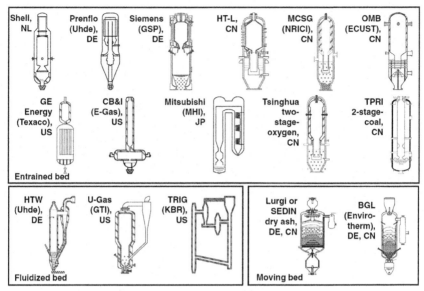

Fig. 5 Commercially available coal gasification technologies [10].

evident in Fig. 5. There is a notable level of diversification in the commercial market for gasifier technology when considering that only five countries are significantly involved in the technological development of gasifiers: the Netherlands, Germany, China, Japan, and the U.S.A (US) [10]. Table 3 provides information for operating IGGC plants with various gasifier technologies regarding the operation of four of these plants [48].

Fig. 6 displays the market share of thermal syngas capacity in GW_e. Shell technology is the current market leader closely followed by GE Energy. It is predicted that the upcoming technology from China, particularly SEDIN and ECUST, will show a significant increase in market share growth [10]. Similarly, circulating fluidized bed gasifiers, such as the TRIG gasifier developed by KBR, have recently become of interest due to the ability to utilize a broad range of low-quality fuel sources and advantages relating to NO_X emissions [49].

3.2 Syngas Cleaning

There are several contaminants that must be removed from the syngas before power generation or the conversion of syngas to different chemical and fuel products can occur. Contaminants removed from syngas generally include *particulate matter* (to avoid downstream fouling, corrosion, and erosion

Table 3 Comparison of four existing IGCC plants

	IGCC Plant			
Characteristic	Tampa electric	Wabash river	Nuon power	ELCOGAS
Gasifier type	Entrained flow	Entrained flow	Entrained flow	Entrained flow
Gasifier technology	GE energy	E-Gas	Shell	Prenflo
Cold gas efficiency (%)	71–76	74–78	80–83	80–83
Carbon conversion (%)	96–98	98	> 98	>98
Feed rate (tonnes per day)	2200–2400	2500	2000	2600
Gasifier availability (%)	82.0	84.4	81.8	76.0
Plant availability (%)	95.0	94.3	96.4	93.0
Designed net efficiency (basis)	41.6 (HHV)	37.78 (HHV)	43.0 (LHV)	43.0 (LHV)
Net power (MW_e)	250	262	253	300

Adapted from Liu K, Cui Z, Chen W, Zhang L. Coal and syngas to liquids. In: Hydrogen and syngas production and purification technologies. Hoboken, NJ: John Wiley & Sons, Inc.; 2009, p. 486–521

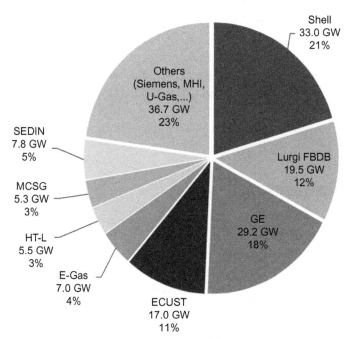

Fig. 6 Global gasification technology in operation or underconstruction [10].

issues [50]), condensable hydrocarbons or tars, sulfur compounds (corrosiveness and contribution to SO_2 release), nitrogen compounds (avoid catalyst poisoning and NOx formation), *alkali metals* (forming metal oxides at lower temperatures, articulating and causing corrosion [21,24,51,52]), and chlorine compounds (significant process threat for corrosion and fouling). Contaminant levels vary greatly and are heavily influenced by the feed material impurities and gasifier operating conditions. The degree of contaminant removal that is required also varies substantially depending on the syngas application [21,24]. There are numerous methods to remove these contaminants, and a significant amount of research has been conducted regarding syngas cleaning, as comprehensively reviewed in [21,24,51,52]. The presence of tars in the syngas is one of the main technical barriers in the development of gasification [16]. Tar is a general term used to represent condensable organic compounds such as mixed oxygenates, phenolic ethers, alkyl phenolic, heterocyclic ethers, and polyaromatic hydrocarbons. Analyzing tar is difficult due to the complex chemical nature of different tar contaminants. Specific tar contaminants will readily condense, even at high temperature, and cause issues such as fouling. Other tar contaminants can generate catalyst poisoning in downstream process operations. Tar contaminants that are water soluble also create secondary issues relating to waste water treatment for water-based cleaning processes. Therefore, removal or decomposition of all variations of tar contaminants is desirable. However, the optimal solution to issues related to tar contaminants would be to reduce or ideally prevent tar formation. This could be achieved through the improvements in the gasification process, notably gasifier technology and optimizing operating conditions [27]. There is also extensive research for high-temperature gas cleaning (HTGC), say $>250°C$, due to overall process efficiency [4].

4 POLYGENERATION

4.1 Power Generation

4.1.1 Combined Cycle

There are two primary sources of power generation in an IGCC plant, gas and steam turbines. A key variable in the operation of a gas turbine is the firing temperature. The current state-of-the-art gas turbines for use in the IGCC process are classified as either F-class or G-class, with firing temperatures that range between 1370 and 1430°C [12]. While increasing the firing temperature improves efficiency, there exists a tradeoff between the firing

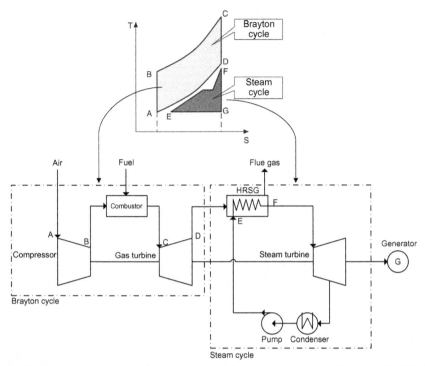

Fig. 7 Combined cycle power generation and corresponding T-S diagram. *(Modified from Higman C, van der Burgt M. Chapter 7: Applications. In: Gasification. 2nd ed. Burlington: Gulf Professional Publishing; 2008, p. 257–321.)*

temperature and the cooling requirements that are necessary to protect the components of the turbine. Modern turbine technologies have between 30 and 35% (LHV) efficiency [53]. To increase the net power efficiency of the process gas turbines are combined with a steam cycle. The exhaust gases from the turbine have the potential to produce power by a steam turbine. Fig. 7 illustrates the temperature-entropy (T-s) curve of a combined cycle power generation plant composed of a Brayton cycle (gas turbine) and Rankine cycle (HRSG). The HRSG is a convective heat exchanger with liquid water and steam on the inside and exhaust gases on the outside. The function of the HRSG is to recover heat from the outlet of the gas turbine to produce steam, which is then directed to steam turbines to generate electricity.

Referring to Fig. 7, air (A) is compressed and now the pressurized air (B) is used for the combustion of the syngas fuel. The resulting hot gases (C) then enter the turbine before being sent to the HRSG (D). Steam is raised in the HRSG (F) that passes through a steam turbine generating work (G) before being condensed and pumped back to the HRSG to complete the cycle (E). It is evident in the T-s diagram that the space below the area of the Brayton cycle is occupied by the steam cycle. The ratio of shaded areas to the area below the upper line of the cycle, which is a measure of the overall efficiency, is subsequently increased.

4.1.2 Fuel Cell Integration

Integrating gasifiers with high-temperature fuel cell systems to enable power generation systems with high efficiencies and low emissions have been attracting significant research attention since the early 1990s. Fuel cells offer an additional source of power as hydrogen produced from syngas is converted to electricity with high power generation efficiency and negligible emissions. The first efforts to integrate gasification and fuel cell systems generally focused on the molten-carbonate fuel cell (MCFC) due to the relative maturity of this technology [54]. The advancement of the solid oxide fuel cell (SOFC) then began to demonstrate promising system efficiency for integration with the IGCC process. The operating conditions of SOFCs enable improved integration with gas turbines to achieve high power production [53]. Therefore, the majority of research and development in the field shifted to the integration of SOFCs with the IGCC process.

Table 4 lists some of the key studies on IGCC-fuel cell integration. The earliest study in the table refers to Jansen et al. [55], 1992, who proposed a system utilizing the entrained flow gasifier developed by Shell, the gas cleaning had an operating temperature of approximately 350°C and concluded that such a system could achieve electricity efficiency of 53.1% (LHV basis). This study further investigated the feasibility of CO_2 capture downstream of the MCFC using shift reaction followed by ceramic CO_2 separation membrane and reported the system efficiency would decrease to 47.5% (LHV basis). In a later study, in 1994, Jansen et al. [56] compared the effects of different gas cleaning processes on the performance of an IGCC system utilizing GE Energy entrained flow gasifier technology and MCFC and demonstrated that the system with HTGC could achieve electricity efficiency of 53.2% (LHV basis) in comparison with conventional LTGC that would decrease the electricity efficiency to 49.2% (LHV basis).

In Table 4, the highest efficiency belongs to the study by Lobachyov et al. [57] in 1997. This study proposed a system composed of E-Gas entrained flow gasifier process and SOFC-gas turbine hybrid system. The system efficiency was estimated to be quite high, 63.1% (HHV basis) though

Table 4 Summary of research literature regarding IGCC fuel cell integration

Gasifier type	Gas cleaning	Fuel cell	CO$_2$ sequestration	Efficiency (basis)	Year	Ref
Entrained flow	HTGC	MCFC	Yes	47.31% (LHV)	2015	[84]
Fluidized bed	HTGC	SOFC	Yes	49.8% (LHV)	2015	[85]
Entrained flow	HTGC	SOFC	Yes	55.0% (LHV)	2014	[86]
Entrained flow	LTGC	SOFC	No	61.5% (HHV)	2010	[54]
Entrained flow	HTGC	SOFC	Yes	56.2% (HHV)	2009	[87]
Entrained flow	HTGC	SOFC	No	61.8% (HHV)	2009	
Fluidized bed	HTGC	SOFC	Yes	50.3% (HHV)	2006	[88]
Entrained flow	HTGC	SOFC	No	variable	2006	[89]
Fluidized bed	LTGC	SOFC	Yes	46.3% (HHV)	2005	[90]
Fluidized bed	HTGC	SOFC	No	60.1% (HHV)	2005	[91]
Fluidized bed	HTGC	SOFC	Yes	49.6% (HHV)		
Entrained flow	LTGC	SOFC	No	46.7% (LHV)	2004	[92]
Entrained flow	LTGC	MCFC	No	53.3% (LHV)	2000	[93]
Entrained flow	None	SOFC	No	63.1% (HHV)	1997	[57]
Entrained flow	HTGC	MCFC	No	53.2% (LHV)	1994	[56]
Entrained flow	LTGC	MCFC	No	49.2% (LHV)		
Entrained flow	HTGC	SOFC	No	49.0% (LHV)		
Entrained flow	HTGC	MCFC	Yes	53.1% (LHV)	1992	[55]
Entrained flow	HTGC	MCFC	No	47.5% (LHV)		

without carbon capture. However, this proposed system relies heavily on the CO_2 acceptor gasification process for not only syngas production but also gas cleaning. This gas cleaning scheme increases the system efficiency significantly as the syngas coming out of the gasifier is used directly in the SOFC with minimal thermal energy loss. The ability to consistently produce a syngas with contaminant concentration able to meet the stringent specifications of an SOFC while utilizing such a gas cleanup process may not be viable in practical applications.

4.2 Chemical and Fuel Synthesis

4.2.1 Hydrogen

Hydrogen currently has significantly diverse applications in the chemical and food industry [58]. Additionally, due to the major advantages in efficiency and environmental benefits, hydrogen as a fuel source and in conjunction with fuel cells has attracted considerable attention within a proposed hydrogen economy. However, the reliable and efficient production of hydrogen is a major issue in preventing the development of this hydrogen economy. Hydrogen production from syngas can be achieved in the water-gas shift WGS reactor followed a sour shift. The WGS reaction typically occurs in a multistaged fixed bed catalytic reactor in order to adjust the CO and H_2 ratio. The WGS reactor can be located either before the sulfur compounds removal step of the syngas cleaning process (sour shift) or after (clean shift). Consequently, the choice of catalyst is dependent on whether the WGS reaction is a sour or sweet shift [59–61]. The reaction proceeds as follows:

$$CO_{(s)} + H_2O \leftrightarrow CO_{2(g)} + H_{2(g)} \ \Delta H^{\circ}_{rxn} = -41.21 \ kJ/mol \qquad (1)$$

In order to drive the reaction equilibrium toward the WGS products, steam is typically introduced at the inlet of the shift reactor. Depending on the desired product and CO_2 sequestration requirements, some quantity of the syngas may bypass the WGS reactor to avoid exceeding the required H_2/CO ratio [59–61]. The CO_2 byproduct from this process can be separated from the hydrogen during sulfur compound removal in the gas cleaning process, in methods such as the Selexol process [48] (see Table 8). If the IGCC process is designed with a sweet shift, then multistage flash drums can be used to extract CO_2 from absorbed solvents. The CO_2-product stream can then be sent for planned utilization/sequestration.

An alternative approach if high purity hydrogen is required is to separate the hydrogen product via palladium diffusion membrane [58]. In this approach, the WGS reaction occurs inside a membrane tube over a suitable catalyst. Another well-documented approach for achieving high purity hydrogen from syngas is pressure swing adsorption (PSA) technology. In the PSA process, gas species are separated at high pressure and low temperature through passing the syngas through a reactor containing a suitable solid sorbent. The pressure is then reduced and this releases the CO_2 from the adsorbent surface.

4.2.2 Methanol

Methanol is an important intermediate product for the manufacture of various other chemicals. Approximately 39% of methanol is converted to formaldehyde, 19% to methyl tert-butyl ether, and 12% to acetic acid [62]. Final product applications include solvents, paints, explosives, and plastics, among numerous other chemicals. Methanol is also increasingly evaluated as an alternate fuel source for internal combustion engines and direct methanol fuel cells. In 2013, global demand for methanol reached 65 million metric tonnes [62]. Demand for methanol has varied substantially causing significant price volatility; however, the average price has historically increased [63].

Methanol synthesis occurs through the reaction of H_2 with CO or CO_2, as follows:

$$CO_{(g)} + 2H_{2(g)} \rightarrow CH_3OH_{(g)} \ \Delta H^{\circ}_{rxn} = -91 \ \text{kJ/mol} \qquad (2)$$

$$CO_{(g)} + 2H_{2(g)} \rightarrow CH_3OH_{(g)} + H_2O \ \Delta H^{\circ}_{rxn} = 0 - 50 \ \text{kJ/mol} \qquad (3)$$

For example, the LPMEOH process design by Air Products and Chemicals Inc. is a prominent syngas-to-methanol conversion system [48]. The process involves a slurry-bubble column reactor wherein a catalyst, typically a mixture of copper, zinc oxide, alumina, and magnesia, is suspended in an inter mineral oil. The mineral oil is a temperature moderator capable of removing the heat of reaction through generating steam in an internal tubular boiler. Operating conditions for this process are between 225 and 270°C and 5–10 MPa [64]. The ability to remove this heat allows the process to operate at isothermally. Therefore, the process is stable and suitable for rapid ramping, idling, shutdown, and startup. This is notably advantageous in a polygeneration energy conversion system such as the IGCC process.

4.2.3 Ammonia

Ammonia as a fertilizer, in the form of either salts, solutions or anhydrously, represents >80% of global ammonia demand [65] with various industrial applications, such as explosives and cleaning agents, the remaining 20%. The price has remained relatively steady at approximately US$500–$600 per metric tonne after a brief peak at $900 per tonne during 2008 [66]. Global ammonia demand has grown at an average annual rate of approximately 2% over the past decade and this growth is forecasted to continue into the foreseeable future.

The production of ammonia from syngas is still similar to the original Haber-Bosch process that first enabled the production of synthetic ammonia [67]. Ammonia is produced from the syngas in a multiple fixed bed converter catalyzed by promoted iron. The catalysts are supplied as magnetite, Fe_3O_4, and reduced in the reactor to produce metallic iron. Removal of oxygen from the solid magnetite pellet produces a porous iron pellet with a surface area between 10 and $20\,m^2/g$ [67]. The catalyst pellets are promoted with small quantities of Al_2O_3 and K_2O. Ruthenium catalysts have proven to be more active than the traditional iron catalysts, allowing the reactor to operate at lower temperatures and pressures. This results in considerable savings in compression and vessel costs; however, ruthenium is a rare and expensive metal, and it is unclear that these savings justify the increased catalyst expenditure that would be required. On each pass over a fixed bed converter, approximately 30% conversion occurs [48]. Therefore, the ammonia is separated from the reactor effluent, and the unreacted syngas is recycled. The separated ammonia is cooled in a series of heat exchangers in order to condense the ammonia is a liquid product. The ammonia synthesis proceeds as follows:

$$N_{2(g)} + 3H_{2(g)} \rightarrow 2NH_{3(g)} \quad \Delta H^{\circ}_{rxn} = -92 \text{ kJ/mol} \tag{4}$$

4.2.4 Synthetic Natural Gas

Natural gas is a vital component of the global energy mix. It is widely used in residential, commercial, and industrial applications. Due to increase in uses, there often exists a geographical mismatch in demand and supply of natural gas [68]. Synthetic natural gas (SNG), also referred to as substitute natural gas, consists primarily of methane (>95% volume), with similar properties to natural gas. SNG is desirable when there is an absence or a shortage of natural gas in a region. SNG can be used in locations that lack the availability natural gas or to augment the gas supply where natural gas is already available to provide a convenient, consistent, and high-quality fuel with combustion characteristics similar to natural gas. The price of substitute natural gas (SNG) is therefore strongly dependent on the competing source of natural gas. It has also been highlighted that producing and transporting natural gas causes the emissions of 2.3 times CO_2-equivalent gas in comparison to mining and delivery coal on per unit of energy basis [69]. This is due to the gas flaring and leaking that occurs when utilizing natural gas and the significant global warming potential of CH_4. Therefore, the production of SNG instead of conventional natural gas may have an environmental advantage.

SNG is synthesized by the reaction of carbon oxides with hydrogen, which occurs in a fixed bed reactor over a nickel or iron catalyst bed. The reactions proceed as follows:

$$CO_{(g)} + 3H_{2(g)} \rightarrow CH_{4(s)} + H_2O_{(g)} \ \Delta H^{\circ}_{rxn} = -206.2 \ kJ/mol \quad (5)$$

$$CO_{2(g)} + 4H_{2(g)} \rightarrow CH_{4(s)} + H_2O_{(g)} \ \Delta H^{\circ}_{rxn} = -165.0 \ kJ/mol \quad (6)$$

It should be noted that SNG catalysts also have a significant water gas shift activity:

$$CO_{(g)} + H_2O_{(g)} \leftrightarrow CO_{2(s)} + H_{2(g)} \ \Delta H^{\circ}_{rxn} = -41.21 \ kJ/mol \quad (7)$$

Conversion to methane will decline with increasing temperature due to this competing water gas shift reaction. However, the conversion will increase with increasing pressure as the four or five moles in Equations IV and V, respectively, react to form two moles of gas. Furthermore, the catalyst is gradually poisoned by sulfur. Subsequently, the catalyst lifetime is directly related to the level of sulfur removal in the gas cleaning process. One of the well-known methanation unit designs is TREMP by Haldor Topsoe [70]. Syngas is mixed with a recycle stream and fed to the first reactor. The effluent is collected and then split into a recycle stream before being sent to the second reactor. The recycle stream is further cooled and fed to the recycle compressor. The purpose of this recycle stream is to dilute the initial syngas feed stream, reduce the temperature rise across the first reactor, ensure complete conversion of CO and H_2 into CH_4. The necessity of this recycle stream results in a notable decrease in process efficiency, requiring significant energy consumption for compression [71].

4.2.5 Fischer-Tropsch Products

Fischer-Tropsch (FT) products predominantly refer to liquid hydrocarbon fuels, and the process is commonly known at gas to liquid (GTL). The GTL option is especially attractive when there is a consistent supply of gas without the infrastructure, in for the form of pipelines or liquefaction methods, to economically transport it to a suitable market. The synthesis of FT products allows for the utilization of the existing and significant infrastructure in place for the transport of liquid hydrocarbons.

There are two conventional synthesis processes for FT products from syngas, ARGE, and Synthol [67]. The ARGE process syngas is converted over a cobalt catalyst with operating conditions of 200°C and 30–40 bars. The reaction occurs in a number of parallel fixed-bed reactors placed in a pressure vessel containing cooling water in order for the reaction to proceed isothermally. In the Synthol process, syngas is converted over an iron catalyst with operation conditions of 250°C and 30–40 bars. The reaction occurs in large fluid-bed reactors. Due to the lack of sulfur and nitrogen contaminants, some FT products from these processes can be hydrogenated for the manufacture of solvents and waxes. Other FT products are ideal as automotive gasoline or for blending to manufacture aviation fuel or automotive diesel fuel. Typically, the main hydrocarbon cut from the ARGE process is Heavy oils and waxes, while the Synthol produces lighter hydrocarbons with gasoline being the main cut [67].

5 PROCESS TECHNOECONOMICS

Technoeconomic analyses on the IGCC process are pivotal in determining aspects such as the overall capital cost of plant construction, operation and maintenance costs, and levelized cost of electricity. Capital cost estimates from a number studies are summarized in Table 5. Note that estimations refer only to base infrastructure of the power generation plant and are therefore exclusive of aspects such as financing fees. The significant disparity that exists between capital cost estimations can be attributed to a number of factors, most notably the feed material and unit cost assumptions necessary for this analysis and the significant variations in process design for the IGCC process.

Optimal IGCC process design requires utilization of efficient modeling and simulation tools. IGCC models that can be found in the literature are usually validated with data from one of the existing plants (see Table 5). In general, research in this area has focused on coal as a feed material and on an entrained bed gasifier as the most viable gasifier technology. In all the work reviewed, Aspen Plus is the most common software used for modeling purposes and there is generally focus on the gasifier section of the plant. Costs

Table 5 Summary of research literature regarding capital cost estimations for the IGCC process

Feed material	Net power (MWe)	CO2 sequestration	Efficiency (basis)	Notable features	Capital cost ($US/ kWe)	Year	Ref
Coal	375	Yes	35.93% (LHV)	None	2707	2017	[94]
Coal	473	Yes	40.70% (LHV)	Additional steam reforming	1904		
Coal	439	No	43.88% (LHV)	None	2605	2014	[95]
Coal	353	Yes	35.26%(LHV)	CO2 sequestration through physical adsorption	5035		
Coal	485	No	46.61% (LHV)	None	2492	2014	[96]
Coal	427	Yes	33.83% (LHV)	Entrained flow gasifier	3645		
Coal	393	Yes	35.09% (LHV)	Fluidized bed gasifier	3075		
Coal	395	Yes	35.57% (LHV)	Moving bed gasifier	2817		
Coal	515	No	46.0% (LHV)	None	1596	2013	[97]
Coal	426	Yes	38.0% (LHV)	CO2 sequestration through physical adsorption	2680		
Coal	808	Yes	40.2% (HHV)	CO2 sequestration through membrane separation	2169	2013	[98]
Coal	561	Yes	35.70% (LHV)	CO2 sequestration through physical adsorption	3529	2012	[99]
Coal	485	No	46.6% (LHV)	Shell gasifier	2410	2012	[100]
Coal	433	Yes	37.1% (LHV)	CO2 sequestration through physical adsorption	3287		
Coal	449	No	43.1% (LHV)		2549		
Coal	420	Yes	36.0% (LHV)		3371		

Continued

Table 5 Summary of research literature regarding capital cost estimations for the IGCC process—cont'd

Feed material	Net power (MWe)	CO2 sequestration	Efficiency (basis)	Notable features	Capital cost ($US/kWe)	Year	Ref
Coal	487	No	35.4% (HHV)	Siemens gasifier CO2 sequestration through physical adsorption	2113	2010	[101]
Coal	444	Yes	30.4% (HHV)	None	2718		
Coal	535	Yes	33.3% (HHV)	CO2 sequestration through physical adsorption	2425		
Coal	572	Yes	36.2% (HHV)	High-temperature gas cleaning	2047		
Coal	691	Yes	38.3% (HHV)	CO2 sequestration through membrane separation	1724		
Coal	502	Yes	40.0% (HHV)	Air separation through ion transport membrane Advanced turbine	1683		
Coal	489	No	44.9% (LHV)	None	1647	2009	[102]
Coal	381	Yes	35.1% (LHV)	CO2 sequestration through physical adsorption	2304		
Coal	396	Yes	36.4% (LHV)	CO2 sequestration through membrane separation	2267		

Feedstock						Year	Ref.
Coal	789	No	46.44% (LHV)	None	1951	2009	[103]
Coal	695	Yes	36.96% (LHV)	CO_2 sequestration through chemical adsorption	2453		
Coal	417	No	43.75 (LHV)	Shell gasifier	1342	2008	[104]
Coal	353	Yes	33.05 (LHV)	Clean shift	1962		
Coal	368	Yes	34.46 (LHV)	Sour shift	1927		
Coal	465	No	39.47 (LHV)	G.E gasifier	1326		
Coal	374	Yes	29.45 (LHV)	Clean shift	1916		
Coal	394	Yes	30.98 (LHV)	Sour shift	1898		
Coal	528	No	37.2% (HHV)	None	1430	2007	[12]
Coal	493	Yes	32.2% (HHV)	CO_2 sequestration through chemical adsorption	1890		
Biomass	149	No	34.0% (HHV)	None	1250	2005	[105]
Biomass	123	Yes	28.0% (HHV)	CO_2 sequestration through physical adsorption	2730		
Coal, petroleum coke	307	No	42.7% (LHV)	ELCOGAS plant design	1386	2000	[74]
Coal	312	No	43.6% (LHV)	ELCOGAS plant design	1210		
Coal	446	No	50.9% (LHV)	Novel plant design	902		

Values are exclusive of inflation and foreign currency conversion is calculated based on annual average exchange rate from the year of publication.

for design purposes, a significant factor regarding IGCC development, are only reported in approximately half of the reviewed literature.

Zheng et al. [72] provide useful analysis through the comparison of different gasifier models in Aspen Plus and confirm that the overall performance of an IGCC plant is significantly influenced by the gasifier type and feedstock characteristics. The model produced by Frey et al. [73] is based on the Tampa IGCC plant and is one of the most exhaustive models found in the literature as it describes the numerous gas cleaning units. The Puertollano IGCC plant is the basis of numerous models, such as the one by Campbell et al. [74], which was utilized for sensitivity analysis of various parameters, Kanniche et al. [75] that evaluated the impact of inclusion of the CO_2 removal train, and Perez-Fortes et al. [76], which assessed feed material. Arienti et al. [77] modeled nine different scenarios in order to choose the best plant configuration. These models take into account a fixed demand of H2 to be produced and also consider that the remaining syngas is converted into power. Descamps et al. [78] also included H_2 purification and study of the efficiency of the whole plant including the H_2 purification units. In this study, integration of CO_2 capture in a complete and detailed IGCC power station simulation model is performed in order to calculate total plant efficiency.

With regards to the future research directions of modeling the IGCC process, a simulation model encompassing a superstructure with multiple technical alternatives should be developed [79]. As explored in this chapter, there are several promising technological advancements, such an alternate technology for air separation and different high- and low-temperature syngas cleaning options that warrant in–depth analysis and comparison. Furthermore, specific boundary conditions including different subsidies, feed material availability, and capital costs of all process operations, among other considerations, must also be deliberated in a more detailed analysis in order to gain a more accurate understanding of the commercial feasibility of the IGCC process [79]. A major challenge in the design of polygeneration systems like the IGCC process will be the determination of the optimal tradeoff between flexibility and capitals costs, as higher flexibility typically involves larger unit process sizes. In the majority of process modeling literature, the performance of all process operations and individual units is assumed to be constant under all operating conditions. In reality, the performance of equipment might significantly decrease when operated below design capacity. Therefore, an analysis that can accurately incorporate this variation in performance would be valuable.

Furthermore, there is minimal literature describing dynamic modeling entire IGCC process due to the lack of experimental data describing the transient behavior of an IGCC system. This is significant as the establishment of dynamic models would provide the time-dependent data that are necessary to develop robust and responsive control systems that can ultimately enable optimized load changing capability [53]. The development of these control systems is therefore a key future challenge.

6 IGCC TECHNOLOGY BARRIERS

The current dependence on coal-based power generation and the infrastructure developed around it acts to support the industry heavily, making it likely that coal will remain a critical energy source for the foreseeable future. There are potential methods to utilize this energy source more sustainably. The significant body of research that has been conducted regarding the IGCC process and the well-outlined benefits of this potentially highly flexible polygeneration energy conversion system convey a promising technical and theoretical perspective. However, there are significant challenges that continue to prohibit the increased establishment of this system.

In the past two decades, the commercial viability of IGCC technology has been explored across various developed countries and power generation capacity of modern IGGC plants has notably increased compared to pioneering plants established in the mid-1990s. This increase in capacity is attributed to the implementation of new technologies, most notably advancements in the gas turbine system. Several IGCC projects have been proposed in Australia, China, the European Union, and the USA, with a number of other countries showing interest. In 2007 around 25 IGCC projects capable of producing on average $500\,MW_e$ each were proposed [4]. However, a significant proportion was canceled, citing *cost escalations* and *uncertainty in legislation* regarding CO_2 emission regulations among the key reasons for discontinuing the projects.

From a practical viewpoint, the operation of an IGCC plant presents several obstacles:

Process operation complexity: For essentially all of the current IGCC demonstration plants, between three and 5 years was required to reach 70%–80% capacity after commercial operation begun [12]. Due to the complexity of the IGCC process, no single unit or process operation within the total system was primarily responsible for the majority of the unplanned shutdowns that these plants have experienced. However, it should be noted that the gasification unit is identified as having the largest factor in reducing IGCC operability. Upon reaching 70%–80% capacity, operational performance for these plants has not typically exceeded 80% on a consistent basis [12].

Load following: IGGC plants are also yet to be designed specifically for high flexibility and the minimum load and ramp rate are generally between 40% and 50% and 3%–5% per minute, respectively [80]. This in comparable with an outdated PC plant.

Different skill set: the operation of an IGCC unit is significantly different from the operation of a PC plant and requires a different operational philosophy. In many regards, construction and operation of an IGCC plant is akin to that of a chemical plant and necessitates a different set of skills from plant designers and operators.

Higher risks: The IGCC process is also still viewed to have unquantified operating risks as the costs for operation and maintenance are relatively uncertain on account of the few reference plants and general lack of operating experience.

Energy-food nexus concern: With regards to the potential for cogasification of coal with biomass there are several concerns, predominantly relating to limitations of land and water, most notably competition with food production. There is the social concern that biomass production for energy conversion systems is nonsustainable or might take precedence over the production of food crops [81]. Under some circumstances, such as harvesting at a rate greater than the rate of natural regeneration, this concern is logical. Essentially there will exist sources of biomass that for a variety of reasons, such as recreation, biodiversity, water cycle management, etc. that should not be utilized for energy conversion systems. The agricultural production of biomass is also relatively land intensive and involves high logistics costs due to the low energy density of biomass. However, for a notable quantity of biomass sources, particularly agricultural waste, acquiring biomass for cogasification could displace costs that would previously be associated with disposal. Therefore, interest is still high as the previously discussed synergy between coal and biomass during gasification has the potential to enable the application of biomass as an energy source on a commercial scale [81].

Legislation barriers: There are also several logistical barriers that arise in attempts to establish a flexible polygeneration energy conversion systems [69]. Any company attempting to adopt a polygeneration system would likely be hesitant of producing chemicals or fuels they have little experience with or knowledge in regards to safety, training, and regulatory issues.

Similarly, the flexible generation of power in large quantities for sale to the grid may find barriers in the perpetual bidding system for power contracts and the extensive grid regulation. Furthermore, barriers caused by polygeneration might arise when attempting to operate in markets that a company has no previous experience or established relationships. Finally, and perhaps more significantly, without the establishment of effective operation strategies, the IGCC process will not have the ability to capitalize on market volatility, a key benefit for this process. The market conditions, for both feed material and product, has an immense impact on the optimal product portfolio and there is considerable risk associated with the speculation of future market prices.

From a broader perspective, there are several notable barriers to the widespread adoption of the IGCC process and indeed any other coal-based power generation process that facilitates CCS:

Robust emissions mitigation policy: A key barrier is a lack of universally applied and robust emissions mitigation policy. Without this, there is no economic driver for CCS. While mitigation policy does exist certainly in countries, in nearly all cases it is simply cheaper to emit CO_2 than to capture and sequester it. Due to the inexpensive nature of coal, it is clear that without additional government intervention, in the form of policy, there will no substantial incentive to minimize CO_2 emissions from coal-based power generation. There are several policy measures that could be established to achieve this such as various renewable energy subsidies and emissions mitigation policies. The selection of the optimal policy combination presents an additional substantial challenge. In the absence of significant charges on CO_2 emissions, developers and utility companies will continue to proceed with conventional coal-based power generation plants over IGCC technology as the cost of electricity generation is lower and the return investment is higher. This choice is logical given that any new plant construction on a commercial scale represents a multiyear, billion-dollar construction venture. Another technological barrier is that CCS is not sufficiently developed to the extent required for coal-based power generation. After capture at a plant, it is envisioned that CO_2 will be transported by pipeline to a suitable geographical storage site and injected for final sequestration. While the technologies required for this process do exist and have been employed at some level for various projects, they have not been demonstrated at the immense scale required for a commercial, high-capacity, coal-based power generation plant.

CAPEX: Perhaps the most significant barrier overall is the capital expenditure required and associated financial risk to establish a commercial scale coal-based power generation plant that facilitates either polygeneration, CCS or both, especially considering the significant change within the power generation sector this would represent. In general, due to the increasing size of investment in new power generation systems and the inherent economic risks, the utility industry is particularly slow to accept new technologies. There are concerns that the environmental benefits of IGCC compared to PC may threaten the continued use of the existing coal-based power generation system, thereby jeopardizing significant current capital. Furthermore, the balance sheet exposure from developing a novel power plant could be substantial, from two perspectives. Firstly, to fund the construction of the plant, and secondly, to manage the integrated power supply risk to the retail business if the operating reliability were less assured than conventional power plants. As a result, current efforts to extend the operating lifetimes of the existing system under the protection of grandfathering provisions in emission regulations essentially block the need for new, expensive, plant construction in the short term.

However, this may not be a viable long-term solution. For example, in Australia coal-based power generation accounts for over 60% of total capacity [82]. By the year 2020, approximately 45% of the coal-based power generation plants will be over 40 years of age [83]. The majority of coal-based power generations plants operational in Australia were built using now-outdated subcritical PC technology, the prevalent technology when they were constructed. As a result, the power generation infrastructure in Australia is one of the least efficient globally, and the electricity supply in Australia is one of the most emissions intensive of any developed nation on a per capita basis. Therefore, in the coming decades, there will be a need to retire and replace a significant portion of this power generation infrastructure in Australia. Should substantial CO_2 mitigation policy be in effect at this stage, the IGCC process then becomes a potentially viable option. There are also concerns that the IGCC process perhaps does not have the competitive advantage over alternative coal-based power generation technologies with CCS, such as oxygen-fired PC. Similarly, it is possible that retrofitting the existing coal-based power generation infrastructure to include CCS could present as the more economically viable option. Without increased commercial-scale establishment of these different technologies, it is difficult to compare and identify an optimal solution with a high degree of confidence.

7 CONCLUSION

The key capability of the IGCC technology is the synthesis of versatile chemical products from various carbonaceous feed material, such as coal, biomass, and byproducts, from the petroleum refining process. This flexibility places the IGCC as a viable alternative for conventional Rankine cycles, which suffer from inflexibility in response to the volatile electricity market. To date, there are few commercial examples of this technology predominantly due to the high capital cost requirement and operation complexity. However, the economic feasibility of the IGCC could be significantly improved with carbon capture obligations. This is due to its lower carbon capture costs as a result of treating high-pressure and high-concentrated CO_2 stream, unlike conventional power generation systems. This chapter provides a comprehensive review of polygeneration IGCC process with multiple-feed and multiple-product flexibility. Then process fundamentals are critically reviewed and technological barriers are discussed.

REFERENCES

[1] Breeze P. Chapter 3: coal-fired power plants. In: Breeze P, editor. Power Generation Technologies. 2nd ed. Boston: Newnes; 2014. p. 29–65.
[2] Khalilpour R, Karimi IA. Investment portfolios under uncertainty for utilizing natural gas resources. Comput Chem Eng 2011;35:1827–37.
[3] Metz B, Davidson O, Coninck HD, Loos M, Meyer L. IPCC special report on carbon dioxide capture and storage. Cambridge: Cambridge University Press, for the Intergovernmental Panel on Climate Change; 2005.
[4] Burnard K, Bhattacharya S. Power Generation From Coal: Ongoing Developments and Outlook. Internation Energy Agency; 2011.
[5] Weil K. Coal gasification and IGCC technology: A brief primer. 2010.
[6] BP. Statistical Review of World Energy; 2014. p. 2014.
[7] Khalilpour R. Flexible operation scheduling of a power plant integrated with PCC processes under market dynamics. Ind Eng Chem Res 2014;53:8132–46.
[8] Cortés C, Tzimas E, Peteves S. Technologies for coal based hydrogen and electricity co-production power plants with CO_2 capture. Joint Research Centre Institute for Energy. European Commission; 2009.
[9] Tsitsiklis J, Xu Y. Pricing of fluctuations in electricity markets; 2015.
[10] Gräbner M. Industrial coal gasification technologies covering baseline and high-ash coal. Wiley-VCH: Weinheim; 2014.
[11] Miller BG. Chapter 2: past, present, and future role of coal. In: Miller BG, editor. Coal energy systems. Burlington: Academic Press; 2005. p. 29–76.
[12] Katzer J, Ansolabehere S, Beer J, Ellerman D, Friedmann J, Herzog H, et al. The future of coal. Massachusetts: Massachusetts Institute of Technology; 2007.
[13] Karimi F, Khalilpour R. Evolution of carbon capture and storage research: Trends of international collaborations and knowledge maps. Int J Greenh Gas Control 2015;37:362–76.

[14] Rong A, Lahdelma R. Role of polygeneration in sustainable energy system development challenges and opportunities from optimization viewpoints. Renew Sustain Energy Rev 2016;53:363–72.

[15] Stucley C, Schuck S, Sims R, Bland J, Marino B, Borowitska M, et al. Bioenergy in Australia: Status and opportunities. Sydney: Bioenergy Australia; 2012.

[16] Ruiz JA, Juárez MC, Morales MP, Muñoz P, Mendívil MA. Biomass gasification for electricity generation: Review of current technology barriers. Renew Sustain Energy Rev 2013;18:174–83.

[17] Kumar A, Jones DD, Hanna MA. Thermochemical biomass gasification: a review of the current status of the technology. Energies 2009;2:556–81.

[18] Farzad S, Mandegari MA, Görgens JF. A critical review on biomass gasification, co-gasification, and their environmental assessments. Biofuel Res J 2016;3: 483–95.

[19] Bridgwater AV. The technical and economic-feasibility of biomass gasification for power-generation. Fuel 1995;74:631–53.

[20] Leung DYC, Yin XL, Wu CZ. A review on the development and commercialization of biomass gasification technologies in China. Renew Sustain Energy Rev 2004;8:565–80.

[21] Abdoulmoumine N, Adhikari S, Kulkarni A, Chattanathan S. A review on biomass gasification syngas cleanup. Appl Energy 2015;155:294–307.

[22] Devi L, Ptasinski KJ, Janssen FJJG. A review of the primary measures for tar elimination in biomass gasification processes. Biomass Bioenergy 2003;24:125–40.

[23] Sutton D, Kelleher B, Ross JRH. Review of literature on catalysts for biomass gasification. Fuel Process Technol 2001;73:155–73.

[24] Woolcock PJ, Brown RC. A review of cleaning technologies for biomass-derived syngas. Biomass Bioenergy 2013;52:54–84.

[25] Emami Taba L, Irfan MF, Wan Daud WAM, Chakrabarti MH. The effect of temperature on various parameters in coal, biomass and CO-gasification: a review. Renew Sustain Energy Rev 2012;16:5584–96.

[26] Emami-Taba L, Irfan MF, Wan Daud WMA, Chakrabarti MH. Fuel blending effects on the co-gasification of coal and biomass – A review. Biomass Bioenergy 2013;57: 249–63.

[27] Brar JS, Singh K, Wang J, Kumar S. Cogasification of coal and biomass: a review. Int J For Res 2012;2012:1–10.

[28] Ahrenfeldt J, Thomsen TP, Henriksen U, Clausen LR. Biomass gasification cogeneration – a review of state of the art technology and near future perspectives. Appl Therm Eng 2013;50:1407–17.

[29] Heidenreich S, Foscolo PU. New concepts in biomass gasification. Prog Energy Combust Sci 2015;46:72–95.

[30] Puig-Arnavat M, Bruno JC, Coronas A. Review and analysis of biomass gasification models. Renew Sustain Energy Rev 2010;14:2841–51.

[31] Prins M, Ptasinski K, Janssen F. From coal to biomass gasification: comparison of thermodynamic efficiency. Energy 2007;32:1248–59.

[32] Valero A, Usón S. Oxy-co-gasification of coal and biomass in an integrated gasification combined cycle (IGCC) power plant. Energy 2006;1643–55.

[33] Sjöström K, Chen G, Yu Q, Brage C, Rosén C. Promoted reactivity of char in co-gasification of biomass and coal: synergies in the thermochemical process. Fuel 1999;78:1189–94.

[34] Fermoso J, Arias B, Plaza MG, Pevida C, Rubiera F, Pis JJ, et al. High-pressure co-gasification of coal with biomass and petroleum coke. Fuel Process Technol 2009;90:926–32.

[35] Hernández JJ, Aranda-Almansa G, Serrano C. Co-gasification of biomass wastes and coal-coke blends in an entrained flow gasifier: an experimental study. Energy Fuel 2010;24:2479–88.

[36] Khosravi M, Khadse A. Gasification of Petcoke and coal/biomass blend: a review. Int J Emerg Technol Adv Eng 2013;3:167–73.

[37] Sofia D, Coca Llano P, Giuliano A, Iborra Hernández M, García Peña F, Barletta D. Co-gasification of coal–petcoke and biomass in the Puertollano IGCC power plant. Chem Eng Res Des 2014;92:1428–40.

[38] Nemanova V, Abedini A, Liliedahl T, Engvall K. Co-gasification of petroleum coke and biomass. Fuel 2014;117:870–5.

[39] Murthy BN, Sawarkar AN, Deshmukh NA, Mathew T, Joshi JB. Petroleum coke gasification: A review. Can J Chem Eng 2014;92:441–68.

[40] Bayram A, Müezzinoğlu A, Seyfioğlu R. Presence and control of polycyclic aromatic hydrocarbons in petroleum coke drying and calcination plants. Fuel Process Technol 1999;60:111–8.

[41] Tamamushi F, Shimojo M, Fujii N. Study of heavy oil gasification for IGCC. JSME Int J Ser B 1998;41:1067–70.

[42] Koukouzas N, Katsiadakis A, Karlopoulos E, Kakaras E. Co-gasification of solid waste and lignite – a case study for western Macedonia. Waste Manag 2008;28:1263–75.

[43] Domenichini R, Gallio M, Lazzaretto A. Combined production of hydrogen and power from heavy oil gasification: Pinch analysis, thermodynamic and economic evaluations. Energy 2010;35:2184–93.

[44] Meratizaman M, Monadizadeh S, Ebrahimi A, Akbarpour H, Amidpour M. Scenario analysis of gasification process application in electrical energy-freshwater generation from heavy fuel oil, thermodynamic, economic and environmental assessment. Int J Hydrogen Energy 2015;40:2578–600.

[45] Holopainen O. Power production from biomass IGCC plant employing heavy-petroleum residues. Bioresour Technol 1993;46:125–8.

[46] Smith AR, Klosek J. A review of air separation technologies and their integration with energy conversion processes. Fuel Process Technol 2001;70:115–34.

[47] Nikrityuk P, Meyer B. Gasification processes: modeling and simulation. Weinheim: Wiley-VCH; 2014.

[48] Liu K, Cui Z, Chen W, Zhang L. Coal and syngas to liquids. In: Hydrogen and syngas production and purification technologies. Hoboken, NJ: John Wiley & Sons, Inc.; 2009. p. 486–521.

[49] Liu Z, Peng W, Motahari-Nezhad M, Shahraki S, Beheshti M. Circulating fluidized bed gasification of biomass for flexible end-use of syngas: a micro and nano scale study for production of bio-methanol. J Clean Prod 2016;129:249–55.

[50] Cortés CG, Tzimas E, Peteves S. Technologies for coal based hydrogen and electricity co-production power plants with CO_2 capture.

[51] Sharma SD, Dolan M, Park D, Morpeth L, Ilyushechkin A, McLennan K, et al. A critical review of syngas cleaning technologies—fundamental limitations and practical problems. Powder Technol 2008;180:115–21.

[52] Mondal P, Dang GS, Garg MO. Syngas production through gasification and cleanup for downstream applications—recent developments. Fuel Process Technol 2011;92:1395–410.

[53] Hossein Sahraei M, McCalden D, Hughes R, Ricardez-Sandoval LA. A survey on current advanced IGCC power plant technologies, sensors and control systems. Fuel 2014;137:245–59.

[54] Li M, Rao AD, Brouwer J, Samuelsen GS. Design of highly efficient coal-based integrated gasification fuel cell power plants. J Power Sources 2010;195:5707–18.

[55] Jansen DAO, van Veen H. CO_2 Reduction potential of future coal gasification based power generation technologies. Energ Conver Manage 1992;365–72.

[56] Jansen D, van der Laag P, Oudhuis A, Ribberink J. Prospects for advanced coal-fuelled fuel cell power plants. J Power Sources 1994;151–65.

[57] Lobachyov K, Richter H. High efficiency coal-fired power plant of the future Energy Conversion and Management; 1997. p. 1693–9.

[58] Shoko E, McLellan B, Dicks AL, da Costa JCD. Hydrogen from coal: Production and utilisation technologies. Int J Coal Geol 2006;65:213–22.

[59] Smith B, Loganathany M, Shanthaz MS. Review of the water gas shift reaction kinetics. Int J Chem React Eng 2010;8.

[60] Grol E, Yang W-C. Evaluation of alternate water gas shift configurations for IGCC systems; 2009.

[61] Platon A, Wang Y. Water-gas shift technologies. In: Hydrogen and syngas production and purification technologies. Hoboken, NJ: John Wiley & Sons, Inc.; 2009. p. 311–28.

[62] Methanol Institute. The Methanol Industry. Methanol Institute; 2011.

[63] ICIS. ICIS pricing; 2014.

[64] Chmielniak T, Sciazko M. Co-gasification of biomass and coal for methanol synthesis. Appl Energy 2003;74:393–403.

[65] Hetland J, Kvamsdal HM, Haugen G, Major F, Karstad V, Tjellander G. Integrating a full carbon capture scheme onto a 450 MWe NGCC electric power generation hub for offshore operations: presenting the Sevan GTW concept. Appl Energ 2009;86: 2298–307.

[66] Chen X. Wholesale ammonia prices have been crashing; 2013.

[67] Higman C, van der Burgt M. Chapter 7: Applications. In: Gasification. 2nd ed. Burlington: Gulf Professional Publishing; 2008. p. 257–321.

[68] Khalilpour R, Karimi IA. Selection of liquefied natural gas (LNG) contracts for minimizing procurement cost. Ind Eng Chem Res 2011;50:10298–312.

[69] Adams TA. Future opportunities and challenges in the design of new energy conversion systems. Comput Chem Eng 2015;81:94–103.

[70] Bell D, Towler B, Fan M. Coal gasification and its applications. Oxford: Elsevier; 2011.

[71] Li S, Gao L, Jin H. Realizing low life cycle energy use and GHG emissions in coal based polygeneration with CO_2 capture. Appl Energy 2017;194:161–71.

[72] Zheng L, Furinsky E. Comparison of Shell, Texaco, BGL and KRW gasifiers as part of IGCC plant computer simulations. Energ Conver Manage 2005;46:1767–79.

[73] Frey C, Akunuri N. Probabilistic modeling and evaluation of the performance, emissions, and cost of texaco gasifier-based integrated gasification combined cycle systems using ASPEN. North Carolina: North Carolina State University; 2001.

[74] Campbell P, McMullan J, Williams B. Concept for a competitive coal fired integrated gasification combined cycle power plant. Fuel 2000;1031–40.

[75] Kanniche M, Bouallou C. CO_2 capture study in advanced integrated gasification combined cycle. Appl Therm Eng 2007;27:2693–702.

[76] Pérez-Fortes M, Bojarski AD, Velo E, Nougués JM, Puigjaner L. Conceptual model and evaluation of generated power and emissions in an IGCC plant. Energy 2009;34:1721–32.

[77] Arienti S, Mancuso L, Cotone P. IGCC plants to meet the refinery needs of hydrogen and electric power. Barcelona: Seventh European Gasification Conference; 2006.

[78] Descamps C, Bouallou C, Kanniche M. Efficiency of an integrated gasification combined cycle (IGCC) power plant including CO_2 removal. Energy 2008;33:874–81.

[79] Chen Y. Optimal design and operation of energy Polygeneration systems. Massachusetts Institute of Technology; 2013.

[80] Henderson C. Increasing the flexibility of coal-fired power plants. IEA Clean Coal Centre; 2014.

[81] Sanchez DL, Kammen DM. A commercialization strategy for carbon-negative energy. vol. 1; 2016. p. 15002.

[82] Ball A, Ahmad S, Bernie K, McCluskey C, Pham P, Tisdell C, et al. Australian energy update. Canberra; 2015.

[83] Stock A. Australia's electricity sector: ageing, inefficient and unprepared. Sydney: Climate Council; 2014.

[84] Duan L, Sun S, Yue L, Qu W, Yang Y. Study on a new IGCC (integrated gasification combined cycle) system with CO_2 capture by integrating MCFC (molten carbonate fuel cell). Energy 2015;87:490–503.

[85] Chen S, Lior N, Xiang W. Coal gasification integration with solid oxide fuel cell and chemical looping combustion for high-efficiency power generation with inherent CO_2 capture. Appl Energy 2015;146:298–312.

[86] Lanzini A, Kreutz TG, Martelli E, Santarelli M. Energy and economic performance of novel integrated gasifier fuel cell (IGFC) cycles with carbon capture. Int J Greenhouse Gas Control 2014;26:169–84.

[87] Gerdes K, Grol E, Kearins D, Newby R. Integrated gasification fuel cell performance and cost assessment. U.S.A Department of Energy; 2009.

[88] Verma A, Rao AD, Samuelsen GS. Sensitivity analysis of a vision 21 coal based zero emission power plant. J Power Sources 2006;158:417–27.

[89] Ghosh S, De S. Energy analysis of a cogeneration plant using coal gasification and solid oxide fuel cell. Energy 2006;31:345–63.

[90] Kuchonthara P, Bhattacharya S, Tsutsumi A. Combination of thermochemical recuperative coal gasification cycle and fuel cell for power generation. Fuel 2005;84:1019–21.

[91] Rao A, Verma A, Samuelsen G. Engineering and economic analyses of a coal-fueled solid oxide fuel cell hybrid power plant. Reno: The American Society of Mechanical Engineers; 2005.

[92] Kivisaari T, Bjornbom P, Sylwan C, Jacquinot B, Jansen D, Degroot A. The feasibility of a coal gasifier combined with a high-temperature fuel cell. Chem Eng J 2004;100:167–80.

[93] Maruyama M, Takahashi S, Iritani J, Miki H. Gasification Technologies Conference. San Francisco; 2000.

[94] Ahmed U, Kim C, Zahid U, Lee C-J, Han C. Integration of IGCC and methane reforming process for power generation with CO_2 capture. Chem Eng Process Process Intensif 2017;111:14–24.

[95] Tola V, Pettinau A. Power generation plants with carbon capture and storage: a techno-economic comparison between coal combustion and gasification technologies. Appl Energy 2014;113:1461–74.

[96] Cormos C-C. Techno-economic and environmental evaluations of large scale gasification-based CCS project in Romania. Int J Hydrogen Energy 2014;39:13–27.

[97] Pettinau A, Ferrara F, Amorino C. Combustion vs. gasification for a demonstration CCS (carbon capture and storage) project in Italy: A techno-economic analysis. Energy 2013;50:160–9.

[98] Siefert NS, Litster S. Exergy and economic analyses of advanced IGCC–CCS and IGFC–CCS power plants. Appl Energy 2013;107:315–28.

[99] Pettinau A, Ferrara F, Amorino C. Techno-economic comparison between different technologies for a CCS power generation plant integrated with a sub-bituminous coal mine in Italy. Appl Energy 2012;99:32–9.

[100] Cormos C-C. Integrated assessment of IGCC power generation technology with carbon capture and storage (CCS). Energy 2012;42:434–45.

[101] Gerdes K. Current and future technologies for gasification-based power generation. In: A pathway study focused on carbon capture advanced power systems R&D using bituminous coal. 2010.

[102] Rezvani S, Huang Y, McIlveen-Wright D, Hewitt N, Mondol JD. Comparative assessment of coal fired IGCC systems with CO_2 capture using physical absorption, membrane reactors and chemical looping. Fuel 2009;88:2463–72.

[103] Martelli E, Kreutz T, Consonni S. Comparison of coal IGCC with and without CO_2 capture and storage: shell gasification with standard vs. partial water quench. Energy Procedia 2009;1:607–14.

[104] Huang Y, Rezvani S, McIlveen-Wright D, Minchener A, Hewitt N. Techno-economic study of CO_2 capture and storage in coal fired oxygen fed entrained flow IGCC power plants. Fuel Process Technol 2008;89:916–25.

[105] Rhodes JS, Keith DW. Engineering economic analysis of biomass IGCC with carbon capture and storage. Biomass Bioenergy 2005;29:440–50.

CHAPTER 8

CO$_2$ Conversion and Utilization Pathways

Ahmad Rafiee*, Kaveh Rajab Khalilpour[†], Dia Milani[‡]
*Cardiff School of Engineering, Cardiff University, Cardiff, United Kingdom
[†]Faculty of Engineering and Information Technology, Monash University, Melbourne, VIC, Australia
[‡]CSIRO Energy Centre, Newcastle, NSW, Australia

Abstract

The transition to net-zero or net-negative emission future implies avoidance of CO$_2$ emissions along with other greenhouse gases. Nevertheless, the research community and industry are progressively converging to a conclusion that CO$_2$ sequestration has serious limitations for the value proposition. Alternatively, creating a demand market and a revenue stream for the recovered almost-pure CO$_2$ may prevail over CO$_2$ sequestration option and improve the economic feasibility of carbon capture technologies. As such, research in the carbon management field is seen to be shifting toward CO$_2$ utilization, directly and indirectly, in energy and chemical industries.

This chapter discusses physical and chemical CO$_2$ utilization pathways and investigates the existing process integration scenarios and the performance assessment benchmarks.

Keywords: Carbon capture and utilization, CO$_2$ conversion, Climate change mitigation, CO$_2$ valorization, CO$_2$ utilization

Abbreviations

Ar	argon
ASU	air separation unit
ATR	autothermal reforming
BiR	bireforming
BTL	biomass to liquid
CCS	carbon capture and storage
CCU	CO$_2$ capture and utilization
CCUS	CO$_2$ capture sequestration and utilization
CFA	carbonated fly ash
CTL	coal to liquid
DMC	dimethyl carbonate
DME	dimethyl ether
DMR	dry-methane reforming
EGR	enhanced gas recovery
EOR	enhanced oil recovery
FPSO	floating production, storage, and offloading
FT	Fischer-Tropsch

Polygeneration with Polystorage
https://doi.org/10.1016/B978-0-12-813306-4.00008-2

GHG	greenhouse gases
GTL	gas to liquid
HEN	heat exchanger network
HFC	hydrofluorocarbons
IPCC	intergovernmental panel on climate change
LCA	lifecycle assessment
LHV	lower heating value
MEA	mono-ethanol-amine
NSoR	nonsolar
OCRM	oxidative CO_2 reforming of methane
OT	once through
POX	partial oxidation
PTG	power to gas
RB	reverse-Boudouard
RC	recycle
RWGS	reverse water gas shift
SMR	steam-methane reforming
Syngas	synthesis gas
TMC	trimethylene carbonate

1 INTRODUCTION

The main options for shifting away from fossil fuels and thus reducing carbon emissions are (1) reduction of energy consumption, e.g., via efficiency improvement, (2) replacing fossil fuels by cleaner resources such as renewables, and (3) capture and storage of CO_2. Carbon capture and storage (CCS) has been viewed as a bridge for decarbonizing future energy [1]. However, after three decades of research, this technology has also been facing barriers as the design and operation costs are still high and unfeasible for niche applications such as enhanced oil/gas recovery. Uncertainty is also associated with the eternity of the stored CO_2 and whether this gas will remain locked underground forever [2–5]. However, creating a demand market and a revenue stream for the recovered almost-pure CO_2 may provide a better alternative for CO_2 sequestration option and improve the economic feasibility of this climate change mitigation approach. As such, research in the carbon management field is seen to be shifting toward CO_2 capture and utilization (CCU) as an alternative direction. The utilization and cofeeding of the captured CO_2 in the synthesis of many products is increasingly in the spotlight for many research organizations and industrial counterparts. This could add an extra benefit for the economics of carbon capture process. Fig. 1 illustrates a possible scenario of process integration where the captured CO_2 can be stored away from geosequestration and be utilized as a commodity or

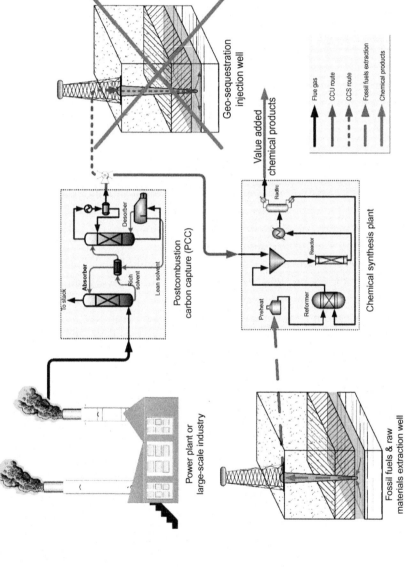

Fig. 1 CO₂ utilization as an alternative to sequestration.

feedstock in another chemical/fuel synthesis process. In this rapidly accelerating field, numerous innovative approaches were articulated. The aim of these studies is either to save on fuel consumption, improve process integration, enhance GHG abatement, or synthesis new materials and composites.

Here, we discuss the fate of the CO_2 molecules in the utilization process and categorize in two main directions, i.e., physical and chemical pathways. Then, we probe the literature in addressing the process/renewable integration scenarios and highlighting the performance assessment benchmarks. The principles of key CO_2 utilization routes, reported in the literature, are identified and discussed. This comprehensive study can serve as an updated criterion for the fellow researchers, engineers, students, and scientists.

2 CO$_2$ UTILIZATION PATHWAYS

Since the Industrial Revolution, and the development of chemical industries, CO_2 has found applications in several industries probably the earliest and the most notable of these being beverage industries for carbonated drinks. Huang and Tan [6] classified different forms of CO_2 valorization into two categories: conversion of CO_2 into fuels and chemicals and direct (physical) utilization of CO_2 [6]. Further to carbonated drinks, CO_2 can be used directly in other applications such as dry ice, fire extinguisher, solvent, refrigerant, process fluid, welding medium, or in algae farms for photosynthesis. These direct applications for CO_2 are limited in scale and have a narrow effect on the overall CO_2 abatement [7]. It can also be used in large-scale industries to indirectly boost/enhance a process as in the enhanced oil recovery (EOR), enhanced gas recovery (EGR), and enhanced geothermal systems (EGS). In all these applications, the CO_2 molecules remain pure or in the dissolved in a mixture and do not react or crack further. The left-hand side of Fig. 2 illustrates some applications and industrial uses for CO_2.

Carbon dioxide can also be used as a feedstock in chemical or fuel syntheses. In those chemical processes and by applying specific heat/pressure conditions, the CO_2 molecules often crack down to the basics (i.e., pure carbon or CO) or react with other components to form longer chains (i.e., hydrocarbons). In the direct utilization, the CO_2 molecule is a building block of a chemical product where the product cannot be formed without CO_2 either as a whole molecule or after the crackdown. While for indirect chemical utilization, CO_2 molecules help to boost/enhance the syntheses

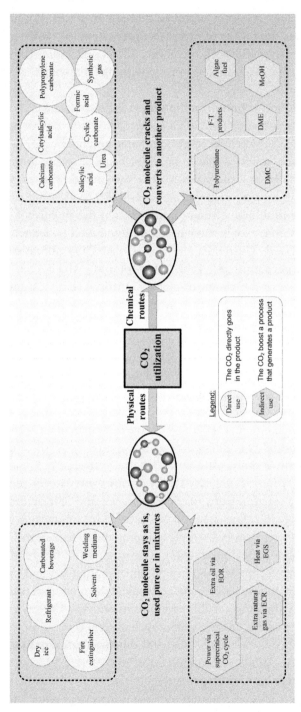

Fig. 2 Various pathways for CO$_2$ utilization.

process, add value, or cut the withdrawal of raw materials. It is noteworthy to mention that the CO_2 utilization in many chemical syntheses (i.e., for synthetic gas) may fulfill both categories (direct and indirect). The right-hand side of Fig. 2 illustrates some chemical/fuel syntheses routes for CO_2 utilization.

Given the huge magnitude of the annual anthropogenic CO_2 emissions, a critical issue in the assessment of CO_2 capture and utilization technologies is to investigate the potential sources and also potential utilizing industries along with their scales. Table 1 provides a list of some anthropogenic CO_2 sources, sinks, and the scales. Unlike the sources, data on CO_2 utilization potentials are very scattered and are subject to a notable uncertainty. Given the historical huge carbon capture costs at the absence of a comprehensive carbon policy, only some necessary limited CO_2-dependent industries have survived. Currently, the total global CO_2 utilization is <200 million tonnes per annum [8] (Table 1, last column), which is relatively negligible compared to the extent of global anthropogenic CO_2 emissions of >32,000 million tonnes per annum [9]. But, as the carbon tax policies are being implemented and the carbon capture becomes inevitable, the CO_2 would be available at low or even negative prices (as an alternative to storage option). This scenario may disrupt the current trend of CO_2 utilization and boost the scale of consumption in the current industries. It could also revive new CO_2-consuming industries that were not viable at the traditional high-cost CO_2 context.

Poliakoff and Leitner [11] defined twelve principles of CO_2 chemistry to assess the practicability of different reactions and processes that utilize CO_2. In a comprehensive study, Otto et al. [12] pre-evaluated 123 reaction paths to valorize CO_2 into chemicals. The reactions were categorized into two scales, i.e., 100 reaction paths for fine chemicals and 23 reaction paths for bulk chemicals [12]. A fine chemical was defined as a substance with a global production of fewer than 10,000 t per annum, while a bulk or basic chemical constitutes higher annual production. The synthesis of fuels and polymers was not considered. They considered potential profitability and CO_2 reduction potential of these reaction paths and ranked the numerous reactions into a list to identify the most suitable reactions in utilizing CO_2 as a feedstock. The ranking list of the 23 bulk chemicals indicated that formic acid, oxalic acid, formaldehyde, methanol, urea, and dimethyl ether mostly fulfill the selection criteria. While the ranking list of the 100 fine chemicals showed that methyl-urethane, 3-oxo-pentanedioic acid, 2-imidazolidinone, ethyl-urethane, 2-oxazolidone, and isopropyl-isocyanate are on the top of the list

Table 1 Anthropogenic CO$_2$ sources, sinks, and the scales

CO$_2$ emissions by economic sector [9]		Large-scale CO$_2$ emission sources (plants ≥0.1 MtCO$_2$/y) [10]			Current CO$_2$ utilization [8]	
Sector	MtCO$_2$ (%)	Process	Number of sources	Emissions (MtCO$_2$/y)	Industry	Quantity (MtCO$_2$/y)
Electricity and heat production	13,655.6 (42.4)	Power	4942	10,539	Urea	114
Transportation	7384.9 (23)	Cement production	1175	932	Methanol	8
Manufacturing industries and construction	6114.8 (19.0)	Refineries	638	798	Dimethyl ether (DME)	3
Residential	1868.7 (5.8)	Iron and steel industry	269	646	Methyl tert–butyl ether (TBME)	1.5
Services	861.9 (2.7)	Petrochemical industry	470	379	Formaldehyde (CH$_2$O)	3.5
Other (agriculture/ forestry, fishing, energy industries other than electricity and heat generation, marine and aviation bunkers)	2303.8 (7.2)	Oil and gas processing	NA	50	Carbonates	0.005
		Other sources	90	33	Polycarbonates	0.01
		Bioethanol and bioenergy	303	91	Inorganic carbonates	50
					Technological	28
					Algae for the production of biodiesel	0.010
Total	32,189.7 (100)	Total	7887	13,466	Total	200

[12]. These chemicals are the most suitable products for future detailed analysis and development [13].

In the following sections, each of the key CO_2 utilization options is discussed in detail, and a critical review of the process features is provided.

3 PHYSICAL CO_2 UTILIZATION

As elaborated in Section 2, there are two CO_2-utilization categories including physical and chemical. In physical utilization, the CO_2 molecules remain pure or suspended in a solution without a chemical reaction. CO_2 can be physically used as dry ice, fire extinguisher, solvent, refrigerant, process fluid, welding medium. It can also be used in large-scale industries to indirectly boost/enhance a process as in the enhanced oil recovery (EOR), enhanced gas recovery (EGR), and enhanced geothermal systems (EGS). Here, we look at the literature addressing physical utilization of CO_2.

Safi et al. [14] simulated CO_2 utilization in an EOR process and optimized the CO_2 injection rate for an operational reservoir in Texas. They considered two scenarios: constant rate and pressure-limited injection. The optimization problem was solved by integrating genetic algorithm with COZSim solver package developed by Nitec, LLC. Bachu [15] investigated 13,000 oil reservoirs in Alberta, Canada, for CO_2 storage and CO_2-enhanced oil recovery potential (CO_2-EOR). The author identified 136 oil reservoirs within 85 fields in the province with suitability for CO_2-EOR or CO_2 storage only. Azzolina et al. [16] carried out a statistical analysis of a database of 31 CO_2-EOR projects to investigate the possible EOR market for CO_2, CO_2-EOR, and pure storage of CO_2. The injection of CO_2 and CH_4 mixture in oil reservoirs was studied by Liu and Zhang [17]. The highest oil recovery and the best efficiency of oil displacement were seen in homogeneous hydrocarbon reservoirs. The mixture of CO_2 and CH_4 was suitable for different rhythm hydrocarbon reservoirs by gas-alternating-water (WAG) flooding.

The CO_2 utilization for EOR in the western section of the Farnsworth Unit (Texas) was simulated by White et al. [18]. The planned net CO_2 injection rate was 0.19 million tonnes per year. Testing for large-scale CO_2-EOR and CO_2 sequestration in the Midwestern USA was also investigated by Gupta and coauthors [19]. They also evaluated the technoeconomic feasibility of CO_2 utilization and storage in Ohio's depleted oil fields [20]. It was found that the depleted gas and oil fields in Ohio had the potential to store 3.4 billion tonnes of CO_2. The results revealed that those fields could be considered as promising candidates for CO_2-EOR. The cumulative net amount of CO_2 stored at 2014 was about 1.4 million tonnes. Boodlal et al. [21] investigated the feasibility of CO_2-EOR and CO_2-storage in Trinidad. The results showed that the primary oil recovery efficiency was 27.2%. Once the oil price slumped to $61 per barrel, the project became economically infeasible. Pingping et al. [22] developed a methodology to evaluate CO_2 storage potential in oil reservoirs in China. For 21 blocks in an oilfield in China, which can be used for CO_2 storage, the total theoretical storage capacity for CO_2 was estimated to be 7.69×10^8 t. Assuming an effective storage coefficient of 0.25, the effective storage capacity became 19.2×10^8 t [22].

The CO_2 utilization through the integration of methanol plant with enhanced gas recovery (EGR) and geosequestration was investigated by Luu et al. [23]. They considered four scenarios: (1) power plant and methanol synthesis chain including natural gas plant (with no EGR) operate independently, (2) CO_2 capture and storage (CCS): captured CO_2 from the power plant is sent to geosequestration while methanol synthesis chain including natural gas plant (with no EGR) operates independently, (3) CO_2 capture and utilization (CCU): the captured CO_2 from the power plant is sent to the methanol reactor, and (4) CO_2 capture sequestration and utilization (CCUS): captured CO_2 is sent to geo-sequestration, EGR, and methanol plant. The results showed that the CCUS scenario could digest a natural gas feed with a maximum CO_2 mole fraction of 0.232. Integration of methanol plant with EGR showed the best performance metrics. In another study, the CO_2 utilization in EGR at the Bahkrabad gas field (Bangladesh) was investigated [24]. About 83.8% of captured CO_2 from the underground coal gasification was used for EGR, and 16.2% of that was utilized for fertilizer production.

Utilizing impure CO_2 as the geofluid for enhanced geothermal systems (EGS) was thermodynamically analyzed by Zhang et al. to extract geothermal energy [25]. Using CO_2 as an alternative for synthetic refrigerants in mobile and stationary air conditioning was discussed in Antonijevic's work [26]. Using CO_2 in automotive air conditioning system showed higher efficiency, which reduced fuel consumption compared to the existing R134a refrigerant. A two-stage stochastic model of CO_2 utilization and disposal infrastructure was developed and applied to an industrial complex on the eastern coast of Korea in 2020 [27]. The uncertainties in the fluctuations of CO_2 emissions were considered. The objective was to maximize the total annual profit of the carbon capture and storage infrastructure. Various aspects of CO_2 utilization for emission reduction in Romania was studied by Dragos et al. [28]. Utilizing CO_2 from thermal power plants for EOR was also addressed. A scalable and comprehensive infrastructure model was developed for CO_2 utilization and disposal [29]. A case study was conducted for an industrial complex on the east coast of Korea in 2020.

4 CHEMICAL CO_2 UTILIZATION

For chemical utilization, the CO_2 molecules often crack down in exothermic/endothermic reactions and can convert to several commodity chemicals, synthetic fuels, or building blocks for other products. This practice will not only significantly reduce the resource intake and carbon emission but will also improve the sustainability image for many chemical processes and industries. The most common routes for chemical CO_2 utilization are highlighted in this section.

4.1 Syngas Production

Reforming is the first step in converting raw materials, including fossil fuels, to value-added chemicals and fuels. Syngas is the core intermediate product of reforming that can lead to the production of several end products. Historically, due to low-cost fossil-fuels and boiler technologies, steam has been used as a high-temperature medium to generate syngas ($H_2O + CH_4 \rightarrow CO + 3H_2$). However, when available, CO_2 can also react with methane to generate syngas with a reaction called dry reforming ($CO_2 + CH_4 \rightarrow 2CO + 2H_2$). Application of CO_2 in dry reforming can indirectly lead to the production of several high-demand products such as diesel, naphtha, etc. [30]. It can also react directly with some other chemicals. Selection of reforming technology is, therefore, the most critical decision in such plants, which is directly relevant to the end product of choice.

Fig. 3 provides a schematic of reforming processes for conversion of low-value materials/fuels to syngas and then value-added chemicals/fuels. The product of reforming is synthesis gas (or "syngas") with main components being hydrogen and carbon monoxide but often contains fractions of water and carbon dioxide. The term "reforming" is generally referred to natural gas conversion to syngas, and solid-feed reforming (biomass, coal, waste, etc.) is called pyrolysis/gasification. The chief reforming technologies for syngas production using natural gas are catalytic SMR (steam-methane reforming), DMR (dry-methane reforming), ATR (autothermal reforming), and POX (partial oxidation) [31–33].

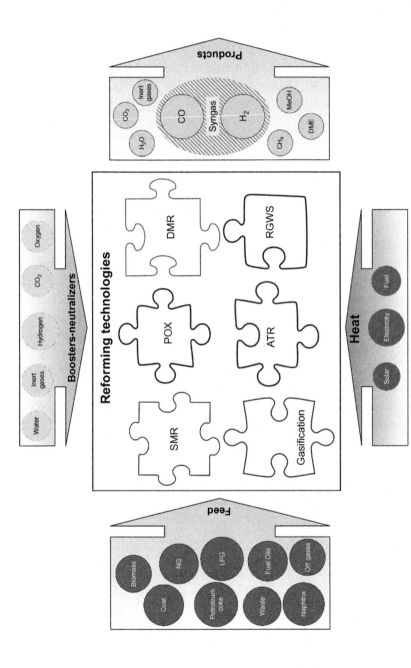

Fig. 3 Schematic of reforming processes for conversion of low-value materials/fuels to syngas and then value-added chemicals/fuels.

The reformer choice would ultimately depend on a compromise between the pros and cons of each technology. Both ATR and POX require O_2 feed for which a costly air separation unit (ASU) is required. They have, however, the advantage of having H_2/CO ratio of around two, which is close to the optimal operating condition of Fischer–Tropsch (FT) gas-to-liquid (GTL) processes. SMR process, in contrast, does not need ASU and thus requires less capital cost and also less land usage. But, it produces syngas with high H_2/CO ratio, being theoretically around 3. Therefore, the syngas before going to GTL process needs to pass through a separation unit to remove excess H_2 or add extra CO to adjust the H_2/CO ratio. The steam reforming of hydrocarbons is the benchmark technology for the production of pure hydrogen in refineries. The SMR reactions can be carried out in several reactor types including adiabatic, tubular, and heat exchanger reformers [33]. The core idea of CO_2 utilization in reforming process is to combine SMR and DMR to obtain the desirable H_2/CO ratio [34].

The optimal design of syngas production with recycled CO_2 utilization was presented by Choi et al. [35]. Their hypothesis was formulated based on an integrated superstructure (Fig. 4) optimization using mixed-integer programming. The optimization results indicated that CO_2 emission decreased by 31% and the annualized profit increased by 14% compared to the base case, which was a widely used syngas production process via reforming of butane or methane [25]. Lim et al. [34] studied a combination of SMR and DMR for CO_2 utilization. The results revealed that net CO_2 emission of the process with combined reformers was reduced by 67% compared to the reference case of SMR only.

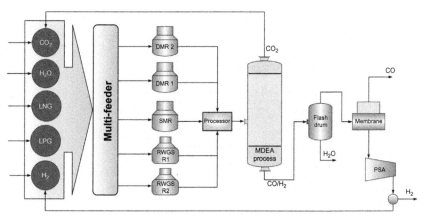

Fig. 4 A superstructure framework for optimal synthesis of gas production process with recycled CO_2 utilization.

CO$_2$ utilization through the integration of a combined heat and power plant (CHP) based on cogasification of biomass and coal feedstock was analyzed by Kuo and Wu [36,37]. The optimum operating conditions such as steam to carbon ratio and CO$_2$ flow rate were determined by maximizing exergy efficiency and energy conversion efficiency. For the case that the feed to the CHP consists of 60 wt% of coal and 40 wt% of biomass, the optimization results showed that the CO$_2$ emission is reduced by 38.23% compared to the case with coal feed only. Feasibility of coal gasification with CO$_2$ utilization to produce liquid fuels and electricity was studied by Yuan et al. [38]. The results showed that CO$_2$ utilization for producing liquid fuels and chemicals was technically feasible.

4.2 Methanol Production

Methanol, the historically known "wood alcohol," has a wide application in various chemical industries and recently as a fuel in some fuel-cell cars. The concept of methanol economy first suggested by 1994 Noble Prize winner George A. Olah claiming that the global dependence on diminishing oil and natural gas sources could be alleviated by replacing these fuels by methanol. Methanol can be produced from chemical recycling of CO$_2$ [39–41].

A schematic of methanol process with various reforming technologies is illustrated in Fig. 5. The production of methanol can use CO$_2$ both indirectly (in reforming stage) and directly in the methanol reactor, which converts syngas to methanol. The governing reactions in methanol synthesis from syngas are:

$$CO + 2H_2 \rightleftharpoons CH_3OH \quad \Delta_{H298K} = -21.7\,kcal/mol$$

$$CO_2 + 3H_2 \rightleftharpoons CH_3OH + H_2O \quad \Delta_{H298K} = -11.9\,kcal/mol$$

CO$_2$ utilization in methanol production is currently in practice at large industrial scale. For instance, Fanavaran Petrochemical plant in Iran produces one million tons of methanol per year through CO$_2$ utilization. The plant receives natural gas and CO$_2$ with flowrates of 610 ktons/y and 268 ktons/y, respectively, from two contiguous petrochemical complexes. The imported natural gas and CO$_2$ together with onsite-generated steam are fed to the SMR system, and the product syngas is then fed to the methanol reactor after pressurization [42].

Currently, CO$_2$ utilization in methanol synthesis is one of the major activities in research and industrial sphere. Chen et al. [43] analyzed four different systems including: (1) integration of coal to liquids and natural gas to liquids, (2) combining nuclear energy with coal to liquids or CO$_2$

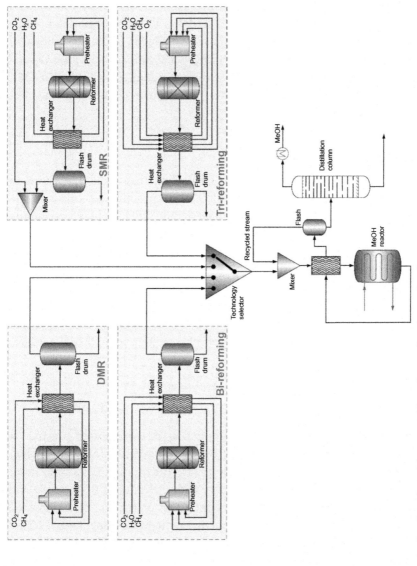

Fig. 5 Schematic of methanol process with various reforming technologies.

hydrogenation to methanol, (3) dry reforming of methane for chemicals/ fuels production, and (4) integration of solar energy with H_2O-CO_2 splitting to produce chemicals/fuels. The results showed that the integration of the processes using fossil fuels as the feedstock with noncarbon energy source would reduce the amount of the CO_2 emitted to the atmosphere. A hybrid of a genetic algorithm and generalized pattern search was used by Alarifi et al. [44] for dynamic optimization of Lurgi-type methanol reactor. The objective was in finding the optimal coolant temperature, and the recycle ratio of CO_2 (recovered from the reformer flue gas) to the methanol reactor. The results showed that with 5% recycle ratio of CO_2, the production rate of methanol increased by $\sim$2.5% compared to an existing operating reference plant. Other studies compared the CO_2 valorization through the integration of a postcombustion power plant and methanol synthesis reactor in six different syngas production routes, i.e., SMR, DMR with surplus CO removal, bireforming, trireforming, the parallel combination of SMR and DMR, and DMR with H_2 addition [45,46]. The results showed that the latter configuration had the lower CO_2 emission per mass of methanol. Roh et al. [47] considered two CO_2 utilization routes in methanol synthesis, i.e., direct hydrogenation and combined reforming. The results showed that replacement or integration of a methanol plant with a combined reformer would represent a sustainable alternative. Furthermore, the synthesis of methanol through direct hydrogenation was a good path to utilize CO_2 when cheap hydrogen sources are available. The effect of feed composition on methanol synthesis using a generalized thermodynamic approach was studied by Iyer et al. [48]. The performance of the methanol reactor under single- and two-phase conditions was analyzed for maximum methanol productivity, $CO + CO_2$ conversion, and CO_2 utilization. Methanol production via indirect hydrogenation of CO_2 and a CO_2-derived formate esters (i.e., methyl, ethyl-, propyl- and butyl-formate) over nanocomposite copper/alumina catalysts was described by Park et al. [49]. The conversion of the formate esters was over 95% and methanol yield of the methyl, ethyl-, propyl-, and butyl-formates were 99, 92, 91, and 91%, respectively.

The production of methanol on a floating production, storage, and off-loading (FPSO) system from stranded natural gas and CO_2 was simulated [50]. The process was optimized with the objective function of maximizing carbon conversion to methanol. Installing an RWGS reactor on the recycle stream increased the methanol production and CO_2 consumption rates by 13.8% and 38.9%, respectively, compared to the case without RWGS [50].

The total consumption of CO_2 was 78.7%. The viability of offshore CO_2 utilization by dry reforming was studied by Lima et al. [51] from a technoeconomic and environmental point of view. Two process configurations were evaluated to produce methanol. In the first case, steam- and dry reforming were assumed to take place in one reactor, while in the second case dry reforming occurred in one reactor and water-gas-shift reaction took place in the subsequent reactor. The results indicated that the performance of the first case was better than the second case as methanol production was reported four times higher, sales revenue 181% higher, capital investment 21.7% lower, and the index for environment impact 47.7% lower. A graphical targeting approach was used to assess the potential environmental route to produce methanol from CO_2 and H_2 [52]. They identified five optimal reaction schemes. The most promising utilization pathway was found to use the captured CO_2 from power plants for methanol production. The CO_2 utilization for the production of syngas and methanol synthesis over Cu-based Zr-containing catalysts was reviewed by Raudaskoski et al. [53]. Whereas CO_2 utilization for production of methanol with coproduction of electricity from an open gas turbine, followed by a separator was studied by Kralj and Glavic [54]. Optimization of the overall process using nonlinear programming yielded an increased annual profit. The first large-scale application of CO_2 reforming of natural gas in a one million tonnes per annum methanol plant in Iran was reported by Holm-Larsen [55]. Carbon dioxide was imported from an ethylene cracker and an ammonia plant. The net CO_2 utilization was ~ 0.12 t per tonne of methanol. Captured CO_2 from the flue gas of a magnesite plant in Liaoning Province (China) was reacted with hydrogen derived from water electrolysis powered by solar energy to produce methanol [56].

Technoeconomic feasibility analysis of polygeneration schemes for generation of fuels, electricity, heat, and chemicals with carbon capture and storage or CO_2 utilization was studied by Ng et al. [57]. They evaluated five process configuration: (1) coal polygeneration with CO_2 precombustion capture and storage to produce methanol, acetic acid, electricity, and hydrogen, (2) coal polygeneration with CO_2 utilization in methane synthesis by Sabatier's reaction to produce methanol, acetic acid, electricity, hydrogen, and methane, (3) coal-integrated gasification combined cycle system with CO_2 capture and storage to produce electricity, (4) coal-integrated gasification combined cycle with trireforming of natural gas to produce electricity and methanol, and (5) this scheme was similar to scheme (1) with bio-oil as a feedstock instead of coal. Economic potentials were increased from cogeneration to polygeneration in addition to producing methanol.

A nonlinear optimization model of the low-pressure Lurgi methanol reactor with CO$_2$ utilization as a reactant for cogeneration of electricity using a gas turbine and fuel cells along with process heat integration was developed by Kralj and Glavic [58]. The yield of product was increased and CO$_2$ emissions reduced by 4800 t per annum. The two routes of methanol production from syngas and CO$_2$ hydrogenation were analyzed by Machado et al. [59]. The effect of process conditions such as pressure, (H$_2$-CO$_2$)/(CO$_2$ + CO), and H$_2$/CO$_2$ ratios on CO and CO$_2$ conversions, selectivity to methanol, and methanol production rate was investigated. In the methanol production route via dry reforming of CO$_2$, the amount of CO$_2$ consumption and CO$_2$ emission was 664,800 t per year and 19.02 t of CO$_2$ per hour, respectively. Methanol production and CO$_2$ management from a bioethanol plant through direct hydrogenation and natural gas bireforming was evaluated by Wiesberg et al. [60]. The consumption amount of CO$_2$ from the upstream bioethanol plant in the direct hydrogenation of CO$_2$ to methanol route was 1000 kmol/h, which was five times greater than that for indirect conversion of CO$_2$ through bireforming route. Production of hydrogen via thermal decomposition of natural gas and further conversion of H$_2$ with CO$_2$ recovered from coal-fired power plant stack gases considerably reduced CO$_2$ emissions by two-thirds from the utility and transportation sectors [61]. Pisarenko et al. [62] developed a process for producing methanol from syngas via DMR. The process consisted of four methanol reactors in series. Recycling of CO$_2$ to the reforming section increased the production rate of methanol. Technoeconomic analysis of a catalytic methanol plant powered by renewable energy sources was performed by Clausen et al. [63]. The syngas was produced by electrolysis of water, biomass gasification, CO$_2$ from postcombustion capture, and ATR of biogas or natural gas.

4.3 Dimethyl Ether (DME) Production

Dimethyl ether (DME) is an organic chemical compound with the formula CH$_3$OCH$_3$. Fig. 6 shows two pathways for DME production: direct and indirect. DME can be produced indirectly by methanol synthesis at first step followed by the dehydration of two methanol molecules (2 CH$_3$OH → (CH$_3$)$_2$O + H$_2$O) in the second step. It can also be produced directly from syngas in a single reactor [64].

Luu et al. studied the CO$_2$ valorization through the integration of a postcombustion carbon capture plant with a parallel combination of BiR and DMR for direct DME production [65]. The results showed that the parallel

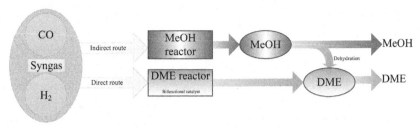

Fig. 6 Two pathways for DME production: direct and indirect.

combination of DMR and BiR could save 22% in methane feed uptake compared to the case where a standalone ATR was used to produce syngas.

Chen et al. [66] used CO_2 to control the reforming reactions in an ATR of a five tonnes/day DME pilot plant. The H_2/CO ratio was adjusted at 1.04 to maximize DME productivity. The effect of methanol dehydration on methanol formation in direct synthesis of DME from syngas was studied by Dadgar et al. [67]. With a dry syngas feed at 50 bar and operating temperature of 210–270°C, CO_2 was consumed totally during the synthesis. This means that conversion of CO_2 to methanol is greater than or equal to its formation via the water gas shift reaction. On the other hand, adding water at two different mole fractions of 6.6% and 9.4% to the syngas feed yielded a higher CO_2 formation via the water gas shift reaction than the CO_2 conversion via hydrogenation reaction to methanol.

Vakili and Eslamloueyan [68] used differential evolution algorithm for optimization of an industrial-scale fixed bed reactor for direct synthesis of DME from syngas. The objective function was chosen to maximize the DME production rate in each tube of the reactor. The design functions were the number of reactor tubes, coolant water, and feed inlet temperatures. The optimization results showed that the proposed optimal reactor configuration was more economical for large-scale purposes compared with the conventional industrial reactors. The conversion of CO and DME productivity increased by 0.73% and 4.84%, respectively. Some other studies have assessed the impact of operating conditions on the selectivity of methanol and DME for direct DME synthesis [69–73].

Manenti et al. [74] applied a systematic staging to a series of fixed bed reactors for direct synthesis of DME. The process was simulated and optimized in order to improve the production yield. Samimi et al. [75] studied the dehydration of methanol to DME in two spherical packed bed reactors connected in series. The decision variables were inlet temperatures and the catalyst distributions for each reactor. The production rate of DME was optimized using differential evolution (DE) method. The optimization results revealed 16.3% higher DME production rate compared with a conventional reactor.

Hu et al. [76] modeled a pipe-shell reactor for direct synthesis of DME from syngas and optimized over a bifunctional catalyst. The authors reported the effect of operating conditions on the selectivity and yield of DME and CO conversion. Peng and coworkers [77,78] studied the optimal DME productivity, minimum CO$_2$ emissions, and several syngas production technologies including DMR, SMR, POX, and coal gasifier. The results showed the optimal H$_2$/CO ratio of 1.0 with recycled stream having a very strong effect on the feed gas composition. Zhang et al. [79] performed a rigorous process simulation of direct synthesis of DME aiming to reduce CO$_2$ emission via a trireformer using Aspen Plus process simulator. The operating conditions were optimized to maximize the productivity of DME. Heat exchanger network (HEN) was designed to minimize utility costs. The results showed that CO$_2$ utilization through trireforming process is economically feasible on an industrial scale.

4.4 Urea Production

Urea or carbamide is an amide compound with chemical formula (NH$_2$)$_2$CO. It plays an important role in mammals' metabolism and has various industrial applications. A schematic of a urea production process is illustrated in Fig. 7. The basic urea process is the so-called Bosch–Meiser developed in 1922, which consists of two main reactions:

$$2\,NH_3 + CO_2 \rightleftharpoons H_2N - COONH_4 \;\; (\text{Carbamate formation})$$

$$H_2N - COONH_4 \rightleftharpoons (NH_2)_2CO + H_2O \;\; (\text{Urea conversion})$$

The first reaction is fast exothermic and happens at high temperature and pressure to produce ammonium carbamate (H$_2$N-COONH$_4$). This component is decomposed through a slow endothermic reaction in the second stage to form ammonia.

For producing one tonne of urea, 0.735–0.750 t CO$_2$ is consumed [7]. One of the key applications of urea is in fertilizers where it reacts with water, releasing CO$_2$ and ammonia to the soil. Furthermore, the amount of CO$_2$ emissions in the urea production process is 2.27 t of CO$_2$-eq. per tonne of CO$_2$ utilized [80]. Consequently, the concept of CO$_2$ utilization in urea production is not recognized as a carbon reduction measure [7]. The analysis of coproduction of urea and power from coal was performed by Bose et al.

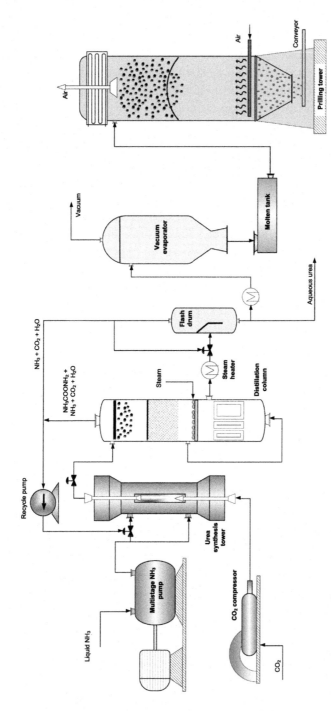

Fig. 7 Schematics of a urea production process.

[81] using Aspen Plus. The performance of the plant was evaluated based on the amount of CO_2 utilized. The results showed that the economic break-even for CO_2 utilization is below 5% and 10% for CO conversion efficiencies at 95% and 90%, respectively, within the water gas shift reactor.

Pérez-Fortes et al. [82] modeled and evaluated urea production from industrial CO_2. They evaluated the process efficiency, the amount of utilized CO_2, the utility requirements, and investment costs. The amount of CO_2 emissions was 0.6 t per tonne of CO_2 utilized.

4.5 Dimethyl Carbonate (DMC) Production

Dimethyl carbonate (DMC) is a carbonate ester with chemical formula of $(CH_3O)_2CO$. Fig. 8 illustrates six different reaction mechanisms for DMC production. It was initially produced by the reaction of phosgene $(COCl_2)$ with methanol $(COCl_2 + 2\ CH_3OH \rightarrow CH_3OCO_2CH_3 + 2\ HCl)$. However, due to the toxicity of phosgene, over time this route was deserted, and other synthesis routes have been used [83]. Garcia-Herrero et al. [84] have explained five other routes including oxidative carbonylation of methanol (Eni), oxidative carbonylation of methanol via methyl nitrite (Ube), ethylene carbonate transesterification (Asahi), urea transesterification, and direct synthesis from CO_2. Currently, the most popular industrial routes are: (1) direct production from methanol and CO_2 [85]; (2) production from methanol, CO_2, and epoxides; (3) production from CO_2 and ortho-ester or acetals [86–95].

Kongpanna et al. [96] developed a systematic framework for sustainable chemical process design with a focus on CO_2 utilization. The methodology was applied to DMC production highlighting some of innovative process designs. In another study, Kongpanna et al. [97] carried out a technoeconomic analysis of four processes for production of DMC with CO_2 utilization target. The evaluated processes were: direct synthesis from methanol and CO_2, indirect synthesis routes including conversion of CO_2 with propylene oxide, ammonia, and ethylene oxide to propylene carbonate, urea, and ethylene carbonate, respectively. The intermediate products then further reacted with methanol to produce dimethyl carbonate. They concluded that the synthesis from ethylene carbonate is the most promising process for DMC production. Bertilsson and Karlsson [98] investigated the CO_2 utilization potential via DMC process through phosgene- and urea-based routes. They reported significant impact of feedstock mixtures (i.e., natural gas, oil, and coal) on total CO_2 emissions.

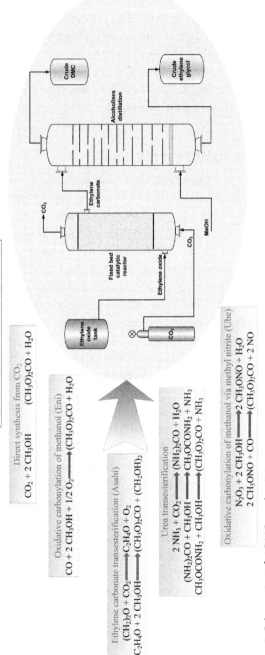

Fig. 8 Schematics of a DMC production process.

4.6 Polyurethane

Polyurethane is an organic polymer with units joined by urethane (carbamate) links (—NH—(C=O)—O—). Polyurethane is produced by reaction of CO_2 and propylene oxide using zinc hexacyanocobaltate catalyst and a multifunctional alcohol as a starter [99]. The key parameters in production of CO_2-based polyurethane materials are: functionality, molecular weight, and physicochemical properties of the poly-ether-carbonate polyols [99]. Since there is a large market for polyurethane, this technology can become one of the largest applications to valorize CO_2 in the polymer industry.

Srivastava [100] investigated the conversion of CO_2 to polyurethane under mild conditions over highly efficient, as-synthesized MCM-41 catalyst and without using an additional cocatalyst or solvent. Carbamates were synthesized with selectivity of 84%–95% without using an additional solvent. Given the CO_2-utilization potential of polyurethane, it is projectable that research and development in this direction will boom.

4.7 Fischer-Tropsch Gas-to-Liquid (FT-GTL) Products

Conversion of syngas to high-chain hydrocarbons has a history of almost a century, invented by Fischer and Tropsch in the early 20th in Germany for securing the country' overseas dependence on liquid fuels for transport. The primary product of the so-called FT reactor is syncrude, which is then upgraded in following separation processes to LPG, naphtha, diesel, wax, etc.

The Fischer-Tropsch GTL process is a technology that first converts natural gas (NG) to syngas by one of the various reformer processes [32]. The syngas is then converted into a liquid mixture of hydrocarbons (syncrude) in the so-called FT reactors and the presence of an iron or cobalt catalysts [101]. Last, this syncrude is upgraded in the following separation processes to target fuels including LPG, naphtha, diesel, and wax (Fig. 9).

Hydrogenation of CO and CO_2 to light olefins over supported iron nanocatalysts was studied by Hu and coworkers [103]. The conversion of CO and CO_2 was reported at 87% and 45%, respectively. Utilizing CO_2

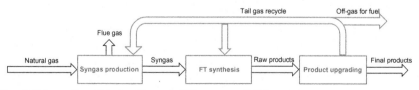

Fig. 9 Schematic of Fischer-Tropsch GTL process [102].

for the production of synthetic hydrocarbons fuels was also studied by Schaub et al. [104]. The simulation and optimization of a GTL process with cobalt-based F-T reactor were studied by Panahi et al. [105]. The hot syngas leaving the ATR was cooled down to ambient temperature before being directed to the CO_2 removal unit. Three expressions for the chain growth probability inside the FT reactor were compared to model the product distribution. Since the WGS reaction over cobalt-based catalysts is negligible, the consumption or production of CO_2 inside the FT reactor with cobalt-based catalyst was zero. The achieved chain growth probability model showed that at the high price of wax, the recycled light components containing mainly H_2, CO, CO_2, CH_4, to the syngas production section had a profound effect on reducing the H_2/CO ratio of fresh syngas. In this case, only 46% of the CO_2 of the stream leaving the ATR was recovered, and the rest of CO_2 was required to reduce the H_2/CO ratio of the stream to the FT reactor. A case study was performed to reduce H_2/CO inside the staged FT reactor with the iron-based catalyst by adding CO_2 to the stage [106]. The reactor temperature was constrained by 250 °C. Optimizing the FT reactor path by extra CO_2 feed showed that adding CO_2 had no effect on reducing the H_2/CO inside the reactor even through RWGS reaction because of the low temperature of the FT reactor. In another study, a GTL plant with different syngas production configurations including ATR, combined reformers, and a series arrangement of gas heated reformers with ATR was optimized [107]. The objective function was the maximization the flowrate of liquid hydrocarbons. The optimization results showed that the combined reformer had the lowest CO_2 emissions, i.e., ~51 kmol/h, while the ATR configuration had the highest CO_2 emission to the atmosphere from the purge stream (~718 kmol/h). Different placements of a CO_2 removal unit in GTL were optimized from an economical point of view [108]. The objective function was selected to be a measure of the annual net cash flow. The optimization results revealed that CO_2 removal unit did not have a considerable influence on carbon and energy efficiencies as well as the annual net cash flow of the process. In a GTL plant with cobalt-based FT reactor, some amounts of CO_2 were needed for the reformer reactions to adjust the H_2/CO ratios of fresh syngas and the H_2/CO to the FT reactor.

The CO_2 utilization through the integration of a postcombustion carbon capture with a GTL process for the objective function of maximizing wax product was studied [30]. For a 300-MW coal-fired power plant, the results indicated that an ATR-based GTL process does not have flexibility for CO_2

intake, while all the captured CO_2 fed to the GTL process via an SMR could be consumed. There was no net utilization of imported CO_2 to the process. For the case of a standalone ATR as a reformer, the chain growth probability inside the FT reactor was 0.966. Alternatively, in the process of a standalone SMR, the probability of chain growth was reduced to 0.92. Ha et al. [109] developed a process model for FT synthesis to evaluate the process efficiency and CO_2 emissions using an SMR and DMR as the syngas production technologies [109]. A combination of SMR and DMR required 20–30% less natural gas feed. Moreover, CO_2 emissions and energy use of three processes including natural gas, coal, and biomass to liquids were compared [110,111].

Furthermore, hybrid coal and natural gas-to-liquid fuels processes were simulated and optimized [112,113]. The CO_2 stream was recycled to the gas reforming block and a reduction in CO_2 emissions was observed from the CTL case.

4.8 Methane Production

Hydrogenation of CO_2 or CO to methane and water is used in ammonia plants to purify the syngas and to reduce the carbon oxides in H_2-rich syngas and polymer electrolyte fuel cell anodes [114]. Sahebdelfar and Takht Ravanchi [115] studied the thermodynamic equilibrium analysis of CO_2 hydrogenation to methane and water. The results showed that high conversion of CO_2 to methane is more feasible at low-temperature ranges (200–300°C). Production of hydrogen and methane powered by the surplus electricity generated from fluctuating renewable sources such as photovoltaic or wind power was investigated by Reiter and Lindorfer [116]. This method is known as power-to-gas technology [117]. Part of the produced hydrogen can be used to convert the captured CO_2 to methane through methanation process via Sabatier reaction. The CO_2 conversion to methane on catalysts containing Ni and Ce supported on a HNaUSY zeolite was investigated by Graca et al. [118]. They reported very high conversion rate of CO_2 and high selectivity of methane by increasing the Ni content.

Different CO_2 sources for power-to-gas applications in Austria were evaluated by Reiter and Lindorfer [119]. They concluded that there is enough amount of CO_2 to be converted to methane via the renewable power sources, i.e., photovoltaics and wind power plants. Also, the best CO_2 sources for power-to-gas (PTG) plants were a bioethanol plant and biogas upgrading facilities.

4.9 Chemical-Looping Dry Reforming

Chemical-looping dry-reforming process consists of three stages: (1) methane reduction, (2) CO_2 reforming to generate CO, and (3) oxidation [120]. This process is recognized as a promising innovative technology for CO_2 valorization. Huang et al. [120] studied the feasibility of using spinel nickel ferrite ($NiFe_2O_4$) in the chemical-looping dry-reforming process. They concluded that $NiFe_2O_4$ is a suitable candidate for this application. The feasibility of chemical looping dry reforming was also studied through thermodynamic screening calculations [121]. The results indicated that high conversion of CO_2 is possible. Aika and Nishiyama [122] studied the oxidative coupling of methane with CO_2 utilization over CaO and BaO-CaO catalysts. The yield of ethane and ethene was higher than the maximum theoretical yield by methane and oxygen reaction.

4.10 Mineralization

The principle and realization of CO_2 mineralization for electricity generation was proposed in HePing's et al. work [123]. The CO_2 mineralization or carbonation reaction is a reaction in which CO_2 is converted into CO_3^{2-} and the energy is released. Different concentrations of CO_2 (mixed N_2/CO_2 gases) were examined. The results showed that higher power density could be achieved at high concentration of the CO_2.

A recent study conducted by Ebrahimi et al. [124] proposed the CO_2 carbonation of fly ash (FA) from a power plant in Australia to be utilized in cement manufacturing in producing "green" building materials. It was concluded that a 10% blending ratio of carbonated fly ash (CFA) resulted in FA uptake of $96.4 \, kg_{FA}/ton_{cement}$ equivalent of $55.4 \, kg_{FA}/GWh$, while the CO_2 utilization was at $72.9 \, kg_{CO2-eq}/ton_{cement}$ equivalent to $41.9 \, kg_{CO2-eq}/GWh$.

5 CONCLUSIONS

CO_2 abatement offers a challenge and an opportunity to the chemical and energy sector. After three decades of research on CCS, this technology is facing barriers as the investment and operation costs are still high and infeasible in majority of cases. As such, research in the carbon management field is shifting toward CO_2 utilization in the synthesis of chemicals and fuels in order to improve the value proposition of carbon capture option.

We critically reviewed the literature on carbon capture, conversion, and utilization routes and assessed the progress in the research and developments in this direction. It is evident from the literature that physical CO$_2$ utilization is a mature industry with limited demand for innovative core academic research. Rather, such routes rely mainly on market dynamics linked with global and local energy prices and climate change mitigation policies such as carbon taxing mechanisms. However, chemical CO$_2$ utilization, except for some limited products, is a developing field with a significant demand for research and development. Research in all aspects from catalysis, to unit operation design, to process configuration, life cycle assessment, environmental benchmarking, and policy analysis is still in early days.

REFERENCES

[1] Karimi F, Khalilpour R. Evolution of carbon capture and storage research: Trends of international collaborations and knowledge maps. Int J Greenhouse Gas Control 2015;37:362–76.

[2] Vidas H, Chikkatur BHA, Venkatesh B. Analysis of the Costs and Benefits of CO$_2$ Sequestration on the U.S. Outer Continental Shelf. Technical report. 2012.

[3] House KZ, et al. The energy penalty of post-combustion CO$_2$ capture & storage and its implications for retrofitting the U.S. installed base. Energ Environ Sci 2009;2 (2):193–205.

[4] Rochelle GT. Amine scrubbing for CO$_2$ capture. Science 2009;325(5948):1652–4.

[5] Benson SM, Cole DR. CO$_2$ sequestration in deep sedimentary formations. Elements 2008;4(5):325–31.

[6] Chih-Hung Huang C-ST. A review: CO$_2$ utilization. Aerosol Air Qual Res 2014;14 (2):480–99.

[7] Muradov N. Industrial utilization of CO$_2$: a win–win solution. In: Liberating energy from carbon: introduction to decarbonization. New York, NY: Springer; 2014. p. 325–83.

[8] Aresta M, Dibenedetto A, Angelini A, The changing paradigm in CO$_2$ utilization. J CO$_2$ Util 2013;3–4:65–73.

[9] IEA. CO$_2$ emissions from fuel combustion. Paris: International Energy Agency; 2015.

[10] Metz B, et al. IPCC special report on carbon dioxide capture and storage. Cambridge: Cambridge University Press; 2005. for the Intergovernmental Panel on Climate Change. x, 431 p.

[11] Poliakoff M, Leitner W, Streng ES. The twelve principles of CO$_2$ chemistry. Faraday Discuss 2015;183:9–17.

[12] Otto A, et al. Closing the loop: captured CO$_2$ as a feedstock in the chemical industry. Energ Environ Sci 2015;8(11):3283–97.

[13] Langanke J, Wolf A, Peters M. Chapter 5: Polymers from CO$_2$—an industrial perspective A2 – Styring, Peter. In: Quadrelli EA, Armstrong K, editors. Carbon dioxide utilisation. Amsterdam: Elsevier; 2015. p. 59–71.

[14] Safi R, Agarwal RK, Banerjee S. Numerical simulation and optimization of CO$_2$ utilization for enhanced oil recovery from depleted reservoirs. Chem Eng Sci 2016;144:30–8.

[15] Bachu S. Identification of oil reservoirs suitable for CO_2-EOR and CO_2 storage (CCUS) using reserves databases, with application to Alberta, Canada. Int J Greenh Gas Control 2016;44:152–65.

[16] Azzolina NA, et al. CO_2 storage associated with CO_2 enhanced oil recovery: a statistical analysis of historical operations. Int J Greenh Gas Control 2015;37:384–97.

[17] Liu P, Zhang X. Enhanced oil recovery by CO_2–CH_4 flooding in low permeability and rhythmic hydrocarbon reservoir. Int J Hydrogen Energy 2015;40(37):12849–53.

[18] White MD, et al. 12th international conference on greenhouse gas control technologies, GHGT-12Numerical simulation of carbon dioxide injection in the western section of the Farnsworth unit. Energy Procedia 2014;63:7891–912.

[19] Gupta N, et al. 12th international conference on greenhouse gas control technologies, GHGT-12Testing for large-scale CO_2-enhanced oil recovery and geologic storage in the Midwestern USA. Energy Procedia 2014;63:6393–403.

[20] Mishra S, et al. 12th international conference on greenhouse gas control technologies, GHGT-12Estimating CO_2-EOR potential and CO-sequestration capacity in Ohio's depleted oil fields. Energy Procedia 2014;63:7785–95.

[21] Boodlal D, Alexander D, Narinesingh J. 12th international conference on greenhouse gas control technologies, GHGT-12The use of reservoir simulation to obtain key parameters for investigating the feasibility of implementing flue gas CO_2-EOR coupled with sequestration in Trinidad. Energy Procedia 2014;63:7517–28.

[22] Pingping S, Xinwei L, Qiujie L. Methodology for estimation of CO_2 storage capacity in reservoirs. Pet Explor Dev 2009;36(2):216–20.

[23] Luu MT, Milani D, Abbas A. Analysis of CO_2 utilization for methanol synthesis integrated with enhanced gas recovery. J Clean Prod 2016;112(Part 4):3540–54.

[24] Nakaten N, Islam R, Kempka T. 12th international conference on greenhouse gas control technologies, GHGT-12Underground coal gasification with extended CO_2 utilization – an economic and carbon neutral approach to tackle energy and fertilizer supply shortages in Bangladesh. Energy Procedia 2014;63:8036–43.

[25] Zhang F-Z, Xu R-N, Jiang P-X. Thermodynamic analysis of enhanced geothermal systems using impure CO_2 as the geofluid. Appl Therm Eng 2016;99:1277–85.

[26] Antonijević DL. Carbon dioxide as the replacement for synthetic refrigerants in mobile air conditioning. Therm Sci 2008;12(3):55–64.

[27] Han J-H, Lee I-B. Two-stage stochastic programming model for planning CO_2 utilization and disposal infrastructure considering the uncertainty in the CO_2 emission. Ind Eng Chem Res 2011;50(23):13435–43.

[28] Dragos L, et al. Aspects of CO_2 utilization toward the goal of emission reduction in Romania. In: Inui MAKISYT, Yamaguchi T, editors. Studies in surface science and catalysis. Amsterdam: Elsevier; 1998. p. 195–200.

[29] Han J-H, Lee I-B. Development of a scalable and comprehensive infrastructure model for carbon dioxide utilization and disposal. Ind Eng Chem Res 2011;50 (10):6297–315.

[30] Rafiee APM, Khalilpour R. CO_2 utilization through integration of post-combustion carbon capture process with fischer tropsch gas-to-liquid (GTL) processes. J CO_2 Util 2016;18:98–106.

[31] Bao B, El-Halwagi MM, Elbashir NO. Simulation, integration, and economic analysis of gas-to-liquid processes. Fuel Process Technol 2010;91(7):703–13.

[32] Wilhelm DJ, et al. Syngas production for gas-to-liquids applications: technologies, issues and outlook. Fuel Process Technol 2001;71(1–3):139–48.

[33] Steynberg A, Dry M, editors. Fischer-Tropsch technology studies in surface science and catalysis. Centi G, editor. vol. 152. Amsterdam: Elsevier; 2004. p. 700.

[34] Lim Y, et al. Optimal design and decision for combined steam reforming process with dry methane reforming to reuse CO_2 as a raw material. Ind Eng Chem Res 2012;51 (13):4982–9.

[35] Choi S, et al. Optimal design of synthesis gas production process with recycled carbon dioxide utilization. Ind Eng Chem Res 2008;47(2):323–31.

[36] Kuo P-C, Wu W. Thermodynamic analysis of a combined heat and power system with CO$_2$ utilization based on CO-gasification of biomass and coal. Chem Eng Sci 2016;142:201–14.

[37] Po-Chih Kuo WW. Optimal process design based on energy and exergy analyses of co-gasification power plant with CO$_2$ utilization, Available from: https://aiche.confex.com/aiche/2015/webprogram/Paper423362.html; 2015.

[38] Wang Y, et al. Modeling and simulation of coal gasification with carbon dioxide reuse system. 2009 Asia-Pacific power and energy engineering conference; 2009.

[39] Olah GAGA, Prakash GKS. Beyond oil and gas: the methanol economy. Weinheim, Germany: Wiley-VCH; 2006.

[40] Olah GA, Goeppert A, Prakash GKS. Chemical recycling of carbon dioxide to methanol and dimethyl ether: from greenhouse gas to renewable, environmentally carbon neutral fuels and synthetic hydrocarbons. J Org Chem 2009;74(2):487–98.

[41] Olah GA. Available from: http://desertec-uk.org.uk/articles/methanol_synthesis.pdf.

[42] Powell CE, Qiao GG. Polymeric CO$_2$/N-2 gas separation membranes for the capture of carbon dioxide from power plant flue gases. J Membr Sci 2006;279(1–2):1–49.

[43] Chen Q, et al. Opportunities of integrated systems with CO$_2$ utilization technologies for green fuel & chemicals production in a carbon-constrained society. J CO$_2$ Util 2016;14:1–9.

[44] Alarifi A, et al. Dynamic optimization of Lurgi type methanol reactor using hybrid GA-GPS algorithm: the optimal Shell temperature trajectory and carbon dioxide utilization. Ind Eng Chem Res 2016;55(5):1164–73.

[45] Luu MT, et al. A comparative study of CO$_2$ utilization in methanol synthesis with various syngas production technologies. J CO$_2$ Util 2015;12:62–76.

[46] Milani D, et al. A model-based analysis of CO$_2$ utilization in methanol synthesis plant. J CO$_2$ Util 2015;10:12–22.

[47] Roh K, et al. A methodology for the sustainable design and implementation strategy of CO$_2$ utilization processes. Comput Chem Eng 2016;91:407–21.

[48] Iyer SS, et al. Generalized thermodynamic analysis of methanol synthesis: Effect of feed composition. J CO$_2$ Util 2015;10:95–104.

[49] Du X-L, et al. Efficient hydrogenation of alkyl Formate to methanol over nanocomposite copper/alumina catalysts. ChemCatChem 2014;6(11):3075–9.

[50] Kim WS, et al. The process design and simulation for the methanol production on the FPSO (floating production, storage and off-loading) system. Chem Eng Res Design 2014;92(5):931–40.

[51] Lima BCDS, de Araujo OQF, de Medeiros JL, Morgado CRV. Technical, economic and environmental viability of offshore CO$_2$ reuse from natural gas by dry reforming. Appl Mech Mater 2016;830:109–16.

[52] Dumont M-N, von der Assen N, Sternberg A, Bardow A. In: Assessing the environmental potential of carbon dioxide utilization: A graphical targeting approach. Symposium on process systems engineering. 2012: Singapore; 2012.

[53] Raudaskoski R, et al. Catalytic activation of CO$_2$: use of secondary CO$_2$ for the production of synthesis gas and for methanol synthesis over copper-based zirconia-containing catalysts. Catal Today 2009;144(3–4):318–23.

[54] Kralj AK, Glavič P. CO$_2$ separation from purge gas and flue gas in the methanol process, using NLP model optimization. Ind Eng Chem Res 2007;46(21):6953–62.

[55] Holm-Larsen H. CO$_2$ reforming for large scale methanol plants: an actual case. Stud Surf Sci Catal 2001;441–6.

[56] Zhao L, et al. Cold energy utilization of liquefied natural gas for capturing carbon dioxide in the flue gas from the magnesite processing industry. Energy 2016;105:45–56.

[57] Ng KS, Zhang N, Sadhukhan J. Techno-economic analysis of polygeneration systems with carbon capture and storage and CO_2 reuse. Chem Eng J 2013;219:96–108.

[58] Kralj AK, Glavič P. H_2 separation and use in fuel cells and CO_2 separation and reuse as a reactant in the existing methanol process. Energy Fuel 2007;21(5):2892–9.

[59] Machado CF, de Medeiros JL, Araújo O, Alves RMB. In: A comparative analysis of methanol production routes: synthesis gas versus CO_2 hydrogenation. International conference on industrial engineering and operations management, 2014. Bali, Indonesia; 2014.

[60] Wiesberg IL, et al. Carbon dioxide management by chemical conversion to methanol: HYDROGENATION and BI-REFORMING. Energ Conver Manage 2016; 125:320–35.

[61] Steinberg M. Production of hydrogen and methanol from natural gas with reduced CO_2 emission. Int J Hydrogen Energy 1998;23(6):419–25.

[62] Pisarenko EV, et al. Power- and resource-saving process for producing methanol from natural gas. Theor Found Chem Eng 2008;42(1):12–8.

[63] Clausen LR, Houbak N, Elmegaard B. Technoeconomic analysis of a methanol plant based on gasification of biomass and electrolysis of water. Energy 2010;35(5):2338–47.

[64] Azizi Z, et al. Dimethyl ether: a review of technologies and production challenges. Chem Eng Process Process Intensif 2014;82:150–72.

[65] Luu MT, et al. Analysis of di-methyl ether production routes: process performance evaluations at various syngas compositions. Chem Eng Sci 2016;149:143–55.

[66] Takashi Ogawa NI, Shikada T, Ohno Y. Direct dimethyl ether synthesis. J Nat Gas Chem 2003;12:219–27.

[67] Dadgar F, et al. Direct dimethyl ether synthesis from synthesis gas: the influence of methanol dehydration on methanol synthesis reaction. Catal Today 2016;270:76–84.

[68] Vakili R, Eslamloueyan R. Design and optimization of a fixed bed reactor for direct dimethyl ether production from syngas using differential evolution algorithm. Int J Chem React Eng 2013;11:147.

[69] Zhen Chen HZ, Ying W, Fang D. Global kinetics of direct dimethyl ether synthesis process from syngas in slurry reactor over a novel Cu-Zn-Al-Zr slurry catalyst. Int Jo Chem Mol Nucl Mater Metall Eng 2010;4(8):501–7.

[70] Zhaoguang Nie HL, Liu D, Ying W, Fang D. Intrinsic kinetics of dimethyl ether synthesis from syngas. J Nat Gas Chem 2005;14:22–8.

[71] Aguayo AT, et al. Kinetic modeling of dimethyl ether synthesis in a single step on a $CuO-ZnO-Al2O3/\gamma-Al_2O_3$ catalyst. Ind Eng Chem Res 2007;46(17):5522–30.

[72] Wang L, et al. Influence of reaction conditions on methanol synthesis and WGS reaction in the syngas-to-DME process. J Nat Gas Chem 2006;15(1):38–44.

[73] Shim HM, et al. Simulation of DME synthesis from coal syngas by kinetics model. Korean J Chem Eng 2009;26(3):641–8.

[74] Manenti F, et al. Systematic staging design applied to the fixed-bed reactor series for methanol and one-step methanol/dimethyl ether synthesis. Appl Therm Eng 2014;70 (2):1228–37.

[75] Samimi F, et al. Mathematical modeling and optimization of DME synthesis in two spherical reactors connected in series. J Nat Gas Sci Eng 2014;17:33–41.

[76] Hu Y, Nie Z, Fang D. Simulation and model design of pipe-shell reactor for the direct synthesis of dimethyl ether from syngas. J Nat Gas Chem 2008;17(2):195–200.

[77] Peng XD, et al. Single-step syngas-to-dimethyl ether processes for optimal productivity, minimal emissions, and natural gas-derived syngas. Ind Eng Chem Res 1999;38 (11):4381–8.

[78] Peng X-D. Kinetic Understanding of the Syngas-To-DME Reaction System and Its Implications to Process and Economics. United States Department of Energy; 2002. Under Contract No. DE-FC22-94 PC93052.

[79] Zhang Y, Zhang S, Benson T. A conceptual design by integrating dimethyl ether (DME) production with tri-reforming process for CO_2 emission reduction. Fuel Process Technol 2015;131:7–13.

[80] Global CCS Institute, P.B. Accelerating the uptake of CCS: Industrial use of captured carbon dioxide, Available from: http://www.globalccsinstitute.com/publications/accelerating-uptake-ccs-industrial-use-captured-carbon-dioxide; 2011.

[81] Bose A, et al. Co-production of power and urea from coal with CO_2 capture: performance assessment. Clean Techn Environ Policy 2015;17(5):1271–80.

[82] Pérez-Fortes M, Bocin-Dumitriu A, Tzimas E. 12th international conference on greenhouse gas control technologies, GHGT-12CO2 utilization pathways: techno-economic assessment and market opportunities. Energy Procedia 2014;63:7968–75.

[83] Tundo P, Selva M. The chemistry of dimethyl carbonate. Acc Chem Res 2002;35 (9):706–16.

[84] Garcia-Herrero I, et al. Environmental assessment of dimethyl carbonate production: comparison of a novel electrosynthesis route utilizing CO_2 with a commercial oxidative carbonylation process. ACS Sustain Chem Eng 2016;4(4):2088–97.

[85] Heyn RH. Chapter 7: Organic carbonates A2 – Styring, Peter. In: Quadrelli EA, Armstrong K, editors. Carbon dioxide utilisation. Amsterdam: Elsevier; 2015. p. 97–113.

[86] Dai W-L, et al. The direct transformation of carbon dioxide to organic carbonates over heterogeneous catalysts. Appl Catal Gen 2009;366(1):2–12.

[87] Razali NAM, et al. Heterogeneous catalysts for production of chemicals using carbon dioxide as raw material: a review. Renew Sustain Energy Rev 2012;16 (7):4951–64.

[88] Aresta M, Dibenedetto A. Carbon dioxide: a valuable source of carbon for chemicals, fuels and materials. In: Catalytic process development for renewable materials. Weinheim: Wiley-VCH Verlag GmbH & Co. KGaA; 2013. p. 355–85.

[89] Danielle Ballivet-Tkatchenko SS. Chapter 10: Linear organic carbonatES. In: - Aresta M, editor. Carbon dioxide recovery and utilization. The Netherlands: Kluwer Academic Publishers; 2003. p. 261–77.

[90] Aresta M, Dibenedetto A. Utilisation of CO_2 as a chemical feedstock: opportunities and challenges. Dalton Trans 2007;28:2975–92.

[91] Laurenczy G, Picquet M, Plasseraud L. Di-n-butyltin(IV)-catalyzed dimethyl carbonate synthesis from carbon dioxide and methanol: An in situ high pressure 119Sn{1H} NMR spectroscopic study. J Organomet Chem 2011;696(9):1904–9.

[92] Quaranta E, Aresta M. The chemistry of N–CO_2 bonds: synthesis of carbamic acids and their derivatives, isocyanates, and ureas. In: Carbon dioxide as chemical feedstock. Weinheim: Wiley-VCH Verlag GmbH & Co. KGaA; 2010. p. 121–67.

[93] Song C. Global challenges and strategies for control, conversion and utilization of CO_2 for sustainable development involving energy, catalysis, adsorption and chemical processing. Catal Today 2006;115(1–4):2–32.

[94] Huang S, et al. Recent advances in dialkyl carbonates synthesis and applications. Chem Soc Rev 2015;44(10):3079–116.

[95] Khavarian M, Chai S-P, Mohamed AR. Carbon dioxide conversion over carbon-based nanocatalysts. J Nanosci Nanotechnol 2013;13(7):4825–37.

[96] Kongpanna P, et al. Systematic methods and tools for design of sustainable chemical processes for CO_2 utilization. Comput Chem Eng 2016;87:125–44.

[97] Kongpanna P, et al. Techno-economic evaluation of different CO_2-based processes for dimethyl carbonate production. Chem Eng Res Des 2015;93:496–510.

[98] Bertilsson F, Karlsson HT. CO_2 utilization options, part II: assessment of dimethyl carbonate production. Energ Conver Manage 1996;37(12):1733–9.

[99] Langanke J, et al. Carbon dioxide (CO_2) as sustainable feedstock for polyurethane production. Green Chem 2014;16(4):1865–70.

[100] Srivastava R, Srinivas D, Ratnasamy P. Syntheses of polycarbonate and polyurethane precursors utilizing CO_2 over highly efficient, solid as-synthesized MCM-41 catalyst. Tetrahedron Lett 2006;47(25):4213–7.

[101] Van der Laan GP, Beenackers AACM. Kinetics and selectivity of the Fischer-Tropsch synthesis: a literature review. Catal Rev Sci Eng 1999;41(3–4):255–318.

[102] Rostrup-Nielsen JR. Syngas in perspective. Catal Today 2002;71(3–4):243–7.

[103] Hu B, et al. Selective hydrogenation of CO_2 and CO to useful light olefins over octahedral molecular sieve manganese oxide supported iron catalysts. Appl Catal Environ 2013;132–133:54–61.

[104] Schaub G, Unruh D, Rohde M. Synthetic hydrocarbon fuels and CO_2 utilization. Stud Surf Sci Catal 2004;153:17–24.

[105] Panahi M, et al. A natural gas to liquids process model for optimal operation. Ind Eng Chem Res 2012;51(1):425–33.

[106] Rafiee A, Hillestad M. Staging of the Fischer–Tropsch reactor with an iron based catalyst. Comput Chem Eng 2012;39:75–83.

[107] Rafiee A, Hillestad M. Synthesis gas production configurations for gas-to-liquid applications. Chem Eng Technol 2012;35(5):870–6.

[108] Rafiee A, Hillestad M. Techno-economic analysis of a gas-to-liquid process with different placements of a CO_2 removal unit. Chem Eng Technol 2012;35(3):420–30.

[109] Ha K-S, et al. Efficient utilization of greenhouse gas in a gas-to-liquids process combined with carbon dioxide reforming of methane. Environ Sci Technol 2010;44(4):1412–7.

[110] Tao R, Patel MK. Basic petrochemicals from natural gas, coal and biomass: energy use and CO_2 emissions. Resour Conserv Recycl 2009;53:513–28.

[111] Ren T. Petrochemicals from Oil, Natural Gas, Coal and Biomass: Energy Use, Economics and Innovation. Utrecht University, Copernicus Institute for Sustainable Development and Innovation, Department of Science, Technology and Society, Faculty of Science; 2009.

[112] Sudiro M, Bertucco A. Production of synthetic gasoline and diesel fuel by alternative processes using natural gas and coal: process simulation and optimization. Energy 2009;34(12):2206–14.

[113] Elia JA, Baliban RC, Floudas CA. Toward novel hybrid biomass, coal, and natural gas processes for satisfying current transportation fuel demands, 2: Simultaneous heat and power integration. Ind Eng Chem Res 2010;49(16):7371–88.

[114] Sivashunmugam Sankaranarayanan KS. Carbon dioxide – a potential raw material for the production of fuel, fuel additives and bio-derived chemicals. Indian J Chem 2012;51A:1252–62.

[115] Sahebdelfar S, Takht Ravanchi M. Carbon dioxide utilization for methane production: a thermodynamic analysis. J Petrol Sci Eng 2015;134:14–22.

[116] Reiter G, Lindorfer J. Global warming potential of hydrogen and methane production from renewable electricity via power-to-gas technology. Int J Life Cycle Assess 2015;20(4):477–89.

[117] Sternberg A, Bardow A. Life cycle assessment of power-to-gas: syngas vs methane. ACS Sustain Chem Eng 2016;4(8):4156–65.

[118] Graça I, et al. CO_2 hydrogenation into CH_4 on NiHNaUSY zeolites. Appl Catal Environ 2014;147:101–10.

[119] Reiter G, Lindorfer J. Evaluating CO_2 sources for power-to-gas applications – a case study for Austria. J CO_2 Util 2015;10:40–9.

[120] Huang Z, et al. Evaluation of multi-cycle performance of chemical looping dry reforming using CO_2 as an oxidant with Fe–Ni bimetallic oxides. J Energy Chem 2016;25(1):62–70.

[121] Najera M, et al. Carbon capture and utilization via chemical looping dry reforming. Chem Eng Res Design 2011;89(9):1533–43.

[122] Aika K-I, Nishiyama T. Utilization of CO$_2$ in the oxidative coupling reaction of methane over CaO based catalysts. Catal Today 1989;4(3):271–8.

[123] Xie HePing WY, Yang H, MaLing G, Tao L, JinLong W, Liang T, Wen J, Ru Z, LingZhi X, Bin L. Generation of electricity from CO$_2$ mineralization: principle and realization. Sci China Technol Sci 2014;57(12):2335–43.

[124] Ebrahimi A, et al. Sustainable transformation of fly ash industrial waste into a construction cement blend via CO$_2$ carbonation. J Clean Prod 2017;156:660–9.

CHAPTER 9

Solar Thermal Energy and Its Conversion to Solar Fuels via Thermochemical Processes

James Hinkley*, Christos Agrafiotis†
*Sustainable Energy Systems, School of Engineering and Computer Science, Faculty of Engineering, Victoria University of Wellington, New Zealand
†Institute of Solar Research, Deutsches Zentrum für Luft- und Raumfahrt/German Aerospace Center (DLR), Linder Höhe, Köln, Germany

Abstract

This chapter reviews the conversion of solar energy to various fuels through the use of thermochemical processes. The chapter begins with an overview of solar thermal technologies capable of providing the necessary energy at appropriate temperatures, before summarizing the sometimes bewildering range of process options. As many hundreds of potential cycles have been proposed over the years, the chapter will necessarily focus on groups of reaction schemes with common characteristics. It will also endeavor to provide the reader with an up-to-date summary of research and demonstration activities around the world, along with a summary of some of the priorities for further work.

Keywords: Solar energy, Solar fuel, Thermochemical cycle, Metal oxides, Biomass, Coal

1 INTRODUCTION

This section is primarily concerned with what is generally referred to by researchers in the solar thermal area as solar chemistry. It is important to note that other researchers can and do use the terms solar chemistry and solar fuels in a completely different sense, e.g., water splitting by photochemical methods or artificial photosynthesis (e.g., [1–3]). Arguably, solar electricity—whether from photovoltaics or concentrating solar power—can also be used to make solar fuels, but the current work is primarily concerned with processes that use solar thermal energy directly in chemical reactions to produce fuels.

Solar thermal fuels rely on the chemical transformation of materials that is possible when they are heated, typically to drive a dissociation reaction. In the case of the thermolysis of water, this results in the production of

Polygeneration with Polystorage
https://doi.org/10.1016/B978-0-12-813306-4.00009-4

247

hydrogen and oxygen. While attractive in its apparent simplicity, thermolysis in practice only occurs at very high temperature, and is considered too challenging from an engineering and efficiency perspective to be a realistic fuel pathway [4].

Research over many decades has sought to identify other materials that can reduce the required maximum temperature to more realistic ranges through, for example, the use of redox cycles. Such cycles require two or more steps to produce hydrogen and oxygen from water, through the sequential reduction of an active material at high temperature and a subsequent oxidation step where the reduced material is exposed to steam and reoxidized to its initial state, producing hydrogen. An advantage of these cycles is that in principle many are directly adaptable to carbon dioxide splitting and/or combined CO_2/H_2O splitting. The production of carbon monoxide from carbon dioxide has been demonstrated as a precursor to the synthesis of carbon-based fuels such as hydrocarbons [5].

Significant research has also been conducted on the solar reforming of methane and the gasification of solids such as coal, biomass, and waste. Reforming is the reaction of water and methane (the principal component of natural gas) at temperatures above 850°C to produce carbon monoxide and hydrogen. This process is a cornerstone of commodity chemical production and is used in both ammonia and methanol production.

Carbothermic reduction reactions are also attractive as the presence of carbon generally leads to a much lower temperature than thermal processes alone. For example, it is possible to reduce ceria at much lower temperatures using methane as a reductant than reduction via thermal dissociation. However, the use of carbon also can affect the sustainability of the overall process, especially if sourced from fossil fuels.

There are also a number of hybrid processes, which use electricity as well as solar thermal energy. These cycles were favored by the nuclear industry, which believed it could provide both the electrical and thermal energy required. However, this was contingent on the development of high temperature reactors in the Generation IV program, and this type of reactor was subsequently removed from the development priorities [6].

2 SOLAR THERMAL TECHNOLOGIES

While solar energy is the world's most abundant renewable energy source, it is also relatively diffuse, with an energy density at the earth's surface of approximately $1\,kW/m^2$. Thermochemical processes require the input of energy at temperatures from as low as 600°C to more than 1800°C; this requires the solar energy to be concentrated. This energy must then be collected by some form of receiver, where the thermal energy can be used in a reactor to drive chemical reactions. Solar thermal reactors are most commonly directly integrated with the receiver, but with some barrier to prevent loss of reactants to ambient.

2.1 Solar Concentration Technologies

The various types of solar collectors (i.e., troughs, towers, linear Fresnel, and dishes) are well described elsewhere and this chapter will instead focus on comparing the different technologies and their relevance for solar fuels production. These technologies are well developed for electricity generation based on steam Rankine cycles, with both parabolic troughs and power towers being commercially implemented at significant scale. In both technologies, solar energy is absorbed by surfaces with special coatings and transferred to a working fluid. This fluid (known as the heat transfer fluid) may be used directly in the case of pressurized water or steam, but is more commonly only used as a heat carrier with subsequent steam generation. Thermal energy is also commonly stored in molten salts to provide dispatchable generation, i.e., electricity can be produced when required.

In *parabolic troughs*, solar energy is concentrated by curved mirrors onto a linear receiver tube, encased in an evacuated glass tube to minimize thermal losses. Parabolic troughs are the most mature and widely deployed CSP technology, accounting for 85% of the approximately 4.9 GW$_e$ of worldwide capacity in 2017. The vast majority of these plants use thermal oils as a heat transfer fluid, and molten nitrate salts are the preferred thermal storage medium. These salts are cheaper and less damaging in the event of a spill. Troughs using oil are limited to an upper temperature of about 390°C to avoid decomposing the oil. The molten salts typically used melt at about 220°C and are stable up to 590°C. Some manufacturers have developed troughs with direct steam generation or molten salt, but these are operationally more challenging due to phase change in the case of steam and the need to avoiding freezing in the case of molten salts.

In *power towers*, or central receivers, a large field of individually controlled mirrors called heliostats track the sun throughout the day to concentrate the solar energy on a tower-mounted receiver. Thermal energy is transferred in the receiver to a molten salt or in some plants steam is produced directly. Molten salts are most commonly used as they can operate at up to 590°C and can readily be integrated with large-scale thermal storage. The solar tower technology also enables higher steam temperatures than is possible with parabolic troughs leading to higher thermal to power conversion efficiency; thus, it is widely expected that towers will become the dominant CSP technology. This is important for solar chemistry applications as power towers are capable of much higher temperatures than currently demonstrated for power generation.

Paraboloidal dishes enable the highest optical concentration and have the highest optical efficiency. The biggest individual dish units built to date are limited to about $500\,kW_{th}$, and as yet they have not been deployed commercially so it is difficult to accurately assess costs. Applying dish systems to chemical processes requires replicating chemical reactors for every receiver to achieve scale. Such an approach is likely to rule out some of the more complex processes that could be contemplated on a tower system. A field of dish receivers supplying thermal energy to a central processing facility is a possibility; however, this would also increase overall complexity and cost of the system [6a].

Concentrated Solar Fuels (CSF) generally require much higher peak temperatures to drive the endothermic (energy-consuming) reactions that enable the storage of solar energy in chemical form than is required for electricity production from steam cycles. As the latter is currently the primary deployment of CSP technologies, some further refinement of concentrating solar technologies will assist but is not essential for solar fuels production.

While the most widely deployed CSP technology currently is parabolic troughs, these are not expected to be useful for many solar fuels applications due to their limited concentration ratio (up to 50–80) and a consequent limit on the peak temperature achievable of around 500–600°C. This is also true for Linear Fresnel collectors.

Towers have higher concentration ratios and therefore can attain higher temperatures. CSP towers are now commercially deployed, but the highest temperature in most commercial applications is still around 600°C due to the thermodynamic cycles used and limitations of the thermal stability of the materials and heat transfer fluids. Open volumetric air receivers have been demonstrated at precommercial scale as high as 700°C and are commercially available.

Existing tower-mounted receivers are typically externally irradiated, which enables surround fields to be used. Beyond a receiver temperature of around 700°C, an externally irradiated receiver loses too much energy through radiative and convective losses and an alternative design is required.

Cavity-type receivers are likely to be required for the highest temperature applications. Achieving higher temperatures becomes a balance between the design of the field and receiver, as the thermal losses from reradiation increase to the fourth power of the temperature difference between the irradiated surface and the surrounds. Different receiver types are discussed in the following sections.

2.2 Solar Thermal Receiver Concepts

2.2.1 Cavity Receivers

In these receivers, the active surface is shielded from the surroundings by partially enclosing it in some form of housing or cavity. Concentrated solar energy enters the receiver through an aperture [7]. Radiation reflected from the walls primarily impinges upon other walls so that only a small amount of the irradiated power is lost by reradiation. Additionally, the fact that the cavity faces primarily downwards reduces convective heat losses and maintains a higher temperature within the cavity. Large cavity receivers require a sector field and relatively high tower height to field radius compared to externally irradiated receivers. Multiple cavities can be installed on a tower making the use of round- or at least multieye fields possible. Therefore, even installations close to the tropics can be efficiently irradiated.

2.2.2 Windowed Receivers

Even higher temperatures can be realized by receivers with cavity apertures sealed with a transparent window. This also opens up the possibility to use gas loops at increased (or reduced) pressures. For example, compressed air can be heated up to be directly used in a gas turbine for power production. Such systems have already been demonstrated at >1000°C [8]. Secondary optics are generally necessary to achieve such high temperatures efficiently. Although secondary optics will result in additional losses (optical and cooling), they reduce the size of the aperture, and therefore reradiation losses, and increase the optical concentration ratio. The windows are made from quartz and need to be dome shaped to withstand the elevated pressures. Even though cavity receivers with particle clouds employed as absorbing media have been proposed [7], in most cases, the heat transfer is realized in porous structures that can be made from metals or ceramics depending on the temperature. The front surface of the structures can in theory be cooler than further into the receiver elements, leading to lower reradiation losses. Such receivers have also been developed for reforming of natural gas. They were operated at $>800°C$, and 15 bar with a thermal power of up to $400\,kW_{th}$. The technology is not commercially available yet. The issues of the technology are mostly related to the window itself. The mechanical strength as well as the sealing capability of the window deteriorates at higher temperatures and therefore the window has to be cooled. It is also limited in size, and therefore multireceiver systems would be required for industrial-scale installations.

2.2.3 Beam Down Receivers

An alternative type of solar concentrating system, the reflective tower or the beam-down system has been developed. It consists of a heliostat field, the "reflective tower," and a ground receiver equipped with/without a secondary concentrator (CPC). The optical path of the reflective tower comprises the heliostat field illuminating a hyperbolic or elliptical reflector. The reflector is placed on a tower below the target point of the field. The upper focal point of the hyperboloid or elliptical reflector coincides with the target point of the field. The reflector directs the beams downward. On the ground, secondary concentrators of the CPC-type can be used to further enhance the concentration of the solar energy. The concentration factor for beam-down concentration systems is in the range of 1000 to 10,000, and the receiver/reactor on the ground can achieve temperatures in excess of $1300°C$. A $300\text{-}kW_{th}$ system with a hyperbolic reflective tower was built at the Weizmann Institute of Science (WIS) in Israel. Recently, $100\text{-}kW_{th}$ and $250\text{-}kW_{th}$ beam-down systems with elliptical reflectors were built in Miyazaki [9] and Nagano, respectively, in Japan.

Advantages of the beam-down system or beam-down receiver/reactor are:
- Heavy or large receiver/reactor at ground level rather than supported on a tower
- No supply of materials (liquid, coal, slurry, etc.) to tower top, eliminating large pumping energy
- Eliminating heat loss from long pipes to/from tower top
- Reduced heat loss by wind
- CPC in beam-down system can receive solar energy from a larger heliostat field compared with that in conventional tower system
- Solid-phase reactants will settle at the bottom of the receiver and reduce the risk of window fouling
- Ideal for liquid-phase and packed bed particle reactors.

But disadvantages are:

- Secondary mirror introduces extra optical losses and also has the effect of increasing the image size
- Secondary reflector & CPC make construction cost higher than conventional tower system
- Higher convective heat loss of cavity receivers for an upward facing cavity (but "windowed" cavity receivers can solve this problem)
- Possible deformation of secondary reflector caused by wind load
- Difficulty to maintain secondary reflector clean.

2.2.4 Hybridization

As distinct from the idea of hybrid solar fuels through the use of fossil fuel reactants or electricity, hybridization of the heat source (i.e., providing heat from combustion of a fuel) for the endothermic reactions can potentially be employed to increase utilization of the plant. The option to hybridize may be an important consideration; this will be dependent on the peak temperature and whether an alternative energy supply can achieve the necessary temperature.

3 SOLAR THERMAL FUELS

We will now briefly review the potential fuels options and the required precursors, as there are a number of possible fuels and they have different feedstock requirements.

3.1 Potential Solar Fuels

Hydrogen can be used as a transport fuel by fuel cell vehicles, or used to generate electricity either using fuel cells or thermally. This technology is developing rapidly in some countries (e.g., Germany, Scandinavia, Japan, Korea, and the USA), with Japan in particular openly promoting the transition to a hydrogen economy as part of its long-term energy policy [10, 11]. While a clean burning fuel, hydrogen is a disruptive technology with a number of challenges around production, storage, transportation, and delivery. Hydrogen should not be dismissed as a longer-term option for transport, however.

Syngas (a mixture of hydrogen and carbon monoxide) can be used as a fuel directly by burning in a gas turbine. Syngas is not practical for transport applications (i.e., aircraft, automobiles) due to the need to compress the gas for storage, and has a comparatively low fuel density compared to conventional hydrocarbon fuels. A gas turbine can be used to generate electricity, however.

Syngas is also readily converted to pure hydrogen by reacting with water to "shift" the carbon monoxide to carbon dioxide, which can be separated.

Synthetic fuels can also be produced from syngas using processes such as Fischer-Tropsch synthesis (FT) and the Methanol-to-Gasoline process, as described in detail elsewhere in this book. Hydrocarbons are attractive fuels as they have very high energy density, are liquids at room temperature, and are compatible with existing infrastructure. Methanol is also an interesting possibility, as it is both a fuel and a commodity chemical widely used in plastic production.

Ammonia is similar to methanol in that it can be both an energy carrier and a commodity chemical. Ammonia is produced today on a massive scale for fertilizer and explosive production using the well-known Haber-Bosch process. Ammonia is relatively easy to transport with conventional infrastructure, but public acceptance may be difficult due to odor and concerns over toxicity.

3.2 Feedstock Requirements for Different Fuels

The feedstock requirements are quite different for different fuels/commodities:

- Hydrogen and ammonia—these only require hydrogen, although both can be produced from syngas as is done conventionally. Hydrogen as a fuel requires purification to remove contaminants such as H_2S and CO, which can poison fuel cells, as well as compression for transport and delivery, which is at high pressure: usually at 350 or 700 bar.
- Hydrocarbons and methanol—synthesis of these fuels requires both hydrogen and a source of carbon. Typically a mixture of H_2 and CO is used, as CO_2 is less reactive than CO. The efficient synthesis of a single product requires a rather exact stoichiometry, but it is possible also to produce multiple products. FT synthesis, covered elsewhere in this book, is particularly suited to polygeneration as the synthesis reaction produces a spectrum of products that can be separated into different final products, e.g., lubrication waxes and transport fuels.

4 SOLAR FUEL PRODUCTION USING CARBON-BASED FEEDSTOCKS

These routes can be split into three major categories: reforming, thermal cracking, and gasification. The first two employ natural gas (methane) as the carbon-based feedstock, whereas the third can process organic feedstocks like coal, biomass, or wastes in the form of solids or slurries.

4.1 Methane Reforming

There are several options for the methane reforming process: steam methane reforming (SMR) and CO_2 (or dry) methane reforming (DMR), represented by the following two equations respectively:

$$CH_4 + H_2O \leftrightharpoons 3H_2 + CO \qquad \Delta H^0_{298\,K} = +206\,kJ/mol$$

$$CH_4 + CO_2 \leftrightharpoons 2H_2 + 2CO \qquad \Delta H^0_{298\,K} = +247\,kJ/mol$$

The two pathways can also be combined in so-called mixed reforming (reaction of methane with a mixture of water and CO_2). The CO in the syngas can be shifted to H_2 via the catalytic water-gas shift reaction ($CO + H_2O \leftrightharpoons H_2 + CO_2$) and the product CO_2 can be separated from H_2 via, e.g., pressure swing adsorption.

Both reactions are highly endothermic and hence favored by high temperatures (industrial reforming processes are carried out between 800°C and 1000°C)—which, though in the solar thermal area, are not among the highest relevant to solar fuel production routes and are considered moderate. The idea in solar reforming is to use concentrated solar thermal energy to provide the heat for the endothermic reforming reaction, embodying solar energy in the product hydrogen or syngas. In conventional reformers, this heat is supplied by combustion of additional natural gas and process waste gas from the downstream hydrogen purification step consuming between 3% and 20% of the feedstock natural gas as fuel [12].

The relevant solar research efforts worldwide have been reviewed in detail recently [13]. Solar reforming was one of the first solar fuel pathways examined, since the early 1980s [14, 15], by the US Naval Research Laboratory that operated a solar tubular reformer in the context of proposing the CO_2/CH_4 reforming-methanation cycle as a mechanism for converting and transporting solar energy. As research efforts evolved worldwide, the process has been extensively studied. Both indirectly and directly solar-heated reactors have been employed. In the former case, the solar receiver heats either the external surface of catalyst-containing tubes within which is fed the reactant gas or a heat transfer fluid, which transfers its enthalpy to the reformer reactor downstream (allothermal heating). Air, sodium (Na) vapors [16], molten salts [17, 18], as well as solid particles [19, 20] have all been proposed and tested as heat transfer fluids for allothermal heating. The concept of allothermal heating via solar-heated air, which is technically the simplest, was first tested by CIEMAT, Spain, and the German Aerospace Research Centre (DLR), Germany, in the late 1980s and early 1990s [21] on

Plataforma Solar de Almeria (PSA), Spain, where solar-heated air (at 980°C and 9 bars) from a receiver was employed to drive a separate steam reformer; the latter achieved final CH_4 conversions between 68% and 93% depending on reaction temperature. Solar reforming of methane by CO_2 in an externally irradiated 480-kW_{th} tubular reformer has been studied at WIS on a solar central receiver [22]; a scaled-up 50-MW_{th} beam-down solar reforming system was designed [23] but never built and tested.

Work in Australia on solar methane reforming has been conducted by the Commonwealth Scientific and Industrial Research Organisation (CSIRO) since 1999, originally in a 25-kW_{th} single-coil reformer (SCORE) coupled to a 107-m^2 dish concentrator and employing two concentric tubes: the catalyst was packed in between the inner and the outer, solar-irradiated tube, whereas the inner tube was for countercurrent heating of the feed water stream (Fig. 1). Production of solar-enriched fuels and hydrogen via steam

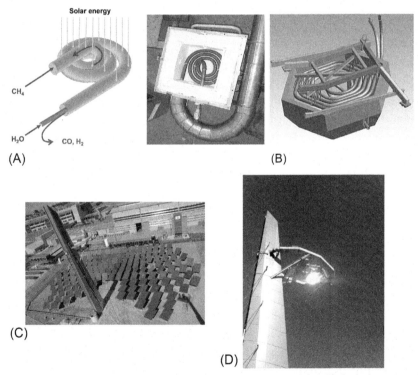

Fig. 1 CSIRO's reformer technology development: (A) the single-coil solar reformer and its operating principle; (B) the double-coil reformer; (C) the single-tower, 500 kW_{th}, heliostat field at Newcastle, Australia; (D) the reformer on the tower during solar operation [24].

reforming of natural gas at temperatures and pressures up to 850°C and 20 bars was demonstrated. Research continued on scaled-up reactors reforming natural gas at 10 bar at 850°C with an absorbed power of $200 kW_{th}$, with outer receiver dimensions of about 2.5 m [25, 26].

Reforming using directly irradiated receivers/reactors has been implemented on reactors/receivers made of porous ceramic honeycombs and foams capable of absorbing concentrated solar irradiation—also called and "volumetric" receivers since they enable the concentrated solar radiation to penetrate and be absorbed within the entire volume of the absorber—and coated with steam-reforming catalysts. In fact, such reactors based on ceramic foams were the first structured reactors to be tested for solar-aided methane reforming by DLR and SANDIA, USA. [27].

Within the successor project SOLASYS between the Weizmann Institute of Science (WIS) in Israel and DLR, receivers were developed for a power input of up to $300 kW_{th}$ [28, 29]. Eventually, within the project SOLREF, an advanced and more compact and cost-effective volumetric receiver/reformer was designed to operate at higher power, pressure, and temperature levels ($400 kW_{th}$, 950°C, 15 bar, respectively) resulting in higher efficiency. The advanced solar reformer was tested and validated under real solar conditions at the WIS, and remains until today as the largest reactor developed and operated (Fig. 2). Structured directly irradiated reactors with fins have been also developed by WIS researchers and tested with respect to CO_2 methane reforming on their solar tower [30, 31].

Large-scale solar thermal reforming will most likely require the economies of scale offered by heliostat fields with central tower receivers that can generate solar thermal fluxes in the MW capacity. The reactor technology of choice will be one, which can be deployed cost effectively at large scale and based on such criteria, "technically simpler" concepts even of lower efficiency, might be more attractive for large-scale implementation and demonstration of the technology (e.g., tubular, indirectly heated, "allothermal" reformers).

4.2 Methane Cracking

Another approach proposed as a carbon dioxide emissions-free route to hydrogen is that of purely thermal decomposition of methane (methane cracking) to hydrogen and elemental carbon as per the reaction below [32]:

$$CH_4 \rightarrow C + 2H_2 \quad \Delta H^0_{298K} = +75.6 \, kJ/mol$$

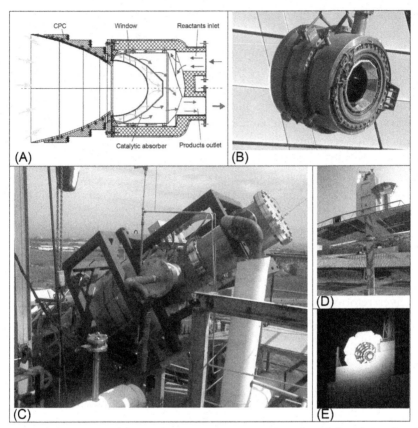

Fig. 2 The "evolution" of the SOLASYS- SOLREF ceramic foam-based directly irradiated solar steam methane reformer: (A) sketch of operation principle; (B) actual reactor; (C) reactor installed on the solar tower of the WIS, Israel; (D), (E) irradiated reformer in operation. *(Images: Courtesy of DLR.)*

This process, being purely thermal and noncatalytic, requires higher temperatures than reforming (in excess of 1500°C); these have been achieved experimentally by either the use of "seeding" dilute carbon particle streams as direct solar absorbing media or by indirectly heated reactors through graphite reactor walls. These indirect "hot–wall" reactors naturally required even higher temperatures imposing further technical and material difficulties.

This route was investigated experimentally by University of Colorado, USA. [33]; ETH, Switzerland [34]; WIS, Israel [35]; and more extensively by Laboratoire PROcédés, Matériaux et Energie Solaire PROMES, France

[36–38] in both types of reactors: carbon-particles-aerosol reactors, some-times in combination with "hot walls." The carbon product can in theory be a useful side product, but in practice has been found not suitable for the carbon black market and sometimes containing toxic polyaromatic hydro-carbons (PAHs). The overall process economics depend strongly on either the ease of disposal or the price and hence the quality of the produced carbon black powder [39]. Neither of them has been proved as yet satisfactory; in combination to the problems related to high temperatures, embarking on process scale-up does not seem likely in the near future.

4.3 Biomass/Coal/Industrial and Municipal Wastes Gasification

Gasification is a process in which carbonaceous materials are reacted with a controlled amount of oxidant like oxygen, CO_2 and/or steam at high tem-peratures ($>700°C$) to produce CO and H_2 (and CO_2 as a by-product). Steam Gasification involves the reaction of steam with solid carbonaceous feedstocks such as coal, coke, biomass, or carbon-containing wastes to pro-duce syngas. The organic component of biomass contains carbon, hydrogen, oxygen, nitrogen, and sulfur; thus, the general highly endothermic steam gasification reaction for stoichiometric water delivery can be written as:

$$C_1H_xO_yS_uN_v + (1 - y) H_2O \rightarrow (x/2 + 1 - y - u) H_2 + CO + u H_2S + v/2 N_2 \quad (3)$$

where x, y, u, and v are the elemental molar ratios of H/C, O/C, S/C, and N/C in the feedstock. Depending on the operating conditions, components such as tars and other volatile organics may also be formed, especially when using coal. Just like reformers, conventional gasifiers use a portion of their product or input stream to drive their reaction, which affects adversely the overall process efficiency. Thus, if this biomass has been obtained from purely renewable sources its gasification via solar thermal energy can lead to emissions-neutral syngas. The ideal syngas composition can be obtained only if the thermodynamic equilibrium above $900°C$ is reached; hence, solar gasifiers have been designed for operation accommodating a range of appli-cations temperatures near $1000°C$ [40].

This reaction involves solid reactant particles; thus, just like the previous case of methane cracking, both directly or indirectly irradiated solar reactors have been used, very often from the same research groups that explored methane cracking. In the first category, the biomass/coal particles are them-selves the solar-absorbing media; in the second, graphite plates or tubular reactor walls are solar-heated and the heat is conducted, radiated, and convected to the particles/slurry/gaseous feedstock. In either case, the particulate feedstock can be stationary in a packed bed reactor configuration or moving, in for example, fluidized bed, spouted bed, "drop-down" (moving bed) or entrained flow/vortex reactor configurations. Irrespective of whether directly or indirectly irradiated, packed bed solar reactors consist of a cavity receiver filled with a batch of feedstock which receives the concentrated solar power directly through a window or indirectly through an emitter plate or the reactor's hot wall. On the one hand, such reactors offer high reaction extent due to the presence of a large quantity of feedstock particles absorbing the radiation and the long solid residence time and can treat a large variety of carbonaceous feedstocks of different compositions. On the other hand, their scale-up involves some technical challenges: ash build-up, temperature gradients, and nonhomogeneous reactions (mass and heat transfer limitations) can occur due to the thermal inertia of the bed [41]. Hence, since the mid-90s and more recently, interest has moved to fluidized bed solar gasifiers in which the feedstock is continuously suspended with inert particles by a neutral gas and/or the oxidizing agent. Several recent reviews describe in detail the progress made in CSP-aided coal/biomass steam gasification [42–44].

Directly irradiated reactors have been used in conjunction with many "bed" concepts. A packed-bed reactor was employed for the gasification of coal, activated carbon, coke and a mixture of coal and biomass, in probably the earliest relevant work published [45]. A solar fluidized coal-bed gasifier contained in a quartz tube was directly exposed to concentrated irradiation for steam gasification [46]. A spouted bed reactor irradiated via a solar simulator from the top through a transparent window was employed to gasify coal coke with CO_2 at temperatures $>1000°C$ [47] and more recently by PROMES for a variety of wood feedstocks [41]. Vortex flow (or entrained flow) petcoke gasifiers were developed by ETH and scaled up from the 5-kW_{th} laboratory scale [48] up to a 500-kW_{th} quartz-windowed cavity reactor with slurry feeding, tested on the PSA solar tower within the so-called SYNPET project. The same group investigated the use of indirectly heated gasifiers in conjunction with a beam-down system [49]; a stationary, 150-kW_{th}, two-cavity, reactor with the upper one functioning as the solar absorber and the lower, packed bed one functioning as the reaction chamber, was irradiated from above at the solar tower of PSA for the gasification of six biomass types [50]. The packed bed was heated by a directly heated SiC plate separating the two cavities. These reactors are actually the largest-scale solar gasification reactors tested so far.

Indirectly irradiated gasifiers are designed in a variety of ways due to the flexibility afforded by not having to design for a window. The University of Colorado group [51] investigated their concept of multitubular "hot wall" reactors (~1150°) inside which biomass was "dropped-down" (i.e., moving bed reactors). The approach was intended to be commercialized by company Sundrop Fuels, but finally the plan has been abandoned.

5 WATER AND CARBON DIOXIDE SPLITTING

Hydrogen production has been widely studied since the 1970s, although in recent years there has been considerable interest in carbon dioxide splitting as well to provide the precursors for synthetic liquid hydrocarbon fuels.

5.1 Thermal Water Splitting

The most conceptually simple method for hydrogen production is the direct thermal dissociation of water, known as thermolysis.

$$H_2O \leftrightarrow H_2 + \frac{1}{2}O_2$$

While apparently simple, this reaction is in fact extremely difficult to realize for two main reasons. The first is that the process requires very high temperatures to achieve a reasonable degree of dissociation, in excess of 2800°C, and therefore a very high temperature heat source is required [52]. A second, related challenge is in materials, for containing the reaction and equally important to separate the product gases before they recombine [4, 52].

One way of avoiding these issues it to separate the net water splitting reaction into two (or more) subprocesses that can be conducted at lower temperatures. In this concept, a series of redox reactions is used in which an active material is internally cycled between oxidized and reduced forms to split water into hydrogen and oxygen. Such cycles are known as thermochemical cycles and hundreds of potential cycles have been proposed to use thermal energy from concentrated solar radiation or nuclear energy. Some cycles also include an electrolysis step and are termed hybrid or thermoelectrochemical cycles.

5.2 Thermochemical Cycles

Thermochemical cycles have received considerable attention in response to global energy crises and challenges such as the oil shocks in the 1970s.

After a hiatus in the 1980s, interest in thermochemical cycles was revitalized in the 1990s and 2000s after the Kyoto Protocol and climate change debate focused attention on decarbonization of the energy chain.

A lot of research was conducted by the nuclear industry in the 1970s, and again from the 1990s, with a view to hydrogen production. Much of this research assumed the successful development of high-temperature nuclear reactors under the Gen IV program, capable of achieving temperatures of up to 950°C. While the initial development roadmap suggested that suitable reactors would be ready for deployment by 2020, the most recent roadmap suggests a timeframe of 2025 to 2030 is more likely [6]. Currently, therefore, only concentrated solar radiation is a feasible energy source for these cycles.

A number of screening programs were conducted alongside the various resurgences in interest in thermochemical cycles. General Atomics, a prominent developer of nuclear technologies, conducted a comprehensive screening analysis for hydrogen production from nuclear power for the US DOE in 2003 [53]. Subsequent interest in solar thermal hydrogen generation led to the STCH program, also funded by the US DOE, which sought to identify the most prominent solar thermochemical hydrogen processes as part of an ongoing merit review process to determine the best allocation of government funding [54]. Other activities include the EU-funded framework project INNOHYP-CA, which was created to identify the most prospective processes for massive hydrogen production for Europe [55]. This work was succeeded by the International Energy Agency's Hydrogen Implementing Agreement (IEA HIA) Task 25 for high-temperature processes for hydrogen production, which developed a number of fact sheets (http://ieahydrogen.org/Activities/Task-25.aspx).

The following is a summary of the most prominent cycles, and includes some notes on the current status and challenges associated with the successful implementation of each. Unfortunately, there are significant challenges in intercomparison of production costs due to inconsistencies in the underlying assumptions, and these will not be reported.

5.3 Sulfur-Based Cycles

These cycles are based on the sequential oxidation and reduction of sulfur between its (IV) and (VI) oxidation states:

$$SO_2 \text{ or } H_2SO_3 + \frac{1}{2}O_2 \leftrightarrow SO_3 \text{ or } H_2SO_4$$

The common reaction in all sulfur cycles is the high-temperature disso-ciation of SO_3 at around 850 to 950°C to produce SO_2 and oxygen. The SO_2 is then used in a subsequent step to remove the oxygen from a water molecule to produce hydrogen and sulfuric acid, which is then recycled to the dissociation step. These cycles are not applicable for CO_2 splitting. They have been described in detail in a recent review [56].

5.3.1 Sulfur-Iodine (General Atomics) Cycle

This cycle, also known as the SI or I-S cycle, was developed by General Atomics in the US in the 1970s and is one of the most extensively studied of all thermochemical cycles. The cycle has 3 basic reactions:

Bunsen reaction:

$$SO_2 + I_2 + 2H_2O \rightarrow 2HI_{(aq)} + H_2SO_{4\,(aq)} \quad 120\,^{\circ}C$$

HI decomposition:

$$2HI \rightarrow H_2 + I_2 \qquad\qquad\qquad 320\,^{\circ}C$$

H_2SO_4 decomposition:

$$H_2SO_4 \rightarrow SO_3 + H_2O \qquad\qquad 450\,^{\circ}C$$

$$SO_3 \rightarrow SO_2 + \frac{1}{2}O_2 \qquad\qquad 850 - 950\,^{\circ}C$$

There are a number of challenges associated with this cycle. The Bunsen reaction requires excess iodine to achieve an acceptable conversion of SO_2 to H_2SO_4, resulting in a considerable inventory of iodine in the cycle. This causes some thermal inefficiency as iodine is a relatively heavy element and additional thermal inventory is associated with the excess iodine. The thermal decomposition of HI is also problematic, as an azeotrope is formed between HI and H_2O, which prevents full decomposition, although alternative reactors and membrane separators have been proposed to overcome this.

Despite decades of research and considerable progress in overcoming many of the issues, the most recent developments have not been favor-able. A major international project was developed between researchers from the French Atomic Energy Agency (CEA) and US researchers at Sandia National Laboratories and at General Atomics in San Diego. The three partners each constructed a skid-mounted plant to carry out one of the three reactions, and the intention was to join the three sections into an integrated pilot plant producing 75 L per hour of hydrogen. While each section was able to operate successfully in an independent mode, integration proved extremely challenging and the project was finalized without a successful outcome [57]. Issues included operating all sections continuously due to iodine flow challenges, the complexity of the reactive distillation step, and materials challenges in the HI section. A number of studies by CEA and Sandia among others have developed unfavorable estimates of the projected cost of hydrogen from the SI cycle, compared to alternative cycles. These factors, together with the delays in development of high temperature reactors, have led to CEA and the US DOE suspending work on the SI cycle, and the decommissioning of the ILS plant.

5.3.2 Hybrid Sulfur (Westinghouse) Cycle

The Hybrid Sulfur (HyS) cycle was developed by the Westinghouse Electricity Corporation in the US and has also been extensively studied. Indeed, the HyS and SI cycles are probably the most widely researched cycles around the world. Key organizations that have been involved in research into this cycle include Westinghouse, CEA, and Savannah River National Laboratories (SRNL) in the US. The chemistry of the cycle is simpler than the SI cycle, with two basic reactions. These are the SO_3 decomposition described, and an electrolysis step in which SO_2 is oxidized at one electrode to produce sulfuric acid while protons are reduced to form hydrogen at the other:

Anode reaction:

$$H_2O + SO_{2(aq)} \rightarrow SO_{3(aq)} + 2H^+ + 2e^- \qquad 85°C$$

Cathode reaction:

$$2H^+ + 2e^- \rightarrow H_{2(g)} \qquad 85°C$$

The advantage of the HyS cycle compared to conventional alkaline electrolysis is that the cell potential required is greatly reduced through the presence of SO_2. This reduction in potential reflects the stored chemical energy in the highly endothermic SO_3 decomposition step. This translates to a theoretical potential for water splitting of 0.158 V compared to 1.23 V for the electrolysis of pure water [58]. In practice, additional overpotential is needed to overcome kinetic and mass transfer limitations to achieve an acceptable current density (and hence economic production rate of hydrogen per unit area of electrolyzer). Efficient electrolyzer operation requires elevated temperature and pressure to reduce the overpotential, which poses significant materials challenges for electrolyzer components and catalysts as the solution needs to be at high concentrations (about 50% H_2SO_4) to minimize the thermal load in acid concentration prior to the decomposition step [59].

The electrolyzer was intensively studied at SRNL between 2004 and 2009 [60] with the aim of leveraging significant advances in proton exchange membrane (PEM) technology in recent years. The electrolyzer consists of two separate compartments separated by a membrane, which allows protons to cross over to complete the electrical circuit, but ideally prevents sulfur dioxide reaching the cathode where it would be reduced to elemental sulfur. In fact, sulfur dioxide crossover remains a significant challenge as elemental sulfur was observed within the MEA. The cell was operated for about 200 h at a current density of and mA/cm^2 with a relatively stable voltage of 760 mV, which is above the program goal of 0.6 V.

Westinghouse developed a bayonet reactor for the H_2SO_4 decomposition with a nuclear heat source, which is similar in concept to that proposed for the SI cycles [61]. However, plans to demonstrate the cycle on Integrated Laboratory Scale (ILS) were suspended indefinitely with the loss of interest in hydrogen production from nuclear energy.

Despite some reduction in interest from the nuclear industry, efforts continued in Europe to develop key cycle components in several projects. DLR successfully demonstrated the evaporation and dissociation of sulfuric acid as part of the HYTHEC project between 2004 and 2007 [62]. A subsequent EU project, HycycleS, investigated the materials and reactor components required to handle the process streams and conduct the reactions [63]. These projects confirmed the technical feasibility of SO_3 splitting, although further development is required for scale up of the solar reactor/receivers.

5.3.3 Sulfur-Bromine (ISPRA Mark 13) Cycles

A number of cycles were developed by the European Joint Research Centre (JRC) at Ispra in Italy utilizing bromine rather than iodine in the SO_2 oxidation step.

Bromine Bunsen reaction:

$$SO_2 + Br_2 + 2H_2O \rightarrow 2HBr_{(aq)} + H_2SO_{4\,(aq)} \quad 120\,^{\circ}C$$

HI decomposition:

$$2HBr \rightarrow H_2 + Br_2 \quad \text{electrochemical}$$

The JRC conducted a series of bench-scale experiments on various thermochemical cycles, which included an evaluation of the comparative merits of the ISPRA Mark 13 and SI cycles (ISPRA Mark 16). The bromine Bunsen reaction is easier than its iodine analogue because the free energy of formation of HBr is higher. This greatly simplifies the first reaction in which SO_2 is oxidized, which is very difficult in the SI cycle due to separation of HI from the excess iodine. However, the decomposition of HBr to form hydrogen is correspondingly more difficult and needs to be conducted electrochemically. The JRC concluded that the Mark 13 cycle was more prospective than the SI cycle, and commissioned a pilot plant with glass and quartz equipment at a scale of $100\,L\,h^{-1}$ in 1978 [64]. This pilot plant operated well, and flow sheet evaluations suggested that a thermal efficiency of around 39% was possible. This was comparable to similar estimates for the HyS cycle of 41%. The production cost of nuclear hydrogen was estimated to be around three times the price of natural gas.

5.4 Volatile Metal Oxides

In these cycles, the oxide is reduced to produce a metal vapor, which is quenched and reacted with steam to produce hydrogen. This has some positive aspects in terms of the reactor configuration as the vapor can be readily removed from the reaction chamber.

5.4.1 Zinc/Zinc Oxide

This is the most widely studied of the volatile metal oxide cycles. The most prominent research groups in the field are the Swiss groups led by A. Steinfeld at ETH Zurich and the Paul Scherrer Institut (PSI), and their collaborators in WIS in Israel. Some work has also been conducted at the University of Colorado. The chemistry of the cycle is quite simple:

Decomposition of ZnO:

$$ZnO \rightarrow Zn + \frac{1}{2}O_2 \qquad 2000\,^{\circ}C$$

Hydrolysis of Zn:

$$Zn + H_2O \rightarrow ZnO + H_2 \quad 450\,^{\circ}C$$

A $10\text{-}kW_{th}$ rotating reactor receiver was been built and tested at PSI and has successfully demonstrated the zinc oxide decomposition step [65]. In this reactor, the walls are lined with zinc oxide powder and irradiated through a quartz window with concentrated solar radiation. Zinc vapor from the decomposition is extracted from the reactor and quenched at extraordinarily high cooling rates to prevent recombination with oxygen. The finely divided metallic powder can be used for zinc batteries or for the production of hydrogen by hydrolysis. A larger reactor was subsequently constructed and evaluated in a high flux solar furnace and also at the solar furnace at Odeillo in France [66, 67].

There are a number of challenges with this cycle. Firstly, the extremely high temperature required for the decomposition requires high precision optics to achieve the necessary concentration of solar radiation, and it is unclear if this is achievable on a tower field even with a secondary concentrator. Secondly, the high quench rates require a large volume of cool gas, and this will affect the energetics of the cycle by making heat recuperation difficult [68]. Finally, closing the cycle for hydrogen production is also challenging, as the ZnO for the reactor needs to be free flowing and finely divided.

An interesting variation of this cycle is the use of carbon to reduce the peak temperature by using a carbothermal reduction of the zinc as per the following equation:

$$ZnO + C \rightarrow Zn + CO \quad 1200\,^{\circ}C$$

This concept was successfully demonstrated at a 300-kW_{th} pilot scale on the WIS tower under the European framework project SOLZINC, using charcoal from biomass to avoid CO_2 emissions [69, 70].

5.4.2 Cadmium/Cadmium Oxide

This cycle is similar to the zinc cycle described here in that the decomposition product is a metal vapor, which must be quenched and then reacted with water to produce hydrogen:

Decomposition of CdO:

$$CdO \rightarrow Cd + \frac{1}{2}O_2 \quad 1450 - 1500\,^{\circ}C$$

Hydrolysis of Cd:

$$Cd + H_2O \rightarrow CdO + H_2 \quad 25 - 350\,^{\circ}C$$

While the peak temperature is much more amenable to concentrated solar radiation, the challenge with this cycle is that the hydrolysis reaction is very slow below the melting point of cadmium (321°C). An alternative process based on cadmium electrolysis and decomposition of the cadmium hydroxide has also been proposed [71]:

Electrolysis of Cd:

$$Cd + 2H_2O \rightarrow Cd(OH)_2 + H_2 \quad 50^{\circ}C$$

Decomposition of $Cd(OH)_2$:

$$Cd(OH)2 \rightarrow CdO + H_2O \, 375^{\circ}C$$

While thermodynamically attractive, cadmium-based processes have been discounted by some because of the extreme biotoxicity of cadmium metal [72].

5.5 Nonvolatile Metal Oxides

Cycles based on nonvolatile metals and their oxides have obvious advantages over the sulfur- and bromine-based cycles in terms of corrosivity and toxicity of the reactants. There has therefore been considerable interest in a number of thermochemical cycles based on such materials, and especially those based on iron oxides. These cycles also have advantages over the volatile metal oxide cycles because both the reduced and oxidized forms can remain in the solid phase, and thus can be supported on nonmoving, fixed-bed structural reactors (monoliths) such as honeycombs and foams that can function simultaneously as receivers. Naturally, with every benefit there are a number of compromises, as is true of all thermochemical cycles.

5.5.1 Fe₃O₄/FeO

One of the first cycles proposed [73] for solar thermochemical water splitting is based on pure FeO and Fe_3O_4:

$$Fe_3O_{4(l)} \rightarrow 3FeO_{(l)} + \tfrac{1}{2}O_{2(g)} \quad > 1600^{\circ}C$$

$$3FeO + H_2O_{(g)} \rightarrow Fe_3O_4 + H_{2(g)} \quad \text{exothermic}$$

This cycle is possible because Fe_3O_4 decomposes at slightly above its melting point to form a liquid phase consisting of a mixture of FeO and Fe_3O_4 [74]. Thermodynamic calculations indicate that as the conversion increases (beyond 40% or 50%), the calculated vapor pressure of $FeO_{(g)}$ also increases. However, the maximum thermal efficiency was achieved at relatively low conversions between 10% and 20%, and vaporization is not expected to be an issue at these conversions. Overall conversion efficiencies of solar energy to electricity produced by a fuel cell of 20.4% to 25.1% have been estimated, at solar concentration ratios of 5000 and 10,000 times, respectively [74].

This process has significant challenges because of the extremely high temperatures required, and the associated difficulties in obtaining sufficiently concentrated solar radiation entering the reactor (requiring secondary concentration). Researchers have consequently identified a number of systems, which utilize the same basic chemistry but using compounds in which some of the iron is substituted by other transition metals to reduce the peak temperatures, discussed later.

5.5.2 Mixed Ferrites

Mixed ferrites have a composition that can be expressed as $M_x(Fe_{3-x})O_4$, where M represents one or more transition metals such as nickel, zinc, manganese, or cobalt. These compounds generally form a spinel-like structure with the substituting element in the (II) oxidation state, for example $Ni_{0.5}Mn_{0.5}Fe_2O_4$ [75], and can be reduced at temperatures in the range 1100–1400°C to form an oxygen-deficient ferrite $Ni_{0.5}Mn_{0.5}Fe_2O_{4-\delta}$, which can be regenerated through reaction with steam in a lower-temperature hydrogen production step. The two reactions can be expressed as:

Oxygen evolution:

$$Ni_{0.5}Mn_{0.5}Fe_2O_{4(s)} \rightarrow Ni_{0.5}Mn_{0.5}Fe_2O_{4-\delta(s)} + \tfrac{\delta}{2}O_{2(g)} \quad 1100 - 1400^\circ C$$

Water splitting:

$$Ni_{0.5}Mn_{0.5}Fe_2O_{4-\delta(s)} + \delta\,H_2O_{(g)} \rightarrow Ni_{0.5}Mn_{0.5}Fe_2O_{4(s)} + \delta H_{2(g)}$$
$$< 800 - 1000^\circ C$$

There has been considerable interest in these materials due to the comparatively moderate reaction temperatures and the fact that the ferrites remain solid throughout the process. A number of organizations have developed reactor concepts in which the ferrite is applied as an active coating on a supporting ceramic matrix. Some of these reactors are described here:

The *Tokyo Institute of Technology* developed a concept where the active coating was deposited on ceramic plates on the outside of a rotating drum [76]. The plates in the drum were sequentially exposed to concentrated solar radiation to reduce the ferrite, and then to a steam atmosphere where the water-splitting reaction takes place. One of the potential issues with this reactor concept is keeping the hydrogen and oxygen streams separate to avoid potentially catastrophic recombination.

Sandia National Laboratories (SNL) developed a "Counter Rotating Ring Receiver Reactor Recuperator" known as the CR5. This also features a rotating design consisting of ceramic discs, which rotate at a speed of around 1 rpm or less [77]. An interesting feature of this reactor is that adjacent plates rotate in opposite directions to aid internal heat recuperation between the high- and low-temperature reactions. Again, separation of the two gas streams was a significant challenge, as is maintaining the tight clearances required between the plates at elevated temperatures. A subsequent design abandoned the rotating plates in favor of a slowly rotating screw conveyor, which elevates active particles in a bed to an irradiated reduction zone [78]. Reduced particles were conceived to be collected in a central annulus, passing through a series of vacuum chambers before exposure to steam in the hydrogen production step, to improve cycle efficiency [79].

APTL/CERTH, Greece and DLR, developed within the EU-funded HYDROSOL Project series (2002-present) a quasicontinuous hydrogen production concept based on stationary reactors. Two or more reactors are used with the oxygen-producing and water-splitting steps carried out in separate reactors to give a semicontinuous flow of oxygen and hydrogen [80]. This concept was tested at a scale of $100\,kW_{th}$ on the solar tower at PSA of CIEMAT in Spain [81]. Two modules were used with the feed and product gas lines being switched over when changing between modes. The thermal load on each reactor was also changed depending on the mode of operation, as the high temperature thermal reduction reaction at 1200°C and above requires significantly more thermal energy than the water splitting reaction at 800°C. This process was managed by dividing the heliostat field into groups, which provided the basic heat requirements of either reactor plus "switch-over" heliostats that provided the extra energy needed for the oxygen producing reaction. Half cycles of 20 to 30 min were used, and hydrogen production was successfully demonstrated, although some degradation of the ferrite was found, as evidenced by falling production rates with cycle number. This was also observed in laboratory-scale experiments and remains an area for ongoing development. Currently, a new scaled-up version of the HYDROSOL reactor to the $750\text{-}kW_{th}$ level has been implemented consisting of three reactors of hemispherical dome cavity configuration including suitable secondary reflectors (Fig. 3). The domes are assembled of foams manufactured entirely of the redox material, nickel ferrite [82]. The hydrogen production campaign is currently in progress.

(A)

(B)

Fig. 3 The new, domed, 750 kW$_{th}$-scale, 3-chamber HYDROSOL reactor installed on the PSA tower: (A) without and (B) with the CPC concentrators, respectively. *(Images: Courtesy of DLR from: DLR. HYDROSOL_Plant: Hydrogen from sunlight. 2017. DLR media release 29/11/2017. Available at: http://www.dlr.de/dlr/en/desktopdefault.aspx/tabid-10081/151_read-25217/#/gallery/29212 [Accessed 19 January 2018].)*

5.5.3 Hercynites

A mixed cobalt ferrite-hercynite system ($CoFe_2O_4/FeAl_2O_4$) has been proposed by the University of Colorado (United States) as a potential redox cycle [83]. The hercynite is produced due to the reaction between $CoFe_2O_4$ and its support material, Al_2O_3. The overall reactions for H_2O splitting can be summarized as:

$$Reduction: \quad CoFe_2O_4 + 3Al_2O_3 \rightarrow CoAl_2O_4 + 2FeAl_2O_4 + \frac{1}{2}O_2 \quad (1)$$

$$H_2O\,Splitting: \quad CoAl_2O_4 + 2FeAl_2O_4 + H_2O \rightarrow CoFe_2O_4 + 3Al_2O_3 + H_2 \tag{2}$$

These reactions could be cycled between 1200°C (reduction) and 1000°C (H_2O splitting) under stable conditions.

The work was extended to CO_2 splitting thus:

$$Reduction: \quad CoAl_2O_4 + 2FeAl_2O_4 + CO_2 \rightarrow CoFe_2O_4 + 3Al_2O_3 + CO \tag{3}$$

Under a total pressure of 100–800 mbar, appreciable amounts of CO were produced after reduction at 1350°C: approximately 100–150 degrees lower than for ceria.

The research group demonstrated that the hercynite cycle can produce H_2 at 1000°C when reduced at 1200°C, with conversions between 14.2% and 18.7% achieved. However, the oxidation of $FeAl_2O_4$ in the presence of steam is less favorable than that for FeO under the same conditions. The group also demonstrated in the laboratory that this cycle could conduct isothermal water splitting, albeit at the somewhat higher temperature of 1350°C, thereby avoiding the need to cycle between reduction temperature and a lower water-splitting temperature [84]. They claim that this has the potential to avoid the thermal and time losses that occur in temperature swing operation, due to the frequent heating and cooling of the metal oxide.

5.5.4 CeO₂/Ce₂O₃

Like the zinc oxide system, the pure cerium oxide system requires a very high temperature for the reduction reaction:

Decomposition of Ce_2O_3:

$$2CeO_{2(s)} \rightarrow Ce_2O_{3(s)} + \frac{1}{2}O_{2(g)} \quad 2000^\circ C, \; 100 - 200\,mbar$$

Water splitting:

$$Ce_2O_{3(s)} + H_2O_{(g)} \rightarrow 2CeO_{2(s)} + H_{2(g)} \quad 400 - 600°C$$

The two reactions for this system have been recently demonstrated on the CNRS-PROMES solar furnace at Odeillo in France [85]. The high-temperature reduction was demonstrated using a pellet of CeO_2, but was only observed at reduced pressures of 100–200 mbar under a nitrogen atmosphere. The duration of the experiment needed to be limited to avoid sublimation of the solid, indicating that volatility is still an issue even with so-called nonvolatile metal oxides—40% to 60% of the CeO_2 was vaporized during the experiments. The hydrolysis reaction (and indeed reaction with oxygen) was found to have rapid kinetics at a temperature of 400°C to 600°C.

The extremely high peak temperature is a concern for practical operation and scale up, as is the relatively large amount of thermal mass that needs to be cycled between the high and low temperatures of the cycle due to stoichiometry—only half a mole of hydrogen is released for every two moles of cerium (IV) oxide decomposed. Doping with other metals has been investigated to overcome issues with sintering [86], with zirconia in particular having been shown to reduce the reduction temperature [87].

There has also been considerable interest in the nonstoichiometric partial reduction of ceria at lower temperatures, at around 1500–1600°C. Despite only partial reduction, it has been claimed that the cycle has comparable productivity per gram of reactant and efficiency to other solid-state thermochemical cycles under investigation [88]. ETH, in collaboration with this group, developed a solar reactor based on reticulated porous ceramic foams manufactured entirely from ceria [89, 90]. Using such foams in a 4-kW_{th} receiver, they performed a total of 291 stable redox cycles yielding 700 standard litres of syngas of composition 33.7% H_2, 19.2% CO, 30.5% CO_2, 0.06% O_2, 0.09% CH_4, and 16.5% Ar, which was compressed to 150 bar and further processed via Fischer-Tropsch synthesis to a mixture of naphtha, gasoil, and kerosene [91] (Fig. 4). Their current work involves the construction of an entirely new 50-kW_{th} solar pilot plant outside Madrid, Spain, incorporating reactor design and operation improvements that will produce liquid fuels via Fischer-Tropsch synthesis of solar produced syngas [92].

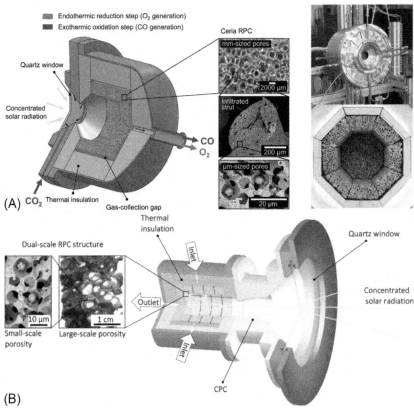

Fig. 4 Technology of ETH's ceria-based solar cavity reactors for two-step, solar-driven thermochemical production of fuels: (A) schematic of the experimental setup, featuring the main system components of the production chain to solar kerosene from H_2O and CO_2; (B) schematic and actual photographs of the solar reactor configuration with the cavity receiver containing a ceria-made reticulated porous ceramic (RPC) structure with dual-scale porosity. ((A) Reproduced with permission from Marxer DA, Furler P, Scheffe JR, et al. Demonstration of the entire production chain to renewable kerosene via solar-thermochemical splitting of H_2O and CO_2. Energy Fuels 2015;29(5):3241–50, Copyright (2015), American Chemical Society; (B) Reproduced with permission from Marxer D, Furler P, Takacs M, Steinfeld A. Solar thermochemical splitting of CO_2 into separate streams of CO and O_2 with high selectivity, stability, conversion, and efficiency. Energy Environ Sci 2017;10:1142–9, Published by The Royal Society of Chemistry.)

5.5.5 Perovskites

While the ceria cycle has become the benchmark for the nonvolatile metal oxide cycles, recent efforts have concentrated on finding new redox materials capable of decreasing the reduction temperature while increasing both the per-cycle H_2/CO yields and thermal stability. Nonstoichiometric perovskites have been identified, principally by research groups at ETH/PSI and SNL, as being potentially superior to ceria in all of these respects.

Earlier work at the Chemical Process Engineering Research Institute, CERTH, Thessaloniki (Greece), demonstrated the technical feasibility of using nonstoichiometric perovskites ($Sr_xLa_{1-x} Mn_yAl_{1-y}O_3$ ($x = 0.4$ or 0.6; $y = 0.4$ or 0.6) for water splitting [93]. Since then, researchers at SNL reported that $LaAlO_3$ doped with Mn and Sr ($La_{0.6}Sr_{0.4}MnO_{3-\delta}$ perovskite) will efficiently split both H_2O and CO_2. They reported H_2 and CO_2 yields that were nine and six times greater, respectively, than those produced from ceria when reduced at 1350°C and reoxidized at 1000°C. The temperature at which O_2 begins to evolve from the perovskite is around 300 degrees lower than that for ceria. The materials also exhibited good thermal stability, maintaining their redox activity over 80 CO_2-splitting cycles [94].

Thermodynamic analysis by ETH/PSI for cycles based on similar perovskite materials has highlighted the effect of temperature and reduction/oxidation thermodynamics on overall cycle efficiency [95]. At a reduction temperature of 1527°C and oxidation temperature of 727°C, the theoretical efficiency of the perovskite cycle (16%) is lower than that of ceria (22%). However, if the reduction temperature is reduced to 1327°C, while still maintaining an oxidation temperature of 727°C, the efficiency of the perovskite cycle (17%) was then greater than that for ceria (13%). This is largely due to the greater degree of reduction of the perovskite at the lower temperature, and its lower sensible heat penalty (due to its higher specific heat) in cycling between reduction and oxidation.

Chinese researchers have tested perovskites of the type $La_xA_{1-x}Fe_yB_{1-y}O_3$ ($A = Sr, Ce, B = Co, Mn; 0 \leq x, y \leq 1$) dispersed in three different commercial support materials (ZrO_2, Al_2O_3 and SiO_2) for the CO_2-splitting reaction. The type of support induced great differences in reaction performance, with a 25-wt% SiO_2 support giving the highest activity [96].

6 OTHER LOW-TEMPERATURE CYCLES

6.1 UT3

The UT-3 cycle is another cycle that has been extensively researched since the 1970s, when it was proposed by the University of Tokyo. The cycle consists of four solid-gas reactions [97]:

$$CaO_{(s)} + Br_{2(g)} \rightarrow CaBr_{2(s)} + \frac{1}{2}O_{2(g)} \quad 500 - 600^{\circ}C$$

$$CaBr_{2(s)} + H_2O_{(g)} \rightarrow CaO + 2HBr_{(g)} \quad 700 - 750^{\circ}C$$

$$Fe_3O_{4(s)} + 8HBr_{(g)} \rightarrow 3FeBr_{2(s)} + Br_{2(g)} \quad 200 - 300^{\circ}C$$

$$3FeBr_{2(s)} + 4H_2O_{(g)} \rightarrow Fe_3O_{4(s)} + 6HBr_{(g)} + H_{2(g)} \quad 550 - 650^{\circ}C$$

The cycle has some attractive features given the relative ease of separating gaseous and solid reaction products and the relatively low peak temperature compared to other cycles. However, the hydrolysis reaction temperatures are close to the melting points of $CaBr_2$ and $FeBr_2$, leading to problems with diminishing reactivity due to melting and sintering of the solid phases [98]. The process was operated at bench scale in a pilot plant named MASCOT (Model Apparatus for Studying Cyclic Operation in Tokyo) designed for a production rate of 3 L/h of hydrogen [99]. As well as the issues of sintering of the solids, there is some debate over the efficiency possible with this cycle. Earlier (1996) estimates using ASPEN-PLUS indicated a thermal efficiency of 49.5% would be possible using concentrated solar radiation [100]. However, later evaluations that considered some of the process difficulties suggested that the process efficiency was more likely to be in the range of 12%–22.5%, with the upper estimate requiring the development of separation membrane technology [98, 101]. It is therefore debatable as to whether the UT-3 cycle offers significant benefits for hydrogen production over conventional electrolysis.

6.2 Hybrid Copper Chloride

This cycle was originally proposed in 1978 as a method to reduce the cell voltage for hydrogen production from water electrolysis by coupling with thermochemical reactions [102]. The initial reaction set proposed is as follows:

$$2CuCl + 2HCl \rightarrow 2CuCl_2 + H_2 \quad \text{Electrolysis, room temp}$$

$$2CuCl_2 + H_2O \rightarrow 2CuCl + 2HCl + \frac{1}{2}O_2 \quad > 600^{\circ}C$$

The second reaction is more precisely expressed by the following subreactions:

$$2CuCl_2 \rightarrow 2CuCl + Cl_2 \quad > 600^{\circ}C$$

$$Cl_2 + H_2O \rightarrow 2HCl + \frac{1}{2}O_2 \quad > 600^{\circ}C$$

The voltage required to run the electrolysis was reported to be around 0.6–1.0 V, comparable with the hybrid sulfur process. More recent development of the cycle has been conducted at Argonne National Laboratories using a modified reaction scheme [103]:

$$2CuCl_{(aq)} + 2HCl_{(aq)} \rightarrow 2CuCl_{2(aq)} + H_{2(g)} \quad 100°C, \ 24 \, bar$$

$$2CuCl_{2(s)} + H_2O_{(g)} \rightarrow Cu_2OCl_{2(s)} + 2HCl_{(g)} \quad 400°C$$

$$Cu_2OCl_{2(s)} \rightarrow 2CuCl_{(s)} + \frac{1}{2}O_{2(g)} \quad 550°C$$

All reactions have been demonstrated in proof-of-concept experiments and a conceptual process design has been developed and simulated in ASPEN PLUS, yielding a relatively high efficiency of about 40% [104]. However, the process has a number of significant challenges, and is rather complicated as it involves a number of unit operations where the reactants pass between aqueous and solid phases. For example, the $CuCl_2$ produced in the electrolyzer must be precipitated out of solution before passing to the hydrolysis section. The $CuCl_2$ is then sprayed into a superheated steam environment where it reacts to form a dry solid powder, which needs to flow under gravity into the decomposition reactor. In the reactor, the Cu_2OCl_2 decomposes to form oxygen and molten CuCl, which spills over a weir for return to the electrolyzer feed tank. While an extremely challenging process engineering problem, this cycle has received considerable interest from the nuclear industry as it is the only cycle that looks suitable for current and foreseeable nuclear reactor designs. Recently, some progress was reported by an international team from 5 countries on various aspects of the cycle including materials and system integration [105].

7 CHALLENGES FOR THERMOCHEMICAL CYCLES

Commercial implementation of thermochemical cycles will require them to outperform alternative methods based on renewable energy such as electrolysis for hydrogen production powered by wind energy, photovoltaics, or concentrating solar power. This has been one of the underlying assumptions in the various screening studies, and there is certainly significant potential to outperform electrolysis due to the high potential efficiencies obtainable with thermochemical cycles [106]. However, the practical implementation of thermochemical cycles remains some years off with a number of challenges yet to be overcome [107]. These include:

- Engineering materials: thermochemical cycles require operation at high temperatures and in some cases with corrosive materials. Ceramic materials such as siliconized silicon carbide are capable of withstanding such temperatures, although thermal cycling places addition stress on materials.

- Active materials: most thermochemical cycles require a redox material. Key attributes include a moderate reduction temperature, good kinetics for both oxidation and reduction, and mechanical stability to withstand both thermal cycling and the changes in crystal structure associated with oxygen loss. Doped ceramics and metal oxides appear the most promising materials but this is still an active research area.

- Reactor design: most cycles have a low and high temperature step as it is thermodynamically nonoptimal to drive the reduction and oxidation reactions at the same temperature although some institutions have investigated the use of pressure swing to influence the equilibrium and kinetics. Recuperation of thermal energy—from reduction to oxidation reactions as well as product and sweep gas streams—remains one of the major challenges to improving the thermal efficiency of reactors.

These challenges are closely intertwined with the transition of thermochemical cycle research from the laboratory to pilot scale, as they reflect some of the engineering issues that need to be resolved.

8 SUMMARY OF RECENT SOLAR THERMAL DEMONSTRATION PLANTS

While many of the most promising cycles have been shown to be technically feasible at laboratory scale, the cost of building pilot plants has meant that a much smaller number have been demonstrated in continuous or semicontinuous operation using solar energy. One of the issues is that building such demonstration plants requires continuous funding over a number of years, with funding agencies often requiring in kind or cash contributions from industrial partners. While this is perfectly understandable, the immature stage of the technology has meant that industry has—with a few exceptions—been generally reluctant to invest in what is regarded as quite risky technology. Despite this, a number of significant demonstration plants have been successfully built and operated (Table 1). The challenge in coming years is to now find industry partners to scale up to precommercial demonstration plants.

Table 1 Recent solar thermochemical fuel demonstration plants

Key features	Cycle family			
	Nonvolatile metal oxide	Volatile metal	Sulfur-based	Methane reforming
Cycle/active material used in demonstration/pilot plant	Doped ferrites, ceria	Zinc oxide	Modified hybrid sulfur "Outotec open cycle"	Conventional catalyst
Project, dates	HYDROSOL-Plant 2014–17 (Preceded by HYDROSOL I, II, 3D; 2002–12)	Solar production of zinc and hydrogen, 2008–15	SOL2HY2 2013–16	Solar thermal reforming with SCORE & DCORE reactors, 1999–2009
Location Demonstration scale (solar step)	PSA, Spain 750 kW$_{th}$	Odeillo, France 100 kW$_{th}$	Jülich, Germany 100 kW$_{th}$ (sulfuric acid cracking)	Newcastle, Australia 200 kW$_{th}$
Key institutions	DLR (Germany), APTL (Greece), CIEMAT (Spain)	PSI (Switzerland)	DLR (Germany), Aalto University (Finland), ENGINSOFT SpA (Italy)	CSIRO
Peak temperature	1200–1400°C	1350–1750°C (observed, design)	750–1200°C	800°C
Challenges	Sintering of ferrites, efficiency	Thermal mass of system (8-h heat up to 1350°C)	Scale up of windowed systems	Scale-up, attractiveness of solar-fossil hybrid
Type of reactor	Stationary monolithic honeycomb with CPC and quartz glass window	Windowed cavity with oxide particles	Stationary catalyst coated monolithic absorber with CPC and quartz glass window	Open cavity with tube-in-tube reactor; catalyst in annulus between tubes
Reference	[82]	[66, 67]	[108, 109]	[13]

9 SUMMARY AND OUTLOOK

The interest in solar chemistry for fuels production has been primarily driven by concerns about energy security and climate change. A large amount of research was conducted in the 1970s by the nuclear industry, followed by a significant hiatus in which interest was minimal. However, growing interest in hydrogen and solar chemistry saw a revitalization of many research programs in the 1990s and 2000s. Despite these 40 years or so of research in thermochemical cycles, no cycle has been demonstrated at an industrial scale, although some cycles have been operated continuously in pilot plants. However, the reality is that these cycles cannot at present compete with the incumbent technology, steam reforming of methane.

A recent review of the processes under IEA-HIA Task 25 by an international consortium of experts concluded that cycles with an electrolysis step were closest to realization. The pure thermochemical cycles have greater challenges primarily due to higher peak temperatures. A number of other cycles are not attractive because of the toxicity of the reactants and/or materials issues to overcome corrosion issues under the extreme conditions many cycles require. Despite all these challenges, the growing maturity of concentrating solar thermal technologies means that a thermal energy source will be available independent of nuclear reactor developments. The potentially high thermal efficiency of thermochemical cycles results in an excellent opportunity to efficiently convert solar energy into high-value solar fuels such as hydrogen.

Meanwhile, valuable lessons can certainly be learned from production of solar fuels through carbon-containing feedstocks like methane, in similar solar-aided processes and reactor designs. These routes can be thought of as a process option for a transition period leading from fossil fuel-based solar fuels to those produced from renewable resources alone.

REFERENCES

[1] Gust D, Moore TA, Moore AL. Solar fuels via artificial photosynthesis. Acc Chem Res 2009;42(12):1890–8. https://doi.org/10.1021/ar900209b.
[2] Marschall R. Solar energy for fuels. In: Solar energy for fuels. Topics in current chemistry, vol. 371. 2015. p. 143–72. https://doi.org/10.1007/128_2015_636.
[3] Nocera DG. Solar fuels and solar chemicals industry. Acc Chem Res 2017;50:616–9. https://doi.org/10.1021/acs.accounts.6b00615.
[4] Yadav D, Banerjee R. A review of solar thermochemical processes. Renew Sust Energ Rev 2016;54:497–532. https://doi.org/10.1016/j.rser.2015.10.026.

[5] Le Gal A, Abanades S, Flamant G. CO_2 and H_2O splitting for thermochemical production of solar fuels using nonstoichiometric ceria and ceria/zirconia solid solutions. Energy Fuels 2011;25(10):4836–45. https://doi.org/10.1021/ef200972r.

[6] OECD Nuclear Energy Agency. In: Technology roadmap update for generation IV nuclear energy systems. Gen IV international forum; 2014. p. 1–66. Available at: https://www.gen-4.org/gif/upload/docs/application/pdf/2014-03/gif-tru2014.pdf.

[6a] Kim J, et al. Methanol production from CO_2 using solar-thermal energy: process development and techno-economic analysis. Energy Sci 2011;4(9):3122–32. https://doi.org/10.1039/c1ee01311d.

[7] Romero M, Steinfeld A. Concentrating solar thermal power and thermochemical fuels. Energy Environ Sci 2012;5:9137–674. https://doi.org/10.1039/c2ee21275g.

[8] Stein WH, Buck R. Advanced power cycles for concentrated solar power. Sol Energy 2017;152:91–105. https://doi.org/10.1016/j.solener.2017.04.054.

[9] Kodama T, et al. Flux me asurement of a new beam-down solar concentrating system in miyazaki for demonstration of thermochemical water splitting reactors. Energy Procedia 2014;49:1990–8. https://doi.org/10.1016/j.egypro.2014.03.211.

[10] Government of Japan. Strategic energy plan, Japan: Ministry of Economy, Trade and Industry; 2014. Available at: http://www.enecho.meti.go.jp/en/category/others/basic_plan/pdf/4th_strategic_energy_plan.pdf. Accessed 1 January 2018.

[11] METI. Summary of the strategic road map for hydrogen and fuel cells, Japan: Japanese Ministry of Economy, Trade and Industry; 2014. Available at: http://www.meti.go.jp/english/press/2014/pdf/0624_04a.pdf.

[12] Mueller-Langer F, et al. Techno-economic assessment of hydrogen production processes for the hydrogen economy for the short and medium term. Int J Hydrog Energy 2007;32(16):3797–810. https://doi.org/10.1016/j.ijhydene.2007.05.027.

[13] Agrafiotis C, et al. Solar thermal reforming of methane feedstocks for hydrogen and syngas production – a review. Renew Sust Energ Rev 2014;29:656–82.

[14] Chubb TA. Characteristics of CO_2-CH_4 reforming-methanation cycle relevant to the solchem thermochemical power system. Sol Energy 1980;24(4):341–5. https://doi.org/10.1016/0038-092X(80)90295-9.

[15] McCrary JH, et al. An experimental study of the CO_2CH_4 reforming-methanation cycle as a mechanism for converting and transporting solar energy. Sol Energy 1982;29(2):141–51. https://doi.org/10.1016/0038-092X(82)90176-1.

[16] Diver RB, et al. Solar test of an integrated sodium reflux heat pipe receiver/reactor for thermochemical energy transport. Sol Energy 1992;48(1):21–30. https://doi.org/10.1016/0038-092X(92)90173-8.

[17] Kodama T, et al. CO_2 reforming of methane in a molten carbonate salt bath for use in solar thermochemical processes. Energy Fuel 2001;15(1):60–5. https://doi.org/10.1021/ef000130t.

[18] Kodama T, et al. Molten-salt tubular absorber/reformer (MoSTAR) project: the thermal storage media of Na_2CO_3-MgO composite materials. J Sol Energy Eng 2009;131(4):41013. https://doi.org/10.1115/1.3197840.

[19] Dahl JK, et al. Solar-thermal processing of methane to produce hydrogen and syngas. Energy Fuel 2001;15(5):1227–32. https://doi.org/10.1021/ef0100606.

[20] Dahl JK, Weimer AW, et al. Dry reforming of methane using a solar–thermal aerosol flow reactor. Ind Eng Chem Res 2004;43:5489–95. https://doi.org/10.1021/ie030307h.

[21] Böhmer M, Langnickel U, Sanchez M. Solar steam reforming of methane. Sol Energy Mater 1991;24:441–8.

[22] Epstein M, et al. In: Solar experiments with a tubular reformer. Proceedings of the 8th int symp solar thermal concentrating technologies, Cologne, Germany. Heidelberg; 1996. p. 1209–29.

[23] Segal A, Epstein M. Solar ground reformer. Sol Energy 2003;75:479–90. https://doi. org/10.1016/j.solener.2003.09.

[24] Stein W, et al. Natural gas: Solar-thermal steam reforming, In: Encyclopedia of electrochemical power sources. 2009. p. 300–12. Available at: http://www. sciencedirect.com/science/article/pii/B978044452745500294X.

[25] McNaughton R, Hart G, Collins M. Solar steam reforming using a closed cycle gaseous heat transfer loop. Marrakech, Morocco: SolarPACES; 2012.

[26] McNaughton R, Stein W. Improving efficiency of power generation from solar thermal natural gas reforming, SolarPACES 2009. Available at: http://solarpaces2009. org/cms/front_content.php; 2009.

[27] Buck R, et al. Carbon-dioxide reforming of methane in a solar volumetric receiver reactor-the caesar project. Sol Energy Mater 1991;24(1–4):449–63. https://doi.org/ 10.1016/0165-1633(91)90082-V.

[28] Tamme R, et al. Solar upgrading of fuels for generation of electricity. J Sol Energy Eng 2001;123(2):160–3. https://doi.org/10.1115/1.1353177.

[29] Worner A, Tamme R. CO_2 reforming of methane in a solar driven volumetric receiver-reactor. Catal Today 1998;46:165–74.

[30] Klein HH, Karni J, Rubin R. Dry methane reforming without a metal catalyst in a directly irradiated solar particle reactor. J Sol Energy Eng 2009;131(May):14. https://doi.org/10.1115/1.3090823.

[31] Rubin R, Karni J. Carbon dioxide reforming of methane in directly irradiated solar reactor with porcupine absorber. J Sol Energy Eng 2011;133:5. https://doi.org/ 10.1115/1.4003678.

[32] Muradov NZ. How to produce hydrogen from fossil fuels without CO_2 emission. Int J Hydrog Energy 1993;18(3):211–5. https://doi.org/10.1016/0360-3199(93) 90021-2.

[33] Dahl JK, Buechler KJ, et al. Solar-thermal dissociation of methane in a fluid-wall aerosol flow reactor. Int J Hydrog Energy 2004;29:725–36. https://doi.org/10.1016/j. ijhydene.2003.08.009.

[34] Trommer D, Hirsch D, Steinfeld A. Kinetic investigation of the thermal decomposition of CH4 by direct irradiation of a vortex-flow laden with carbon particles. Int J Hydrog Energy 2004;29:627–33. https://doi.org/10.1016/j.ijhydene. 2003.07.001.

[35] Kogan M, Kogan A. Production of hydrogen and carbon by solar thermal methane splitting. I. The unseeded reactor. Int J Hydrog Energy 2003;28:1187–98. https:// doi.org/10.10360-3199/03.

[36] Abanades S, Flamant G. Production of hydrogen by thermal methane splitting in a nozzle-type laboratory-scale solar reactor. Int J Hydrog Energy 2005;30:843–53. https://doi.org/10.1016/j.ijhydene.2004.09.006.

[37] Abanades S, Flamant G. Solar hydrogen production from the thermal splitting of methane in a high temperature solar chemical reator. Sol Energy 2006;80:1321–32. https://doi.org/10.1016/j.solener.2005.11.004.

[38] Rodat S, et al. Hydrogen production from solar thermal dissociation of natural gas: development of a 10 kW solar chemical reactor prototype. Sol Energy Pergamon 2009;83(9):1599–610. https://doi.org/10.1016/J.SOLENER.2009.05.010.

[39] Rodat S, et al. Characterisation of carbon blacks produced by solar thermal dissociation of methane. Carbon 2011;49(9):3084–91. https://doi.org/10.1016/j.carbon. 2011.03.030.

[40] Trommer D, et al. Hydrogen production by steam-gasification of petroleum coke using concentrated solar power-I. Thermodynamic and kinetic analyses. Int J Hydrog Energy 2005;30(6):605–18. https://doi.org/10.1016/j.ijhydene.2004.06.002.

[41] Bellouard Q, Abanades S, Rodat S. Biomass gasification in an innovative spouted-bed solar reactor: experimental proof of concept and parametric study. Energy Fuel 2017;31(10):10933–45. https://doi.org/10.1021/acs.energyfuels.7b01839.

[42] Loutzenhiser PG, Muroyama AP. A review of the state-of-the-art in solar-driven gasification processes with carbonaceous materials. Sol Energy 2017;156:93–100. https://doi.org/10.1016/j.solener.2017.05.008.

[43] Piatkowski N, et al. Solar-driven gasification of carbonaceous feedstock-a review. Energy Environ Sci 2011;4:73–82. https://doi.org/10.1039/c0ee00312c.

[44] Puig-Arnavat M, et al. State of the art on reactor designs for solar gasification of carbonaceous feedstock. Sol Energy 2013;97:67–84. https://doi.org/10.1016/j.solener.2013.08.001.

[45] Gregg DW, et al. Solar gasification of coal, activated carbon, coke and coal and biomass mixtures. Sol Energy Pergamon 1980;25(4):353–64. https://doi.org/10.1016/0038-092X(80)90347-3.

[46] von Zedtwitz P, Steinfeld A. Steam-gasification of coal in a fluidized-bed/packed-bed reactor exposed to concentrated thermal radiationmodeling and experimental validation. Ind Eng Chem Res 2005;44(11):3852–61. https://doi.org/10.1021/ie050138w.

[47] Kodama T, et al. Coal coke gasification in a windowed solar chemical reactor for beam-down optics. J Sol Energy Eng 2010;132:041004. https://doi.org/10.1115/1.4002081.

[48] Z'Graggen A, et al. Hydrogen production by steam-gasification of petroleum coke using concentrated solar power-II reactor design, testing, and modeling. Int J Hydrog Energy 2006;31:797–811. https://doi.org/10.1016/j.ijhydene.2005.06.011.

[49] Piatkowski N, Steinfeld A. Solar-driven coal gasification in a thermally irradiated packed-bed reactor. Energy Fuel 2008;22:2043–52. https://doi.org/10.1021/ef800027c.

[50] Wieckert C, et al. Syngas production by the thermochemical gasification of carbonaceous waste materials in a 150 kWth packed-bed solar reactor. Energy Fuel 2013;27:4770–6. https://doi.org/10.1021/ef4008399.

[51] Lichty P, et al. Rapid high temperature solar thermal biomass gasification in a prototype cavity reactor. J Sol Energy Eng 2010;132(February)011012https://doi.org/10.1115/1.4000356.

[52] Steinfeld A. Solar thermochemical production of hydrogen – a review. Sol Energy 2005;78:603–15. https://doi.org/10.1016/j.solener.2003.12.012.

[53] Brown LC, et al. High efficiency generation of hydrogen fuels using nuclear power. Final technical report for the period August 1, 1999 through to September 30, 2002. General Atomics; 2003.

[54] Perret R. Solar thermochemical hydrogen production research (STCH) thermochemical cycle selection and investment priority, SANDIA REPORT SAND2011-3622. Albuquerque, NM: SANDIA National Lab; 2011. Available at: https://energy.gov/sites/prod/files/2014/03/f9/solar_thermo_h2.pdf.

[55] Ewan B, et al. INNOHYP CA – final report. Grenoble, France: Commissariat à l'énergie atomique; 2007.

[56] Sattler C, et al. Solar hydrogen production via sulphur based thermochemical water-splitting. Sol Energy 2017;156:30–47. https://doi.org/10.1016/j.solener.2017.05.060.

[57] Pickard PS, Russ B. II.G.2 Sulfur-Iodine Thermochemical Cycle. Albuquerque, NM, Available at: https://www.hydrogen.energy.gov/pdfs/progress09/ii_g_2_pickard.pdf; 2009.

[58] Gorensek MB, et al. A thermodynamic analysis of the SO_2/H_2SO_4 system in SO_2-depolarized electrolysis. Int J Hydrog Energy 2009;34(15):6089–95. https://doi.org/10.1016/j.ijhydene.2009.06.020.

[59] Gorensek MB, Summers WA. Hybrid sulfur flowsheets using PEM electrolysis and a bayonet decomposition reactor. Int J Hydrog Energy 2009;34(9):4097–114. https://doi.org/10.1016/j.ijhydene.2008.06.049.

[60] Summers WA, et al. Hybrid sulfur thermochemical cycle. DOE hydrogen program FY 2009 annual progress report. Aiken, SC; 2009 Available at: https://www.hydrogen.energy.gov/pdfs/progress09/ii_g_3_summers.pdf.

[61] Gorensek MB, Edwards TB. Energy efficiency limits for a recuperative bayonet sulfuric acid decomposition reactor for sulfur cycle thermochemical hydrogen production. Ind Eng Chem Res 2009;48(15):7232–45. https://doi.org/10.1021/ie900310r.

[62] Noglik A, et al. Solar thermochemical generation of hydrogen: development of a receiver reactor for the decomposition of sulfuric acid. J Sol Energy Eng 2009; 131(1):11003. https://doi.org/10.1115/1.3027505.

[63] Roeb M, et al. Sulphur based thermochemical cycles: development and assessment of key components of the process. Int J Hydrog Energy 2013;38(14):6197–204. https://doi.org/10.1016/j.ijhydene.2013.01.068.

[64] Beghi GE. A decade of research on thermochemical hydrogen at the Joint Research Centre, Ispra. Int J Hydrog Energy 1986;11(12):761–71. https://doi.org/10.1016/0360-3199(86)90172-2.

[65] Steinfeld A. Solar hydrogen production via a two-step water-splitting thermochemical cycle based on Zn/ZnO redox reactions. Int J Hydrog Energy 2002;27:611–9.

[66] Meier A, Koepf E. Solar production of Zinc and Hydrogen – 100 kW solar pilot reactor for ZnO dissociation. Villigen, Switzerland, Available at: https://www.aramis.admin.ch/?DocumentID=34853; 2015.

[67] Villasmil W, et al. Pilot scale demonstration of a 100-kWth solar thermochemical plant for the thermal dissociation of ZnO. J Sol Energy Eng 2013;136(1):011016–1/11. https://doi.org/10.1115/1.4025512.

[68] Weidenkaff A, et al. Direct solar thermal dissociation of zinc oxide: condensation and crystallization of zinc in the presence of oxygen. Sol Energy 1999;65(1):59–69. https://doi.org/10.1016/S0038-092X(98)00088-7.

[69] Epstein M, et al. Towards the industrial solar carbothermal production of Zinc. J Sol Energy Eng 2008;130(1):14505. https://doi.org/10.1115/1.2807214.

[70] Wieckert C, et al. A 300 kW solar chemical pilot plant for the carbothermic production of Zinc. J Sol Energy Eng 2007;129(2):190. https://doi.org/10.1115/1.2711471.

[71] Whaley T, et al. 'Status of the cadmium thermoelectrochemical hydrogen cycle. Int J Hydrog Energy 1983;8(10):767–71. https://doi.org/10.1016/0360-3199(83)90206-9.

[72] Abanades S, et al. Screening of water-splitting thermochemical cycles potentially attractive for hydrogen production by concentrated solar energy. Energy 2006;31:2805–22. https://doi.org/10.1016/j.energy.2005.11.002.

[73] Nakamura T. Hydrogen production from water utilizing solar heat at high-temperatures. Sol Energy 1977;19(5):467–75. https://doi.org/10.1016/0038-092x(77)90102-5.

[74] Steinfeld A, Sanders S, Palumbo R. Design aspects of solar thermochemical engineering—a case study: two-step water-splitting cycle using the fe3o4/feo redox system. Sol Energy Pergamon 1999;65(1):43–53. https://doi.org/10.1016/S0038-092X(98)00092-9.

[75] Tamaura Y, et al. Production of solar hydrogen by a novel, 2-step, water-splitting thermochemical cycle. Energy 1995;20(4):325–30. https://doi.org/10.1016/0360-5442(94)00099-O.

[76] Kaneko H, Miura T, Fuse A, et al. Rotary-type solar reactor for solar hydrogen production with two-step water splitting process. Energy Fuel 2007;21(4):2287–93. https://doi.org/10.1021/EF060581Z.

[77] Diver R, et al. Solar thermochemical water-splitting ferrite-cycle heat engines. J Sol Energy Eng 2008;130(4)041001. https://doi.org/10.1115/1.2969781.

[78] Ermanoski I, Siegel N, Stechel E. A new reactor concept for efficient solar-thermochemical fuel production. J Sol Energy Eng 2013;135(3)031002. https://doi.org/10.1115/1.4023356.

[79] Ermanoski I. Cascading pressure thermal reduction for efficient solar fuel production. Int J Hydrog Energy 2014;29:13114–7.

[80] Agrafiotis C, et al. Solar water splitting for hydrogen production with monolithic reactors. Sol Energy 2005;79:409–21. https://doi.org/10.1016/j.solener.2005.02.026.

[81] Roeb M, et al. Test operation of a 100 kW pilot plant for solar hydrogen production from water on a solar tower. Sol Energy Pergamon 2011;85(4):634–44. https://doi.org/10.1016/J.SOLENER.2010.04.014.

[82] Säck J-P, et al. High temperature hydrogen production: design of a 750KW demonstration plant for a two step thermochemical cycle. Sol Energy 2016;135:232–41. https://doi.org/10.1016/j.solener.2016.05.059.

[83] Scheffe J, Li J, Weimer A. A spinel/hercynite water-splitting cycle. Int J Hydrog Energy 2010;35:3333–40.

[84] Muhich C, et al. Efficient generation of hydrogen by splitting water with an isothermal redox cycle. Science 2013;341:540–2.

[85] Abanades S, Flamant G. Thermochemical hydrogen production from a two-step solar-driven water-splitting cycle based on cerium oxides. Sol Energy 2006;80(12):1611–23. https://doi.org/10.1016/j.solener.2005.12.005.

[86] Kaneko H, Miura T, Ishihara H, et al. Reactive ceramics of CeO_2–MOx (M = Mn, Fe, Ni, Cu) for H_2 generation by two-step water splitting using concentrated solar thermal energy. Energy 2007;32(5):656–63. https://doi.org/10.1016/j.energy.2006.05.002.

[87] Le Gal A, Abanades S. Catalytic investigation of ceria-zirconia solid solutions for solar hydrogen production. Int J Hydrog Energy 2011;36(8):4739–48. https://doi.org/10.1016/j.ijhydene.2011.01.078.

[88] Chueh W, Haile S. A thermochemical study of ceria: exploiting an old material for new modes of energy conversion and carbon dioxide mitigation. Philos Trans R Soc A Math Phys Eng Sci 2010;368(1923):3269–94.

[89] Furler P, et al. Thermochemical carbon dioxide splitting via redox cycling of ceria reiculated foam strucutures with dual-scale porosities. Phys Chem Chem Phys 2014;16: 10503–11.

[90] Furler P, Scheffe J, Steinfeld A. Syngas production by simultaneous splitting of water and carbon dioxide via ceria redox reactions in a high temperature solar reactor. Energy Environ Sci 2012;5:6098–103.

[91] Marxer D, et al. Demonstration of the entire production chain to renewable kerosene via solar thermochemical splitting of H_2O and CO_2. Energy Fuel 2015;29: 3241–50.

[92] Marxer D, et al. Solar thermochemical splitting of CO_2 into separate streams of CO and O_2 with high selectivity, stability, conversion, and efficiency. Energy Environ Sci 2017;10(5):1142–9. https://doi.org/10.1039/C6EE03776C.

[93] Evdou A, Zaspalis V, Nalbandian L. La(1Lx)SrxMnO3Ld perovskites as redox materials for the production of high purity hydrogen. Int J Hydrog Energy 2008;33: 5554–62.

[94] McDaniel A, et al. Sr- and Mn-doped LaAlO3-[small delta] for solar thermochemical H_2 and CO production. Energy Environ Sci 2013;6(8):2424–8. https://doi.org/10.1039/c3ee41372a.

[95] Scheffe J, Weibel D, Steinfeld A. Lanthanum-strontium-manganese perovskites as redox materials for solar thermal splitting of water and carbon dioxide. Chem Fuel 2013;27:4250–7. https://doi.org/10.1021/ef301923h.

[96] Jiang Q, et al. Thermochemical carbon dioxide splitting reaction with Cex M1-x O2-d (M = Sn, Hf, Zr, La, y and Sm) solid solutions. Sol Energy 2014;99:55–66.

[97] Aihara M, et al. Reactivity improvement in the UT-3 thermochemical hydrogen production process. Int J Hydrog Energy 1992;17(9):719–23. https://doi.org/10.1016/0360-3199(92)90093-C.

[98] Lemort F, et al. Physicochemical and thermodynamic investigation of the UT-3 hydrogen production cycle: a new technological assessment. Int J Hydrog Energy 2006;31(7):906–18. https://doi.org/10.1016/j.ijhydene.2005.07.011.

[99] Nakayama T, et al. MASCOT-a bench-scale plant for producing hydrogen by the UT-3 thermochemical decomposition cycle. Int J Hydrog Energy 1984;9 (3):187–90. https://doi.org/10.1016/0360-3199(84)90117-4.

[100] Sakurai M, et al. Solar UT-3 thermochemical cycle for hydrogen production. Sol Energy 1996;57(1):51–8. https://doi.org/10.1016/0038-092X(96)00034-5.

[101] Teo ED, et al. A critical pathway energy efficiency analysis of the thermochemical UT-3 cycle. Int J Hydrog Energy 2005;30(5):559–64. https://doi.org/10.1016/j.ijhydene.2004.08.003.

[102] Dokiya M, Kotera Y. Hybrid cycle with electrolysis using CuCl system. Int J Hydrog Energy 1976;1(2):117–21. https://doi.org/10.1016/0360-3199(76)90064-1.

[103] Lewis MA, et al. Evaluation of alternative thermochemical cycles - Part III further development of the Cu-Cl cycle. Int J Hydrog Energy 2009;34(9):4136–45. https://doi.org/10.1016/j.ijhydene.2008.09.025.

[104] Lewis MA, Masin JG. The evaluation of alternative thermochemical cycles. part II: the down-selection process. Int J Hydrog Energy 2009;34(9):4125–35. https://doi.org/10.1016/j.ijhydene.2008.07.085.

[105] Naterer GF, et al. Advances in unit operations and materials for the Cu–Cl cycle of hydrogen production. Int J Hydrog Energy 2017;42(24):15708–23. https://doi.org/10.1016/j.ijhydene.2017.03.133.

[106] Kolb GJ, Diver RB. In: Screening analysis of solar thermochemical H$_2$ concepts. IEA task 25 meeting general atomics, USA; 2008.

[107] Bulfin B, et al. Applications and limitations of two step metal oxide thermochemical redox cycles; a review. J Mater Chem A 2017;5(36):18951–66. https://doi.org/10.1039/C7TA05025A.

[108] Guerra Niehoff A, et al. In: Thermodynamic model of a solar receiver for superheating of sulfur trioxide and steam at pilot plant scale. ASME. Energy sustainability, volume 1: Biofuels, hydrogen, syngas, and alternate fuels; CHP and hybrid power and energy systems; concentrating solar power; energy storage; environmental, economic, and policy considerations of advanced energy systems; Geot, (50220), Charlotte, North Carolina, USA, June 26–30, 2016; 2016. https://doi.org/10.1115/ES2016-59167. p. V001T10A001.

[109] Odorizzi S. (ENGINSOFT S). SOL2HY2 – solar to hydrogen hybrid cycles project publishable summary. Available at: http://cordis.europa.eu/docs/results/325/325320/final1-sol2hy2-publishable-summary-final-report-24-2.pdf; 2016.

CHAPTER 10

Polygeneration Systems in Iron and Steelmaking

Hamid Ghanbari

Technology Specialist, Chemical Engineering, COMSOL Inc., Stockholm, Sweden

Abstract

Iron and steel manufacturing is one of the main industries subject to sustainability concerns. In recent years, several sites have been closing in the OECD (Organization for Economic Co-operation and Development) countries due to environmental regulations and pollution prevention, while the business has been growing fast in China, India, and other developing countries.

This chapter evaluates the polygeneration operation concept of future steelmaking processes with the aim of minimizing emission footprint while sustaining the profitability. One possible approach to increase efficiency and reduce carbon dioxide emissions is integration of steelworks with chemical plants for possible utilization of available off-gases in the system as chemical products. It is shown that through a polygeneration superstructure framework, it is possible to develop the most efficient process integration with alternative routes for off-gases pretreatment and further utilization of electricity, district heat, and methanol.

Keywords: Polygeneration, Steelmaking, Blast furnace, Carbon dioxide, Electricity, District heat, Methanol

1 INTRODUCTION

1.1 Steelmaking Technologies

Iron and steel are fundamental components of almost all infrastructures within which we live. In 2017, around 1.69 billion tons of crude steel were manufactured with China responsible for approximately half of this production [1]. This number divided by the world population of 7.55 billion translates to annual 224 kg/capita of crude steel consumption. There will be continuing growth in the volume of steel produced, particularly in developing areas such as Latin America, Asia, Africa, and the Indian subcontinent, where steel will be vital in raising the welfare of the developing societies [2].

There are currently four routes available for steel production worldwide [3]. These routes are shown in Fig. 1:

- Blast Furnace/Basic Oxygen Furnace (BF–BOF)
- Electric Arc Furnace (EAF)
- Direct Reduction (DR)
- Smelting Reduction (SR)

Conventional steelmaking (BF-BOF) is known as an energy-intensive sector, a remarkable source of carbon dioxide emission among primary industries and it is responsible for more than 65 percent of the steel manufactured worldwide.

Electric arc furnace as the second most common steelmaking route accounts for approximately 30 percent of steel production [5]. This route is primarily based on scrap, i.e., recycled steel.

Direct reduction involves reduction of oxygen from iron ores by using natural gas as a reducing agent in a solid-state process. The unit process is carried by shaft furnaces technology such as MIDREX [6] and HYL [7], rotary kilns technology such as the SL/RN process [8], rotary hearth furnace technology such as Fastmet/Fastmelt [9] and ITmk3 [10], or fluidized bed reactors technology such as Circofer [11].

Smelting reduction is an alternative to the BF to produce liquid iron. Like in the DR, there is no longer need of a coke oven plant; hence, SR is aimed at using a wide range of coals and iron fines. Smelting reduction has two stages, where firstly iron ore is heated and partially reduced by gases generated in the smelter. In the second stage, the smelter, further iron reduction takes place in the liquid state in contact with coal and oxygen. The FINEX process [12], an improved version of the COREX process [13], is the main SR technologies and HISARNA technology has developed under ULCOS program [14].

The primary motivation behind the DR and SR processes compared to the BF is their lower environmental impacts, such as lower carbon dioxide emission, due to gaseous direct reduction. Furthermore, these processes do not need a coke oven plant; they cause less dust emission and have a better water treatment. However, the economy and operational condition of these processes are the local challenges to compete with conventional steelmaking [7].

1.2 Conventional Steelmaking

Conventional BF–BOF steelmaking (Fig. 2) ranks foremost among all the steelmaking processes mainly due to cost, energy efficiency, and high

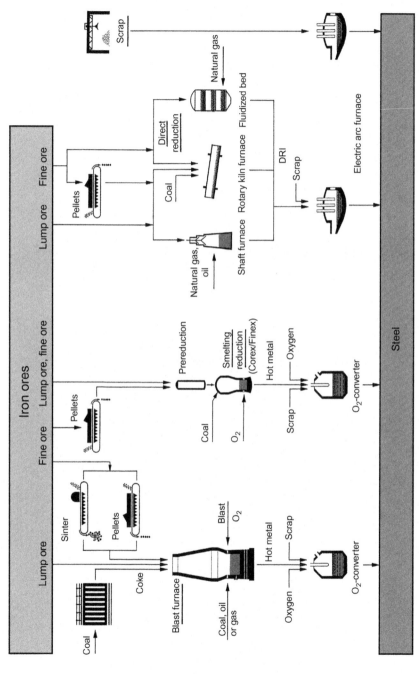

Fig. 1 Flow diagram of steel production routes [4].

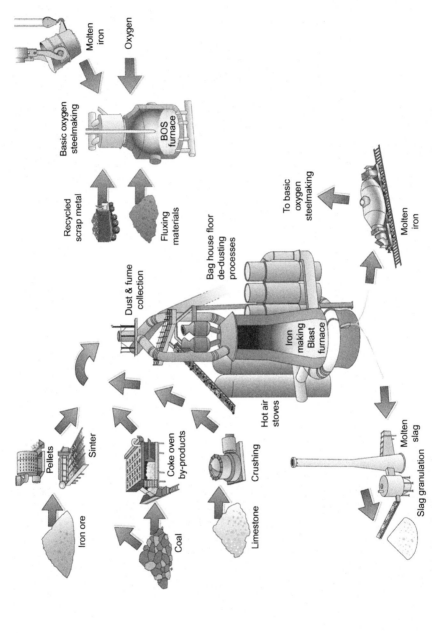

Fig. 2 Conventional steelmaking process [26].

production rate as well as the high degree of heat utilization [15]. This primary steelmaking process is well established and has already evolved into a mature state. Therefore, it is difficult to reduce substantially energy demand and the emissions especially in the blast furnace, which operates very close to its theoretical minimum in terms of reductants and energy consumption as it has approached its physical limits with respect to energy efficiency [16,17]. In the following, a short summary of each unit and advances in process technologies is provided.

Sinter plant: One of the first processes involved in primary steelmaking is the sinter plant (SP), which converts a raw material mixture, with iron oxides as the main constituent, into agglomerated particulate form, sinter, which is fed to the blast furnace. A bed of sinter feed mix travels under an ignition hood where hot combustion gases ignite coke blended into the sinter mix to start the sintering process, which is maintained by sucking large volumes of air through the bed from below. Therefore, the process is a source of CO_2 emissions (around 12% of integrated steelmaking) that can be estimated from the mass balance of the sinter plant [18].

Coke plant: In the Coking Plant (CP) operation, coal is dry-distilled to coke, which is a strong particulate matter of low reactivity at moderate temperatures suitable as a feed to the blast furnace. Cokemaking products are coke, Coke Oven Gas (COG), tar, and residual fuel oil. The main part of the coke goes to the blast furnace, while a smaller amount of the coke breeze goes to the sinter plant [18]. COG as a byproduct of the coking plant is a hydrogen-rich gas and appoint of high interest to enhance energy efficiency and reduce greenhouse gas (GHG) emissions [19]. It can be used as fuel in different processes in steelmaking such as coke oven, in preheating furnaces, as a reducing agent in the BF [20], in power plants [21] or feedstock to other chemical and metallurgical plants such as methanol [22–24] and DRI [25] production. The waste heat from the COG can also be used to dry the coal used for cokemaking, which may reduce the fuel consumption up to $0.3\,GJ/t$.

Blast furnace: Evolution of hot metal production in the blast furnaces reported to the European Blast Furnace Committee (EBFC) since 1990 shows that the average production per blast furnace has increased by approximately 48% while the average working volume of the furnace increased only by 26.6%. This demonstrates that apart from enlargement of the furnaces, the measures to increase furnace productivity enabled required hot metal production with fewer furnaces [27].

Blast furnace as the heart of steel plant acts as a large shaft-like counter-current heat exchanger and chemical reactor, where the agglomerated iron-bearing burden is charged with coke in alternate layers. The combustion of coke, which is maintained by the supply of preheated air (blast), provides CO to reduce iron oxides to iron and provides energy in form of heat to melt the iron and impurities. The hot metal (hm) and byproduct, slag, are tapped intermittently through tap holes from the lower furnace, while the top gas leaves the furnace top through uptakes.

The main improvement in the BF operation is regarding reducing agents and top gases utilization. Other potential energy mitigation options for the blast furnace are charging carbon composite agglomerates, application of top pressure recovery turbines, improvement of blast furnace control system (up to $0.4\,GJ/t_{hm}$), and slag heat recovery (up to $0.35\,GJ/t_{hm}$) [28].

1.2.1 Reducing Agents

Various reducing agents are available as injectants in the BF. Carbon/hydrogen/hydrocarbons in the form of granular or pulverized coal, heavy fuel oil, oil residues, used oils, fats and emulsions, animal fats, eco-oil, natural gas, coke oven gas, BOF gas, BF gas, waste plastics, coal tar, and biomass products are generally available in sufficient quantities at reasonable cost, however varying greatly between different regions. Hence, the choice among several reducing agents is determined by cost and operation constraints of the blast furnace. Coke, as the main reductant, also serves as a physical carrier of the bulk column in the BF, without which the BF operation would not be possible [3].

Coke: Coke is the primary fuel and reducing agents in the BF process. Depending on auxiliary reducing agents used, the coke consumption level is around 350–$400\,kg/t_{hm}$ in modern blast furnaces. The metallurgical coke in the blast furnace acts as a reducing agent, energy carrier, and support medium for the burden material. By implementation of new concepts of blast furnace operation, the coke consumption may be decreased to $200\,kg/t_{hm}$ [29].

Oil injection: Heavy oil or waste oil has been used as an auxiliary reducing agent in the BF to partially replace coke. The main advantage of oil is effective injectable hydrogen-carrying reducing agent, which may lead to reduce CO_2 emission. During many years, the Finnish steelmaking company Ruukki reported a consumption of about $360\,kg/t_{hm}$ coke and about $100\,kg/t_{hm}$ oil [27, 30].

Natural gas injection: Natural gas (NG) injection is an alternative injectant for medium-size furnaces. Its selection depends on the price of natural gas versus coal and its availability. It can be also injected simultaneously with pulverized coal. Increasing the natural gas injection rate may require increasing oxygen enrichment to keep the flame temperature and bosh gas volume within the operating bounds.

Along with natural gas injection, the utilization rate of CO enhances while that of H_2 decreases. Permeability, H_2 indirect reduction, and productivity of the blast furnace also increases [31]. In USA, injection of natural gas up to $155 \, kg/t_{hm}$ has been reported [32].

Pulverized coal injection: Pulverized coal (PC) is the most used auxiliary reducing agents in the BF process. There is a practical upper limit to the scale of pulverized coal injection, depending on coal types and raw material qualities among other variables. Pulverized coal injection rates above $200 \, kg/t_{hm}$ are considered massive and may not be sustained for long periods especially for large furnaces [28] even though rates up to $250 \, kg/t_{hm}$ have been reported as a monthly average.

Off-gas injection: Large volumes of off-gases from coke plant, blast furnace, and basic oxygen furnace (COG/BFG/BOFG) are available in integrated steelmaking. These gases contain mainly CO, CO_2, CH_4, H_2, and N_2, which are used as fuel in the hot stoves, preheating furnaces, and power plant. However, the gases also have the potential to be used as reducing agents in BF. This concept has been investigated and implemented in pilot and semi-industrial units [33, 34]. For applying Top Gas Recycling (TGR) in the BF, a sufficient level of oxygen enrichment is necessary to burn carbon, CO, and H_2 in lower part to produce reduction gases at sufficient temperature. By implementation of TGR, the total BF operation fuel consumption is estimated to be around $300 \, kg/t_{hm}$ in interchange of top gas utilization unit operation costs.

Biomass injection: Biomass as low-carbon or carbon-neutral carrier has been studied to replace fuels in blast furnace. Most of the efforts have been on solid fuel such as Charcoal BioMass (CBM) [35] and up to $150 \, kg/t_{hm}$ has been used in practice in small blast furnaces in Brazil [32, 36]. The heating value of biomass is low compared to fossil fuels. Thermochemical conversion processes ranging from torrefaction to pyrolysis may enhance the biomass properties to make it useful in the form of solid-, liquid-, and gaseous-reducing agents. Some processes may produce valuable byproducts that can be utilized in other chemical and energy sectors [37–39].

Replacement of fossil carbon by renewable biomass-based carbon is an effective measure to mitigate carbon dioxide emission intensity from the blast furnace. Besides the characteristic of the biomass as feasible reductant such as volatile matter and ash contents [40], the availability and economic competitiveness of biomass treatment plays an important role. The possible use of Finnish biomass in the integrated steel plant, particularly as auxiliary reducing agents in the blast furnace, after preprocessing it to decrease the oxygen content and increase the heating value has been investigated [39, 41, 42].

A key challenge is to develop efficient conversion technology, which can make the product compete with fossil fuels economically, considering the environmental benefit. The availability of large amounts of low-temperature gases increases the potential of the torrefaction process, which produces Torrefied BioMass (TBM), compared to higher degree of pyrolysis (e.g., charcoal) [39, 43–45]. The results indicate that based on typical cost of today, biomass products may not be economically competitive compared to fossil fuels (particularly coal). However, introduction of a carbon trading scheme or high carbon tax is expected to increase the motivation and interest in using biomass to partly replace coal in steelmaking [37, 46].

Table 1 shows some properties of introduced reducing agents [47]. Coke oven gas, heavy oil, natural gas, pulverized coal, and biomass products were studied as different reducing agents to be injected in the blast furnace for partial replacement of coke. In the reference plant, up to 17.6 t/h of coke oven gas was assumed to be available according to the coke production limitations. COG, which contains carbon monoxide, carbon dioxide, hydrogen, oxygen, nitrogen and methane, either can be injected into blast furnace or sent to the polygeneration system. For the sake of simplicity, in the BF all the injectants studied were taken to have an upper limit for the injection rate of 120 kg/t_{hm}.

Table 1 Reducing agents and their composition

Reducing agents	C	H_2	CH_4	CO	CO_2	N_2	O	HHV (MJ/kg)
PC	73.2	4.7	–	–	–	1	9	29.8
Oil	85.5	11.2	–	–	–	0.8	–	43.1
COG	0	12.3	42.1	17.2	7.8	19.8	–	42.4
NG	1.55	0.35	96.3	–	0.3	1.5	–	54.5
CBM	87.69	3.39	–	–	–	–	9	33.5
TBM	Estimated as f(Temp,time)							

Hot stoves: Hot stoves (HS), also known as cowpers, are used to preheat air (blast) required for combustion of coke in the BF. They work as a countercurrent regenerative heat exchanger. Low-cost and low-calorific value residual gases from the BF operation with small amount of other fuels, such as COG or natural gas, are commonly used in the stoves to increase the hot blast temperature up to 1523 K. It is known that the coke consumption decreases by 10–15 kg/t_{hm} with an increase of 100 K in the hot blast temperature [48]. Many BFs have three hot stoves. While two of them are being heated, the blast passes through the regenerative chamber of the third stove on its way to the blast furnace. Hot stove automation can reduce energy consumption by optimal operational condition up to 17 percent. Another potential for energy saving is by heat recovery from flue gases for preheating of air. This concept can reduce fuel consumption by 0.085 GJ/t_{hm} and yield energy savings of up to 0.35 GJ/t_{hm}. Improvement of combustion condition through more efficient burners can lead to energy saving up to 0.04 GJ/t_{hm} [28]. In order to reduce overall fuel consumption in industrial heating, oxygen enrichment of combustion air can be very effective. The application of oxygen enrichment in hot stoves will lead to lower fuel rate and increase hot blast stoves efficiency [49–52]. In this study, an upper bound of 32% enrichment was considered in hot stoves due to physical restriction.

Basic oxygen furnace: The Basic Oxygen Furnace (BOF) process converts the molten iron from the blast furnace with limestone and up to 30% steel scrap by injecting oxygen at supersonic speed, resulting in oxidation of carbon and impurities, producing liquid crude steel with typically 0.1–0.5 weight-% of carbon. High-purity oxygen is blown through the molten bath to lower carbon, silicon, manganese, and phosphorus content of iron, while various fluxes are used to reduce the sulfur and phosphorus level. BOF can operate either in open-hood or closed-hood vessels. This makes opportunities to recover heat or fuel from off-gases, which is rich in CO. Closed-hood BOF offers the best potential for both. This technology has been extensively implemented in Western Europe and Japan. This technology would reduce unavoidable CO_2 generation up to 0.16 ton per ton liquid steel, resulting in energy saving in the range of 0.53–0.92 GJ/ton liquid steel.

Air separation unit: In the system, there is a large demand for oxygen to inject into the BF (either as oxygen-enriched blast or as cold pure oxygen) as well as in the BOF. Cryogenic air separation (ASU) is the most efficient way to produce high volumes of oxygen, but it is an energy intensive process. The carbon dioxide emission from the air separation unit is estimated from energy required and available fuel in polygeneration system [53].

2 NOVEL PATHWAYS FOR SUSTAINABLE STEELMAKING

The considerable contribution of integrated steelmaking to greenhouse gas emissions, particularly of carbon dioxide, has already initiated research and development programs among main iron and steel producers [2,54]. Research and development efforts on the conventional BF-BOF steelmaking route have been carried out, e.g., in the Ultra-Low CO_2 Steelmaking (ULCOS) programme within EU [34,55], and in the CO_2 Ultimate Reduction in Steelmaking Process by Innovation Technology for Cool Earth 50, which is recognized as the COURSE50 programme, in Japan [56,57].

Carbon dioxide emission reduction methods in conventional steelmaking may be categorized into (1) use of alternative reducing agents such as neutral or low-carbon carriers, and (2) reduction of CO_2 emissions that are inevitable in the process by capturing CO_2 followed by sequestration.

Fig. 3 shows a schematic of the blast furnace operation under generalized carbon dioxide emission reduction methods in ULCOS and COURSE50. The main improvement in the technology is recycling CO_2-stripped blast furnace top gases back to the tuyeres. To implement the top gas recycling concept external investment needed to recover and purify the gases, including a capturing unit process and technology for transporting the separated CO_2, utilizing or sequestrating.

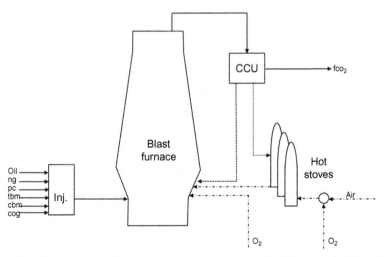

Fig. 3 Blast furnace operation under top gas recycling and different reducing agents.

Many researches have been undertaken in postcombustion CO_2 capturing, where exhaust gas from coal or gas-fired power plants is captured and prepared to be sent for geological storage. The most common technologies under development are amine-based chemical absorption, pressure swing adsorption, and membrane separation [58–60].

Off-gases in an integrated steel plants have some important differences compared to flue gases from power plants, such as higher CO_2 content, lack of oxygen, and different total pressure, which may result in other optimal solutions for off-gas treatment [17,61,62]. Researchers have provided comparative case studies for different scenarios of operation. The possibility for further utilization of residual gases in a polygeneration system has been considered by [47,63,64]. Next, brief descriptions of main unit processes available for off-gas utilization in a steel plant integrated with a polygeneration system are presented.

2.1 Carbon Capture and Utilization Units

Depending on BF operation under different emerging technologies concepts, several gas separation units such as alternative carbon-capturing process are considered for off-gas treatment and utilization. To recover CO and H_2 from the stripped gas streams, the most common methods employed are liquefaction, chemical absorption, and selective swing adsorption. Liquefaction may not be suitable in the conventional BF operation due to relatively high nitrogen concentration in the blast furnace top gas, as CO and N_2 are very similar in nature, which makes their separation by physical means difficult.

Residual gases can be used for top gas recycling, as fuel in the power plant or to produce chemicals in a chemical plant. COG can be used in the integrated plant either in BF as reductant or in the power plant. Another option would be further separation of COG contaminants to recover methane and to produce syngas through gasification. Syngas, in turn, can be used to produce chemicals such as hydrogen and methanol [22,23].

Carbon dioxide after compression is assumed to be sent to the pipeline at 110 atm for further transport and sequestration [65].

2.1.1 Separation Unit Operations

An integration of carbon dioxide capturing unit with a polygeneration system could be a long-term solution for suppressing CO_2 emissions from steel plants. In conventional operation, the blast furnace produces a top gas that contains more than 20 vol-% of carbon monoxide, which could be used to

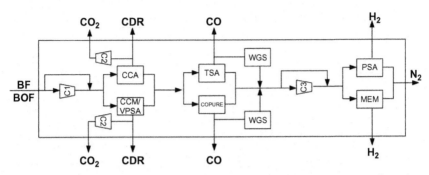

Fig. 4 Carbon capturing unit in the superstructure. *TSA*, temperature swing adsorption; *COPURE*, chemical absorption unit; *WGS*, water gas shift reactor; *CCA*, CO_2 chemical absorption; *CCM*, CO_2 capturing membrane; *(V)PSA*, (vacuum) pressure swing adsorption; *MEM*, membrane separation.

produce high-value chemical byproducts, such as methanol. It also contains more than 20 vol-% carbon dioxide, which could be captured and seques-trated. Fig. 4 shows the unit processes for gases treatment from the blast fur-nace and basic oxygen furnace. Three major units with different CO, CO_2, and H_2 separation technologies are included.

Chemical Absorption: Chemical absorption is a separation technology suited for large volumes of gases. For CO_2 capturing, the development of technology for BF gas has resulted in alternatives with an energy consump-tion of about $2\,GJ/t_{CO2}$, which is a half of the value required by the con-ventional chemical CO_2 amine-based solvent process [66]. The CCA process includes gas treatment, CO_2 removal, solvent regeneration, condi-tioning, and compression steps. The lowest consumption of energy per ton carbon dioxide was achieved by combining the 2-amino-2-methyle-1-propanol (AMP) with intercooling [67,68].

The COPure^{SM} process can be used to apply selective separation of CO from the blast furnace top gas. Low-pressure and low-temperature operation and noncorrosive solvent make the process economical in terms of capital and operation costs. Reported carbon monoxide recovery and purity exceed 98% and 99%, respectively. The approximate utility requirements, e.g., electrical power, reboiler and cooling duties, steam needed, and investment cost used in the present work were provided through private communica-tion with Rockey Costello, R.C. Costello & Assoc., Inc, [69].

Selective Swing Adsorption: The Vacuum Pressure Swing Adsorption (VPSA) as a technology for CO_2 capturing has been developed and tested in the experimental blast furnace in Lulea, Sweden, under the ULCOS pro-ject [14,55,70].

The Temperature Swing Adsorption (TSA) process may be applied for selective adsorption of CO from gas streams with an adsorbent mass comprising of crystalline zeolite molecular sieves. Pressure is not a critical factor and temperature changes in 273–573 K with a reported CO recovery of 99%. This process can purify a gas stream containing as little as 10 ppm by volume of carbon monoxide, but it is preferred to utilize the process to make bulk separation of CO from gas streams containing at least 5 vol-% [71,72].

An efficient H_2 recovery could be achieved by operating a Pressure Swing Adsorption (PSA) in a high-pressure ratio of the feed over the residue or by a membrane in a high-pressure ratio of the feed over the product, which may result in extra compression costs for both investment and operation [73].

Membrane Technology: For CO_2 recovery (CCM), a fixed carrier site membrane with amine groups is suggested to selectively separate CO_2 from the blast furnace top gas. In this process, water in the feed gas is an advantage rather than a problem since the membrane should be humidified during the operation, which makes it a proper choice after sulfur scrubbing of the feedstock. The process is in two stages with low-temperature feedstock.

For H_2 recovery, a Prism separator system, developed by Air Product and Chemical Inc, is chosen. The unit is controlled by pressure and flow adjustment of gas streams. The separator utilizes the principle of selective permeation through a gas-permeable membrane that has specially designed hollow fibers. Permeability coefficients of gases through a multicomponent membrane are used to estimate the flow rate and operational pressure of the system [74,75].

2.1.2 Gasification Unit Operations

The composition of the coke oven gas is assumed constant and it contains high amounts of hydrogen and methane, which can be used directly in the BF, polygeneration system, or as feedstock to the gas reforming plant. Fig. 5 shows the alternative unit processes for the gasification route. Pressure swing adsorption and membrane technologies are used to separate mainly hydrogen from methane and Steam Methane Reforming (SMR), Carbon Dioxide Reforming (CDR), and Partial Oxidation Reforming (POR) technologies are applied for methane gasification to produce more hydrogen. In the gas-reforming units, the optimal values of the critical parameters (temperature, pressure, and reactant ratio) have been taken from literature based on standalone processes.

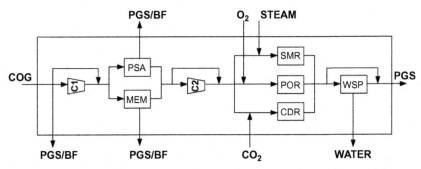

Fig. 5 Methane gasification units in superstructure. *PSA*, pressure swing adsorption; *MEM*, MEMbrane adsorption; *SMR*, steam methane reforming; *POR*, partial oxidation reforming; *CDR*, carbon dioxide reforming; *PGS*, polygeneration system; *WSP*, water separation unit.

Steam Methane Reforming: Production of synthesis gas from methane can be realized through steam methane reforming. This process is highly endothermic, but has low carbon deposition and high H_2/CO ratio, which makes it suitable for methanol synthesis. The steam-methane reaction takes place with the water gas shift reaction

$$CH_4 + H_2O = CO + 3H_2 \tag{1}$$

$$CO + H_2O = CO_2 + H_2 \tag{2}$$

The feedstock to the SMR unit is mixed with saturated steam in a methane-to-steam ratio of 3.681 kmol H_2O/kmol CH_4, and the reaction takes place at 20 bar and 1153–1300 K. The conversion of CH_4 and CO is $x_{CH_4}^{SMR} = 81.5\%$ and $x_{CO}^{SMR} = 40.2\%$ [76]. The heat of reaction at typical reformer operating conditions for the steam-methane reaction and water-gas shift reaction is 234.7 kJ/mol and -34.6 kJ/mol, respectively.

Carbon Dioxide Reforming: Carbon dioxide reforming (CDR) of methane can be expressed as

$$CO_2 + CH_4 = 2CO + 2H_2 \tag{3}$$

and has a great potential to be used in chemical industry to gain more environmental benefits by suppressing carbon dioxide emissions. It has been shown that the optimal operating condition would be 1143–1313 K at a pressure of 1 bar, equal ratio of methane to carbon dioxide, and the conversion of methane in the reaction is $x_{CH_4}^{CDR} = 0.90$. The heat of reaction has been reported as 247 kJ/mol at 298 K [77].

Partial Oxidation Reforming: The third alternative technology for synthesis gas formation is the exothermic methane partial oxidation reaction with standard heat of reaction of $-35.9\,kJ/mol$ and is expressed as

$$CH_4 + 0.5\,O_2 = CO + 2\,H_2 \qquad (4)$$

This process shows a high yield of hydrogen, but the oxygen stream makes it costly. It has been determined that by increasing temperature, the selectivity of carbon monoxide and hydrogen increases, but pressure has a negative effect on methane conversion and hydrogen production. The investigators [78] also reported that a CH_4/O_2 ratio of 2, a temperature between $1073-1473$ K at 1 bar are ideal conditions for an oxy-reforming reaction to get a high yield of the synthesis gas. The conversion of methane is $x_{CH_4}^{POR} = 0.95$.

2.1.3 Torrefaction Unit

Biomass as energy source has some characteristics such as potential contaminants that make it a complicated fuel in terms of wide range of resources and waste streams. Furthermore, it has a low volumetric heating value and its moisture content may vary considerably. A key challenge is to develop efficient conversion technology, which can make biomass compete with fossil fuels economically, preserving the benefits regarding environmental aspects. Availability of large volumes of low-temperature gases as source of heat increases the potential of the torrefaction process in comparison with a higher degree of pyrolysis. In this study, a torrefaction process [47], including dryer, torrefaction, cooling, and grinding were added to the system to find optimum operational condition for the torrefaction process in this environment [43].

Mass and energy yield, *ME* and *EY*, on a dry ash-free basis is represented by correlations as function of temperature and time. The residence time of the torrefaction process considered was between 0.2 and 1 h and temperature of torrefaction may vary between $200\,^{\circ}C$ and $300\,^{\circ}C$ [44]. Mass and energy balance for the torrefaction reactor can be estimated from its Moisture Content (*MC*), High Heating Value (HHV), torrefied biomass composition, and energy required for grinding are given as a function of torrefaction temperature [79].

2.1.4 Water-Gas Shift Reactor

Water-gas shift reaction (WGS) is an important industrial reaction that is used in the manufacturing of ammonia, hydrocarbons, methanol, and hydrogen.

The WGS reactor is considered to provide a carbon monoxide to hydrogen ratio required for methanol synthesis. It provides a source of hydrogen at the expense of carbon monoxide.

$$CO + H_2O = CO_2 + H_2 \qquad (5)$$

It is assumed that the reaction takes place at 473 K with CO conversion greater than 0.9 and heat of reaction of -41.2 kJ/mol [80].

2.1.5 Polygeneration System

Integrated steel plants in the EU have often an onsite power plant where the process gases, such as BF, BOF, and COG, is used to produce heat and power. Investigations have shown (Fig. 6) there is a high potential of energy saving in steel sectors by considering the Best Available Technologies (BATs) and state-of-the-art power plant [81].

It has been proposed that the primary steelmaking should be integrated with a polygeneration system to solve the problems of energy utilizing efficiency and to reduce emissions. An idea would be to remove carbon dioxide and use the residual gases as feedstock to a polygeneration system to produce district heat, electricity, and methanol. Polygeneration systems are claimed to be more energy efficient than standalone processes, as they have higher flexibility to switch between different forms of energy products depending

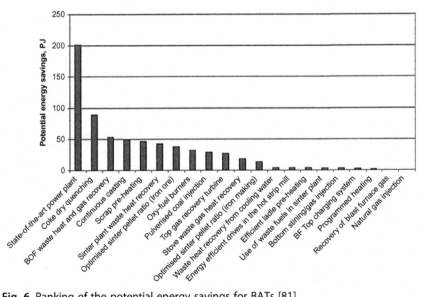

Fig. 6 Ranking of the potential energy savings for BATs [81].

on regional and seasonal demands. The benefit of liquid methanol in comparison with other forms of energy is its higher volumetric heating value and the fact that it can be stored and used as a substitution for traditional energy carriers or as a feed material for small-to-medium scale chemical industries. Additionally, the steel plant could avoid paying for emission or sequestration of CO_2 by converting the residual carbon to methanol, which extends the life cycle of carbon. However, the energy required for this upgrading must be considered carefully.

There have been several studies on optimal design of polygeneration systems. Liu [82] studied a system with different feedstocks and technologies that coproduce electricity and methanol. Their study showed that conversion rate of technologies, price of feedstock, capital investment, and the fixed operating cost have strong influence on the net present value. They also presented a multiobjective mixed-integer nonlinear programming formulation of a typical polygeneration process operating over a time horizon, where both profitability and environmental impacts are considered.

Chen [83] proposed a coal and biomass polygeneration system to produce power, liquid fuels, and chemicals under different economic scenarios using nonlinear programming. He also performed a simultaneous optimization, analyzing the design and the operational decision variables. The results showed that higher net present values could be obtained with increasing operational flexibility.

2.1.5.1 Methanol Unit Operations

All the syngas from gas utilization units are distributed and used in the polygeneration system to produce methanol, electricity, and district heat. Methanol can be produced via two different technologies:

- Gas-Phase Methanol (GPMEOH)
- Liquid-Phase Methanol (LPMEOH)

In the GPMEOH production, methanol is synthesized in a gas-phase reaction over a heterogeneous catalyst from a synthesis gas that consists primarily of hydrogen, carbon monoxide, and carbon dioxide. Newer processes focus on the use of CO-rich synthesis gas instead of H_2-rich synthesis gas, thereby utilizing cheaper synthesis gas to produce methanol. One of the promising technologies utilizing CO-rich synthesis gas is the LPMEOH synthesis process, but the single-pass conversion of syngas in the LP reactor is still limited [84].

Fig. 7 shows methanol production and purification route considered in this study. Main reactions take place in the synthesis of methanol are:

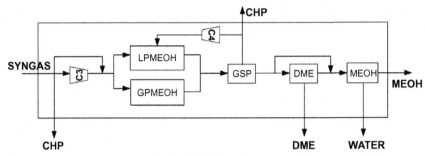

Fig. 7 Methanol production units in superstructure. *CHP*, combined heat and power plant; *LPMEOH*, liquid-phase methanol reactor; *GPMEOH*, gas-phase methanol rector; *GSP*, gas separation unit; *DME*, dimethyl ether purification; *MEOH*, methanol purification.

$$CO + 2\,H_2 = CH_3OH \tag{6}$$

$$CO_2 + 3\,H_2 = CH_3OH + H_2O \tag{7}$$

$$2CO + 4H_2 = C_2H_6O + H_2O \tag{8}$$

The heat of reaction for the first and the second reactions at standard temperature and pressure is -90.79 kJ/mol and -49.50 kJ/mol, respectively. In addition, the third reaction takes place as a side reaction in the GP, producing dimethyl ether (DME) to a limited extent with a reaction heat of -204.94 kJ/mol at standard conditions.

In the GP reactor carbon monoxide, carbon dioxide and hydrogen are catalytically converted to methanol and dimethyl ether. Typical operating conditions are 50 atm and 533 K and all reactions are exothermic and the excess heat must be removed to maintain optimum operational conditions. The conversion depends on temperature, pressure, hydrogen-to-carbon monoxide ratio, space velocity, catalyst composition, and carbon dioxide content.

The overall conversion of the carbon monoxide and carbon dioxide in the syngas-to-methanol is typically 0.95, and methane and nitrogen are considered as inert. The amount of dimethyl ether produced is 2 wt-% of methanol produced [76, 84].

The product stream is cooled down to 318 K to condense all methanol, dimethyl ether, and water. Unreacted H_2, CO, CO_2, CH_4, and N_2 do not condense at the conditions of the exchanger and must be recovered in a flash drum and sent to utility to be burned or released.

Dimethyl ether can be separated from methanol by extractive distillation at 11.2 atm and a reflux ratio of 20 mol recycled liquid/mol distillate. In this column, almost complete recovery of DEM is assumed, which is accomplished as a top product, while methanol and water leave at the bottom. DME can be sold to be used as an additive to diesel.

The LPMEOH process has some advantages. Although the operation conditions such as ability to control temperature, achieving higher conversion per pass with the same H_2/CO ratio (≥ 2) are like those of the gas phase process, heat of reaction can be more effectively used to generate high-pressure steam. Catalyst can also be added and withdrawn from the system while on stream without the necessity of shut down the process. However, the conversion per pass in the liquid phase reactor in CO-rich syngas is low and therefore the methanol yield is low. LPMEOH operational conditions depend on reactor pressure, temperature, the composition of the feed syngas, which per pass conversion of syngas to methanol may vary from 15% to as high as 60%. Eqs. (6), (7) are considered as the main reactions, and the reactor typically operates at 523 K and 50 atm. The conversion of carbon dioxide in Eq. (7) is assumed to be fixed at 8.9% and carbon monoxide conversion is estimated to be 30.6% [76, 84].

The product stream is sent to heat exchanger to condense methanol and water from the unreacted gases. In the methanol separator, a simple phase separation takes place, the bottom product is sent to the methanol distillation column, and the unreacted syngas is recovered to be recycled or sent to power generation plant. At the final stage, in the methanol distillation column, water and methanol are separated at 3.4 atm and 318 K with 99.9% of methanol purity as a top product at a reflux ratio of 1.5. Water with balance methanol leaves as the bottom product of the distillation column and is sent to a wastewater treatment facility [76, 84].

2.1.5.2 Combined Heat and Power Plant

In the combined heat and power plant, the syngas is modeled to be burned to release heat at high temperature to produce high-pressure steam for a turbine, with given efficiency factors in the turbine and in the generator. The low-pressure steam is finally condensed, releasing heat for district heat production. To estimate the amount of electricity and district heat, which could be sold, energy balances for the main processes such as compressor, reactors, and purification columns are used to calculate the internal power requirement. The energy used in the compressor calculated by determining the work required for compressing from inlet to outlet pressure. The reference case of the compressor is assumed to operate isentropically, and the true operation is estimated with adiabatic, motor drive, and mechanical efficiencies of $\eta_{ad} = 0.9$, $\eta_{md} = 0.9$, and $\eta_{mech} = 0.85$, respectively.

2.2 Superstructure Development

Primary steelmaking is characterized by regional constraints as each blast furnace has a unique behavior and its operation depends on the availability of feed materials, and other local conditions such as economy and environmental aspects, as well as social impacts.

In this study, process integration techniques have been investigated for future steelmaking concepts and their possibilities for enhanced sustainability. The studies were carried on in two steps: (1) focus on blast furnace operation with top gas recycling (Fig. 8), (2) integration of steelwork with a polygeneration system where difficult decision-making situations arise. As blast furnace is the heart of the system, at first step, a nonlinear semiempirical mathematical model was developed based on over a decade of industrial data from Ruukki steelwork and the feasibility of the concept was approved.

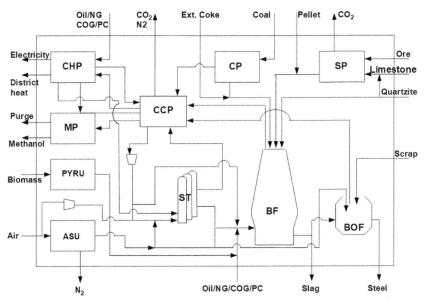

Fig. 8 Integrated steel plant. *CP*, coke plant; *SP*, sinter plant; *ST*, hot stoves; *CCP*, CO_2 capturing plant; *BF*, blast furnace; *BOF*, basic oxygen furnace; *CHP*, combined heat and power plant; *PYRU*, PYRolysis Unit; *ASU*, air separation unit; *MP*, methanol plant.

Table 2 Some of the variables and constraints of the model

Variable	Range	Variable	Range
BF production rate	130–160 t_{hm}/h	Bosh gas volume	150–220 km^3n/h
Recycled top gas	0–220 km^3n/h	Solid residence time	6.0–9.5 h
Blast oxygen content	21–99 vol-%	Slag rate	≥ 0 kg/t_{hm}
Specific fuel rate in BF	0–120 kg/t_{hm}	Top gas volume	≥ 0 km^3n/h
Blast temperature	250–1200°C	Top gas CO content	≥ 0 vol-%
Specific pellet rate	0–600 kg/t_{hm}	Top gas CO_2 content	≥ 0 vol-%
Pyrolysis temp	150–500°C	Top gas H_2 content	≥ 0 vol-%
Specific coke rate	≥ 0 kg/t_{hm}	Top gas N_2 content	≥ 0 vol-%
Flame temperature	1800–2300°C	Top gas heating value	≥ 0 MJ/m^3n
Top gas temperature	115–250°C	Sinter feed flow	0–160 t/h
Own coke feed flow	0–55 t/h		

Table 2 shows the BF input and some of the output variables and their constraints, as well as sinter and coke mass production rate constraints of the plant. Limitations in terms of mathematical formulations and computational time-imposed restrictions on the level of details in the studies. Therefore, a surrogate model based on previous model for blast furnace was generated to include feasible combination scenarios of top gas recycling (Table 3 and Fig. 9), blast (BL) oxygen enriching and different injectants and methanol production.

At second step, a superstructure (Fig. 10) using the surrogate model for blast furnace was introduced and a mixed-integer nonlinear mathematical model was developed as a tool for pre-engineering feasibility study of integration of steelwork with a polygeneration systems. The objective of study covers economic and environmental aspects as well as key operating conditions for the system.

Table 3 Regions of operation used to generate BF surrogate model

Oxygen enrichment of the blast (%)	Top gas recycling rate (km^3n/h)		
	0	80–100	180–200
21–32	States 1, 2, and 3	States 1, 2	–
55–65	–	States 1, 3	States 1, 3
84–99	–	States 1, 2, and 3	States 1, 3

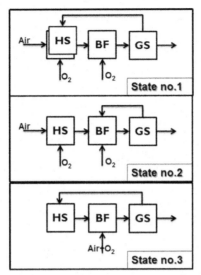

Fig. 9 Different states of blast preheating and oxygen enrichment: State 1: preheating TGR + BL in two sets of hot stoves, State 2: preheating BL in one set of hot stoves, and State 3: preheating of TGR in one set of hot stoves. GS indicates the gas separation unit processes.

3 POLYGENERATION SUPERSTRUCTURE EVALUATION

3.1 Economic Evaluation

Fig. 11 shows the structure of a typical economic evaluation model. The first step in the evaluation procedure is to map design decisions into flow rates and process equipment specifications using process models. The flow rates can be converted into recurring cash streams (e.g., revenues and operating costs) by multiplying them by unit prices. Process equipment specifications can be translated into purchased equipment costs by cost correlations or equipment fabrication cost models.

At the conceptual design stage, all other capital costs are estimated as a function of the purchased equipment cost. Additional recurring cash streams (e.g., operating labor and maintenance costs) are also a function of the equipment specifications. The last step of the analysis is the combination of capital costs and recurring cash streams into a single measure of economic performance. Recurring cash streams occurring in future periods can be discounted and added to the capital cost to obtain a net present value (NPV), or the capital costs can be annualized and added to the recurring cash streams to obtain a total annualized profit [85].

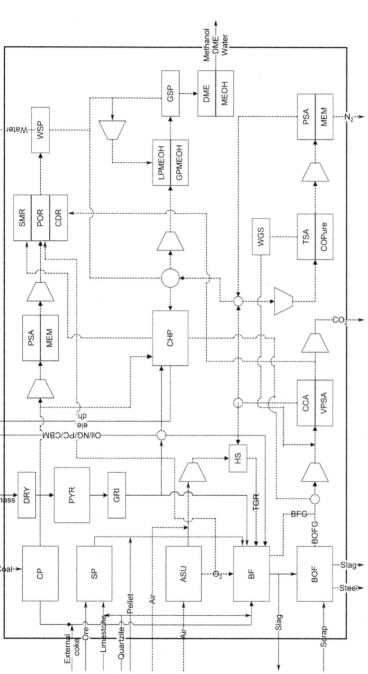

Fig. 10 Superstructure for suggested integrated steel plant. *Lines* depict solid- and liquid-phase material flow and *dashed lines* are residual gas network. It includes coke plant (CP), sinter plant (SP), hot stoves (ST), blast furnace (BF), basic oxygen furnace (BOF), combined heat and power plant (CHP), air separation unit (ASU), and available technologies for carbon capturing and sequestration plant, which are: pressure swing adsorption (PSA), membrane adsorption (MEM), steam methane reforming (SMR), partial oxidation reactor (POR), carbon dioxide reforming (CDR), water separation (WSP), liquid-phase methanol reactor (LPMEOH), gas-phase methanol rector (GPMEOH), gas separation unit (GSP), dimethyl ether purification (DME), methanol purification (MEOH), temperature swing adsorption (TSA), chemical absorption unit (COPURE), water gas shift reactor (WGS), CO_2 chemical absorption (CCA), vacuum pressure swing adsorption (VPSA) and compressors.

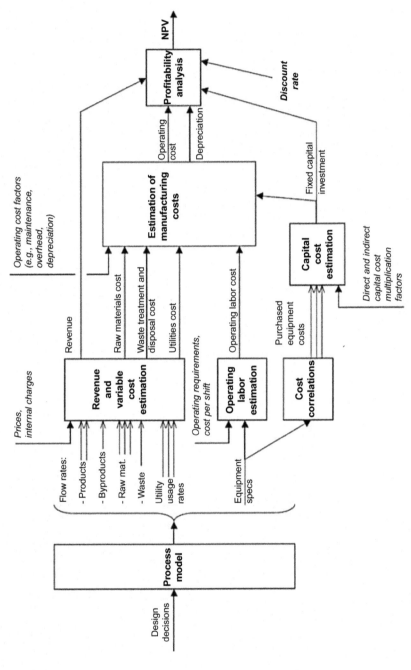

Fig. 11 Element of an economic valuation model [85].

Table 4 Economic parameter considered in this study ($[€/\$]_{index} = 1.3$)

Parameter	Value	Parameter	Value	Parameter	Value
c_{ore}	104 \$/t	c_{dme}	200 \$/t	c_{O_2}	65 \$/km^{3}n
c_{pel}	156 \$/t	c_{ls}	550 \$/t	c_{scrap}	130 \$/t
c_{coal}	143 \$/t	c_{el}	65 \$/MW	c_{meoh}	325 \$/t
$c_{coke,\ ext}$	390 \$/t	c_{dh}	13 \$/MW	c_{dme}	200 \$/t
c_{oil}	195 \$/t	$c_{emi.}$	0 − 150 \$/t	c_{pc}	230 \$/t
$c_{limestone}$	39 \$/t	$c_{seq.}$	0 − 150 \$/t	c_{bm}	65 \$/t
c_{quartz}	39\$/t	c_{ng}	260 \$/t	c_{cbm}	340\$/t

In this project, economic parameters set to 40% tax rate, life, and depreciation time (30 and 10 years) of the project, respectively, and 12% the annual discount rate. Fixed capital investment was assumed to be the sum of manufacturing and nonmanufacturing cost and is estimated as 1.4 times by bare module cost with 25% contingency. Working capital cost and direct expenses considered to be 19.4% and 4% of fixed capital investments, respectively [45]. The economic parameters used in this study are presented in Table 4.

3.2 Emerging BF Technologies

The system under TGR and oxygen enrichment and oil injection is investigated for two different cases: Case 1 refers to operation without any external obligation: The integrated plant can be completely flexible to distribute byproducts according to the price to reach the maximum net present value. In Case 2, an external demand for electricity (40 MW) was imposed on the system. This case projects possibility of local demands of electricity and district heat from some Nordic steelworks.

Fig. 12 show the NPV and estimated steelmaking costs for both cases applying different blast furnace technologies, compared to conventional blast furnace operation. On the horizontal axis, the first number refers to blast furnace states (1–3, cf. Fig. 9), the second number refers to the top gas recycling rate (0 = no, 1 = intermediate, and 2 = high level of recycling, cf. Table 3) and the last to the oxygen enrichment (1 = normal, 2 = intermediate level of enrichment, and 3 = oxyBF, cf. Table 3). Thus, e.g., 213 means BF state 2 with intermediate top gas recycling rate and cold oxygen injection.

Scenario 323 has the highest NPV for Case 1, while the external electricity demand has a strongly reducing effect on the estimated value, also in comparison with the optimized conventional BF operation. In addition,

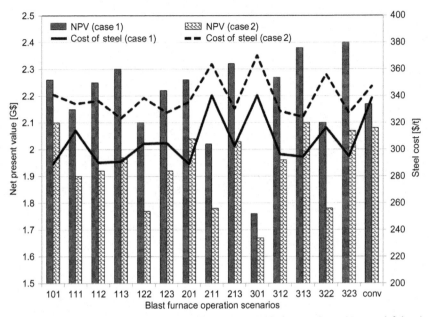

Fig. 12 NPV and steel production cost of the integrated plant without (Case 1, *left bars*) and with (Case 2, *right bars*) external energy demand considering different scenarios of blast furnace operation. The number codes on the abscissa express the BF operation scenario, the extent of top gas recycling, and oxygen enrichment. Conv represents conventional steelmaking without integration with CCU and MEOH plants. Number code: First number expresses BF states (1–3, cf. Fig. 9), second the top gas recycling rate (0 = no, 1 = intermediate and 2 = high recycling), and the third the oxygen enrichment (1 = normal, 2 = intermediate enrichment and 3 = cold oxygen injection, i.e., oxyBF).

scenario 313 (with intermediate top gas recycling rate) turns out to be promising in Case 1. It is also perceived how the combination of recycling and enrichment affects the final cost of steel production.

3.2.1 Sensitivity Analysis

In order to analyze the sensitivity of the suggested system with oil (ISP-323), Fig. 13 shows the changes in the net present value of the optimized system as perturbations of ±50% in the price of the feed material and byproducts were introduced. Quite naturally, the price of ore has the highest influence (up to 25%), followed by emissions, pellets, and coal. Electricity has the lowest effect on the net present value due to the small power production in the initial state.

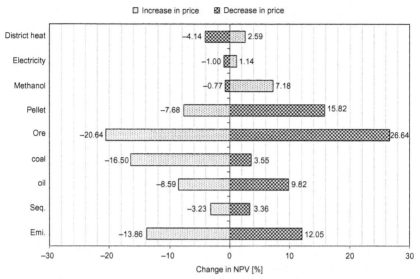

Fig. 13 Sensitivity analysis of ±50% change in the price of key feed materials and byproducts on net present value for integrated steelmaking (ISP-323).

As environmental restrictions increase, it becomes more challenging to deliver low-price electricity from the steelmaking sector. On the other hand, social acceptance of electricity produced from nuclear reactors has decreased. This could be an initiative toward the concept of integration in steel plants to suppress emission from the system by replacing carbon carriers in the blast furnace, such as coke, also converting the residual off-gases to fuels, which may replace fossil fuels and therefore increase the life cycle of carbon.

Fig. 14 shows the sensitivity of the NPV and specific emission based on changes in the cost of CO_2 emission from the system. The results were obtained after maximizing the net present value for the superstructure with defined costs for feed materials and sequestration and price of products.

For all cases, the system has proportionally higher NPV and specific emission for a decrease in the costs of emissions. This effect is shown less for the system with biofuel due to minimized use of necessary fossil carbon.

For the system with COG/NG, CBM, and PC, the oxyBF (nitrogen-free) operation has shown economic benefit while decreasing the specific emission due to the higher cost of fuel and external coke price.

For the system with CBM and PC, extra coke is available to sell from the integrated plant. For other fuels, the effect of specific emission price, at first is

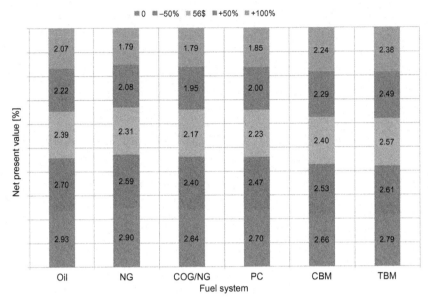

Fig. 14 Sensitivity of net present value of carbon dioxide emission cost for different fuel systems.

on moving toward top gas recycling and oxygen enrichment and the second shift is for changing in an optimal state of operation in plant particularly in PG and MeOH production. This can be seen for the system with TBM gradually effect on the decrease in emission level. Among fossil fuels, the system with oil is more sensitive to change in emission costs. In higher emission costs, all fossil fuels have shown a lower boundary ($\sim 0.5 t_{CO2}/t_{ls}$) for suppressing of CO_2 from the system.

3.3 Investment Cost Distributions

Fig. 15 shows the bare module cost of unit operations selected in an integrated system with different fuel supply systems by maximizing NPV over one period of operation. Results are shown for the maximum production rate of the steel plant, considering investment costs for the new processes.

The integrated system with torrefaction shows the highest NPV; it also has the second highest investment cost among the alternative fuels studied for the system. Swing adsorptions, partial oxidation of methane, and gas-phase methanol production are selected as the main unit processes.

CBM, PC, and COG/NG are selected to operate as oxyBF with no external energy production. Therefore, main investment cost is for CO_2 capturing and sequestration units. Vacuum pressure swing adsorption is

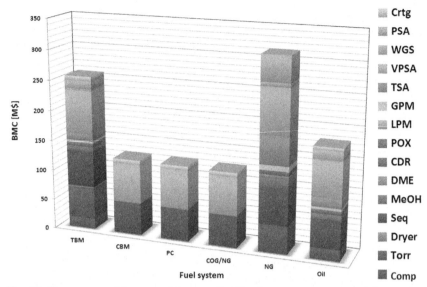

Fig. 15 Percentage of bare module cost (BMC) of unit operations in different fuel systems for the optimum operational state.

the main CO_2-capturing unit process. Conventional BF operation has highest NPV for the system with NG. The system has highest investment cost due to treatment of off-gases for high production of MEOH.

By comparing each scenario for NPV and emissions, there is a tradeoff between process economics, environmental impacts, and energy demands. Multiobjective optimization technique can be used to get the Pareto frontier for a specific plant to investigate the tradeoff [86]. For instance, for the system with NG, OxyBF and conventional operation are competitive in the amount of emission from the system. This is balanced by the amount of carbon, which leaves from the system as methanol varying between 30-50 t/h. This results to higher NPV for conventional operation with a higher amount of methanol production for approximately 40% increase in investment costs.

System with NG shows lower specific emission for most of the scenarios of operation. For the system with PC, CBM, and COG/NG, conventional BF operation has shown higher NPV compared to medium degree of cold top gas recycling, and preheating would be beneficial.

For the system with TBM, NPV increases by applying top gas recycling and oxygen enrichment up to oxyBF operation, but different scenarios of top gas recycling and oxygen enrichment do not have a drastic change on emission level.

Using charcoal may have a comparative economic with fossil fuels particularly for oxyBF operation while it has effective reduction in specific emission of the system by 45%. System with TBM has shown highly competitive economic profit and reducing emission by 60% in cost of further investment. The main difference between system with TBM and CBM system is due to methanol production benefits, which may have further effect toward sustainability. In case of oxyBF operation, higher environmental effect and economic profit may achieve.

The optimal operational conditions for the torrefaction unit are estimated as drying biomass from 40% initial moisture content to 15% at 167 (°C) and dried biomass is sent to torrefaction reactor at 227 (°C) for half an hour with a final moisture content of 3.5%. Mass and energy yield are estimated to be 0.88 and 0.929, respectively [47].

3.4 Environmental Evaluation

3.4.1 Effect of Production Rate and Fuels

The effect of production rate and fuel systems on carbon dioxide emission is investigated by maximizing the net present value. Fig. 16 presents the specific emission for the integrated system for different fuels and hot metal

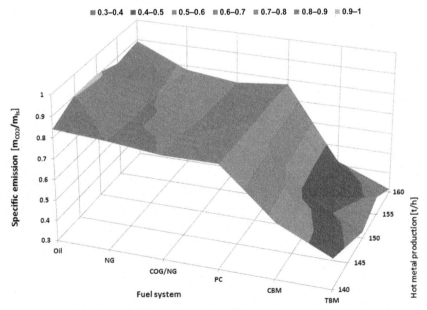

Fig. 16 Specific carbon dioxide emission for different hot metal production rate and fuel systems.

production. The optimization results show that for biofuels, by increasing the capacity of the plant, the specific emission may decrease slightly while for other fuels, a marginal increase in emission is expected. The results express that usage of coke oven gas can reduce emissions from the sector, although it has lower net present value compared to the oil and natural gas systems. The integrated system with pulverized coal shows a medium behavior in terms of economic and environmental impacts; it has lower NPV and specific emission compared to the other fossil fuel-based systems.

3.4.2 Carbon Flow in the System

Fig. 17 shows the percentage of carbon flow based on fossil resources in the integrated system and different fuel supplies after maximizing the NPV. In all cases, coal is the main source of carbon in the blast furnace steelmaking, varying between 30% and 47%. The results show that the production of methanol can affect the carbon flow in the system, while it gives a wide flexibility to distribute energy depending on demand and price toward a more sustainable operation. In case of fossil fuels, there is a tradeoff between carbon dioxide sequestration and emission. This depends on the economics and development of blast furnace technology in the future. In the case of CBM and PC use, the system produces more coke than it needs, which leaves the system while in the other cases an external supply for coke is needed.

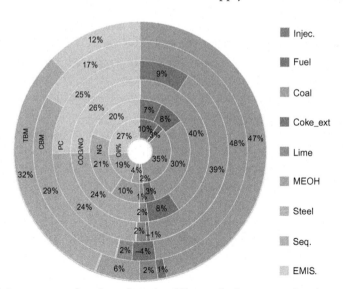

Fig. 17 Percentage of carbon flow in different fuel systems for the optimum operational state.

For the system with CBM, PC, and COG/NG, no methanol production is expected. The optimal operation state is for oxyBF with the utilization of off-gases to minimize extra fuel requirements for the system. This could be as a result of higher price for these fuels. For the system with NG, conventional BF operation is chosen in combined with high MeOH production, which decreases share of coal in the system down to 30% due to export of fossil-based carbon as MEOH (10%) from the system.

This effect was seen by forcing the system with the oxyBF operation, which exchanges 7% of carbon flow in the system between coal and MEOH. The results show top gas recycling will change the injection rate from its upper bound ($120 \, kg/t_{hm}$) in conventional BF operation to a lower level depends on the price of fuel and cost of emission and sequestration. This will affect the energy balance between preheating in hot stoves, oxygen enrichment, the degree of top gas recycling, and utilization of off-gases in polygeneration system considering external demands. For the system with biofuel, injection rate has observed at its upper boundary and for the system with COG/NG is 60% of available COG that it is included as carbon in coal flow.

3.5 Blast Furnace Operation Evaluation

The effect of methanol as carbon and energy carrier, produced from byproducts of a future steel plant with a blast furnace operated under top gas recycling (TBF) and blast oxygen enrichment, uses different costs of CO_2 emission and sequestration is investigated. The results indicate that the oxygen blast furnace (OBF) could be a promising approach due to a suitable composition of the top gases and the eco-environmental effects of carbon capturing and sequestration. It is important to stress that for low emission costs the optimized system operates without top gas recycling and for the oxygen blast furnace; the degree of recycling is not on its highest level at the economic optimum. Table 5 shows the effect of integration of a steel plant with a methanol plant on some key process variables for the TBF and OBF concepts. The optimal TBF states are seen to apply high, but not maximum, oxygen levels in the blast, higher coke rates, and clearly higher oil rates. Furthermore, the BF top gas temperature is at its lower limit. The OBF concepts, in turn, apply practically full top gas recycling and "pure" oxygen injection.

Effect of different reducing agent injection on blast furnace operation is shown in Table 6. Pyrolysis of the biomass was considered to examine the effect of integration by minimizing the manufacturing costs of steel. The

Table 5 Optimal process variables for a steel plant (with top gas recycling BF or oxygen BF operation of the blast furnace) integrated with methanol production steel production rate of 180 t/h, costs of emissions $c_{CO_2} = 50$ €/t, and capturing $c_{storage} = 10$ €/t

Variable	TBF	TBF-INT	OBF	OBF-INT
Blast volume (km^3n/h)	51.6	52.6	27.0	26.3
Oxygen enrichment (vol-%)	86.6	84.4	99.0	99.0
Blast furnace top gas volume (km^3n/h)	192	193	203	215
Blast furnace recycling gas volume (km^3n/h)	109	108	201	212
Sinter feed rate (t/h)	160	160	160	160
Coal feed rate (t/h)	79.1	79.1	79.1	79.1
Specific coke rate (kg/t_{hm})	312	311	280	297
Specific oil rate (kg/t_{hm})	**120**	**120**	26.0	10.5
Specific pellet rate (kg/t_{hm})	513	513	513	513
Flame temperature (°C)	**1800**	**1800**	**1800**	**1800**
Blast temperature (°C)	**1200**	**1200**	**1200**	**1200**
Bosh gas volume (km^3n/h)	197	197	197	204
Top gas temperature (°C)	115	115	178	193
Burden residence time (h)	7.0	7.0	7.6	7.3
Slag rate (kg/t_{hm})	215	215	198	199
Coke oven gas volume (km^3n/h)	17.6	17.6	17.6	17.6
Basic oxygen furnace gas volume (km^3n/h)	6.5	6.5	6.5	6.5
Oil needed for other units than BF (t/h)	–	9.0	–	7.2
Bought coke (t/h)	1.49	1.23	–	–
Sold coke (t/h)	–	–	3.66	0.91
Sold methanol (t/h)	–	26.2	–	18.0
Sold electricity (MW)	35.0	–	–	–
Sold district heat (MW)	178	–	36.4	–
Specific emission (t_{CO2}/t_{ls})	1.17	1.13	0.54	0.48
Specific steel cost (€/t_{ls})	288.1	274.7	251.6	232.7

Variable values at their bounds are written in bold face.

effect of partial replacement of coke with biomass can be more effective when the cost of emission is high and its effect will decrease by minimizing amount of coke requirement in the blast furnace. In all optimized states, the sinter plant operates at maximum production (160 t/h) and the remaining required iron ore is brought into the BF as pellets while the blast or recycled gas is heated to the maximum temperature (1200°C). In the cases where biomass is used, it is always processed at the maximum temperature, i.e., to maximum carbonization. For the TBF (except operation with NG) and OBF concepts, the flame temperature is at its minimum due to the injection of cold gases.

Table 6 Optimal process variables for the system with a steel production rate of 170 t_{ls}/h, costs of CO_2 emissions $c_{CO_2} = 40$ €/t, and storage $c_{storage} = 20$ €/t for conventional BF, with top gas recycling BF, or oxygen BF operation of the blast furnace

Variable	CBF-BM	CBF-NG	CBF-OIL	TBF-BM	TBF-NG	TBF-OIL	OBF-BM	OBF-NG	OBF-OIL
Blast volume (km^3n/h)	150.4	129.9	126.8	40.0	123.6	47.5	25.4	25.7	26.0
Blast oxygen (vol %)	22.0	30.1	28.0	98.4	31.9	88.1	**99.0**	**99.0**	**99.0**
BFG volume (km^3n/h)	221.5	217.2	203.0	178.4	216	179.0	183.8	187.0	179.3
BF TG rate (km^3n/h)	–	–	–	130.6	7.4	102.2	179.5	183.3	173.8
Sinter feed rate (t/h)	**160.0**	**160.0**	**160.0**	**160.0**	**160.0**	**160.0**	**160.0**	**160.0**	**160.0**
Coal feed rate (t/h)	79.1	79.1	79.1	79.1	79.1	79.1	79.1	79.1	79.1
Specific coke rate (kg/t_{hm})	326.7	330.8	320.7	299.5	330.4	308.6	220.4	289.4	256.0
Specific fuel rate (kg/t_{hm})	**120.0**	**120.0**	**120.0**	**120.0**	**120.0**	**120.0**	91.7	16.18	48.3
Specific pellet rate (kg/t_{hm})	457.6	457.6	457.6	457.6	457.6	457.6	457.6	457.6	457.6
Flame temperature (°C)	2265	1881	2246	**1800**	1862	**1800**	**1800**	**1800**	**1800**
Blast temperature (°C)	**1200**	**1200**	**1200**	**1200**	**1200**	**1200**	**1200**	**1200**	**1200**
Pyrolysis temperature (°C)	**500.0**	–	–	**500.0**	–	–	**500.0**	–	–
Bosh gas volume (km^3n/h)	193.7	**220.0**	186.0	174.5	**220.0**	183.3	178.6	184.6	179.4
Top gas temperature (°C)	194.6	221.9	143.0	**115.0**	221.0	**115.0**	184.1	155.2	156.5
Burden residence time (h)	7.2	7.1	7.3	7.7	7.2	7.5	**9.5**	7.9	8.6
Slag rate (kg/t_{hm})	210.7	211.0	216.4	209.4	211.0	214.7	205.3	208.9	203.9
COG volume (km^3n/h)	17.6	17.6	17.6	17.6	17.6	17.6	17.6	17.6	17.6
BOFG volume (km^3n/h)	6.5	6.5	6.5	6.5	6.5	6.5	6.5	6.5	6.5
Aux. fuel excluding BF (t/h)	10.1	10.5	9.3	12.4	10.5	10.1	11.6	6.3	8.1
Bought/sold coke (t/h)	5.8	9.8	0.0	−3.1	9.4	−2.0	−14.7	−4.5	−9.5
Sold methanol (t/h)	21.8	40.1	28.6	20.1	39.9	25.6	18.2	18.2	18.4
Specific emission (t_{CO2}/t_{ls})	1.1	1.4	1.5	0.6	1.4	1.2	0.4	0.6	0.7
Specific steel cost (€/t_{ls})	234.9	246.2	257.4	219.7	246.2	252.5	198.0	224.5	225.0

Boldface numbers indicate values at their constraints.

CBF and TBF apply (practically) maximum injection of auxiliary fuel, while the level for OBF is lower. However, for TBF with natural gas, the optimal top gas-recycling rate is very low and the operation resembles that of CBF. The OBF concepts with oxygen injection apply more than 95% recycling irrespective of the auxiliary fuel used. OBF-BM shows the lowest specific emission, due to a nonfossil carbon source combined with carbon capture and storage, but also the lowest steel production cost of the nine cases studies. The effect of reducing agent injection in the oxygen blast furnace as a main potential to minimize the emission levels requires a closer focus. In all cases studied, injecting auxiliary reductants to the blast furnace lowers the steel production cost, but this positive effect decreases at increasing specific emission costs. At the highest emission cost, there is no economic benefit of oil or natural gas injection. In all scenarios, injecting auxiliary reductants increases the specific emission rate mainly due to lower optimal top gas recycling rate. This effect becomes less prominent at increasing emission costs.

3.6 Conceptual Design and Operation

Fig. 18 displays the suggested integrated steel plant (ISP) configuration and Table 7 shows the optimal values of some key variables in the integrated

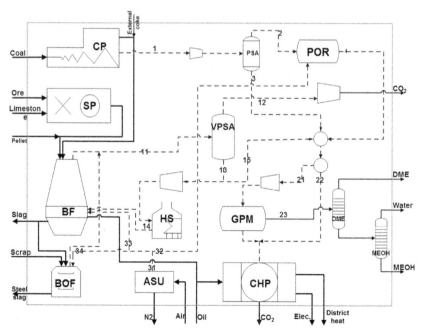

Fig. 18 Suggested integrated plant (ISP-323) and the main streams.

Table 7 Main streams (in kmol/h) and BF variables in the ISP-323 system

	CO	CO_2	H_2	O_2	N_2	CH_4	H_2O	MEOH	DME
1	41.4	13.9	456	1.1	50.7	222			
2	41.4	13.9	89.5	1.1	50.7	222			
3			366						
4	252	13.9	511		50.7	11.1			
11	3910	2787	731		997				
12		2647							
13	3910	139	731		997				
14	3618	135	703		989				
15	292	3.4	27.4		7.3				
21	439	13.9	877		50.7	11.1			
22	105	3.4	27.4		7.3				
23	21.9	0.7	4.3		50.7		13.2	438	8.7
31				1484					
32				105					
33				1076					
34				302					

Oxygen volume [km³n/h]	33.2	Blast volume [km³n/h]	23.9
Specific coke rate [kg/t_{hm}]	261	Flame temperature [°C]	1800
Specific oil rate [kg/t_{hm}]	42.1	Recycled top gas temp. [°C]	1200
Specific pellet rate [kg/t_{hm}]	458	Recycled top gas volume [km³n/h]	180
Coal flow rate [t/h]	81.6	Bosh gas volume [km³n/h]	183
Ore flow rate [t/h]	154	Top gas temperature [°C]	193
Limestone rate [t/h]	21.3	Burden residence time [h]	8.8
Scrap rate [t/h]	37.1	Slag rate [kg/t_{hm}]	260
Sinter flow rate [t/h]	160	COG volume [km³n/h]	17.6
External (sold) Coke flow rate [t/h]	9	BOFG volume [km³n/h]	6.2
Average extra oil needed [t/h]	30	Carbon dioxide send out [t/h]	116

system as result of superstructure periodic optimization (maximizing NPV) for the modified model covering all scenarios of TGR and oxygen enrichment with oil as auxiliary reducing agent under seasonal external energy demand.

Process synthesis shows tradeoff between economic and environmental impacts and different topologies may achieve. Swing adsorption processes are the main gas separation units.

For carbon dioxide capturing, depending on utility requirements and availability, VPSA and CCA processes can be more competitive than membrane technology. Due to availability of in-house oxygen, selection of POR technology seems to be preferable among other methane reforming technologies. Methanol technology depends on fuel, BF operation and top gas composition may vary.

Coke oven gas (stream-1) goes to the gasification plant. First, mainly hydrogen is separated from it in a pressure swing adsorption (PSA) unit, which operates at 10 atm, and is sent to the polygeneration system (stream-3). The rest of the gas, which has high methane content, is sent to the gasification reactor. The partial oxidation (POR) process was selected, which operates at low pressure and at its optimal temperature (1143 K). A mixture of blast furnace and basic oxygen furnace gases (stream-11), taken to be mostly CO (46%), CO_2 (34%), H_2 (9%), and N_2 (11%) after scrubbing, goes to a vacuum pressure swing adsorption and after separation unit for CO_2, is pressurized to 110 atm and sent out of the system for sequestration. The residual gases are rich in carbon monoxide and are assumed to be divided between the polygeneration system and the blast furnace, before which it is preheated (stream-14). The solution has omitted the TSA unit based on the composition of the gases in this state ($\sim$88% CO). The residual gases from capturing and gasification are pressurized and sent to the polygeneration system to produce methanol, electricity, and district heat according to the local demand, considering the economics.

3.7 Performance of Polygeneration System

Fig. 19 depicts the performance of PG under different level of top gas recycling and external energy (electricity) demand. Electricity and district heat production are on their lower-bound external energy demands in all

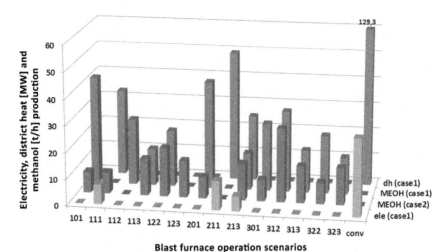

Fig. 19 Electricity, district heat, and methanol production of integrated plant without fixed external demand of electricity (Case 1) and methanol production for the case with fixed external energy demand (Case 2), considering different scenarios of blast furnace operation. The scenario number codes are defined in the caption of Fig. 12. The constraint considered as lower bound (external energy demand) is active for all scenarios in Case 2.

scenarios in Case 2 (and therefore not shown), but the production of methanol may vary up to 35 t/h from case to case (e.g., in scenario 101).

The conventional scenario is seen to have a high production of electricity and heat in Case 1, while the values are clearly lower for the other scenarios, except scenarios 211 and 111 (where oxygen enrichment is low). Overall, the production of methanol is lower in Case 2 compared to Case 1. The promising scenario 323 has a moderate methanol production of about 10 t/h.

Fig. 20 illustrates methanol production rate for the systems studied with different fuels and hot metal production rates by maximizing NPV for case 1. Methanol production may vary because of the flexible operation of the polygeneration system, which adapts itself to satisfy the demands of steam, electricity, district heat and methanol. The integrated system with torrefaction is seen to show a stable behavior in terms of methanol production despite varying steel production, by contrast to the case for other fuels studied. This is due to lower direct price of biomass as feed material and top gas composition. It is noticeable that maximum NPV does not happen in oxyBF operation for all fuel systems.

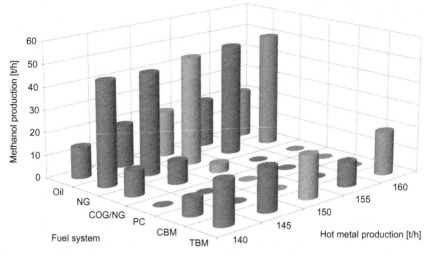

Fig. 20 Methanol production for different hot metal production rate and fuels.

4 CONCLUSION

Primary steelmaking is characterized by regional constraints as each blast furnace has a unique behavior and its operation depends on the availability of feed materials and other conditions in the steel plants such as economy and environmental aspects, as well as social impacts. Improvements on feed material and operation, competitive energy market, environmental restrictions, and the role of Nordic steelworks as energy supplier (electricity and district heat) make a great motivation behind integration among industries toward more sustainable operation, which could increase the overall energy efficiency and decrease environmental impacts.

In this study, through different steps, a model is developed for primary steelmaking, with the Finnish steel sector as a reference, to evaluate future operation of a steelmaking site regarding sustainability by integration with a polygeneration system concept. A tool developed, which can provide useful information concerning techno-environmental and economic aspects for decision making and estimate optimal operational condition of current and future primary steelmaking under alternative scenarios.

The results demonstrated that by integration with a polygeneration system, it is possible to decrease the specific CO_2 emissions of current primary steelmaking from fossil fuels ($1.6\,t_{CO2}/t_{ls}$) to a level of 0.75–$1.0\,t_{CO2}/t_{ls}$ and further by more than 85 % for use of biofuels in considered scenarios. This requires a capital investment for polygeneration. However, it can be economical in the long term, which depends on global environmental restriction and cost of emissions.

REFERENCES

[1] World Steel Association. Press Release (January), worldsteel. Available from:https://www.worldsteel.org/media-centre/press-releases/2018/World-crude-steel-output-increases-by-5.3–in-2017.html; 2018.

[2] World Steel Association. Steel's contribution to a low carbon future, worldsteel. .

[3] Remus R, Monsonet MAA, Roudier S, Sancho LD. Best available techniques (BAT) reference document for iron and steel production, industrial emissions directive. JRC Reference Report. 2010/75/EU.

[4] German Steel Federation. Hot metal and crude steel production. Available from: https://www.stahl-online.de/; 2018.

[5] World Steel Association. Steel statistical yearbook 2013, worldsteel. .

[6] Klawonn R, Siuka D. Current status and future of the Midrex direct reduction technology. Stahl Und Eisen 2006;126:23–9.

[7] Huitu K, Kekkonen M, Holappa L. Novel steelmaking process-liturature study and critical review. Helsinki University of Technology Publications; 2009. TKK-MT-207.

[8] Erwee MW, Pistorius PC. Nitrogen in SL/RN direct reduced iron: origin and effect on nitrogen control in EAF steelmaking. Ironmak Steelmak 2012;39:336–41.

[9] Joyner K. Fastmet (R)/Fastmelt (R): the final steps in waste recovery. Rev Metall/Cah d'Inf Tech 2000;97:461–9.

[10] Tsuge O, Kikuchi S, Tokuda K, Ito S, Kobayashi I, Uragami A. In: Successful iron nuggets production at ITmk3 pilot plant. 61st ironmaking conference proceedings, vol. 61. 2002. p. 511–9.

[11] Husain R, Sneyd S, Weber P. Circored((R)) and Circofer((R)) two new fine ore reduction processes. vol. 99. 129.

[12] Eberle A, Bohrn W, Milionis K, Tessmer G, Reidetschlager J. Smelting reduction and direct reduction of iron ores VAI technologies for scrap substitutes. COREX (R), FIN-MET (R), FINEX (R)vol. 99. ; 1999. p. 211.

[13] Wieder K, Boehm C, Schmidt U, Grill W. COREX (R) prepared for present and future iron making challenges. J Iron Steel Res Int 2009;16:1238–44.

[14] Birat JP. In: Steel and CO_2-the ULCOS program, CCS and mineral carbonation using steelmaking Slag. 1st international slag valorisation symposium. Leuven, Belgium, 6–7 April; 2009.

[15] Biswas AK. Principales of blast furnace ironmaking. 1st ed. Brisbane, Australia: Cootha Publishing House; 1981.

[16] Lungen HB, Schmöle P. Blast furnace operation without coke and carbon? Stahl Eisen 2004;124:63.

[17] Xu C, Cang D. A brief overview of low CO_2 emission technologies for iron and steel making. J Iron Steel Res Int 2010;17:1–7.

[18] Ghosh A, Chatterjee A. Ironmaking and steelmaking: theory and practice. 1st ed. New Delhi, India: Prentice-Hall of India Private Limited; 2008.

[19] Wang C, Larsson M, Ryman C, Grip C, Wikström J, Johnsson A, et al. A model on CO_2 emission reduction in integrated steelmaking by optimization methods. Int JEnergy Res 2008;32:1092–106.

[20] Richlen S. Using coke oven gas in a blast furnace saves over 6$ million anually at a steel mill. Office of Industrial Technologies Energy Efficiency and Renewable Energy; 2000. EE-20 U.S. Department of Energy.

[21] Modesto M, Nebra SA. Exergoeconomic analysis of the power generation system using blast furnace and coke oven gas in a Brazilian steel mill. Appl Therm Eng 2009;29:2127–36.

[22] Arvola J, Harkonen J, Mottonen M, Haapasalo H, Tervonen P. Combining steel and chemical production to reduce CO_2 emissions. Low Carbon Econ 2011;2:115–22.

[23] Bermúdez JM, Arenillas A, Luque R, Menéndez JA. An overview of novel technologies to valorise coke oven gas surplus. Fuel Process Technol 2013;110:150–9.

[24] Guo W, Wang J, Gao W, Wang H. Study on the preparation of higher alcohols using blast furnace gas and coke oven gas. Adv Chem Mater Metall Eng 2013;634-638:842–5.

[25] Johansson MT, Söderström M. Options for the Swedish steel industry–energy efficiency measures and fuel conversion. Energy 2011;36:191–8.

[26] Steel Stewardship Forum. Bluescope steel, basic oxygen steel making process; 2014.

[27] Luengen H, Peters M, Schmöle P. Ironmaking in Western Europe. Iron Steel Technol 2012;9:63–76.

[28] EPA. Available and emerging technologies for reducing greenhouse gas emissions from the iron and steel industry. Sector Policies and Program Division, Office of Air Quality Planning and Standards, U.S. Environmental Protection Agency; 2012.

[29] Hooey PL, Bodén A, Wang C, Grip C, Jansson B. Design and application of a spreadsheet-based model of the blast furnace factory. ISIJ Int 2010;50:924–30.

[30] Slaby S, Andahazy D, Winter F, Feilmayr C, Bürgler T. Reducing ability of CO and H$_2$ of gases formed in the lower part of the blast furnace by gas and oil injection. ISIJ Int 2006;46:1006–13.

[31] Guo T, Chu M, Liu Z, Tang J, Yagi J. Mathematical modeling and exergy analysis of blast furnace operation with natural gas injection. Steel Res Int 2013;84:333–43.

[32] Babich A, Gudenau H, Mavrommatis K, Froehling C, Formoso A, Cores A, et al. Choice of technological regimes of a blast furnace operation with injection of hot reducing gases. Rev Metal 2002;38:288–305.

[33] Tseitlin MA, Lazutkin SE, Styopin GM. A flow-chart for iron making on the basis of 100-percent usage of process oxygen and hot reducing gases injection. ISIJ Int 1994;34:570–3.

[34] Zuo G, Hirsch A. The trial of the top gas recycling blast furnace at LKAB's EBF and scale-up. Rev Metall/Cah d'Inf Tech 2009;106:387–92.

[35] Faleiro RMR, Velloso CM, de Castro LFA, Sampaio RS. Statistical modeling of charcoal consumption of blast furnaces based on historical data. J Mater Res Technol 2013;2:303–7.

[36] Lampreia J, De A, Muylaert MS, De Campos CP, Freitas MAV, Rosa LP, Solari R, et al. Analyses and perspectives for Brazilian low carbon technological development in the energy sector. Renew Sust Energ Rev 2011;15:3432–44.

[37] Norgate T, Langberg D. Environmental and economic aspects of charcoal use in steelmaking. ISIJ Int 2009;49:587–95.

[38] Suopajärvi H, Pongrácz E, Fabritius T. The potential of using biomass-based reducing agents in the blast furnace: a review of thermochemical conversion technologies and assessments related to sustainability. Renew Sust Energ Rev 2013;25:511–28.

[39] Suopajärvi H, Pongrácz E, Fabritius T. Bioreducer use in Finnish blast furnace ironmaking – analysis of CO$_2$ emission reduction potential and mitigation cost. Appl Energy 2014;124:82–93.

[40] Jahanshahi S, Deev A, N Haque LL, Mathieson J, Norgate T, et al. Current status and future direction of low-emission integrated steelmaking process. Cham: Springer; 2014. p. 303–16.

[41] Suopajärvi H, Fabritius T. Towards more sustainable ironmaking—an analysis of energy wood availability in finland and the economics of charcoal production. Sustain For 2013;5:1188–207.

[42] Helle H. Towards sustainable Iron-and steelmaking with economic optimization. Åbo Akademi University; 2014.

[43] Batidzirai B, Mignot APR, Schakel WB, Junginger HM, Faaij APC. Biomass torrefaction technology: techno-economic status and future prospects. Energy 2013;62:196–214.

[44] Saari J, Zakri B, Sermyagina E, Vakkilainen E. In: Integration of torrefaction reactor with steam power plant. The 8th international black liquor colloquium–black liquor and biomass to bioenergy and biofuels, Federal University of Minas Gerais (UFMG), Brazil, May 19–23; 2013.

[45] Ghanbari H. Sustainable steelmaking by process integration. Thermal and Flow Engineering Laboratory, Department of Chemical Engineering, Abo Akademi Univesity; 2014.

[46] Fick G, Mirgaux O, Neau P, Patisson F. Using biomass for pig iron production: a technical, environmental and economical assessment. Waste Biomass Valoriz 2014;5:43–55.

[47] Ghanbari H, Pettersson F, Saxén H. Sustainable development of primary steelmaking under novel blast furnace operation and injection of different reducing agents. Chem Eng Sci 2015;129:208.

[48] Moon J, Kim S, Sasaki Y. Effect of preheated top gas and air on blast furnace top gas combustion. ISIJ Int 2014;54:63–71.

[49] Bisio G, Bosio A, Rubatto G. Thermodynamics applied to oxygen enrichment of combustion air. Energy Convers Manag 2002;43:2589–600.

[50] Wang C, Karlsson J, Hooey L, Bodén A. In: Application of oxygen enrichment in hot stoves and its potential influence on the energy system in an integrated steel plant. International confernce of world renewable energy congress, IEEE, Linköping, Sweden, 8–13 May; 2011.

[51] Wang C, Cameron A, Bodén A, Karlsson J, Hooey L. In: Hot stove oxygen-enriched combustion in an iron-making plant. Oral presentation at Swedish-Finnish Flame Days; 2011.

[52] Sandberg J, Wang C, Larsson M. Analysis of oxygen enrichment and its potential influences on the energy system in an integrated steel plant using a new solution space based optimization approach. Inte J Energy Eng 2013;3:28.

[53] Castle W. Air separation and liquefaction: recent developments and prospects for the beginning of the new millennium. Int J Refrig RevInt Froid 2002;25:158–72.

[54] Fu J, Tang G, Zhao R, Hwang W. Carbon reduction programs and key technologies in global steel industry. J Iron Steel Res Int 2014;21:275–81.

[55] Danloy G, Berthelemot A, Grant M, Borlée J, Sert D, Van der Stel J, et al. ULCOS-Pilot testing of the low-CO_2 Blast Furnace process at the experimental BF in Luleå. Revue de Métallurgie Int J Metallurgy 2009;108:1.

[56] Tonomura S. Outline of course 50. Energy Procedia 2013;37:7160–7.

[57] Watakabe S, Miyagawa K, Matsuzaki S, Inada T, Tomita Y, Saito K, et al. Operation trial of hydrogenous gas injection of COURSE50 project at an experimental blast furnace. ISIJ Int 2013;53:2065–71.

[58] Quintella CM, Hatimondi SA, Musse APS, Miyazaki SF, Cerqueira GS, Moreira AA. CO_2 capture technologies: an overview with technology assessment based on patents and articles. Energy Procedia 2011;4:2050–7.

[59] Kuramochi T, Ramírez A, Turkenburg W, Faaij A. Comparative assessment of CO_2 capture technologies for carbon-intensive industrial processes. Prog Energy Combust Sci 2012;38:87–112.

[60] Li B, Duan Y, Luebke D, Morreale B. Advances in CO_2 capture technology: a patent review. Appl Energy 2013;102:1439–47.

[61] Oda J, Akimoto K, Sano F, Homma T. Diffusion of CCS and energy efficient technologies in power and iron & steel sectors. Energy Procedia 2009;1:155–61.

[62] Kuramochi T, Ramírez A, Turkenburg W, Faaij A. Techno-economic assessment and comparison of CO_2 capture technologies for industrial processes: preliminary results for the iron and steel sector. Energy Procedia 2011;4:1981–8.

[63] Liu P, Pistikopoulos EN, Li Z. A mixed-integer optimization approach for polygeneration energy systems design RID C-4913-2011 RID E-7840-2011. Comput Chem Eng 2009;33:759–68.

[64] Ghanbari H, Pettersson F, Saxén H. Optimal operation strategy and gas utilization in a future integrated steel plant. Chem Eng Res Des 2015;102:322.

[65] Lie JA, Vassbotn T, Hägg M, Grainger D, Kim T, Mejdell T. Optimization of a membrane process for CO_2 capture in the steelmaking industry. Int J Greenhouse Gas Control 2007;1:309–17.

[66] Hayashi M, Mimura T. Steel industries in Japan achieve most efficient energy cut - off chemical absorption process for carbon dioxide capture from blast furnace gas. Energy Procedia 2013;37:7134–8.

[67] Tobiesen FA, Svendsen HF, Mejdell T. Modeling of blast furnace CO_2 capture using amine absorbents. Ind Eng Chem Res 2007;46:7811–9.

[68] Hooey L, Tobiesen A, Johns J, Santos S. Techno-economic study of an integrated steelworks equipped with oxygen blast furnace and CO_2 capture. Energy Procedia 2013;37:7139–51.

[69] Costello R. COPureSM process. R.C. Costello & Assoc., Inc.; 2011.

[70] Torp TA. In: Drastik reduksjon av drivhusgasser fra stalproduksjon med CO_2 Fangst & Lagring (CCS)–ULCOS prosjektet. Technical committee, ULCOS steering committee meeting, Rome, Italy; 2005.

[71] Rabo JA, Francis JN, Angell CL. Selective adsoption of carbon monoxide from gas streams; 1977.

[72] Roark SE, White JH. Selective removal of carbon monoxide. US Patent. 6787118; 2004.

[73] Ruthven DM, Farooq S, Knaebel KS. Pressure swing adsorption. 1st ed. New York, USA: VCH Wiley; 1994.

[74] Henis JMS, Tripodi MK. Multicomponent membrane for gas separations; 1980.

[75] Porter MC. Handbook of industrial membrane technology: the separation of gases by membrane. 1st ed. Saddle River, NJ: Noyes Publications; 1990.

[76] Van Dijk CP, Solbakken A, Rovner JM. Methanol from coal and natural gas; 1983.

[77] S Wang GL, Millar G. Carbon dioxide reforming of methane to produce synthesis gas over metal-supported catalysts: state of the art RID C-5507-2008 RID A-2859-2008. Energy Fuel 1996;10:896–904.

[78] Zhu J, Zhang D, King K. Reforming of CH_4 by partial oxidation: thermodynamic and kinetic analyses. Fuel 2001;80:899–905.

[79] Phanphanich M, Mani S. Impact of torrefaction on the grindability and fuel characteristics of forest biomass. Bioresour Technol 2011;102:1246–53.

[80] Cheng W, Kung HH. Methanol production and use. 1st ed. Boca Raton, FL: CRC Press; 1994.

[81] Moya JA, Pardo N. The potential for improvements in energy efficiency and CO_2 emissions in the EU27 iron and steel industry under different payback periods. J Clean Prod 2013;52:71–83.

[82] Liu P. Modelling and optimisation of polygeneration energy systems. [Doctor of Philosophy]. Imperial College London; 2009

[83] Chen Y. Optimal design and operation of energy polygeneration systems. Massachusetts Institute of Technology; 2012.

[84] Vaswani S. Development of models for calculating the life cycle inventory of methanol by liquid phase and conventional production process. Raleigh, NC: Department of Civil Engineering; 2000.

[85] Cano Ruiz JA. Decision support tools for environmentally conscious chemical process design. Massachusetts Institute of Technology; 1999.

[86] Ghanbari H, Saxén H, Grossmann IE. Optimal design and operation of a steel plant integrated with a polygeneration system. AICHE J 2013;59:3659–70.

CHAPTER 11

Renewable Hybridization of Oil and Gas Supply Chains

Ahmad Rafiee*, Kaveh Rajab Khalilpour[†]
*Cardiff School of Engineering, Cardiff University, Cardiff, United Kingdom
[†]Faculty of Engineering and Information Technology, Monash University, Melbourne, VIC, Australia

Abstract

The oil and gas sector is not only a major energy producer but also a major consumer of energy. Energy requirements in the form of heat, power, or shaft work during onshore/offshore production, enhanced recovery operation, transportation, processing, and upgrading are mainly provided by fossil-fuel-based conventional technologies. Adaptation of sustainable and eco-friendly renewable energies such as hydropower, photovoltaics, solar thermal energy, wind power, biomass, geothermal energy, and wave power in different industries including the oil and gas sector is now globally examined with some successful stories. Renewable hybridization of the conventional oil and gas sector would reduce energy costs as well as greenhouse gas (GHG) emissions. This chapter investigates recent developments in the integration of renewable energy resources with the oil and gas industry and highlights the prospects.

Keywords: Renewable hybridization, Oil and gas, Supply chains, Marine transport, Marine energy, Transport fuels, Oil refineries

1 INTRODUCTION

Global energy demand projections show an increasing trend, with annual consumption predicted to reach around 778 Etta Joule by 2035 [1]. This will present major challenges for the oil and gas industry sector, which is a major producer and consumer of energy. With the reduction of the so-called easy oils and the move toward unconventional oil and gas, the emissions intensity of the oil and gas supply chains, which include onshore/offshore production, enhanced recovery, refining, and transportation (road, rail, or marine), is increasing rapidly. Some oil and gas production facilities are located in remote areas, which make the supply of energy demands for the facilities a big challenge. Hence, onsite renewable-based energy generation can not only improve the sustainability of such processes but also reduce the high

Polygeneration with Polystorage
https://doi.org/10.1016/B978-0-12-813306-4.00011-2

cost of fossil fuel delivery to these sites. Such solutions are also applicable to remote mining industries, which heavily rely on diesel generators.

Renewable energies (also called green power or clean energy) do not contaminate the air or the water and are derived from renewable resources, which are continually replenished by nature. These energies may originate directly from the sun (such as photoelectric, photochemical, and thermal energy), or indirectly from the sun (such as hydropower, wind, and bio-mass), or derived from other natural mechanisms (such as tidal energy and geothermal heat). Available renewable energy technologies convert these forms of natural energy to four distinct types of usable energy, which can replace conventional fossil fuels: electricity, water and air cooling/heating, transport fuels, and offgrid (rural) energy areas. The potential of the world's renewable energy resources is much higher than the current global energy demand. To increase energy security, renewables will have a significant share in the future energy mix. For example, the target of 20% of final electricity consumption from renewables by 2020 has been set for the European Union (EU) members [2]. In 2014, the EU countries had achieved a 16% share of renewables [3].

There are over 65,000 oil and gas fields around the world [4], around 33% of which are located offshore [5]. In 2015, the total amount of crude oil produced and refined globally was 4416 and 4189 million tons (Mt), respectively [6, 7]. The amount of energy consumption in the chemical and petrochemical industries in 2015 was 438 Mt of oil equivalent (Mtoe). Oil and gas industries consume refined petroleum products for energy generation. Also, some amounts of steam or electricity are purchased and imported into the plants. For example, US refineries processed 844 Mtoe in 2015 [6], and the amounts of purchased steam, electricity, and natural gas were 128,339 million pounds, 46,860 million kWh, and 852,067 million ft^3, respectively [8].

In this study, we address recent developments in renewable energy technologies worldwide and the endeavors of the oil and gas sector over the past years in adopting renewable energies.

2 AN OVERVIEW OF OIL AND GAS PRODUCTION SYSTEMS

2.1 Upstream Processes

Fig. 1 is a schematic of upstream oil and gas production processes. The well fluid is received by a test separation unit, which performs primary separation of solid, oil, gas, and water. Oil passes through a number of separation units

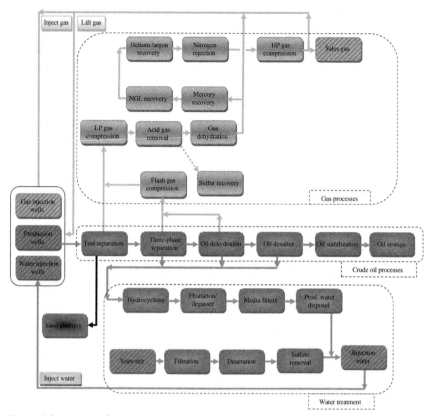

Fig. 1 Schematics of upstream oil and gas processes [10].

to attain export quality in terms of critical components such as water, gas (vapor pressure), and salt content. The gas stream is passed through multiple units including natural gas liquids (NGL) recovery, acid removal, and dehydration. When applicable, the gas goes through other processes for the removal of mercury, helium/argon, and nitrogen. The treated gas is then compressed before export via pipelines or reinjection into the reservoir for enhancement of oil recovery. Gas can also be converted to other forms such as liquefied natural gas (LNG), gas to liquids (GTL), and natural gas hydrate (NGH) [9].

The life of an oil and gas well has three distinct phases in which various techniques are employed to maintain production at maximum levels [11]. The primary importance of these techniques is to force oil into the wellhead where it can be pumped to the surface.

Primary recovery: In this phase, oil is forced out by pressure generated from the gas present in the oil.

Secondary recovery: In this phase, the reservoir is subjected to gas injection or water flooding to maintain a pressure that continues to move oil to the surface.

Tertiary recovery: This phase, also known as enhanced oil recovery (EOR), introduces fluids that reduce viscosity and improve the flow of well fluid. These fluids may consist of gases such as carbon dioxide, which are miscible with oil. They can also include steam, air, polymer solutions, gels, surfactant-polymer formulations, alkaline-surfactant-polymer formulations, or microorganism formulations [12].

During the primary recovery, up to 15% of a reservoir's original oil can be produced while the secondary recovery methods extend an oil field's productive life by injecting gas or water to move the reservoir's oil to a production wellbore. Secondary recovery techniques result in the recovery of up to 35% of the original oil in place. The disadvantage of this approach is high water cut of the well fluid. Tertiary or enhanced oil recovery (EOR) methods enable extraction of up to 70% of a reservoir's oil [13]. Three major classes of EOR are commercially in use [11]:

1. *Thermal recovery*: In this approach, heat is introduced to an oilfield to lower the viscosity. As an example, over 40% of the US EOR (mainly in California) projects utilize the thermal method.

2. *Gas injection*: In this method, gases such as CO_2, nitrogen, or natural gas expand in the oil reservoir and push additional oil to the surface. Other gases can also be used to lower oil viscosity. Around 60% of US EOR projects use gas injection.

3. *Chemical injection*: In this scheme, injection of dilute solutions of various chemical materials (such as emulsifiers, surfactants [14], acids, polymers, solvents, and dispersants) reduces surface tension and aids fluid mobility. Chemical injection techniques comprise about 1% of the US EOR projects.

2.2 Downstream Processes

Crude oil requires further processing to be converted to standard market fuels and chemicals. Fig. 2 is a process flow diagram of a refinery, which carries out crude oil upgrading.

The oil, after desalter (see Fig. 1), enters the atmospheric distillation unit. Distillation columns include furnaces, which provide thermal energy for

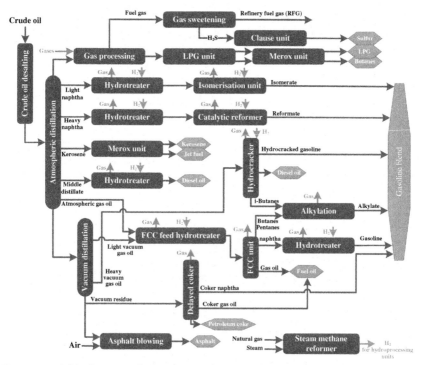

Fig. 2 A simplified process flow diagram of an oil refinery.

boiling the crude oil. In this unit, the oil is separated into various "cuts" based on predefined boiling ranges. Obviously, light hydrocarbons are separated from the top of the column, and the heaviest cuts are at the bottom. The most typical cuts include fuel gas, light naphtha, heavy naphtha, kerosene and jet oil, middle distillate, and gas oil. Any cut heavier than gas oil is generally sent to another distillation column, which operates under a vacuum condition. This process provides further separation and produces some more gas oil. The residue of the vacuum distillation column is used for the production of asphalt and petroleum coke. Given market standards and environmental constraints, the distillation column cuts are sent to various processes for quality improvement. For instance, both gas and liquid need to be treated to remove CO_2, sulfur components (such as H_2S and mercaptans), and nitrogen components, which lead to avoidance or minimization of SO_x, NO_x, and other emissions, which are environmentally concerning. Hydroprocessing is also a critical stage in the refining system for quality improvement of end products, as elaborated in Section 6.2.

In a simple description, the complex refining processes include thermal and hydrogen treatment where both require energy resources. Thermal energy can be supplied from the refinery's in-house products. However, hydrogen is generally supplied by natural gas reforming. Renewable energy resources could, therefore, be integrated with refinery systems to supply both thermal and hydrogen energy requirements.

3 AN OVERVIEW OF RENEWABLE ENERGY RESOURCES AND TECHNOLOGIES

It has been estimated that renewable sources could provide 3078 times the current global energy requirements, see Fig. 3 [15]. Each number in Fig. 3 represents the ratio of the corresponding energy resource to the current energy needs of the world. Solar energy and wind power can provide around 2850 and 200 times the global energy requirements, respectively.

Fig. 4 depicts various sources of renewable energy. Table 1 also summarizes the advantages and disadvantages of these renewable energy technologies and compares their levelized cost of energy (LCOE). The LCOE is an approximation for the electricity price at which the product would need to be sold to reach the breakeven point. Recently, in several jurisdictions, some of these technologies have reached so-called *grid parity* and are competitive with conventional energy resources. As such, the share of renewables in the energy mix is increasing.

The global total installed renewable capacity (including pumped storage) in 2016 amounted up to 2128.7 GW [20]. Fig. 5 illustrates the installed renewable capacity by subtechnology in 2016. From this figure, it can be seen that after the obvious hydro obviously leads in the renewable market,

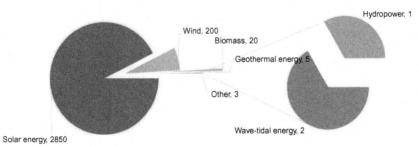

Fig. 3 Worldwide energy resources and comparison with current energy requirements [15].

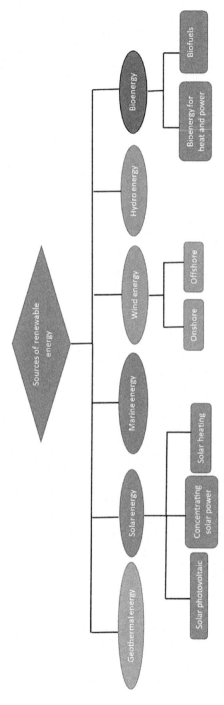

Fig. 4 Diverse sources of renewable energy.

Table 1 The advantages and disadvantages of renewable energy sources [16, 17]

Renewable energy	Advantages	Disadvantages	LCOE range ($/kWh) [18]
Solar	• Abundant • Environment friendly • Available all over the world • Reduces electricity costs • Various applications including offgrid locations, power satellites in space, and distilling water in Africa • Community shared solar gardens • Silent as there is no moving part in most applications • Financial support from governments • Minimal maintenance requirement • Technological advances • Single dwellings with their own electricity supply	• Intermittent • Expensive energy storage • Transportation and installation of solar systems indirectly create pollution • Exotic materials • Space requirement	0.06–0.56 (solar PV) 0.06–0.25 (CSP)
Wind	• Enormous potential (200 times worldwide needs) • Wind turbines space-efficient • Associated capital costs keep decreasing • Low operational costs • Standalone wind turbines offer energy savings and protect	• Intermittent source of energy • Wind turbines can be a potential threat to wildlife (bats, birds) • Unsightly covering of landscape with wind turbines • Noisy operation	0.04–0.12 (offshore wind) 0.1–0.21 (onshore wind)

Table 1 The advantages and disadvantages of renewable energy sources [16, 17]—cont'd

Renewable energy	Advantages	Disadvantages	LCOE range ($/kWh)
	inhabitants' power outages • Proper method to supply energy to remote areas • Land beneath wind turbines usable for farming		
Hydro	• Reliable compared to wind, solar power • Flexible • Safe as no fuel involved • Low operating cost • Peak demands can be coped with just by storing water above the dam • Hydropower plants can go to full power very quickly Constant electricity production	• Environmental effects • Related to interferences in nature • Depends on the amount of water available and droughts can affect electricity generation • Dam-building very expensive • Finding an appropriate site can be problematic	0.02–0.26 (small hydropower) 0.01–0.28 (large hydropower)
Geothermal	• Environmentally friendly • Massive potential (5 times current global energy needs) • Excellent to meet the base energy demand (in contrast with other renewables such solar or wind) • Even small households can benefit from geothermal heating	• In extreme cases, geothermal powergeneration plants can cause earthquakes • Most geothermal resources are not cost competitive	0.03–0.14

Continued

Table 1 The advantages and disadvantages of renewable energy sources [16, 17]—cont'd

Renewable energy	Advantages	Disadvantages	LCOE range ($/kWh)
	• Stable geothermal electricity prices and low-cost fluctuations • Small footprint on land as facilities can be built partly underground • Available everywhere, but some resources are cost-effectively exploitable • Technological improvements make more resources exploitable and lower costs		
Tidal	• Does not release GHG • Does not occupy much space • Highly predictable • Effective at low water speeds • Long lifespans	• Tidal barrages need manipulation on ocean levels, which has environmental effects • Expensive at small scales	0.13–0.28 [19]
Wave	• Enormous energy potential • Reliable energy source • Area efficient	• Onshore wave farms can cause conflicts with local acceptance and tourism services • Costly due to early stages of development	0.12–0.47 [19]
Ocean thermal	• No impact on the environment	• Expensive • Location specific	0.15–0.28 [19]

Table 1 The advantages and disadvantages of renewable energy sources [16, 17]—cont'd

Renewable energy	Advantages	Disadvantages	LCOE range ($/kWh)
energy conversion (OTEC)	• Produces fresh water along with energy • High capacity factor	• Requires expensive transportation options	
Bioenergy	• Carbon neutral • Cost effective • Abundant	• Requires space • Extraction of biomass materials can be expensive • Burning biomass causes air pollution	0.01–0.16

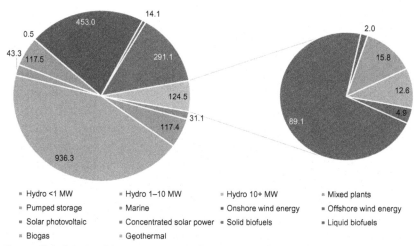

Fig. 5 Global renewable energy installed capacity (MW) [20].

■ Hydro <1 MW   ■ Hydro 1–10 MW   ■ Hydro 10+ MW   ■ Mixed plants
■ Pumped storage   ■ Marine   ■ Onshore wind energy ■ Offshore wind energy
■ Solar photovoltaic ■ Concentrated solar power ■ Solid biofuels ■ Liquid biofuels
■ Biogas   ■ Geothermal

wind and solar photovoltaic (PV) take second and third place, respectively, while other alternatives are still under further development.

4 ENERGY CONSUMPTION IN OIL AND GAS INDUSTRIES

Various phases of oil and gas extraction, processing, transport, and end-use require a significant amount of energy. GHGs originate from different sections of these industries including onshore/offshore oil and gas extraction, pipeline, rail and marine transportation of crudes and refined products, gathering and boosting stations, natural gas compressor stations, LNG storage, petroleum refineries, natural gas processing plants, and LNG import/export facilities [21, 22].

The world crude oil and natural gas consumption in 2016 was 4418 Mt and 3204 Mtoe, respectively [23]. In the same year, the worldwide refinery throughput was 4121 Mt. Fig. 6 shows the breakdown of energy consumption in the chemical and petrochemical industries,[1] pipeline transport,[2] as well as textile and leather industries.[3] The total energy consumption in chemical and petrochemical industries in 2015 was 438 Mtoe, with 70 Mtoe for the mining and quarrying sector.[4]

Based on the available data, the shares of the refining, bulk chemicals, and mining sectors in industrial energy consumption in 2016 were 18%, 25%, and 12%, respectively [24]. Another study reveals that these industries consumed about 6.9% of the total energy produced by the oil and gas industries [25].

From a cost perspective, the energy input cost of the oil and gas industry in the United States was around $720 M in 2016 [26]. Assuming an average crude oil price of $50/barrel, the density of crude oil of $900 kg/m^3$, the energy consumption of the US oil and gas industry in 2016 would have been approximately 2.06 Mtoe. The total energy consumption in the US mining and quarrying sector in 2015 was 8 Mtoe [6].

Energy intensity (EI) is the total amount of energy requirement in an oil and gas production process as a percentage of the total exported products from the production facility. As production rates of an oil and gas field decline, the associated energy demand of that field does not reduce, implying an increased EI. Several techniques can be deployed to increase production

---

[1] Includes (1) manufacture of basic pharmaceutical products and pharmaceutical preparations, (2) manufacture of human-made fibers, (3) manufacture of basic chemicals, fertilizers and nitrogen compounds, plastics and synthetic rubber in primary forms, and (4) manufacture of other chemical products (ISIC Divisions 20 and 21).

[2] Comprises the amount of energy used to support and operate pipelines transporting gases, liquids, slurries, and other commodities, together with the energy used for maintenance of the pipeline and pump stations.

[3] Manufacture of textiles, wearing apparel, leather and related products, according to International Standard Industrial Classification (ISIC) Divisions 13 to 15.

[4] Includes (1) extraction of crude petroleum and natural gas, (2) mining support service activities, (3) mining of metal ores, (4) mining of coal and lignite, and (5) other mining and quarrying (ISIC Divisions 07 & 08 and Group 099).

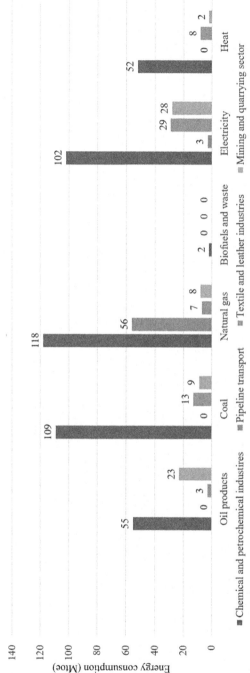

Fig. 6 Breakdown of energy consumption in various sectors in 2015. It includes LPG, aviation gasoline, motor gasoline, refinery gas, ethane, gas/diesel oil, fuel oil, jet fuels, white spirit, lubricants, paraffin waxes, petroleum coke, kerosene, naphtha, bitumen, and other oil products. *(Data from Offshorewind. IN DEPTH TECHTALK: Floating wind-powered water injection. [cited 06 Jan. 2018]. Available from https://www. offshorewind.biz/2016/11/25/in-depth-techtalk-floating-wind-powered-water-injection/.)*

Table 2 Key events in the lifetime of an oil/gas field (sorted based on the life of the field) [27]

Event	Impact on energy intensity (GJ/GJ of product)
Reduced operating pressure (more evident in gas fields)	↑
Gas lift	↑
Modification of facilities	No effect
Power sharing between installations and facilities	↓
Water injection to the field	↑
Suboptimal throughput	
Water cut	↑
Switch off train	↓
Switch off spinning reserve	
Increased downtime	↑
Compressor recycling	↑
Real-time optimization	↓
Fuel gas import	No effect
Gas-to-oil ratio	↓

rates of a field. The energy consumption relative to the production rate of a field increases later in the lifetime of an oil and gas field, leading to a declining shift in energy efficiency indicators. The key events in the lifetime of a field, which drive around 90% of the total energy requirements, are summarized in Table 2.

The key parameters affecting the energy demand as well as the energy performance of an installation are summarized in Table 3 [27].

The energy intensity (EI) of five oil fields studied by Vanner [27] varied from 0.7% to 2.7% (mean EI = 1.6%). However, the EI of seven oil and gas fields ranged from 1.4% to 3.7% (mean EI = 2.0%), whereas the EI of three gas fields varied from 2.3% to 6.5% (mean EI = 4.7%). In addition, the EI of all cases was low and increased during the first 15 years of production life.

Fig. 7 is a schematic of an oil extraction site [28]. The energy and mass flows of 306 fields in California are shown in Table 4. From this table, it can be seen that gross crude oil flow declined from 2488 to 1446 Peta Joule/year (41.88% depletion). However, the quantity of external energy increased by 4.78% from 1985 to 2005. Self-consumption of crude oil in 2005 was zero.

The offshore oil and gas industries are significant emitters of GHG. In the United Kingdom, for example, the sector emitted 14.64 Mt CO_{2e} in 2016 [27]. Significant GHG emissions from Norwegian petroleum extraction

Table 3 Parameters that affect energy demand as well as energy performance of oil/gas fields

	Parameter or element of oil/gas installation
Main energy factors of an oil field	• Ratio of field pressure to export pressure • Field uptime • Energy requirement for activation of gas lift or water injection
Main energy factors of a gas field	• Ratio of field pressure to export pressure • Field production rate • Ratio of field production rate to original design capacity (recycle percentage)
Main process issues	• Ratio of field pressure to export pressure • Change in gas-to-liquid as well as gas-to-oil ratios with time • Ratio of devices throughput to original design capacity • Unplanned outages
Equipment issues	• Turbines, motors, compressors • Submersible pumps • Water injection and gas lift facilities • Need for backup devices (e.g., spinning reserve on generators) • Field modifications and upgrades (modern technologies) • Availability of control systems for compressors and turbines

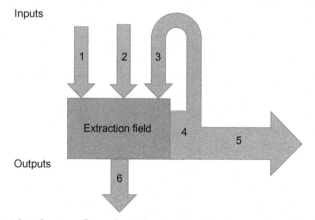

Fig. 7 Material and energy flow of an oil extraction site, streams: 1: external energy, 2: refined products, 3: self-consumption of crude oil, 4: gross crude oil flow from the field, 5: net crude oil flow from field to the refinery, 6: wasted heat + wasted gas (venting + flaring).

Table 4 Energy flow in California oil extraction (Peta Joule/year) [28]

Flow	1985	2005
Gross crude oil flow	2488	1446
Net crude oil flow to the refinery	2342	1446
External energy	169	243
Refined products	1	0
Self-consumption of crude oil	146	0
Wasted heat + wasted gas	0	0

come from gas turbines located on offshore platforms. These turbines are much less efficient than large-scale gas turbines [29]. In Norway, the amount of CO_2 emissions per unit of oil and gas from 1997 to 2012 increased significantly as a field's extraction rate declined [29]. A field's share of oil in the total original oil and gas reserves also increased the CO_2 emission intensity.

All these statistics highlight the energy and emission intensity of the oil and gas supply chain. If these operations could be integrated with clean energy sources, not only could their emission footprint decline, but they may also be able to reduce their production costs when technoeconomically alternative energy sources are available.

5 UPSTREAM: APPLICATION OF RENEWABLE ENERGY IN OIL AND GAS EXTRACTION

New applications of renewable energy in the oil and gas fields are being introduced at an increasing pace. Choi et al. [30] reviewed applications of photovoltaic, wind, geothermal, and solar thermal energy in the oil and gas fields, summarized here in Table 5.

Among these, offshore wind is receiving marked industrial attention both for export power or onsite applications (see Fig. 8). MacDonald [31] determined the energy demands of a sample oil rig living quarter. He examined the ability of several renewable sources including wind and wave power to satisfy energy demands (instead of fossil fuels transferred either via pipeline or fueling ships). Renewable sources (together with storage devices) were able to produce sufficient power for the offshore oil platforms evaluated.

Wei et al. [32] examined the technical feasibility of deploying an offshore wind farm to supply supplementary power for offshore oil and gas platforms and to export the surplus energy to an onshore grid. Three wind farm cases with the rated capacities of 20, 100, and 1000 MW were studied. The foci of the study were the reduction of CO_2 and NO_x emission, the stability of the

Table 5 Application of renewable energy in oil and gas production industries [30]

Renewable energy technology	Oil/gas field (OF/GF)	Rated capacity (kW)	Name of field (country)	Operator
Photovoltaic	OF	500	Midway-Sunset Oil Field (USA)	Chevron Texaco
		750	Kern River Oil Field (USA)	Chevron Texaco
		17.85	Louisiana Bayou Oil Field (USA)	Kyocera Solar
Wind	OF	1500	Suizhong 36-1 Oil Field (China)	CNOOC
		5000	Beatrice Oil Field (Scotland)	Scottish and Southern Energy, Talisman Energy
		6000	Utsira Nord Oil Field (Norway)	DNV GL of Norway
Geothermal	OF	217	Rocky Mountain Oil Field Testing Centre (USA)	–
	OF/GF	700–1000	Fort Liard (Canada)	Thompson and Dunn
Solar thermal	OF	300	McKittrick Oil Field (USA)	Glass Point Solar
		29,000	Coalinga Oil Field (USA)	Bright Source
		7000	Amal Oil Field (Oman)	Glass Point Solar

electrical grid, and the technical feasibility. All three cases were found to be theoretically feasible. However, logistical, operational, and economic issues must be taken into account.

Kolstad et al. [33] investigated the possibility of providing electricity for offshore oil and gas platforms from offshore wind power and an onshore electricity grid. A system including five oil and gas platforms was considered. The results demonstrated the feasibility of the system configuration. In 2015, BG group began the first oil production from the Mukta-B platform (Bombay Offshore Basin, India) powered by solar panels and wind turbines [34]. Design and performance validation of an offshore wind and wave energy floating platform were performed by Hanssen et al. [35], who analyzed two sites, 20 km off the coast of the Iberian Peninsula at the depth of

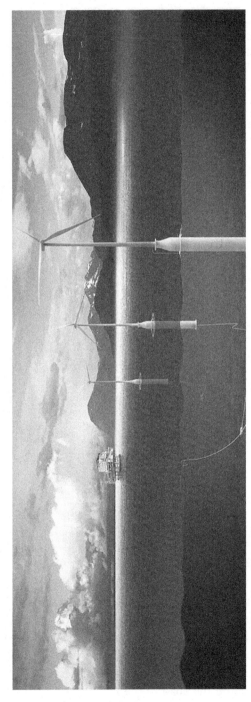

Fig. 8 Illustration of wind power for the onsite power consumption of oil and gas production systems. The DNV GL-led WINd powered Water INjection, WIN WIN, project. *(Illustration by Edmond Yang, Copyright DNV GL.)*

around 200 m, and a North Sea (west of Norway) location some distance between platforms in the Norwegian Continental Shelf and the mainland. Favorable economic results as well as encouraging technical performance were reported.

Shell designed a Monotower Platform with two wind turbines as well as two arrays of PV panels on board [36] (see Fig. 9). The configuration led to reduced cost and environmental impacts compared with subsea power cable. Two platforms of this type were installed in 2005 (K17 in The Netherlands and Cutter in the United Kingdom). Cornelia and Davies provided a rank/ value for offshore power sources from both conventional-energy and

Fig. 9 Shell's Monotower Platform with solar, wind and battery. *(Courtesy of TSS4U B.V.)*

renewable-energy sources [37]. The renewable energies assessed had solar, wind, and ocean sources. The authors claimed that although the wind and solar technologies were mature, ocean thermal energy conversion technology offered good potential in some offshore locations. Another study addressed the potential for producing power from the water produced from oilfields for facility consumption [38]. A pilot plant based on this concept came into operation at the Rocky Mountain Oilfield Testing Centre (RMOTC). The initial unit was a 250-kW Organic Rankine Cycle (ORC) power plant. Gong et al. studied the impact of the temperature of injected water and the water injection rate on reservoir temperature during electricity generation from the LB oil reservoir (located in Hebei province, northern China) [39]. Their results indicated that an increase in the water injection rate increased oil production, which would be helpful in the conduct of electricity generation projects. The application of geothermal energy sources in the oil and gas industries was addressed in [40, 41].

A more recent example of such developments is the WIN WIN (WINd powered Water INjection) joint industry project led by DNV GL, which has justified the feasibility of floating wind-powered water injection, particularly for marginal fields requiring enhancement of recovery [5]. The power generated is used for pumping and basic water treatment [42]. In 2016, DNV GL and partners claimed a saving of $3/barrel of oil using a 6-MW wind turbine driving two units of 2MW water injection pumps into offshore oil wells [43]. It was an autonomous system, and the injection of 44,000 barrels of processed water was performed through risers. For each well, the annual amount of CO_2 abatement was reported to be 16,500 tons.

Another application of renewable energy in oil extraction is the use of solar energy for enhanced oil recovery (EOR). Solar energy is used to heat water and produce steam (see Fig. 10). According to the IEA [44], CSP technology could be widely used in providing high-temperature process heat or steam for EOR. The generated steam is then injected into an oil reservoir to reduce the viscosity of the fluid, facilitating its flow to the surface. Solar EOR is a viable alternative to conventional gas-fired steam generation for the oil industry. Gupta and Laumert [45] investigated the market potential for solar thermal EOR while reviewing 12 major ongoing steam injection EOR projects globally. They also carried out a technoeconomic analysis for Issaran Oil Field in Egypt. The levelized cost of solar steam for that oilfield (oil production rate of 5000–6000 bbl/day) was estimated to be 17.3 $/ton. The internal rate of return was found to be 27.4% with a payback period of 6 years.

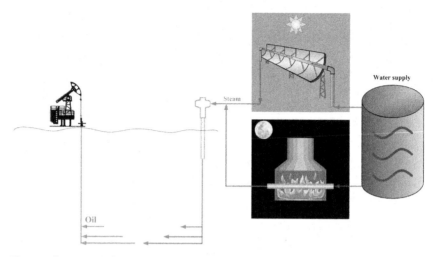

Fig. 10 Illustration of CSP-integrated EOR.

In early 2011, the GlassPoint company revealed the first commercial solar EOR project with a capacity of $300\,kW_t$ in McKittrick, California [46]. The same year, Chevron Technology Ventures and BrightSource built the world's largest 29-MW_t solar EOR demonstration facility in Coalinga, California [47, 48]. A recent large-scale industrial example of the use of solar energy is the production of 850-MW_t plus $26.5\,MW_e$ energy in Belridge oilfield in Kern County, California, by GlassPoint [46]. The project is anticipated to produce electricity and steam as early as 2020. The solar thermal facility generates 12 million barrels of steam per year, while the 26.5-MW PV system facility produces electricity to power oilfield onsite operations. The generated steam from sunshine as well as the PV system results in a saving of 4.87 bcf (billion ft^3) of natural gas per year. Moreover, the CO_2 abatement potential by the Belridge Solar project is >376,000 tons/year, equal to offsetting emissions from 80,000 cars annually. The GlassPoint company with Petroleum Development Oman (PDO) commissioned a pilot in 2013 for the EOR [46]. The success of the pilot paved the way for the Miraah 1.021-GW solar thermal project, which is now under construction. The Miraah project will produce 6000 tons of steam per day. Solar-generated steam would enable Oman to save up to 80% of the natural gas currently used for EOR operations.

For obvious reasons, ocean energy (wave and tidal) is of notable interest for providing the energy demands of offshore operations. Though most ocean technologies are still in the early stage of their learning curves and relatively distant from market competitiveness, some successful implementations exist. For instance, OgWAVE is a deep-water device to harness wave energy, which can be attached either to the legs of oil and gas platforms or moored near the platform [49]. A single OgWAVE device can provide 500 kW or more of electrical power. Alternatively, an aquaWAVE device can be used for 10–15 m of water depth [50]. One aquaWAVE device would provide 1 MW of electrical power.

It is important to note that both CO_2 and nitrogen can be used for enhanced gas recovery (EGR) [13, 51, 52]. Nitrogen can be sourced from air separation units (ASU). However, the energy consumption in ASU plants is 400 kWh/ton of oxygen [53]. CO_2 requires more compression than nitrogen, and greater amounts of CO_2 are required to boost pressure in gas reservoirs. Consequently, the EGR method using pressurized nitrogen is more efficient than the CO_2 alternative in terms of energy consumption. The CO_2 sourced from underground wells is an economical alternative for EGR in the United States. Alternatively, in jurisdictions with a carbon tax, CO_2 captured from other industries may be viable for EOR [54].

Water desalination systems are among the major energy-intensive processes. Water desalination in the oil and gas industries is used for the following reasons:

1. production of drinking water onshore/offshore
2. removal of sulfate during waterflooding to improve oil recovery
3. enhancement of oil recovery with a steam flood, polymer flood, or low-salinity waterflood. Injection of low-salinity water to oilfields increases crude oil recovery [55].
4. facilitation of hydraulic fracture flow recycling using salt-sensitive polymers

The desalination market shares in the oil/gas and petrochemical industries are 14.2% and 17.1%, respectively [55]. According to Adham [55], an average of 3–4 barrels of water is produced with one barrel of crude oil from a field. The role of water is critical for unconventional oil production industries [55]. An average of 2–4 barrels of water is required per barrel of oil produced for oil sands and around 90% of the water is recycled ($\sim$10% make-up water is required) [55]. However, the estimated amount of water required for unconventional reservoirs is 50,000–150,000 barrels for hydraulic fracturing with about 15%–80% of the water recycled.

The application of renewable energy sources such as solar, wind, and geothermal in desalination processes was investigated by Bourouni [56]. The major desalination technologies include the multistage flash process

(MSF), reverse osmosis (RO), and multieffect distillation (MED) [56–62]. MED and MSF are considered thermal desalination technologies, whereas RO is classified as a membrane technology. Some other desalination technologies, which have limited applications, are electrodialysis (ED), mechanical vapor compression (MVC), and humidification and dehumidification of air (HDA) [56]. The energy requirements for water desalination are categorized as electrical energy to drive mechanical equipment such as pumps, compressors, auxiliary equipment, and thermal energy to heat the water. The thermal energy consumption of MSF and MED technologies for seawater desalination is 190–290 and 150–290 kJ/kg, respectively [63]. However, the electrical energy requirement for seawater desalination with MSF, MED, MVC, or RO with/ without heat recovery is 4–6, 2.5–3, 8–12, 7–10/3–5 kWh/m^3, respectively. On the other hand, the electrical energy requirements for brackish water desalination with RO with/without heat recovery and electrodialysis are 1–3/1.5–4 and 1.5–4 kWh/m^3, respectively. Breakdown of the global desalination processes powered by renewables indicated that PV/RO had the highest share with 32% while PV/ED had a 6% share. On the other hand, wind/RO share was 19%, followed by solar/MED (13%), solar/MSF (6%), wind/MVC (5%), hybrid (4%), and other (15%).

Natural gas flaring is a common practice in exploration, production, and processing operations. In 2015, around 147 bcm (billion cubic meters) of natural gas was flared, which represented about 4% of the global natural gas production. Gas flaring in 2015 showed an increase from the 2014 level (145 bcm) [64]. Russia, Iraq, Iran, the United States, and Venezuela flared 21, 16, 12, 12, and 9 bcm, respectively. Flare gases from oil and gas production industries offshore can be converted to methanol (known as a green methanol project) to reduce CO_2 emissions [65]. The project continues to be developed within Maersk floating production storage and offloading (FPSO).

6 DOWNSTREAM: APPLICATION OF RENEWABLE ENERGY IN OIL REFINING

6.1 Energy and Emission Intensive Processes of Refineries

Several aspects of our daily life rely on refinery products. The average worldwide demand for refined products in 2016 was 96.5 million barrels per day (Mbd) [23]. Oil refining is the most energy-intensive sector within the oil and gas supply chain [25]. The energy efficiency of refineries can be improved, leading to lower energy intensity. For instance, by 2013, the energy intensity of refineries in OECD countries had reduced by 13% since 1980 [66].

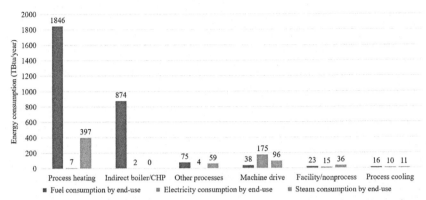

Fig. 11 Breakdown of fuel, electricity, and steam consumption in the US petroleum refining by end use, 2010 [67].

Fig. 11 shows the breakdown of fuel, electricity, and steam consumption in the US petroleum refining sector by end use in 2010 [67]. As evident from the figure, most of the fuel (64%) was used for process heating equipment including heated reactors, fired heaters, and heat exchangers. The most common electricity use in refineries was for machine-driven devices including compressors, pumps, fans, and materials handling, and processing equipment. A majority of the onsite- and offsite-generated steam was used for process heating, steam turbines, process cooling and refrigeration, and other process end uses.

Obviously, the amount of CO_2 emissions from refineries is affected by various factors such as crude oil quality, end product mix, and local standards. The average global refinery CO_2 emissions originate from the combustion of refinery fuels (48%), emissions from the catalytic cracking processes (28%), power generation (13%), and hydrogen plants (11%) [68]. For example, according to Ref. [69], for a typical crude oil refinery, the CO_2 emission distribution showed that three crude oil distillation units accounted for ca. 21.1% of the total CO_2 emission. The CO_2 emitted from the distillation units was related to fuel combustion to preheat feed streams. Steam plant, along with cogenerator, was the next largest emitter with 20.5% and 18.4% of the total refinery emissions, respectively. The rest of the CO_2 emissions came from hydrotreater/continuous catalyst regenerator (7.9%), reformer (7.5%), flare (6.6%), hydrocracker (4.4%), H_2 plant (4%), vis-breaker (1.6%), diesel hydrotreating unit (0.3%), and sulfur removal unit (0.2%). Indirect emissions (purchase of electricity) contributed to 7.6% of the total CO_2 emission. In a steam plant, however, around 30% of the total fuel gas is consumed.

The key challenge of contemporary oil refining is sustainable operation through improvement of energy efficiency and reduction of emission footprint. Brueske and Fisher [67] utilized four different energy measures (bands) to describe onsite energy consumption as well as potential saving opportunities in the petroleum refining industries:

- Energy consumption in a base year (say 2010)
- State of the art (SOA): the minimum energy requirements when deploying the best available technologies and practices
- Practical minimum (PM): the minimum energy requirements when employing applied R&D technologies, which are under development globally
- Thermodynamic minimum (TM): the minimum energy requirements under ideal conditions—unattainable in real applications

Nine processes (see Fig. 2) were selected for the analysis:

- Atmospheric crude distillation
- Vacuum crude distillation
- Alkylation
- Catalytic hydrocracking
- Fluid catalytic cracking (FCC)
- Isomerization
- Catalytic reforming
- Coking/visbreaking
- Hydrotreating

The energy intensity of refinery processes is illustrated in Fig. 12 [67]. Two potential energy-saving opportunities (bandwidths) were defined: (1) current opportunity (OP1 = the 2010 energy consumption, SOA), and (2) R&D opportunity (OP2 = SOA-PM). For the nine selected processes, the amount of energy savings for the OP1 and OP2 opportunity bandwidths was estimated to be 286 (14% saving) and 540 (26% saving) TBtu[5] per year, respectively. It is important to note that the thermodynamic minimum energy consumption of the nine selected processes was 115 TBtu/year.

However, the estimated energy savings for the entire US crude oil refining sector of OP1 and OP2 were 420 and 793 TBtu/year, respectively (Fig. 13). The thermodynamic minimum energy consumption of petroleum refining sector-wide was 169 TBtu/year.

---

[5] Trillion Btu.

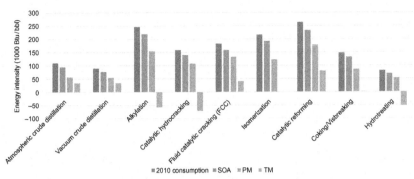

Fig. 12 Energy intensity of refinery processes [67].

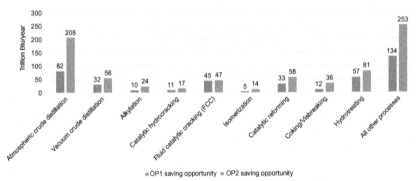

Fig. 13 OP1 and OP2 energy-saving opportunities for the US petroleum refining sector-wide [67].

6.2 Hydrogen Management in Oil Refineries

As shown in Fig. 2, several processes in an oil refinery rely on hydrogen as feedstock. Traditionally, this demand has been supplied by natural gas reforming. Despite hydrogen abundancy, access to elemental hydrogen is cumbersome. It is not a constituent of air and the only pathway to obtain it is through chemical reactions. Even the early discovery of hydrogen occurred only when Robert Boyle was experimenting combining iron with acid ($Fe + H_2SO_4 \rightarrow Fe^{2+} + SO_4^{2-} + H_2$) and observed an inflammable gas later called hydrogen [70].

Obviously, today the simplest pathway for H_2 production is water splitting. However, that approach has not been traditionally favored due to its high energy demand. The most widely used approach has been fossil fuel reforming for syngas (CO/H_2) generation followed by water-gas shift

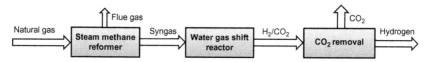

Fig. 14 Schematic of hydrogen production from natural gas.

(WGS) reactions $(CO + H_2O \rightleftharpoons CO_2 + H_2)$, and CO_2 removal (Fig. 14). Though coal, biomass, and oil gasification are also possible approaches for syngas generation, natural gas is a preferred feed due to its high H/C ratio. The chief reforming technologies for syngas production using natural gas are catalytic SMR (steam-methane reforming), DMR (dry-methane reforming), ATR (autothermal reforming), and POX (partial oxidation) [71–74]. Among these options, SMR is the prevalent technology for hydrogen production as it leads to syngas with the highest theoretical H_2/CO ratio $(CH_4 + H_2O \rightarrow 3H_2 + CO)$.

The most critical application of hydrogen has been in refining and ammonia (followed by urea) production. Crude oil refining is constituted of several processes for separation of certain so-called petroleum cuts such as LPG, petrol, naphtha, diesel, and fuel oil. Given the various quality standards for each of these products, several treatment processes are required for removing polluting elements such as mercury, sulfur, nitrogen, and aromatics and breaking long hydrocarbon chains. As such, hydrotreatment and hydrocracking are two essential components in any refinery.

Hydrogen is used in refinery processes such as hydrocrackers and desulfurization units [75]. The production of hydrogen is energy intensive, using natural-gas-fueled reformers and naphtha reformers.

A small fraction of the hydrogen demand of refineries is supplied by their byproducts, mainly from the fluid catalytic cracking (FCC) unit and catalytic reforming. Some secondary units that are not necessary components exist in refineries, which could be installed when justified by the economics of the process. Fluid catalytic cracking (FCC) is one such secondary unit, used for generating much a finer boiling range with cracking of large-chain molecules (e.g., gas oil). Unlike hydrocracking, which is important in locations with high diesel demand (such as Asia), FCC is installed when there is a high demand market for additional gasoline and distillate fuels (such as in the United States). The exiting gas stream (fuel gas, methane, ethane, etc.) from the FCC unit contains some H_2, which can be removed with various approaches for feeding into hydroprocessing units. Naphtha is an intermediate refining product with molecular weight properties close to that of gasoline but with a poor octane number. Catalytic reforming converts isomeric paraffins (linear hydrocarbons) into isoparaffins (branched alkanes) and cyclic naphthenes. It also aromatizes paraffins and naphthenes to aromatics, a process that generates a significant amount of hydrogen. The total demand of refineries with the subtraction of the byproducts from FCC and catalytic reforming has been conventionally supplied by fossil fuel reformers.

With increasing sustainability concerns, interest in renewable hydrogen has been accelerating. It can be produced from biomass through gasification, or renewable thermal and electrical energies can be utilized for water splitting. The intermediate product of biomass gasification is syngas, which can also be converted to liquid fuels in a Fischer-Tropsch synthesis reactor [53, 76–82]. Trainham et al. [83] investigated critical issues of four solar fuel technologies including photoelectrochemical (PEC) systems, solar thermochemical cycles (STC), solar-powered water electrolysis, and solar biomass gasification. They found that solar fuel technologies could not compete with the oil and natural gas prices in 2009 [83]. The costs of hydrogen production from steam methane reforming, STC, PEC, water electrolysis, and solar biomass gasification were $1.14, $3.4–3.8, $5.52, $5.58–6.91, and $1–1.3/kg, respectively. For synthesis gas production, the solar biomass gasification process was introduced as the more competitive route versus steam reforming of methane. The steam and hydrogen (or H_2/CO) produced from a solar reforming process can be used in oilfields and petroleum refineries [83]. However, heavy residual oil from petroleum refineries can be cracked in a solar cracking process, and the cracked products are pumped back to the refinery for further upgrading and processing (Fig. 15).

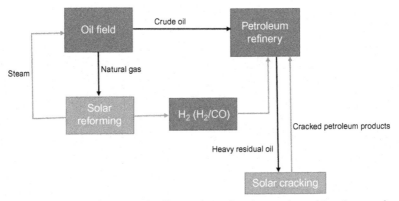

Fig. 15 Potential application of offsite solar reforming and cracking in petroleum refining [83].

The cracked residue of a heavy oil upgrader can be fed to a solar gasification process. The steam and CO_2 of the solar gasifier can be used in EOR, while the H_2 (or H_2/CO) effluent can be used in a refinery [83].

The solar thermal cracking of methane to produce hydrogen and carbon was investigated by Maag et al. [84]. Sheu et al. reviewed solar reforming systems including dry reforming with no catalyst, dry reforming of methane using porous ceramics and metal foam, dry reforming with honeycomb absorber, dry reforming with metal oxide catalyst in liquid bed, dry reforming with heat pipe, steam reforming in packed bed, and use of a tubular reactor [85].

The steam gasification of two types of petroleum coke, Flexicoke and Petrozuata Delayed coke, via concentrated solar radiation was investigated by Trommer et al. [86]. A hybrid combination of solar/petroleum coke thermochemical processes reduced emissions as well as conserved petroleum resources. The molar ratio of H_2 to CO was around 1. Z'Graggen and Steinfeld designed a solar reactor for steam gasification of carbonaceous feedstocks (e.g., biomass, petroleum coke, coal) [87]. The solar reactor was a 5-kW prototype. The mole fractions of effluent H_2, CO, and CO_2 were 0.66, 0.25, and 0.09, respectively. Gálvez et al. conducted a kinetic and thermodynamic analysis of a two-step thermochemical cyclic process for hydrogen production (see Fig. 16) [88].

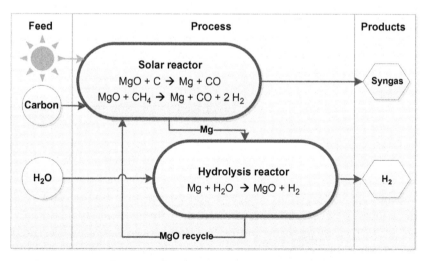

Fig. 16 Schematics of solar hydrogen production with MgO/Mg thermochemical cyclic process [88].

In the first step, petroleum coke and wood charcoal were fed to a solar reactor. CSP technology was used to produce Mg by endothermic methanothermal $(MgO + CH_4 = Mg + CO + 2H_2)$ or carbothermal $(MgO + C = Mg + CO)$ reduction of MgO. In the next step, Mg and water were fed to a hydrolysis reactor to produce hydrogen $(Mg + H_2O \rightarrow MgO + H_2)$. The second step was exothermic, and MgO was recycled to the solar reactor. Frost et al. investigated synthesis gas production via solar concentrator PV (SCPV) and high-temperature coelectrolysis (HTCE) of H_2O and CO_2 [89]. The electricity and heat generated from the SCPV system was used by the HTCE technology to electrolyze H_2O and CO_2. The synthesis gas produced could then be converted to a variety of products including chemicals and FT-derived liquid fuels.

Mohsin et al. investigated the economic viability of wind-generated hydrogen for four proposed sites in Pakistan [90]. The production rate of H_2 in the summer season in Baghan, Golarchi, Nooriabad, and DHA Karachi was 2.519, 2.460, 3.162, and 2.404 tons/day. All the sites were commercially viable for renewable hydrogen production. The supply cost of hydrogen at all sites was $5.30 − $5.80/kg$_{H_2}$. Hydrogen production from excess renewable energy sources in California through commercial electrolysis for fuel cell electric vehicles was studied in Ref. [91]. The 12,000 GWh of excess renewable energy in 2030 could produce 243,000 tons of hydrogen from electrolysis. The benefit-to-cost ratio was estimated to be >1. An optimization model was developed by Kim and Kim for an integrated renewable hydrogen supply in Korea [92]. The primary renewable sources for hydrogen production were solar, wind, and biomass. The production, storage, as well as transportation of H_2 was considered. The potential for hydrogen production from solar, wind, and biomass sources in Argentina was addressed by Sigal et al. [93]. The potential of renewable hydrogen production was found to be around 1 billion tons/year. Hydrogen production from a methanol reforming process powered by solar energy was considered in Ref. [94]. The heat requirements of the methanol reforming to hydrogen were supplied by nonconcentrating solar collectors. The highest total efficiency of 43% was reported. In another work, Serna et al. detailed the production of renewable hydrogen by water electrolysis in an offshore plant [95]. The required power came from wave energy converters and wind turbines. The production of methane and hydrogen-powered by excess electricity sourced from renewable sources including photovoltaic or wind power was addressed in Ref. [96].

7 RENEWABLE ENERGY IN CHEMICAL INDUSTRIES

For cleaner production practices, some studies have investigated the integration of renewable energy with carbon capture and utilization (CCU) processes (Chapter 8). The production of methanol from renewable hydrogen and CO_2 was investigated by Boretti [97]. Technoeconomic analysis of methanol production from biomass gasification and water electrolysis was performed by Clausen et al. [98], who studied six plant configurations. The oxygen to the gasifier and autothermal reforming of all plant configurations was sourced from a water electrolyzer. The exergy efficiency of methanol plants ranged between 59% and 72%. Masel et al. [99] investigated three possible ways of supplying energy for CO_2 conversion to chemicals, namely, including fossil fuels, concentrated solar thermal, and renewable electricity. In another study, a new process for CO_2 utilization through conversion to fuels and chemicals was introduced by Masel et al. [100]. The solar-energy-based process began on the cathode of an electrolyzer to convert CO_2 to CO and formic acid (HCOOH), which then underwent further reactions on mixtures of imidazolium compounds and active metals to produce propylene, gasoline, formaldehyde, etc. The conversion of CO_2 to CO was over 98% at an overall energy efficiency of 80%. The authors claimed that the production of formic acid, formaldehyde, carbon monoxide, and acrylic acid was economically feasible at CO_2 separation costs of $60/ton and without a tax on emissions. A life cycle analysis (LCA) of electrochemical reduction of CO_2 to produce formic acid was conducted by Dominguez-Ramos et al. [101]. Sustainability evaluation indicated that energy and materials consumption was too high. Therefore, by integrating renewable energy, reduction in the consumption of chemicals and a formic acid alternative purification to conventional distillation were proposed. Catalytic conversion of propane in the presence of CO_2 was studied by Du et al. [102] in the temperature range of 300–500°C, using solar thermal energy. The application of renewable energy and catalytic aspects of some reactions, including the production of ethylene/propylene from CO_2, conversion of CO_2 to syngas, DME, methanol, formic acid, FT synthesis, and methane was addressed by Centi et al. [103]. The concept of power to methanol was developed by the Carbon Recycling International (CRI) company. The world's largest CO_2 methanol plant located in Svartsengi, Iceland, began operation in late 2011 [104]. In 2015, CRI boosted the plant production capacity from 1300 to >5000 tons/year.

Today (2017), the plant recycles 5500 tons of CO_2 per year, which would otherwise be released into the atmosphere. The plant energy requirements come from renewables, i.e., geothermal and hydro sources. Hydrogen from water electrolysis is converted into methanol in a catalytic reaction with CO_2.

8 RENEWABLE ENERGY APPLICATIONS IN THE SHIPPING SECTOR

Globally, around 90% of all trades in goods, raw materials, and energy in recent years has been transported by the shipping industry [105]. According to the United Nations Conference on Trade and Development (UNCTAD), estimated international seaborne trade increased from 2.6 to 10.05 billion tons loaded between 1970 and 2015 [106]. The trend in shipping demand is anticipated to grow even further. Breakdown of the international shipping trade in 2015 implies that the share of oil and gas cargoes by weight in the international trade was about 2.95 billion tons (29.3%) [106]. A more precise measure of the global demand for seaborne trade capacity is obtained by multiplying the weight of cargoes by the distance traveled. In 2015, the total of shipped cargoes by weight and distance was 86.24 trillion tons × km. The share of shipped oil, gas, and chemicals cargoes by weight and distance traveled was 27% of the total shipped cargoes.

The required energy for the propulsion of ships has been evolving significantly over the past decades. The shipping industry relies mainly on oil as its energy source. The estimated amount of fuel consumed by the world's marine fleet between 2007 and 2012 ranged from 247 to 325 million tons [107]. The global demand for marine fuel in 2016 was met mainly by fuel oil (194.499 million tons) and gasoil (31.225 million tons), which are the two fuels with the highest environmental impacts. Data of 2012 show that international shipping accounted for 2.2% of global GHG emissions (796 million tons of CO_2, to which can be added 12.7 and 9.8 million tons of N_2O and CH_4, respectively). The breakdown of CO_2 emissions from different shipping sectors by fuel type is illustrated in Fig. 17. International oil tankers, chemical tankers, and liquefied gas tankers emitted 124, 55, and 46 million tons of CO_2 in 2012, respectively [107]. This means that about 28.3% of the total CO_2 emitted from international ships originated from oil tankers, chemical tankers, and liquefied gas tankers.

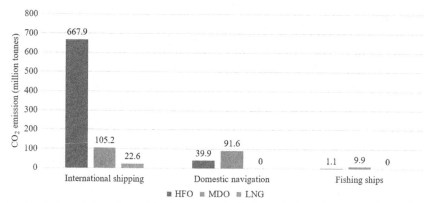

Fig. 17 CO_2 emissions from shipping sectors by fuel type in 2012 [107], HFO (heavy fuel oil), MDO (marine diesel oil), LNG (liquefied natural gas).

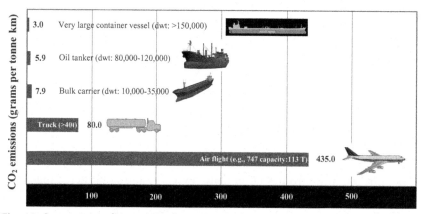

Fig. 18 Comparison of typical CO_2 emissions between modes of transport [109].

It is important to note that the shipping industry, compared to other means of transport, emits the lowest CO_2 emissions per kilometer per ton [108]. Fig. 18 compares typical CO_2 emissions between various modes of transport. Whereas on average, air flight emits $435\,gCO_2/km/ton$, and the value for large trucks is $80\,gCO_2/ton$-km, tankers consume $<7.9\,gCO_2/ton$-km. The emission footprint of ships reduces as their size or deadweight increases. Deadweight tonnage is defined as the total weight of cargo, fuel, fresh water, ballast water, provisions, passengers, and crew.

Mandatory operation and technical measures have been stipulated by the International Convention for the Prevention of Pollution from Ships (MARPOL), aiming for CO_2 emission reduction and efficient energy use. Considering 2008 as the base year, it has been estimated that the shipping sector can reduce its CO_2 emissions per ton per km traveled by at least 20% and 50% until 2020 and 2050, respectively [105]. To meet these targets, cleaner power and fuel options must be considered by ship operators.

Renewables can be used in all types of ships for primary, hybrid, and/or ancillary propulsion, along with onboard or onshore energy use. Using a hybrid combination of renewable sources increases the complementarity and availability of energy sources.

Potential renewable energies for shipping applications are solar PV, wind (e.g., fixed sails, soft sails, kite sails, Flettner rotors, as well as conventional wind turbines), wave energy, biofuels, and use of electricity capacitors charged by renewables. It is noteworthy that small ships (i.e., DWT <10,000) are more predominant globally. Small ships compared to larger ships transport a lower amount of the total cargo while emitting more GHG per kilometer per weight of load [108].

The following benefits can be achieved by using renewable energy sources in the shipping sector [108]:

1. Direct economic benefits:
   * fuel cost substantially reduces
   * security of supply and stability in energy prices increase
   * wear and tear leading to increased stability reduce
   * port fees, as well as local levies, potentially reduce
2. Indirect economic benefits:
   * resilience to externalities increases
   * uneconomical routes potentially revive, resulting in increasing trade system access together with motivating regional growth
   * new or expanded existing industries are created
   * competition in energy supply chain increases
   * death toll related to shipping emissions reduces, in turn lowering healthcare costs
3. Social/environmental benefits:
   * GHG emissions significantly decrease
   * risk of fuel spills and consequently environmental damage diminishes
   * marine noise decreases
   * health benefits arise
   * the danger of exposure of passengers and crew to flammable fuels drops
   * new challenges are encountered by deploying new renewable devices and technologies
   * new employment opportunities are created

According to the International Renewable Energy Agency (IRENA) [108], advanced biofuels have high potential to supply energy choices of the shipping sector. Also, hydrogen fuel cells can be used for power supply. However, the production of hydrogen (e.g., via electrolysis of seawater using electricity sourced from either offshore wind farm or wind-powered generators on ships), along with reliable and low-cost storage options, is a critical issue that must be considered. IRENA addressed the viability of several biofuels (as drop-in fuels) for seaborne transport, including straight vegetable oil (SVO), dimethyl ether, biomass-based Fischer-Tropsch diesel, biomethanol, bioethanol, biodiesel, hydrotreated vegetable oil (HVO), liquefied biomethane (LBM), and pyrolysis oil [108]. Furthermore, marine gas oil or marine diesel oil can be replaced by pyrolysis oil, biomethanol, dimethyl ether, as well as conventional and advanced biodiesel (best option is to blend fuel with maximum 20% biodiesel). In addition, intermediate or heavy fuel oil can be replaced by HVO and SVO. However, liquefied natural gas can be replaced by LBM or biomethane and dimethyl ether. Bio-hydrogen fuel cells can be used for electricity production in ships. It is important to note that if LBM is to be used as fuel in ships, some cargo space must be allocated and dual fuel cryotanks (i.e., storage tanks for supercold fuels) operating at low temperatures are needed. Nevertheless, DME also needs cargo space for storage. If using HVO, SVO, or biodiesel, some indirect costs are associated, including the nexus across water, energy, land, and food (WELF), which can impose serious impediment [108]. A full list of ships (either newly built or retrofit of existing ships) categorized by (1) renewable source, (2) market status of each renewable technology (currently in use, proven application, concept proved, design stage, concept, and uncertain), (3) vessel weight (<400, $400-10,000$, $10,000-50,000$, and $>50,000$ tons), and (4) a crude assessment of the overall potential, is given in Table 6 [108]. The overall potential was assessed based on three metrics: economic, environmental, and social factors. A high, medium, or low potential application was identified for ships which scored on three, two, or one of the three metrics.

The Modec company offers up to $125,000\,m^3$ LNG storage (known as Medium Solution) with power generation range of $83-480\,MW$ [110]. Medium Solution benefits from integration with intermittent wind, solar, hydropower. Supplemental power is dispatched as and when required.

Table 6 Application and potential of renewable energy for the shipping sector

Renewable type		New build (NB) or retrofit of existing (RF)	Weight (tonnes)			
			< 400[a]	400-10,000[b]	10,000-50,000[c]	> 50,000[d]
Wind energy	Fixed wings	NB	**	***	***	**
		RF	**	**	**	*
	Soft sails	NB	***	***	***	**
		RF	***	***	***	**
	Kites	NB/RF	**	**	**	*
	Rotors	NB	***	***	***	**
		RF	**	**	**	**
	Turbines	NB/RF	*	*	*	*
Solar PV	Main	NB	*	NA	NA	NA
		RF	NA	NA		
	Auxiliary	NB	**	NA	*	NA
		RF	**	NA	*	NA
	Ancillary	NB/RF	**	NA	*	NA
Biofuels	1st Generation	NB	**	**	**	*
		RF	**	**	**	*
	2nd Generation	NB	***	***	***	**
		RF	NA	NA	NA	NA
	3rd Generation	NB	***	***	***	**
		RF	NA	NA	NA	NA
Wave energy	Main	NB	*	NA	*	NA
	Auxiliary	NB	*	NA	*	NA
Electricity	Super capacitor	NB	*	*	NA	NA

Key: ***, high potential; **, medium potential; *, limited; NA, not available.
[a]For example, recreation, tourism, small fishery, coastal patrol and security, research, passenger, break.
[b]Including medium fishery, large landing craft, tramp, domestic RollOn-RollOff (Ro-Ro), container, tanker, etc.
[c]For example, deep sea fishery, Ro-Ro, cruise liner, car carrier.
[d]Including very large crude oil carriers, large container ships, Aframax, Panamax.

Color indicators: ▢, currently in use; ▢, proven application; ▢, concept proved; ▢, design stage; ▢, concept; ▢, uncertain.

9 CONCLUSIONS

The oil and gas sector is not only a major energy producer but also a major consumer of energy. The energy requirements of the oil and gas sector in the forms of heat, power, or shaft work are supplied mainly from conventional resources. Adaptation of green energies such as hydro, solar, wind, biomass, and ocean energy is the focus of interest to reduce fuel costs as well as GHG emissions.

The current status of renewable energy resources, including installed capacity, pros and cons, costs, and trends in future energy supply *is discussed*. The amount of energy consumption in oil and gas industries during extraction, transmission, and processing is outlined. Recent technological developments in deploying renewables in the oil and gas industry are covered. In addition, the production of renewable hydrogen, methane, and chemicals is addressed.

REFERENCES

[1] OPEC. World oil outlook 2013. Secretariat, Organization of the Petroleum Exporting Countries; 2013. [cited 2018 15 January]. Available from: http://www.opec.org/.

[2] European Renewable Energy Council. Mapping renewable energy pathways towards 2020. [cited 2018 15 January]. Available from: http://www.eufores.org/fileadmin/eufores/Projects/REPAP_2020/EREC-roadmap-V4.pdf; 2018.

[3] EC. Renewable energy: moving towards a low carbon economy. Available from: https://ec.europa.eu/energy/en/topics/renewable-energy.

[4] Janco C. The political ecology of oil and gas activities in the nigerian aquatic ecosystem. London: Academic Press; 2017.

[5] Offshorewind. In depth techtalk: floating wind-powered water injection; n.d. [cited 2018 06 January]. Available from: https://www.offshorewind.biz/2016/11/25/in-depth-techtalk-floating-wind-powered-water-injection/.

[6] IEA. Sankey diagram, energy balance. [cited 2018 15 January]. Available from: https://www.iea.org/Sankey/; 2015.

[7] IEA. Key world energy statistics. Paris: International Energy Agency; 2017. [cited 2018 10 January]. Available from: https://www.iea.org/publications/freepublications/publication/KeyWorld2017.pdf.

[8] EIA. Fuel consumed at refineries; n.d. [cited 2018 15 January]. Available from: https://www.eia.gov/dnav/pet/PET_PNP_CAPFUEL_DCU_NUS_A.htm.

[9] Khalilpour R, Karimi IA. Investment portfolios under uncertainty for utilizing natural gas resources. Comput Chem Eng 2011;35(9):1827–37.

[10] Khalilpour R. In: Produced water management: an example of a regulatory gap. SPE oilfield water management conference and exhibition. Kuwait City, Kuwait: Society of Petroleum Engineers; 2014.

[11] DOE. Enhanced oil recovery. US DOE; 2017. [cited 2018 12 January]. Available from: https://energy.gov/fe/science-innovation/oil-gas-research/enhanced-oil-recovery.

[12] Speight JG. Introduction to enhanced recovery methods for heavy oil and tar sands. Amsterdam: Elsevier Science; 2016.

[13] LindeGroup. Enhanced oil and gas recovery (EOR/EGR). n.d. [cited 2018 05 January]. Available from: http://www.the-linde-group.com/en/clean_technology/clean_technology_portfolio/enhanced_oil_gas_recovery/index.html.

[14] Hakiki F, Maharsi DA, Marhaendrajana T. Surfactant-polymer coreflood simulation and uncertainty analysis derived from laboratory study. J Eng Labor Stud 2015;47:706–25.

[15] EverythingConnects. Clean, renewable energy. [cited 2018 15 January]. Available from: http://www.everythingconnects.org/renewable-energy.html; 2017.

[16] BBC. Renewable energy sources; n.d.. [cited 2018 11 January]. Available from: http://www.bbc.co.uk/schools/gcsebitesize/geography/energy_resources/energy_rev2.shtml.

[17] Energyinformative n.d. [cited 2017 28 December]. Available from: http://energyinformative.org/tidal-energy-pros-and-cons/.

[18] OpenEI. Transparent cost database; n.d. [cited 2018 23 January]. Available from: https://openei.org/apps/TCDB/-blank.

[19] IEA-OES. International levelised cost of energy (lCOE) for ocean energy technologies. Edinburgh: IEA Technology Collaboration Programme for Ocean Energy Systems (OES), 2015. https://www.ocean-energy-systems.org/documents/65931-cost-of-energy-for-ocean-energy-technologies-may-2015-final.pdf/.

[20] IRENA. Renewable energy technologies. [cited 2018 05 January]. Available from: http://resourceirena.irena.org/gateway/dashboard/?topic=4&subTopic=19; 2017.

[21] EPA. Greenhouse gas reporting program (GHGRP); n.d. [cited 2018 01 January]. Available from: https://www.epa.gov/ghgreporting/ghgrp-and-oil-and-gas-industry.

[22] Khalilpour R, Karimi IA. Selection of liquefied natural gas (LNG) contracts for minimizing procurement cost. Ind Eng Chem Res 2011;50(17):10298–312.

[23] Cooper J, editor. Statistical report. Brussels, Belgium: FuelsEurope; 2017.

[24] EIA. Industry uses many energy sources. [cited 2018 20 January]. Available from: https://www.eia.gov/energyexplained/index.cfm?page=us_energy_industry; 2017.

[25] Singh SPaS. Towards an energy efficient oil & gas sector. New Delhi: The Energy & Resources Institute (TERI); 2015.

[26] Statista. U.S. oil and gas extraction industry energy inputs from 1997 to 2016 (in million U.S. dollars). [cited 2018 20 January]. Available from: https://www.statista.com/statistics/196369/us-oil-and-gas-extraction-energy-inputs-since-1997/; 2017.

[27] Vanner R. Energy use in offshore oil and gas production: Trends and drivers for efficiency from 1975 to 2025. London: Policy Studies Institute; September 2005.

[28] Brandt AR. Oil depletion and the energy efficiency of oil production: the case of California. Sustainability 2011;3(10):1833.

[29] Gavenas E, Rosendahl KE, Skjerpen T. CO2-emissions from Norwegian oil and gas extraction. Ås, Norway: Statistics Norway Research Department; 2015.

[30] Yosoon Choi CL, Song J. Review of renewable energy technologies utilized in the oil and gas industry. Int J Renew Energy Res 2017;7(2):592–8.

[31] MacDonald J. Providing scope for reducing the carbon footprint of an offshore oil rig. In: Sustainable engineering: renewable energy systems and the environment. Glasgow: University of Strathclyde; 2014.

[32] He W, et al. Case study of integrating an offshore wind farm with offshore oil and gas platforms and with an onshore electrical grid. J Renew Energy 2013;2013:10.

[33] Kolstad ML, et al. Grid integration of offshore wind power and multiple oil and gas platforms. Bergen: MTS/IEEE OCEANS; 2013.

[34] OffShoreEnergyToday. BG's renewable energy-powered platform produces first oil; n.d. [cited 2018 10 January]. Available from: https://www.offshoreenergytoday.com/bgs-renewable-energy-powered-platform-produces-first-oil/.

[35] Hanssen JE, et al. In: Design and performance validation of a hybrid offshore renewable energy platform. 2015 Tenth international conference on ecological vehicles and renewable energies (EVER); 2015.

[36] Mundheim K. In: Monotower platform with renewable power. Offshore mediterranean conference and exhibition. Ravenna, Italy: Offshore Mediterranean Conference; 2007.

[37] Cornelia Noel RD. In: Renewables, ready or not?. Offshore technology conference. Houston, TX: Offshore Technology Conference; 2012.

[38] Johnson LA, Walker ED. In: Oil production waste stream, a source of electrical power. Thirty-Fifth workshop on geothermal reservoir engineering. California: Stanford; 2010.

[39] Bin Gong HL, Xin S, Li K. In: Effect of water injection on reservoir temperature during power generation in oil fields. Thirty-sixth workshop on geothermal reservoir engineering. California: Stanford; 2011.

[40] Gioia Falcone CT. In: Oil and gas expertise for geothermal exploitation: the need for technology transfer. Europec/EAGE conference and exhibition. Rome, Italy: Society of Petroleum Engineers; 2008.

[41] Syed Zahoor Ullah SRSB. In: Geothermal reservoirs: a renewable source of energy and an extension of petroleum engineering. CIPC/SPE gas technology symposium 2008 joint conference. Calgary, Alberta, Canada: Society of Petroleum Engineers; 2008.

[42] DNVGTL. Wind powered oil recovery concept moves closer to implementation; n.d. [cited 2018 06 January]. Available from: https://www.dnvgl.com/news/wind-powered-oil-recovery-concept-moves-closer-to-implementation-90002.

[43] Sandberg J. The third generation of wind power—floating turbines. DNV GL, Oslo, 2017.

[44] IEA. Technology roadmap—solar thermal electricity. Paris, France: International Energy Agency; 2014

[45] Sunay Gupta RG, Laumert B. In: Market potential of solar thermal enhanced oil recovery-a techno-economic model for Issaran oil field in Egypt. AIP conference proceedings; 2017.

[46] GlassPoint. Belridge solar. [cited 2018 15 January]. Available from: https://www.glasspoint.com/belridgesolar/; 2017.

[47] BrightsourceEnergy. Chevron/brightsource solar-to-steam demonstration facility. [cited 2017 25 December]. Available from: http://www.brightsourceenergy.com/coalinga-.Wkk9IEpl-Uk; 2011.

[48] BrightsourceEnergy. COALINGA project facts. 2011.

[49] AquanetPower. ogWAVE—offshore application; n.d. [cited 2018 05 January]. Available from: https://www.aquanetpower.com/ogwave.

[50] AquanetPower. aquaWAVE—shallow water application; n.d. [cited 2018 05 January]. Available from: https://www.aquanetpower.com/aquawave.

[51] Muhammad Hasan ME, Mahmoud M, Elkatatny S, Shawabkeh R. In: Enhanced gas recovery (EGR) methods and production enhancement techniques for shale & tight gas reservoirs. Annual technical symposium and exhibition, 24–27 April, Dammam, Saudi Arabia; 2017.

[52] Torsten Clemens KW. In: CO_2 enhanced gas recovery studied for an example gas reservoir. SPE annual technical conference and exhibition, San Antonio, TX; 2002.

[53] Panahi M, et al. A natural gas to liquids process model for optimal operation. Ind Eng Chem Res 2012;51(1):425–33.

[54] Karimi F, Khalilpour R. Evolution of carbon capture and storage research: trends of international collaborations and knowledge maps. Int J Greenhouse Gas Control 2015;37:362–76.

[55] Adham S. In: Desalination needs and opportunities in the oil & gas industry. International conference on emerging water desalination technologies in municipal and industrial applications; 2015 [cited 2018 12 January]. Available from: http://www.desaltech2015.com/assets/presenters/Adham_Samer.pdf.

[56] Bourouni K. Optimization of renewable energy systems: the case of desalination. In: Sahin AS, editor. Modeling and optimization of renewable energy systems. 2012, https://doi.org/10.5772/2283. ISBN: 978-953-51-0600-5.

[57] VeoliaWaterTechnologies. Desalination technologies; n.d. [cited 2018 07 January]. Available from: http://www.veoliawatertechnologies.co.za/water-solutions/desalination/desalination-technologies/.

[58] Eltawil MA, Zhengming Z, Yuan L. A review of renewable energy technologies integrated with desalination systems. Renew Sustain Energy Rev 2009;13(9):2245–62.

[59] Mbarga AA, et al. Integration of renewable energy technologies with desalination. Curr Sustain Renew Energy Rep 2014;1(1):11–8.

[60] Bourouni K, Chaibi MT, Tadrist L. Water desalination by humidification and dehumidification of air: state of the art. Desalination 2001;137(1):167–76.

[61] Ali MT, Fath HES, Armstrong PR. A comprehensive techno-economical review of indirect solar desalination. Renew Sustain Energy Rev 2011;15(8):4187–99.

[62] Absi Halabi M, Al-Qattan A, Al-Otaibi A. Application of solar energy in the oil industry—current status and future prospects. Renew Sustain Energy Rev 2015;43:296–314.

[63] Bilton AM, et al. On the feasibility of community-scale photovoltaic-powered reverse osmosis desalination systems for remote locations. Renew Energy 2011;36(12): 3246–56.

[64] WorldBank. New data reveals uptick in global gas flaring. 2016.

[65] MaerskDrilling. Environmental report. 2008.

[66] IPIECA. Saving energy in the oil and gas industry: climate change. London: IPIECA; 2013.

[67] Sabine Brueske CK, Fisher A. Bandwidth study on energy use and potential energy savings opportunities in U.S. petroleum refining. Energetics 2015;100.

[68] Abdul-Manan AFN, Arfaj A, Babiker H. Oil refining in a CO_2 constrained world: effects of carbon pricing on refineries globally. Energy 2017;121:264–75.

[69] Kosan Roh HL, Yoo H, Imran H, Al-Hunaidy AS, Lee JH. Technical and strategic approaches for CO2 management in refining businesses. Saudi Aramco J Technol 2017;41–50.

[70] Boyle R. Tracts, containing new experiments, touching the relation betwixt flame and air: and about explosions. An hydrostatical discourse occasion'd by some objections of Dr. Henry More against some explications of new experiments made by the author of these tracts: to which is annex't, an hydrostatical letter, dilucidating an experiment about a way of weighing water in water. New experiments, of the positive or relative levity of bodies under water. Of the air's spring on bodies under water. About the differing pressure of heavy solids and fluids. London: Printed for Richard Davis; 1673.

[71] Bao B, El-Halwagi MM, Elbashir NO. Simulation, integration, and economic analysis of gas-to-liquid processes. Fuel Process Technol 2010;91(7):703–13.

[72] Wilhelm DJ, et al. Syngas production for gas-to-liquids applications: technologies, issues and outlook. Fuel Process Technol 2001;71(1–3):139–48.

[73] Steynberg A, Dry M, editors. Fischer-Tropsch technology studies. In: Surface science and catalysis. vol. 152. Amsterdam: Elsevier; 2004. p. 700.

[74] Dincer I, Joshi AS. Solar hydrogen production. In: Solar based hydrogen production systems. New York, NY: Springer New York; 2013. p. 27–71.

[75] Worrell E, Galitsky C. Energy efficiency improvement and cost saving opportunities for petroleum refineries. In: An ENERGY STAR® guide for energy and plant managers. Berkeley, CA: University of California; 2005.

[76] Rafiee A, Hillestad M. Synthesis gas production configurations for gas-to-liquid applications. Chem Eng Technol 2012;35(5):870–6.

[77] Rafiee A, Hillestad M. Techno-economic analysis of a gas-to-liquid process with different placements of a CO_2 removal unit. Chem Eng Technol 2012;35(3):420–30.

[78] Rafiee A, Panahi M. Optimal design of a gas-to-liquids process with a staged Fischer-Tropsch reactor. Chem Eng Technol 2016;39(10):1778–84.

[79] Rafiee A, Panahi M, Khalilpour KR. CO_2 utilization through integration of post-combustion carbon capture process with Fischer Tropsch gas-to-liquid (GTL) processes. J CO2 Util 2017;18:98–106. https://doi.org/10.1016/j.jcou.2017.01.016.

[80] Panahi M, Yasari E, Rafiee A. Multi-objective optimization of a gas-to-liquids (GTL) process with staged Fischer-Tropsch reactor. Energ Conver Manage 2018;163:239–49.

[81] Fazeli H, Panahi M, Rafiee A. Investigating the potential of carbon dioxide utilization in a gas-to-liquids process with iron-based Fischer-Tropsch catalyst. J Nat Gas Sci Eng 2018;52:549–58.

[82] Wood DA, Nwaoha C, Towler BF. Gas-to-liquids (GTL): a review of an industry offering several routes for monetizing natural gas. J Nat Gas Sci Eng 2012;9:196–208.

[83] Trainham JA, et al. Whither solar fuels? Curr Opin Chem Eng 2012;1(3):204–10.

[84] Maag G, Zanganeh G, Steinfeld A. Solar thermal cracking of methane in a particle-flow reactor for the co-production of hydrogen and carbon. Int J Hydrogen Energy 2009; 34(18):7676–85.

[85] Sheu EJ, Mokheimer EMA, Ghoniem AF. A review of solar methane reforming systems. Int J Hydrogen Energy 2015;40(38):12929–55.

[86] Trommer D, et al. Hydrogen production by steam-gasification of petroleum coke using concentrated solar power—I. Thermodynamic and kinetic analyses. Int J Hydrogen Energy 2005;30(6):605–18.

[87] Z'Graggen A, Steinfeld A. Heat and mass transfer analysis of a suspension of reacting particles subjected to concentrated solar radiation—application to the steam-gasification of carbonaceous materials. Int J Heat Mass Transfer 2009;52(1):385–95.

[88] Gálvez ME, et al. Solar hydrogen production via a two-step thermochemical process based on MgO/Mg redox reactions—thermodynamic and kinetic analyses. Int J Hydrogen Energy 2008;33(12):2880–90.

[89] Lyman Joseph Frost JH, Elangovan S. Formation of synthesis gas using solar concentrator photovoltaics (SCPV) and high temperature co-electrolysis (HTCE) of CO_2 and H_2O. Offshore technology conference, Houston, Texas, USA; 2010.

[90] Mohsin M, Rasheed AK, Saidur R. Economic viability and production capacity of wind generated renewable hydrogen. Int J Hydrogen Energy 2018;43(5):2621–30.

[91] Schoenung SM, Keller JO. Commercial potential for renewable hydrogen in California. Int J Hydrogen Energy 2017;42(19):13321–8.

[92] Kim M, Kim J. Optimization model for the design and analysis of an integrated renewable hydrogen supply (IRHS) system: application to Korea's hydrogen economy. Int J Hydrogen Energy 2016;41(38):16613–26.

[93] Sigal A, Leiva EPM, Rodríguez CR. Assessment of the potential for hydrogen production from renewable resources in Argentina. Int J Hydrogen Energy 2014; 39(16):8204–14.

[94] Real D, Dumanyan I, Hotz N. Renewable hydrogen production by solar-powered methanol reforming. Int J Hydrogen Energy 2016;41(28):11914–24.

[95] Serna Á, et al. Predictive control for hydrogen production by electrolysis in an offshore platform using renewable energies. Int J Hydrogen Energy 2017;42 (17):12865–76.

[96] Reiter G, Lindorfer J. Global warming potential of hydrogen and methane production from renewable electricity via power-to-gas technology. Int J Life Cycle Assess 2015;20(4):477–89.

[97] Boretti A. Renewable hydrogen to recycle CO_2 to methanol. Int J Hydrogen Energy 2013;38(4):1806–12.

[98] Clausen LR, Houbak N, Elmegaard B. Technoeconomic analysis of a methanol plant based on gasification of biomass and electrolysis of water. Energy 2010;35 (5):2338–47.

[99] Masel R, et al. CO_2 conversion to chemicals with emphasis on using renewable energy/resources to drive the conversion. In: Commercializing biobased products: opportunities, challenges, benefits, and risks. Cambridge, UK: The Royal Society of Chemistry; 2016. p. 215–57 [Chapter 10].

[100] Masel R, et al. 12th international conference on greenhouse gas control technologies, GHGT-12unlocking the potential of CO_2 conversion to fuels and chemicals as an economically viable route to CCR. Energy Procedia 2014;63:7959–62.

[101] Dominguez-Ramos A, et al. Global warming footprint of the electrochemical reduction of carbon dioxide to formate. J Clean Prod 2015;104:148–55.

[102] Du X, et al. Catalytic dehydrogenation of propane by carbon dioxide: a medium-temperature thermochemical process for carbon dioxide utilisation. Faraday Discuss 2015;183:161–76.

[103] Centi G, Quadrelli EA, Perathoner S. Catalysis for CO_2 conversion: a key technology for rapid introduction of renewable energy in the value chain of chemical industries. Energ Environ Sci 2013;6(6):1711–31.

[104] CRI. C.R.I. [cited 2018 04 January]. Available from: http://carbonrecycling.is/george-olah/2016/2/14/worlds-largest-co2-methanol-plant; 2017.

[105] ICS. 2017 annual review. London: International Chamber of Shipping; 2017.

[106] UN. Review of maritime transport, United Nations conference on trade and development. New York/Geneva: United Nations Publication; 2016 ISBN 978-92-1-112892-5, eISBN: 978-92-1-057410-5, ISSN 0566-7682.

[107] Organization IM. Third IMO greenhouse gas study 2014. .

[108] IRENA. Renewable energy options for shipping. International Renewable Energy Agency; 2015. Available from: http://www.irena.org/-/media/Files/IRENA/Agency/Publication/2015/IRENA_Tech_Brief_RE_for-Shipping_2015.pdf.

[109] ICS, I.C.O.S. Shipping, world trade and the reduction of CO_2 emissions. United Nations Framework Convention on Climate Change (UNFCCC); 2014.

[110] MODEC. Floating storage regasification water-desalination & power-generation; n.d. [cited 2018 06 January]. Available from: http://www.modec.com/fps/fsrwp/pdf/fsrwp.pdf.

CHAPTER 12

Biorefinery Polyutilization Systems: Production of Green Transportation Fuels From Biomass

Pankaj Kumar, Mohan Varkolu, Swarnalatha Mailaram, Alekhya Kunamalla, Sunil K. Maity
Department of Chemical Engineering, Indian Institute of Technology Hyderabad, Sangareddy, India

Abstract

The green transportation fuels (green gasoline, green diesel, and green jet fuel) (GTF) derived from biomass are quite similar to the petroleum-derived transportation fuels and compatible with existing petroleum refinery infrastructures and combustion engines. The present chapter provides an outline of the various routes for the production of GTF from biomass. In general, the biomass is converted to GTF through thermochemical, chemical, biochemical, and platform chemical-based routes. The current chapter presents an outline of thermochemical conversion processes such as biomass gasification, liquefaction, and pyrolysis and chemical conversion process such as hydrodeoxygenation of triglycerides. The ethanol to gasoline and butanol to gasoline are two important biochemical conversion processes for the production of GTF and discussed in the present chapter. The present chapter also provides an outline of the production of GTF and fuel additives from the platform chemicals such as 5-hydroxymethylfurfural, furfural, and levulinic acid.

Keywords: Hydrocarbon biorefinery, Green transportation fuels, Fast pyrolysis, Hydrodeoxygenation of triglycerides, Ethanol to gasoline, Butanol to gasoline, Platform chemicals

1 INTRODUCTION

The transportation fuels (gasoline, diesel, and jet fuel) are one of the primary energy-consuming sectors in the world. At present, the transportation fuels contribute about 28% of the world's energy consumption (Fig. 1). These transportation fuels are primarily derived from the fossil fuels, more specifically from petroleum. The petroleum reserves are depleting continuously and expected to be exhausted in <50 years based on the estimates of the

Polygeneration with Polystorage
https://doi.org/10.1016/B978-0-12-813306-4.00012-4

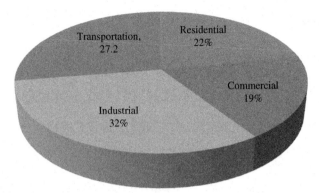

Fig. 1 Energy consumption by different sectors.

proven oil reserves in the world [1]. Moreover, the large-scale use of fossil fuels is continuously increasing the concentration of greenhouse gases in the earth atmosphere leading to global warming. The production of transportation fuels from renewable sources is thus highly desirable for sustainability of civilized society. The biomass is the only renewable carbon source in the world with the tremendous potential to produce a wide range of bio-based products including transportation fuels (called as biofuels). The integrated facility for the conversion of biomass to fuels, power, and chemicals is generally known as biorefinery [2, 3].

The wide ranges of biomass are generally processed in the biorefinery. The biomass used in the biorefinery is classified into three broad categories based on their chemical nature: (i) triglycerides, (ii) sugar and starchy biomass, and (iii) lignocellulosic biomass. The triglycerides consist of vegetable oil, animal fats, waste cooking oils, and microalgal oils. In triglycerides, one molecule of glycerol is bonded with three molecules of fatty acids by ester bonds. The sugar biomass consists of sugarcane, sugar beet, and sweet sorghum. The sugar is a disaccharide of glucose and fructose. The corn, wheat, barley, etc. are included in the starchy biomass. The starch is a polymer of glucose. On the other hand, the lignocellulosic biomass consists of stem, branches, needles, straws, stalks, and leaves of the plant. The agriculture and forest are the main sources of this biomass. The lignocellulosic biomass is composed of cellulose (40%–50%), hemicellulose (20%–30%), and lignin (10%–25%). The cellulose is a crystalline polymer of glucose, whereas hemicellulose is composed of both hexose (glucose, galactose, and mannose) and pentose (xylose and arabinose) sugars. The lignin is a three-dimensional network of three different phenylpropane units: coumaryl alcohol, coniferyl alcohol, and sinapyl alcohol.

The biomass conversion processes are generally developed based on the chemical nature of the biomass. So the biorefinery is broadly classified into three different types: (i) triglycerides biorefinery, (ii) sugar and starchy biorefinery, and (iii) lignocellulosic biorefinery. The outline of possible opportunities of these biorefineries is shown in Fig. 2–4 [2]. These traditional

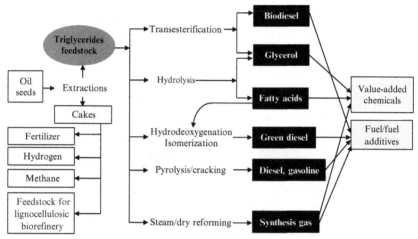

Fig. 2 Possible opportunities of triglyceride biorefinery [2].

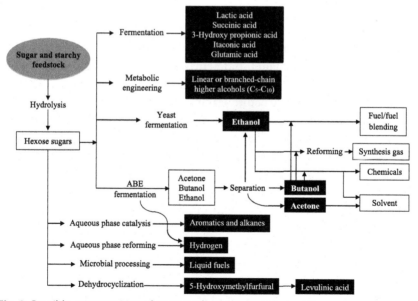

Fig. 3 Possible opportunities of sugar and starchy biorefinery [2].

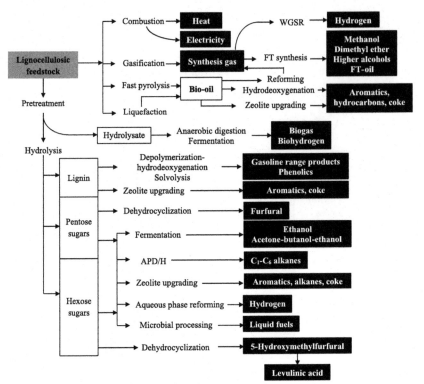

Fig. 4 Possible opportunities of lignocellulosic biorefinery [2].

biorefineries have potentials to produce a wide range of biofuels such as bio-diesel, bioethanol, biobutanol, biomethanol, dimethyl ether, etc. to meet the needs of transportation fuels in our society. These biofuels, however, contain oxygen in their structure leading to lower fuel mileage compared to petroleum-derived transportation fuels. Moreover, these biofuels are incompatible with internal combustion engines due to their unfavorable properties such as cold flow properties, vapor pressure, miscibility with water, etc. The applications of these biofuels are thus limited to blending with petroleum-derived transportation fuels. For example, biodiesel can be blended with diesel to the extent of 20% by volume. The blending of bioethanol to gasoline is limited to 15% by volume. These biofuels are also incompatible with existing petroleum refinery infrastructures including transportation, storage, and fueling station.

The new biomass processing concepts are thus developed to produce hydrocarbon biofuels analogous to the petroleum-derived transportation fuels. This concept is generally known as the hydrocarbon biorefinery [2, 3]. The hydrocarbon biofuels produced in the hydrocarbon biorefinery are generally called as green transportation fuels such as green gasoline, green diesel, and green jet fuel. These green transportation fuels are fully compatible with unmodified combustion engines and petroleum refinery infrastructures. This concept thus alleviates the problem of developing completely new infrastructures for combustion engines, transportation, storage, and fueling station. The possible routes for the production of green transportation fuels from biomass are schematically represented in Fig. 5. The processing of biomass in the hydrocarbon biorefinery is broadly classified as (i) thermochemical conversion processes, (ii) chemical conversion processes, (iii) biochemical conversion processes, and (iv) platform chemical-based conversion processes. These conversion processes are briefly discussed in this chapter.

2 THERMOCHEMICAL CONVERSION PROCESSES

The thermochemical conversion processes are the chemical transformation of biomass to useful fuels and chemicals using heat energy. The liquefaction, gasification, and pyrolysis are three major thermochemical conversion processes as described here.

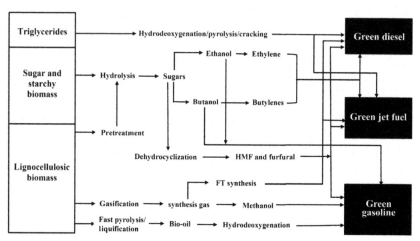

Fig. 5 The possible routes for the production of green transportation fuels from biomass in the hydrocarbon biorefinery.

2.1 Gasification

The gasification converts organic carbonaceous materials (biomass, coal etc.) into gaseous products in presence of air or oxygen. When air is used in the gasification, the product gas is called producer gas [4]. The producer gas thus contains nitrogen and finds application in the generation of heat and electricity. On the contrary, the synthesis gas (SG) is mainly composed of CO and H_2 and used to produce transportation fuels, fuel additives, and chemicals by various routes as shown in Fig. 6 [5]. The SG is used for the production of gasoline, jet fuel, and diesel range hydrocarbons by Fischer-Tropsch synthesis (FTS), hydrogen by water-gas shift reaction (WGSR), and methanol. The methanol can be further transformed into gasoline range hydrocarbons, olefins, and other chemicals. The composition of SG, however, depends on the types of feedstock, which in turn affects the economics of the process. The natural gas is the most economical feedstock for the production of transportation fuels through SG. The production of SG contributes >50% of the production cost of transportation fuels. The feedstock is an important factor for the economic feasibility of the overall process for the production of transportation fuels from SG. The production of transportation fuels through biomass gasification is generally uneconomical due to huge capital investment for feed handling as well as complex SG purification process.

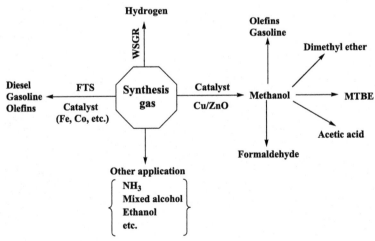

Fig. 6 Applications of SG.

2.2 Methanol to Gasoline

The methanol is currently produced from SG using FTS. The methanol cannot be used as fuel or fuel additives due to its high volatility and oxygen content (50 wt%). It is, therefore, converted to either dimethyl ether by dehydration reaction for application as fuel additive or gasoline range hydrocarbons for application in the gasoline engine. In 1970s, the conversion of methanol to hydrocarbons was first reported by Chang and Silvestri using shape-selective zeolite catalyst [6]. This led to the development of Exxon Mobil's methanol-to-gasoline (MTG) process [7] (Fig. 7). The world's first MTG plant was commercialized in New Zealand in 1986. In this process, the reaction is performed in a fixed-bed reactor over Mobil proprietary HZSM-5 catalyst at 623–639 K and 18–22 bar [8, 9]. The yield of various products is as follows: 86 wt% gasoline range hydrocarbons, 12.7 wt% LPG, and 1.3 wt% fuel gas. The gasoline has research octane number (RON) and motor octane number (MON) of 92.2 and 82.6, respectively, and density of 730 kg/m^3 at 288 K. In this reaction, the methanol is partially dehydrated to an equilibrium mixture dimethyl ether and water [7, 10–12] (Fig. 8). The dimethyl ether and methanol react further to form C_2-C_4 range light olefins followed by higher olefins. The higher olefins then converted to paraffins, aromatics, and naphthenes. Some researchers also proposed a hydrocarbon pool mechanism for MTG process [13–15]. The product distribution in MTG depends on the topology, Si/Al ratio, and acidity of the

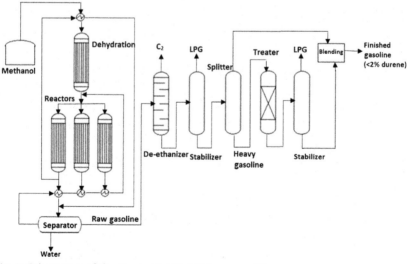

Fig. 7 Schematics of the Exxon Mobil's MTG process [7].

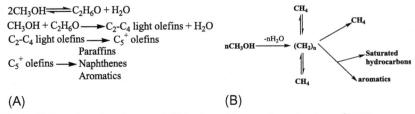

(A)   (B)

Fig. 8 (A) Reaction chemistry and (B) hydrocarbon pool mechanism of MTG process.

catalyst. As the Si/Al ratio in HZSM-5 increases, the amount of acid sites decreases and Brönsted/Lewis acidity (B/L) ratio increases. At lower temperatures, the formation of C_2-C_4 olefins is mainly controlled by increasing B/L ratio [16]. The modified SAPO-34 with transition metals (Co, Mn, Ni) and HZSM-5 with metal oxides showed improved conversion and yield of hydrocarbons [17, 18]. Doping of CuO and ZnO into HZSM-5 increases the acidity of the catalyst that results in the increase in conversion from 38% to 97% and yield of hydrocarbons from 15.2% to 42.16% [19].

2.3 Liquefaction

In liquefaction, the lignocellulosic biomass is converted to the liquid product called bio-oil or biocrude [20]. The char, water-soluble components, and gas are the byproduct of this process. It is a high-pressure (100–250 bar) and medium-temperature (553–643 K) process. In this process, the water acts as a reactant as well as the catalyst. Unlike fast pyrolysis and gasification, the wet biomass is processed in this process. Moreover, the biocrude has relatively high heating value compared to the pyrolysis. The hydrothermal liquefaction is, however, facing serious challenges in the commercialization due to the high-pressure operation and corrosive nature of the biocrude. The high operating pressure requires expensive equipment like feed pump for continuous, smooth, and safe operation. This process is thus highly capital intensive. Moreover, the biocrude is corrosive in nature and hence the processing of biocrude needs expensive alloys. The process is thus mainly limited to the lab- and bench-scale operation so far.

2.4 Fast Pyrolysis

The pyrolysis is a thermochemical decomposition of carbonaceous organic materials at an elevated temperature in absence of oxygen or air. It is an old technology and generally used for the production of charcoal. Since last few decades, the fast pyrolysis technology is developed to maximize the yield of liquid fuels from biomass. In the fast pyrolysis, the lignocellulosic biomass is rapidly decomposed at a high temperature in absence of air or oxygen to produce pyrolysis vapors with a small quantity of char. The pyrolysis vapor is then separated from char and quenched to produce a low viscous dark brown colored liquid called bio-oil or biocrude. The pyrolysis is categorized based on the reaction conditions and desired product as shown in Table 1 [21]. The fast pyrolysis operates at about 773–923 K with a short residence time (<2 s) and produces a liquid up to 75 wt% yield. The process performance is highly influenced by operating conditions, types of feedstock, heating rate, and vapor and solid residence time. The importance of these features on the yield of liquid products is described here.

- The particle size of the biomass should be below 5 mm. The small particle size of biomass helps rapid heating of biomass, which in turn leads to rapid devolatilization and high yield of bio-oil.
- A good amount of water is produced in the fast pyrolysis. Therefore, the feed moisture content of the biomass should be below 10 wt% to avoid excessive amount of water in the bio-oil.
- The optimum operating temperature is crucial to minimize the formation of char and maximize volatile hydrocarbons. The low temperature favors formation of char, while high temperature accelerates thermal cracking of pyrolysis vapors resulting in a high yield of volatile hydrocarbons.
- The rapid heating is essential for minimal exposure of biomass at the lower temperatures to reduce the formation of char. However, bringing the biomass to the desired temperature with minimal exposure at the lower temperature is quite challenging in the fast pyrolysis.

Table 1 Types of pyrolysis based on operating conditions and their major products [21]

	Temperature (K)	Heating rate	Residence time	Major product	Yield (%)
Carbonization	673	Very slow	Days[a]	Solid (charcoal)	35
Torrefaction	563	–	5–60 min[a]	Solid (charcoal)	80
Gasification	1023–1173	Fast	Very high[b]	Gas	85
Intermediate	773	Fast	10–30 s[b]	Liquid	50
Fast	773–923	Fast	1–2 s[b]	Liquid	75

[a]Solid residence time.
[b]Vapor residence time.

- The short vapor residence time below 2 s is also essential to minimize secondary reaction. So, heat and mass transfer, phase transition, and reaction kinetics play a crucial role in fast pyrolysis.
- Rapid separation of char to avoid further cracking of pyrolysis vapors.
- Rapid quenching of pyrolysis vapor after char separation to maximize the yield of bio-oil.

The various types of reactors used in fast pyrolysis of biomass are summarized in Table 2. Among various types of reactors, the bubbling fluidized bed and circulating fluidized bed reactor are the most attractive ones because of the high yield of liquid product. These reactors are commonly used in fast pyrolysis [22]. The augur reactors are simple in design. Ensyn has been producing bio-oil since 25 years in North America using Rapid Thermal Processing (RTP) technology using circulating fluidized bed reactor [23]. Empyro BV is another player for the production of bio-oil by fast pyrolysis. The company built a plant in the Netherlands with the capacity of 24,000 tons per annum using rotating cone reactor. The bio-oil is used to co-fire with natural gas in the boiler. At present, the fast pyrolysis is facing following technical challenges.

- Preparing complete dry, finely divided biomass particles
- Maintaining high bio-oil yields
- Limited demand of bio-oil
- Expensive transportation
- Incompatible with oil refinery metallurgy
- Improving bio-oil stability
- Determining optimal-scale facility

Table 2 Comparison of fast pyrolysis reactors [115]

Reactor	Liquid yield (Wt%)	Complexity	Carrier gas	Feed size	Scale-up	Status
BFB	75	Medium	Yes	Very small	Easy	Commercial
CFB	75	High	Yes	Very small	Easy	Commercial
Ablative	75	High	No	Large	Difficult	Lab scale
Rotary cone	70	High	No	Very small	Medium	Demonstration
Augur	60	Medium	No	Medium	Medium	Pilot scale

BFB, bubbling fluidized bed; *CFB*, circulating fluidized bed.

The bio-oil has high oxygen content, low calorific value, and low pH. It is highly unstable and prone to condensation and polymerization reaction. The bio-oil thus needs improvement in properties for application as the transportation fuel. Two different upgrading techniques (hydroprocessing and catalytic cracking) are generally used for removal of oxygen from bio-oil to produce gasoline and diesel range hydrocarbons as discussed here.

2.4.1 Bio-Oil Upgrading by Hydroprocessing

The catalytic hydroprocessing is the removal of oxygen from bio-oil in the form of water in presence of hydrogen (200 bar) at about 673 K. The hydroprocessing is an existing refinery process to remove sulfur from various fractions of crude oil. The possibility of using existing refinery facility is the added advantage of this route. The complete removal of oxygen from bio-oil results gasoline range hydrocarbons predominantly [21]. The typical yield of gasoline range hydrocarbons in hydroprocessing of bio-oil is about 25 wt% of biomass. The physicochemical properties of the hydrocarbons produced by hydroprocessing of bio-oil using commercial hydrotreatment catalyst are shown in Table 3 [24, 25]. Almost entire oxygen was removed from bio-oil. As a result, the higher heating value (HHV) of bio-oil was enhanced from 18.8 to 41.4 MJ/kg. The HHV of the upgraded bio-oil is close to the HHV of conventional diesel and hence it can be used as the transportation fuel in the existing combustion engines. The hydroprocessing of bio-oil is generally carried out using commercial hydrotreating catalysts such as sulfided CoMo and NiMo supported on alumina and aluminosilicates. These catalysts are, however, suffering from the instability in the high water environment with high level of coking. The stripping of sulfur is another drawback of the sulfided catalysts. The carbon and titania are considered as the alternative support to overcome the catalyst instability [26]. The Pt and other metallic group catalysts were reported as more active at the lower temperatures than sulfided NiMo and CoMo [26].

2.4.2 Upgrading of Bio-Oil by Catalytic Cracking

The catalytic cracking is carried out over zeolites in absence of hydrogen [27, 28]. It is quite similar to fluidized catalytic cracking (FCC) used in petroleum refinery for the production of gasoline from heavy fractions. In this process, the oxygen is removed mainly in the form of CO_2 and H_2O. The aromatic and aliphatic hydrocarbons are generally observed as the products in this process. This process is commercialized by Kior. The maximum hydrocarbons yield was 27.9 wt% of bio-oil over HZSM-5 at 563–683 K and atmospheric pressure [27].

The aromatics content was found to be higher for HZSM-5 and H-Mordenite compared to HY, silicalite, and silica-alumina. Toluene, xylenes, and trimethylbenzenes are the main aromatics during catalytic fast pyrolysis of bio-oil. The aliphatic hydrocarbon was composed of C_6-C_9 range hydrocarbons.

2.5 Catalytic Fast Pyrolysis

In catalytic fast pyrolysis (CFP), the lignocellulosic biomass is converted to hydrocarbons (mostly aromatics and olefins) over zeolites at 773–873 K and atmospheric pressure [29]. The aromatics formed in CFP are compatible with the gasoline and hence can be blended with gasoline for enhancing the octane number. The CFP involves several basic reactions such as cracking, dehydrogenation, dehydration, decarbonylation, decarboxylation, cyclization, aromatization, and condensation to produce aromatics, olefins, CO, CO_2, and H_2O. The aromatics undergo condensation to polyaromatics and further to coke over the zeolite catalyst. The selection of catalyst, process conditions (temperature, catalyst/oil ratio, residence time, etc.), and high heating rates play the crucial role in the quality and yield of liquid. The properties of the liquid in CFP of biomass over HZSM-5 are shown in Table 3. The CFP enriched the aromatics in liquid products with the simultaneous

Table 3 The comparison of properties of bio-oil and hydroprocessed bio-oil with diesel and fuel oil [24, 25]

	Bio-oil[a]	Bio-oil[b]	Diesel	Fuel oil	Bio-oil upgrading[c]
O, wt%	40.28	14.69	0.01	1	0.5
N, wt%	0.82	1.88	65 ppm	0.3	0.4
Aromatic, wt%	7.62	74.22	nd	nd	nd
H/C mole ratio	1.61	1.51	1.84	1.55	1.56
Empirical formula	$CH_{1.61}O_{0.58}N_{0.01}$	$CH_{1.51}O_{0.15}N_{0.02}$	$CH_{1.84}$	$CH_{1.55}O_{0.01}N_{0.003}$	$CH_{1.22}O_{0.03}N_{0.004}$
HHV (MJ/kg)	18.8	34.6	45.5	40	41.4
pH	2.8	5.2	nd[d]	nd	nd
Specific gravity	1.18	0.95	nd	0.94	0.93
Char, wt%	0.3	0.2	nd	1	nd

[a] Fast pyrolysis.
[b] CFP with HZSM-5.
[c] Hydrocarbons by hydroprocessing of bio-oil using commercial hydrotreatment catalyst.
[d] nd, not determined.

decrease in other oxygenated hydrocarbons. The decrease in oxygen content increased the HHV from 18.8 to 34.6 MJ/kg. Hence, the liquid product can be considered as the transportation fuel [20]. Kior developed the CFP for the production of liquid from lignocellulosic biomass. A plant was built in the USA with biomass processing capacity of 20,833 kg/h using circulating fluidized bed reactor [15]. The liquid oil is further converted to hydrocarbons by hydrodeoxygenation (HDO).

For CFP of cellulose in a pyroprobe analytical pyrolyzer over ZSM-5, the highest aromatic yield (30 wt%) and lowest char yield were observed at the highest heating rate (1273 K/s) [30]. The yield of aromatics decreased to half while the yield of char increased from 35% to 40% for decreasing heating rate from 1273 to 274 K/s [31]. The aromatics were increased with increasing the reaction time from 1 to 200 s. The increase in the temperature up to 873 K increased aromatic yield (30 wt%) and the aromatics yield remained unchanged beyond 873 K [32]. Various types of zeolites have been reported for the CFP of biomass such as ZSM-5, SAPO-34, Y-zeolite, Beta, MCM-44, MCM-22, etc. The maximum $\sim$35 wt% yield of aromatics was observed over ZSM-5 with an intersecting 10-membered ring pore system composed of straight (5.3–5.6 Å) and sinusoidal (5.1–5.5 Å) channels, while yield of aromatics was 25 wt% only over ZSM-11 with two intersecting straight channels (5.3–5.4 Å) [33]. The silicalites have pore architecture similar to ZSM-5, but it favors coke formation with lesser aromatics compared to ZSM-5. The mesoporous ZSM-5 was reported to yield more aromatics and less coke than microporous ZSM-5 [34]. The metal-incorporated zeolites enhance the yield of aromatics due to the bifunctional behavior of catalyst [35]. The Ga/ZSM-5 enhanced the yield of aromatics by 40% compared to ZSM-5.

3 CHEMICAL CONVERSION PROCESS: PRODUCTION OF GREEN TRANSPORTATION FUELS FROM TRIGLYCERIDES

The triglycerides are composed of three long carbon chains of fatty acids on a glycerol backbone. Generally, C_8-C_{24} fatty acids are present in the triglycerides with the majority being C_{16} and C_{18} fatty acids [36]. The fatty acids are either saturated or unsaturated. The microalgal oils are generally composed of both lighter and heavier fatty acids depending on the source [3]. In general, the vegetable oils are two types: edible oil (mustard oil, soybean oil, peanut oil, coconut oil, sunflower oil, etc.) and nonedible oils (jatropha oil, karanja oil, mahua oil, etc.).

The early use of vegetable oils as diesel was reported by Rudolf Diesel in 1900 [37]. The direct use of vegetable oils as diesel is, however, limited due to its unfavorable properties such as high viscosity and low volatility. The vegetable oils are thus upgraded to fatty acid methyl esters (biodiesel) by transesterification reaction [4]. In transesterification reaction, the triglycerides react with an alcohol (commonly methanol) in presence of a catalyst to produce biodiesel with glycerol as the byproduct [38, 39]. The alkali-catalyzed transesterification reaction is quite fast and most commonly used. The soybean, rapeseed, and palm oils are predominant feedstock for the production of biodiesel. The biodiesel contains about 15% oxygen in the structure resulting in poor fuel mileage and unfavorable properties (high viscosity and poor cold flow properties). The biodiesel is thus blended with diesel up to 20 vol% only for application in unmodified diesel engines. The production of (oxygen-free) green diesel and jet fuels from triglycerides is thus needed for direct use in unmodified diesel engines.

The removal of oxygen from triglycerides can be done by pyrolysis and catalytic cracking. In pyrolysis, the triglycerides undergo thermal decomposition to remove oxygen in the form of CO, CO_2, and water. The triglycerides can also be cracked into gasoline and diesel-range hydrocarbons using zeolites [40]. The pyrolysis and catalytic cracking of triglycerides, however, lead to a low yield of liquid hydrocarbons. The HDO is another process to remove oxygen from the triglycerides for the production of diesel/jet fuel range hydrocarbons [41]. There are two possible routes for HDO of vegetable oils. (i) The vegetable oils directly undergo HDO in this route over the supported metal catalyst to produce diesel/jet fuel range hydrocarbon. (ii) In this route, the vegetable oils undergo hydrolysis to corresponding fatty acids and glycerol. The fatty acids then undergo HDO to produce green diesel/jet fuel range hydrocarbons.

The HDO of triglycerides is carried out under high hydrogen pressure (25–150 bar) in the temperature range of 523 to 773 K depending on the catalysts, hydrogen/oil ratio, and feedstock [36, 42]. Unlike alkali-catalyzed transesterification, the triglycerides with high free fatty acids content can be used as feedstock for the HDO. Moreover, the HDO of triglycerides blended with crude oil fraction can also be performed in the existing hydrodesulfurization unit of petroleum refinery called coprocessing. In HDO, the unsaturated triglycerides are first saturated in presence of hydrogen over the supported metal catalyst. The saturated triglycerides are then converted to respective fatty acids with the release of propane. The resultant fatty acids then undergo reduction to corresponding fatty alcohol. The HDO of fatty

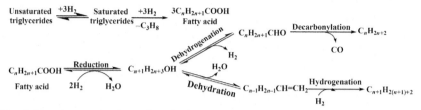

Fig. 9 Reaction mechanism of HDO of triglycerides.

alcohol follows two different routes: dehydrogenation/decarbonylation and dehydration/hydrogenation as shown in Fig. 9 [43]. The dehydrogenation followed by decarbonylation is predominating route over supported metal catalysts. The dehydration/hydrogenation route is mainly followed over acidic catalysts.

The alumina supported transition metals (Ni, Co, NiMo, CoMo, and NiW) are used in the commercial hydrotreatment units. The sulfided, non-sulfided, and phosphided form of these catalysts are inexpensive, highly active, and quite stable. These catalysts were thus extensively studied for HDO of triglycerides [44–46]. Tiwari et al. studied hydroprocessing of soya oil and gas oil mixtures over sulfided $NiMo/\gamma-Al_2O_3$ and $NiW/SiO_2-Al_2O_3$ catalysts at 613–653 K. Almost complete conversion (>99%) of soya oils with about 85%–95% yield of $C_{15}-C_{18}$ hydrocarbons was reported over sulfide $NiMo/\gamma-Al_2O_3$ compared to 80% over $NiW/SiO_2-Al_2O_3$ [47]. >80 wt% diesel-range hydrocarbons were observed over Pd and Ni catalyst (decarbonylation pathway) [36, 48]. The dehydration route is favored over NiMo, and CoMo catalysts.

The metals are generally dispersed over various supports such as $\gamma-Al_2O_3$, activated carbon, SiO_2, and SBA-15, etc. Among the available supports, the better dispersion of metals can be obtained for Al_2O_3 compared to SiO_2 and TiO_2 [49]. The acidic sites and strength of 7 wt% Ni supported catalysts follow the order of $SiO_2 < SAPO-11 < \gamma-Al_2O_3 < HZSM-5 < HY$ [45]. The maximum of 93% of diesel-range hydrocarbons was observed for HDO of methyl palmitate over 7 wt% $Ni/SAPO-11$ at 493 K and 20 bar due to medium acidity. The higher acidity of the catalyst leads to cracking with the high yield of lighter hydrocarbons. HDO of vegetable oils/fatty acids was studied in the different type of reactors such as the high-pressure batch reactor, semibatch reactor, fixed-bed reactor, and trickle-bed reactor [44].

UOP developed an integrated two-stage hydrorefining process called Ecofining process for the production of green diesel from vegetable oils or fatty acids feedstock as shown in Fig. 10. In this process, the vegetable oils

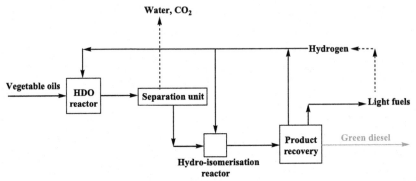

Fig. 10 Schematic diagram of Ecofining process [50].

or fatty acids are first deoxygenated by catalytic HDO to produce diesel-range hydrocarbons. Then, the hydrocarbons are mixed with additional hydrogen and sent to the second stage catalytic hydroisomerization process to improve the cold flow properties, where branched paraffin-rich diesel fuel is produced. The fuel properties of green diesel are shown in Table 4 [43, 50]. The commercial unit of UOP/Eni Ecofining technology becomes operational in Italy. Another plant with >130 million gallon/annum production capacity is operational at Diamond green diesel facility in Norco, Louisiana. Another company, Neste oil, is operating two units in Finland and Singapore with a combined capacity of 170,000 tons/annum of green diesel and Rotterdam with capacity of 800,000 tons/annum [42].

Table 4 Comparison of properties of green diesel (via HDO) with biodiesel and diesel [43, 50]

	Diesel	Biodiesel	Green diesel
Specific gravity	0.84	0.88	0.78
Viscosity at 313 K, cP	–	–	3.81
Sulfur, ppm	<10	<1	<1
Heating value, MJ/kg	43	38	44
Cloud point, K	268	268 to 288	253 to 293
Cetane	40	50–60	70–90
Flash point, K	–	–	411
Fire point, K	–	–	418
Pour point, K	–	–	282
Stability	Good	Marginal	Good
O, wt%	0	11	0.53
H/C	–	–	1.97

4 BIOCHEMICAL PROCESSES

4.1 Ethanol to Gasoline

The bioethanol is currently produced by fermentation of either sugar or starchy biomass depending on their availability. For example, it is mainly produced from sugarcane in Brazil and India and starchy biomass in the USA. The sugar and starchy biomass are generally edible in nature and hence it leads to food-vs-fuel conflict. The lignocellulosic biomass such as energy crops, agricultural waste and residues, forestry waste and residues is abundant and inedible and hence preferable for the bioethanol production [3, 51]. However, the excessive production cost of bioethanol from lignocellulosic biomass inhibits the commercialization of the process so far. The bioethanol can also be produced by fermentation of SG using *Clostridium ragsdalei* as microorganism [52]. The bioethanol is recognized as a most promising biofuel because of its favorable properties. The bioethanol is, however, suffering from the limitations of partial compatibility with the existing internal combustion engines and transportation infrastructure, corrosiveness and hygroscopic in nature, low energy density, and high solubility in water. These factors limit bioethanol as gasoline blend up to 15 wt% [3]. Therefore, there is an increasing tendency to convert bioethanol to gasoline–range hydrocarbons. The ethanol to gasoline (ETG) is usually carried out in a fixed-bed reactor using zeolites as the catalyst in the temperature range of 573–723 K. The yield of liquid hydrocarbons was about 52 wt% at 673 K, $0.5 h^{-1}$ weight hourly space velocity (WHSV) over HZSM-5 [53]. The yield of liquid hydrocarbons decreases with increasing temperature due to cracking reactions. The yield of liquid hydrocarbons was enhanced with increasing pressure and about 70 wt% yield of liquid hydrocarbon was observed at 20.26 bar [53]. The HZSM-5 promoted with iron oxide (and mixed with zinc salts, copper, tin, boronic acid and later molded with binding aluminum oxide) yielded 77%–85% liquid hydrocarbons at 648–673 K [54]. The energy consumption for the separation of ethanol can be reduced by the conversion of aqueous ethanol to liquid hydrocarbons [55]. In ETG process, the ethanol is first dehydrated to ethylene over the acidic catalyst. The ethylene was then converted to gasoline-range hydrocarbons consisting of C_5^+ alkanes and C_7-C_{10} monocyclic aromatics [3].

The ETG was demonstrated by Mobil Oil over HZSM-5. Predominantly C_5^+-range aliphatic hydrocarbons were observed at 583 K [56]. The heavy fraction consisting of mainly higher aromatics was noticed at 638 K [56]. The heat-integrated catalytic conversion of bioethanol to gasoline range hydrocarbons over zeolites was proven to be self-sufficient in energy [57]. Initially, the fermentation broth containing 8–10 wt% ethanol was concentrated to 90 wt%. The solution was then converted to gasoline-range hydrocarbons with 70 wt% yield containing 45–50 wt% aromatics [57]. The exothermic heat in the conversion reactor was used in ethanol recovery from the fermentation broth. The consumption of energy for the whole plant was 1800 Btu/gal, which is much lower compared to the use of ethanol as an additive consuming 21,000 Btu/gal [58]. The two-stage process was also demonstrated to obtain a true engine fuel with allowable aromatics content (Fig. 11) [54]. In this process, bioethanol was converted to paraffins, aromatics, and C_3-C_4 olefins over zeolite or modified zeolite. The olefins and water were separated from liquid hydrocarbons. The liquid hydrocarbons were then hydrogenated at 523–573 K and 100 bar over Re-Pt catalyst to reduce the aromatics content to 12–35 wt%. The separated olefins can also be converted to gasoline-range hydrocarbons by oligomerization reaction.

4.2 Butanol to Gasoline

The bio-n-butanol is produced by ABE fermentation (acetone: butanol: ethanol = 3:6:1) of sugar and starchy or lignocellulosic biomass using clostridia acetobutylicum bacteria. The biobutanol has octane rating similar to gasoline and is compatible with existing internal combustion engines. These characteristics make biobutanol as a superior biofuel over bioethanol. The biobutanol is, however, relatively more expensive than bioethanol. This factor alone inhibits the commercial-scale production of biobutanol so far. Similar to bioethanol, the biobutanol or ABE mixture can also be converted to gasoline-range hydrocarbons over acidic catalysts especially zeolites. Only a few studies have been reported on butanol to gasoline

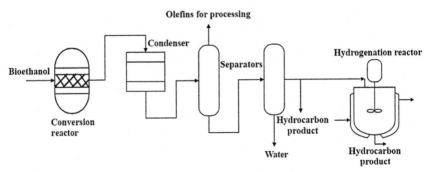

Fig. 11 Schematics of two-stage ETG process [54].

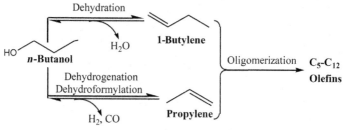

Fig. 12 Reaction mechanism of BTG process [61].

(BTG) so far [59–61]. The HZSM-5 showed a high yield of gasoline range of hydrocarbons [60, 61]. The optimum reaction condition for the production of aromatics free gasoline-range hydrocarbons was reported as 523 K with low WHSV [61]. The BTG reaction follows two pathways: dehydration to butylenes and dehydrogenation to propylene [61] (Fig. 12). The dehydration route is predominant over the solid acid catalyst such as HZSM-5. The olefins are further converted to higher olefins (C_5-C_{12}) by oligomerization reaction. These olefins are then hydrogenated to gasoline-range hydrocarbons.

5 PRODUCTION OF TRANSPORTATION FUELS FROM PLATFORM CHEMICALS

5.1 Platform Chemicals

The research laboratories, Pacific Northwest National Laboratory (PNNL) and National Renewable Energy Laboratory (NREL), identified twelve biomass-derived chemicals from about 300 potential candidates based on their derivative potentials [62]. These chemicals are known as platform chemicals. The list of platform chemicals has been revised in 2010 based on the fresh information as shown in Table 5 [63]. These chemicals are the building blocks in the biorefinery and have potentials to produce a diverse range of organic chemicals, commodity products, and transportation fuels or fuel additives.

5.2 Green Transportation Fuels From Platform Chemicals

The 5-hydroxymethylfurfural (HMF) and furfural are the two most important platform chemicals with diverse derivative potentials including transportation fuels. The HMF and furfural are produced by dehydrocyclization of biomass-derived C_6 and C_5 sugars, respectively, in presence of acid

Table 5 Revisited list of platform chemicals [63]

Ethanol	Biohydrocarbons
Furfural	Succinic acid
5-Hydroxy methyl furfural	Hydroxypropionic acid/aldehyde
Furan dicarboxylic acid	Levulinic acid
Glycerol and derivatives	Sorbitol
Lactic acid	Xylitol
Isoprene	

catalysts. These platform chemicals, however, contain oxygen in the structure. The removal of oxygen from these platform chemicals by HDO is thus necessary to produce green transportation fuels. The direct HDO of HMF and furfural, however, results in C_5-C_6 range volatile hydrocarbons. These hydrocarbons are unsuitable as liquid fuels or fuel blends. The increasing molecular weight by the C—C bond forming reactions is an important strategy to produce gasoline- or diesel-range transportation fuels. The aldol-condensation, hydroxyalkylation-alkylation (HAA), and oligomerization are the widely used C—C bond forming reactions in the biorefinery as briefly described here. These reactions are generally favorable at mild reaction conditions.

5.2.1 Aldol Condensation
The aldol condensation is the reaction between two carbonyl compounds with a reactive α-hydrogen on at least one of the carbonyls. The HMF and furfural, however, do not contain α-hydrogen in their structure and hence cannot undergo the self-aldol condensation. The HMF and furfural can be either partially hydrogenated to generate α-hydrogen in their structure for self-aldol condensation or reacted with α-hydrogen containing carbonyl compounds such as acetone (coproduct of ABE fermentation), hydroxyacetone, glyceraldehydes, etc. The aldol condensation product is subsequently converted to C_7-C_{15} range liquid alkanes by HDO. The distribution of products and quality of the fuel depends on the overall yield of aldol condensation, which in turn depends on the reaction temperature, reactant mole ratio, and nature of the catalyst.

The aldol condensation can be carried out using either basic or acidic catalyst (homogeneous or heterogeneous) as shown in Fig. 13. The homogeneous acid catalysts are rarely used because of lower yields and difficulty in product separation. The NaOH is most commonly used as the homogeneous base catalyst for aldol condensation [64, 65]. The homogeneous base

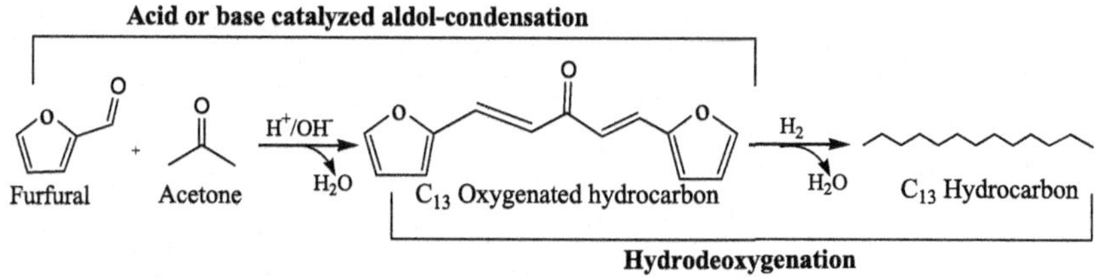

Fig. 13 Aldol condensation of furfural with acetone followed by HDO of condensation product.

catalysts are generally not preferred because of corrosiveness to process equipment, generation of large quantities of wastewater, and difficulty in separation of the catalyst from the product [66]. The heterogeneous base catalysts such as hydrotalcite-like materials and other mixed oxides are generally employed for the aldol condensation [67, 68]. These catalysts, however, deactivate quickly as they are sensitive toward ambient CO_2 [69]. Recently, the solid base catalyst $Zr(CO_3)_x$ was reported to be reused for many cycles without significant loss of catalytic activity [70]. The protonated titanate nanotube was also found to be a highly efficient catalyst and can be used several times without a significant drop in catalytic activity [71]. The zeolites with the different type of structures were also employed for aldol condensation and displayed a good catalytic activity compared to other solid acid catalysts [72]. However, the catalytic activity of the zeolites decreases during the reaction because of the formation of carbonaceous deposits inside their micropores.

From the bibliographic standpoint, the aldol condensation of HMF and furfural with acetone was achieved in the monophasic reactor using water, ethanol/water or methanol/water as the solvent. To overcome the drawback of solvent separation, the biphasic reactor has been employed for continuous separation of hydrophobic condensed products [73]. The furanic aldehydes were also reported to undergo aldol condensation with cyclic ketones under solvent-free conditions and produce aldol adducts with 90% selectivity [74]. With increasing the temperature from 298 to 423 K, the selectivity to products slowly increased [75]. For Mg-Al hydroxides catalyst, the increasing reaction temperature reduces the carbon yield due to the formation of coke [76]. For aldol condensation of furfural with acetone, the product selectivity was increased by 43% for increasing furfural:acetone ratio from 1:1 to 2:1 [77]. Recently, a process has been patented for the production of fuel components and chemicals from levulinic acid (LA) [78]. This process consists of the conversion of cellulose to LA followed by aldol condensation with aldehydes/ketones and HDO of condensed product over alumina-supported metal catalysts (Pd, Pt, Ru, Fe, and Ni).

5.2.2 Hydroxyalkylation-Alkylation Reaction

The HAA reaction occurs between α-hydrogen containing furan with a carbonyl compound (Fig. 14) [79–81]. The oxygen heteroatoms are then removed from the HAA product by HDO to produce diesel/jet fuel range hydrocarbons. This reaction takes place under mild reaction condition (313–363 K) in presence of acid catalysts. The furfural is a potential candidate of α-hydrogen containing furan compound. The furfural, however, does not undergo self or cross HAA reaction with another carbonyl compound because of electron withdrawing effect of aldehyde group. On the other hand, the 2-methylfuran is highly active for HAA reaction. The 2-methylfuran is produced by selective hydrogenation of furfural. The HAA reaction of 2-methylfuran with furfural, HMF, acetoin, cyclopentanone, butanal, acetone, angelica lactone, and cyclohexanone has been reported. The wide ranges of acid catalysts were also tested for HAA reaction: mineral and organic acids, zeolites, heteropoly acids, acidic resins, sulfated ZrO_2, Lewis acids, graphene oxide, and $NbOPO_4$. The zeolites are inactive for HAA reaction [82]. The heteropoly acids showed poor selectivity to HAA product due to over alkylation [83]. The use of mineral acids led to the formation of char during HAA reaction.

The yield of HAA product decreased in the following sequence: Nafion-212 > Nafion-115 > H_2SO_4 > amberlyst-15 > amberlyst-36 > CMK-3-SO_3H > AC-SO_3H > MC-SO_3H > ZrP > HZSM-5 > HY [84]. The catalytic activity of the mineral acids for HAA reaction was in the order of H_2SO_4 > H_3PO_4 > CH_3COOH [82]. The catalytic activity was decreased in the following order for HAA of 2-methylfuran with angelica lactone and ethyl levulinate: Brønsted acids: trifflic acid > sulfuric acid > phosphoric acid > acetic acid, Lewis acids: $FeCl_3$ > $SnCl_4$ > $ZnCl_2$ > $TiCl_4$, and acidic resins: Nafion-212 > Amberlyst-15 > Amberlite IRC [85]. The conversion

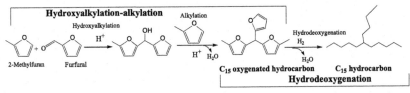

Fig. 14 The HAA reaction of 2-methylfuran with furfural followed by HDO of HAA product.

of 2-methylfuran and yield of HAA product decreased in the following order: butanal > furfural > acetone > mesityl oxide over protonated titanate nanotubes [86].

The HDO of HAA products is necessary to produce diesel/jet-fuel-range hydrocarbons. The HAA product of cyclopentanone and 2-methylfuran further hydrodeoxygenated over M/SiO_2-Al_2O_3 (M = Fe, Co, Ni, Cu) catalysts. Among the tested catalysts, Ni/SiO_2-Al_2O_3 demonstrated the best performance [82]. For HAA of 2-methylfuran and angelica lactone followed by HDO, the higher carbon yields of diesel and jet-fuel-range alkanes were achieved over 5 wt% Pt/C (81.0%) and 5 wt% Pd/C (81.6%) than 5 wt% Ru/C (60.4%) [85]. For HDO of HAA product (furfural with 2-methyl furan) over $Pd/NbOPO_4$ catalyst, the 89.1% yield of diesel-range alkanes (C_{10}-C_{15}) was observed under the optimal reaction conditions (473 K, 40 bar, and 12 h) [87].

5.2.3 Oligomerization

The oligomerization of olefins such as ethylene and butylenes is another route for increasing the carbon number for the production of motor fuels. The ethylene and butylenes are derived by dehydration of bioethanol and biobutanol, respectively, in presence of solid acid catalysts. In this process, the monomers combine to form dimers, trimers, tetramers, etc. over the acidic catalyst. The ethylene oligomerization is the well-studied reaction. The butylene oligomerization is thus briefly discussed in this section. The oligomerization reaction mainly proceeds through carbenium ion mechanism [30]. The butylene oligomerization has been studied over the wide range of acid catalysts such as mineral acids, cation exchange resins, zeolites, metal modified zeolites, etc. The cation exchange resins are highly active for oligomerization reaction. It was reported that the monomer absorption constant is greater than the dimers and trimers, etc. and hence the presence of monomer on catalytic sites inhibits the formation of dimers, trimers, and tetramers [88]. The $NiSO_4/\gamma$-Al_2O_3 was found to be a better catalyst than H_2SO_4 and HF for oligomerization of isobutylene (323–363 K and 21.7 bar) and the reaction was mainly directed toward dimerization without deactivation [89]. For oligomerization butylene-CO_2 mixture in a fixed-bed reactor over HZSM-5 and Amberlyst 70 at 443–525 K and 20–35 bar, >90% butylene conversion with 95% selectivity toward liquid alkenes (C_8+), was reported [90]. For Ni-doped HZSM-5, it was reported that the certain amount of acidic sites with suitable B/L ratio is necessary for selective trimerization of 1-butylene [91]. The high pressure favors conversion of olefins but results in the deactivation of catalyst [88]. UOP developed a unit for catalytic polymerization of olefin containing gases over solid phosphoric acid at 448–508 K (usually 473–493 K), 27–35 bar, and 2.5 L/kg-h [92]. In this process, the yield of gasoline boiling range hydrocarbons is 90–97 wt% of olefin or 0.7 barrel of gasoline per barrel of olefin.

5.3 Transportation Fuels From Platform Molecules

5.3.1 Levulinic Acid Derivatives as the Transportation Fuels

The LA is an important platform chemical due to its multifunctionality (keto and acid group). It has also potential to produce a number of fuel additives such as γ-valerolactone (GVL), 2-methyltetrahydrofuran (MTHF), and ethyl levulinate. Similar to bioethanol, these compounds can be blended with the petroleum-derived transportation fuels without engines modification. The fuel properties of these compounds are compared with ethanol as shown in Table 6. The LA is derived by dehydration of HMF in presence of acid catalyst with the elimination of one mole of formic acid. Hanna and co-workers reported 32.6% yield of LA from 10% kernel flour at 473 K, 8% H_2SO_4 [93]. Heeres et al. achieved 60 mol% yield of LA at 423 K from 1.7 wt% cellulose using 1 M H_2SO_4 [94]. Biofine renewables, LLC developed a continuous process for the production of LA with >70% yield [95]. Recently, GF Biochemicals developed world's largest commercial production of LA at Caserta, Italy. The company targeted the production of 10,000 MT/annum of LA by 2017 [96]. The following section describes the production of the fuel additives from LA.

Table 6 The comparison of fuel properties of GVL, MTHF, and ethyl levulinate with ethanol [116]

	Ethanol	GVL	MTHF	EL
Molecular weight, gmol^{-1}	46.07	100.12	86.13	144.17
Boiling point, K	351	480	353	479.2
Melting point, K	159	242	137	–
Flash Point, K	286	369.1	261.9	468
Density, gL^{-1}	0.789	1.0485	0.86	1.014
Solubility in water, mgL^{-1}	Miscible	≥100	13	Soluble
RON	108.6	–	80	–
Cetane number	5	–	23.5	<10

GVL, γ-valerolactone; MTHF, 2-methyltetrahydrofuran; EL, ethyl levulinate.

5.3.2 Production of γ-Valerolactone

The GVL is typically synthesized through the hydrogenation of LA using either pure hydrogen or transfer hydrogen sources. The formic acid is commonly used as the transfer hydrogen source. The hydrogenation of LA to GVL was performed using both homogeneous and heterogeneous catalysts [97]. These catalysts have been tested in the batch and continuous reactor [97]. The supported ruthenium-based catalysts have been used commonly for the hydrogenation of LA to GVL. The cost of the catalysts, metal leaching, volatile nature of solvents, high pressures, and harsh reaction conditions, however, limit their application at large scale. The hydrogenation of LA proceeds through two different pathways as shown in the Fig. 15. The hydrogenation of LA follows Path-I over the acidic catalysts or supports [98, 99]. The catalysts with dominant metallic sites, however, divert the reaction toward Path-II. In reaction Path-I, the LA undergoes dehydration reaction over acidic sites of the catalyst followed by hydrogenation to GVL. Following Path-II, the LA is hydrogenated to 4-hydroxy levulinic acid over metallic sites of the catalyst followed by cyclization to GVL [100, 101].

For hydrogenation of LA at 538 K and 10 bar over 5 wt% Cu/SiO_2, the yield of GVL was 99.9% [102]. The catalyst showed a decrease in activity with time-on-stream. The stability of the catalyst was, however, improved for 100 h time-on-stream by the addition of promoter, nickel. For hydrogenation of aqueous LA, the catalytic activity was in the order of $Ru > Pt > Cu > Pd > Ni/HAP$ [103]. Among various supports (SiO_2, TiO_2, Al_2O_3, ZrO_2), the Al_2O_3 was reported as the best one for hydrogenation of LA [104]. Among SiO_2, Al_2O_3, and ZrO_2, the SiO_2 was found to be the best support [105]. For the reaction study over various supported (SiO_2, Al_2O_3, ZrO_2, TiO_2, and ZnO) Ni catalysts, the complete conversion

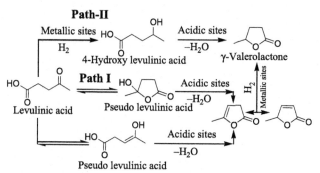

Fig. 15 Routes of the production of GVL from LA.

of LA with 87% yield of GVL was observed over 30 wt% Ni/SiO$_2$ [100]. Mohan et al. studied the hydrogenation of LA over a bifunctional catalyst (metallic and acidic), 30 wt% Ni/HZSM-5 and 67.5% yield of GVL was reported at 523 K and H$_2$/LA mole ratio of 5.5 [98]. For vapor-phase hydrogenation of LA over Ni/MgO, Ni/Al$_2$O$_3$, and Ni/MgO-Al$_2$O$_3$ using formic acid as the hydrogen source, about 87% yield of GVL was observed over Ni/Al$_2$O$_3$ with LA/formic acid mole ratio of 1:5 at 523 K [106]. A drastic decrease in conversion of LA with time-on-stream was observed over Ni/MgO and Ni/MgO-Al$_2$O$_3$ due to the transition of the brucite-periclase structure of MgO [106]. The hydrogenation of LA using formic acid as hydrogen source over Ni/HZSM-5 and Ni/SiO$_2$ showed 63.8% and 75.7% GVL yield, respectively [106]. The SiO$_2$ thus seems to be an effective support for the hydrogenation of LA with formic acid as the hydrogen source. Among the various catalyst preparation methods, the Ni/SiO$_2$ prepared by citric acid assisted method and calcined under N$_2$ flow showed the best catalytic performance [107]. The 20Ni60Cu/SiO$_2$ showed the best catalytic performance for the hydrogenation of LA using formic acid as the hydrogen source [108].

5.3.3 Production of Ethyl Levulinate

The alkyl levulinates especially ethyl levulinate can be blended to diesel up to 5 wt% [109] (Table 6). In general, the ethyl levulinate is synthesized through the esterification of LA, ethanolysis of furfuryl alcohol, and ethanolysis of hexose and pentose sugars derived from carbohydrates in the presence of an acid catalyst. The various catalysts such as mineral acids, ionic liquids, sulfonic acid functionalized catalysts, zeolites, heteropoly acids, sulfated metal oxides, cation exchange resins, etc. were reported for the production of ethyl levulinate. Among the various catalysts, the ionic liquid and sulfonic acid functionalized catalysts are the most promising ones with high selectivity to ethyl levulinate. The separation from the reaction mixture and reuse are, however, the challenges associated with the use of ionic liquids. For the vapor-phase esterification of LA with ethanol over Zr-SBA-15 at 523 K, almost complete conversion with 50% selectivity to ethyl levulinate was observed [110]. The ethyl levulinate synthesis was reported over C-SO$_3$H derived from glycerol and achieved 99% ethyl levulinate yield at 180 min [111]. The catalyst was recycled up to five cycles without significant change in activity. Among the various tested catalysts, the catalysts with Brönsted acidity were often used for the esterification of LA with ethanol [109].

5.3.4 Production of 2-Methyltetrahydrofuran

The MTHF is considered as a potential replacement of solvent, tetrahydrofuran (THF). The MTHF is also a choice of gasoline substitute or blending component of gasoline due to its favorable fuel properties such as hydrophobic nature, high energy density, lower flammability, low toxicity, and high specific gravity [112, 113]. About 70% of MTHF can be blended with gasoline. The MTHF can be produced by two different routes: sequential catalytic hydrogenation of LA and hydrogenation of 2-methylfuran derived from furfural (Fig. 16). Various catalysts (Cu, Ni, and noble metal-based) have been reported for the production of MTHF. For example, for the hydrogenation of LA to MTHF at 538 K and 10 bar using 80 wt% Cu/ SiO_2, 64% MTHF yield was reported [102]. The addition of Ni as a promoter (8wt%Ni72wt%Cu/SiO_2), the MTHF yield was increased to 89%, and the catalyst remained stable for 320 h time-on-stream. For hydrogenation of LA and GVL independently over Ni/KIT-6, Ni/COK-12, and Ni/SBA-16 catalysts, the formation of MTHF was noticed whenever Ni particles were confined in mesopores [101]. Recently, Zheng et al. investigated the ethyl levulinate hydrogenation over Cu/SiO_2 and Cu/Al_2O_3-SiO_2 catalyst [114]. About 65% selectivity to MTHF was observed over Cu/Al_2O_3-SiO_2 at 523 K.

6 SUMMARY AND CONCLUSIONS

The hydrocarbon biorefinery provides the plenty of opportunities for the production of green transportation fuels from various types of biomass. Followings are the summary of the present chapter.

- The thermochemical conversion processes use inexpensive and abundant lignocellulosic biomass as the feedstock. In this route, the biomass

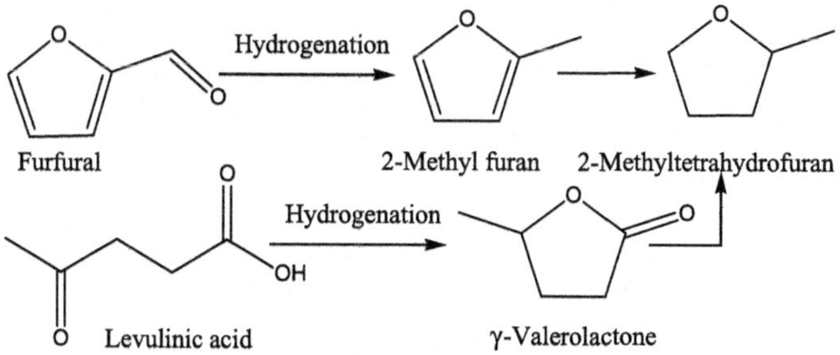

Fig. 16 Routes for the production of MTHF.

is directly converted to liquid fuels. This route is thus most attractive in hydrocarbon biorefinery. The fast pyrolysis and CFP are the most attractive thermochemical conversion processes because of their simplicity and high yield of transportation fuels.

- The triglycerides are generally edible in nature and hence expensive. The availability of the feedstock is one of the major challenges of the HDO of triglycerides. The technological advancement of microalgae is necessary for the abundant availability of triglycerides in near future.

- The MTG and ETG processes provide gasoline-range hydrocarbons with high aromatic content, which is unacceptable as per the current gasoline specification. The further processing of these hydrocarbons is necessary for the reduction of the aromatics content. Alternatively, further studies are needed to develop suitable catalysts for the one-step production of aromatics-free gasoline range hydrocarbons from biomethanol and bioethanol. The studies on BTG are, however, limited and further studies are required for the production of aromatics-free gasoline-range hydrocarbons.

- The aldol condensation, HAA, and oligomerization of olefins are the extensively studied C—C bond forming reactions in the biorefinery to produce green transportation fuels. The platform chemicals also provide the opportunities to produce various fuel additives such as GVL, MTHF, ethyl levulinate, etc. These routes, however, involve multistep resulting in high production cost.

- There is a need for further research on design and development of catalysts and process improvement to make the hydrocarbon biorefinery successful in near future.

REFERENCES

[1] International Energy Statistics. U.S. Energy Information Administration 2013, http://www.eia.gov/cfapps/ipdbproject/IEDIndex3.cfm?tid=5&pid=5&aid=2; 2013.

[2] Maity SK. Opportunities, recent trends and challenges of integrated biorefinery: Part I. Renew Sustain Energy Rev 2015;43:1427–45. https://doi.org/10.1016/j.rser.2014.11.092.

[3] Maity SK. Opportunities, recent trends and challenges of integrated biorefinery: Part II. Renew Sustain Energy Rev 2015;43:1446–66. https://doi.org/10.1016/j.rser.2014.08.075.

[4] Huber GW, Iborra S, Corma A. Synthesis of transportation fuels from biomass: chemistry, catalysts, and engineering. Chem Rev 2006;106:4044–98. https://doi.org/10.1021/cr068360d.

[5] Spath PL, Dayton DC. Preliminary screening – technical and economic assessment of synthesis gas to fuels and chemicals with emphasis on the potential for biomass-derived syngas. Natl Renew Energy Lab 2003;1–160. https://doi.org/10.2172/15006100.

[6] Chang CD, Silvestri AJ. The conversion of methanol and other 0-compounds to hydrocarbons over zeolite catalysts. J Catal 1977;47:249–59.

[7] EMRE. Methanol to gasoline (MTG) technology: an alternative for liquid fuel production. World CTL Conference; 2010.

[8] Allum KG, Williams AR. Operation of the world's first gas-to-gasoline plant. Stud Surf Sci Catal 1988;36:691–711. https://doi.org/10.1016/S0167-2991(09)60566-8.

[9] Yurchak S. Development of Mobil's fixed-bed methanol-to-gasoline (MTG) process. Stud Surf Sci Catal 1988;36:251–72. https://doi.org/10.1016/S0167-2991(09)60521-8.

[10] Haw JF, Song W, Marcus DM, Nicholas JB. The mechanism of methanol to hydrocarbon catalysis. Acc Chem Res 2003;36:317–26. https://doi.org/10.1021/ar020006o.

[11] Chang CD. Mechanism of hydrocarbon formation from methanol. Stud Surf Sci Catal 1988;36:127–43. https://doi.org/10.1016/S0167-2991(09)60507-3.

[12] Kolboe S. On the mechanism of hydrocarbon formation from methanol over protonated zeolites. Stud Surf Sci Catal 1988;36:189–93. https://doi.org/10.1016/S0167-2991(09)60512-7.

[13] Martinez-Espin JS, Mortén M, Janssens TVW, Svelle S, Beato P, Olsbye U. New insights into catalyst deactivation and product distribution of zeolites in the methanol-to-hydrocarbons (MTH) reaction with methanol and dimethyl ether feeds. Cat Sci Technol 2017;7:2700–16. https://doi.org/10.1039/C7CY00129K.

[14] Olsbye U, Bjørgen M, Svelle S, Lillerud KP, Kolboe S. Mechanistic insight into the methanol-to-hydrocarbons reaction. Catal Today 2005;106:108–11. https://doi.org/10.1016/j.cattod.2005.07.135.

[15] Svelle S, Joensen F, Nerlov J, Olsbye U, Lillerud KP, Kolboe S, et al. Conversion of methanol into hydrocarbons over zeolite H-ZSM-5: ethene formation is mechanistically separated from the formation of higher alkenes. J Am Chem Soc 2006;128:14770–1. https://doi.org/10.1021/ja065810a.

[16] Benito PL, Gayubo AG, Aguayo AT, Olazar M, Bilbao J. Effect of Si/Al ratio and of acidity of H-ZSM5 zeolites on the primary products of methanol to gasoline conversion. J Chem Technol Biotechnol 1996;66:183–91. https://doi.org/10.1002/(SICI)1097-4660(199606)66:2<183::AID-JCTB487>3.0.CO;2-K.

[17] Obrzut DL, Adekkanattu PM, Thundimadathil J, Liu J, Dubois DR, Guin JA. Reducing methane formation in methanol to olefins reaction on metal impregnated SAPO-34 molecular sieve. React Kinet Catal Lett 2003;80:113–21. https://doi.org/10.1023/A:1026088327000.

[18] Freeman D, Wells RPK, Hutchings GJ. Methanol to hydrocarbons: enhanced aromatic formation using a composite Ga_2O_3/H-ZSM-5 catalyst. Chem Commun 2001;0:1754–5. https://doi.org/10.1039/b104844a.

[19] Zaidi HA, Pant KK. Catalytic conversion of methanol to gasoline range hydrocarbons. Catal Today 2004;96:155–60. https://doi.org/10.1016/j.cattod.2004.06.123.

[20] Toor SS, Rosendahl L, Rudolf A. Hydrothermal liquefaction of biomass: a review of subcritical water technologies. Energy 2011;36:2328–42. https://doi.org/10.1016/j.energy.2011.03.013.

[21] Bridgwater AV. Review of fast pyrolysis of biomass and product upgrading. Biomass Bioenergy 2012;38:68–94. https://doi.org/10.1016/j.biombioe.2011.01.048.

[22] Butler E, Devlin G, Meier D, McDonnell K. A review of recent laboratory research and commercial developments in fast pyrolysis and upgrading. Renew Sustain Energy Rev 2011;15:4171–86. https://doi.org/10.1016/j.rser.2011.07.035.

[23] Oasmaa A, Van de Beld B, Saari P, Elliott DC, Solantausta Y. Norms, standards, and legislation for fast pyrolysis bio-oils from Lignocellulosic biomass. Energy Fuel 2015;29:2471–84. https://doi.org/10.1021/acs.energyfuels.5b00026.

[24] Elliott DC. Transportation fuels from biomass via fast pyrolysis and hydroprocessing. Wiley Interdiscip Rev Energy Environ 2013;2:525–33. https://doi.org/10.1002/wene.74.

[25] Zhang H, Xiao R, Huang H, Xiao G. Comparison of non-catalytic and catalytic fast pyrolysis of corncob in a fluidized bed reactor. Bioresour Technol 2009;100:1428–34. https://doi.org/10.1016/j.biortech.2008.08.031.

[26] Elliott DC, Hart TR. Catalytic hydroprocessing of chemical models for bio-oil. Energy Fuels 2009;23:631–7. https://doi.org/10.1021/ef8007773.

[27] Adjaye JD, Bakhshi NN. Production of hydrocarbons by catalytic upgrading of a fast pyrolysis bio-oil. Part I: conversion over various catalysts. Fuel Process Technol 1995;45:185–202. https://doi.org/10.1016/0378-3820(95)00034-5.

[28] Adjaye JD, Bakhshi NN. Production of hydrocarbons by catalytic upgrading of a fast pyrolysis bio-oil. Part II: comparative catalyst performance and reaction pathways. Fule Process Technol 1995;45:185–202. https://doi.org/10.1016/0378-3820(95)00040-E.

[29] Zheng A, Jiang L, Zhao Z, Huang Z, Zhao K, Wei G, et al. Catalytic fast pyrolysis of lignocellulosic biomass for aromatic production: chemistry, catalyst and process. Wiley Interdiscip Rev Energy Environ 2017;6:1–18. https://doi.org/10.1002/wene.234.

[30] Carlson TR, Tompsett GA, Conner WC, Huber GW. Aromatic production from catalytic fast pyrolysis of biomass-derived Feedstocks. Top Catal 2009;52:241–52. https://doi.org/10.1007/s11244-008-9160-6.

[31] Carlson TR, Vispute TP, Huber GW. Green gasoline by catalytic fast pyrolysis of solid biomass derived compounds. ChemSusChem 2008;1:397–400. https://doi.org/10.1002/cssc.200800018.

[32] Carlson TR, Jae J, Lin Y-C, Tompsett GA, Huber GW. Catalytic fast pyrolysis of glucose with HZSM-5: the combined homogeneous and heterogeneous reactions. J Catal 2010;270:110–24. https://doi.org/10.1016/j.jcat.2009.12.013.

[33] Jae J, Tompsett GA, Foster AJ, Hammond KD, Auerbach SM, Lobo RF, et al. Investigation into the shape selectivity of zeolite catalysts for biomass conversion. J Catal 2011;279:257–68. https://doi.org/10.1016/j.jcat.2011.01.019.

[34] Li J, Li X, Zhou G, Wang W, Wang C, Komarneni S, et al. Catalytic fast pyrolysis of biomass with mesoporous ZSM-5 zeolites prepared by desilication with NaOH solutions. Appl Catal Gen 2014;470:115–22. https://doi.org/10.1016/j.apcata.2013.10.040.

[35] Cheng YT, Jae J, Shi J, Fan W, Huber GW. Production of renewable aromatic compounds by catalytic fast pyrolysis of lignocellulosic biomass with bifunctional Ga/ZSM-5 catalysts. Angew Chemie Int Ed 2012;51:1387–90. https://doi.org/10.1002/anie.201107390.

[36] Kumar P, Yenumala SR, Maity SK, Shee D. Kinetics of hydrodeoxygenation of stearic acid using supported nickel catalysts: effects of supports. Appl Catal Gen 2014;471:28–38. https://doi.org/10.1016/j.apcata.2013.11.021.

[37] Knothe G. Historical perspectives on vegetable oil based diesel fuels. Inform 2001;12:1103–7.

[38] Zhang Y, Dubé MA, McLean DD, Kates M. Biodiesel production from waste cooking oil: 1. Process design and technological assessment. Bioresour Technol 2003;89:1–16. https://doi.org/10.1016/S0960-8524(03)00040-3.

[39] Sales EA, Ghirardi ML, Jorquera O. Subcritical ethylic biodiesel production from wet animal fat and vegetable oils: a net energy ratio analysis. Energ Conver Manage 2017;141:216–23. https://doi.org/10.1016/j.enconman.2016.08.015.

[40] Dupain X, Costa DJ, Schaverien CJ, Makkee M, Moulijn JA. Cracking of a rapeseed vegetable oil under realistic FCC conditions. Appl Catal Environ 2007;72:44–61. https://doi.org/10.1016/j.apcatb.2006.10.005.

[41] Furimsky E. Catalytic hydrodeoxygenation. Appl Catal Gen 2000;199:147–90. https:// doi.org/10.1016/S0926-860X(99)00555-4.

[42] Glisic SB, Pajnik JM, Orlović AM. Process and techno-economic analysis of green diesel production from waste vegetable oil and the comparison with ester type biodiesel production. Appl Energy 2016;170:176–85. https://doi.org/10.1016/j.apenergy.2016.02.102.

[43] Yenumala SR, Maity SK, Shee D. Hydrodeoxygenation of karanja oil over supported nickel catalysts: influence of support and nickel loading. Cat Sci Technol 2016;6:3156–65. https://doi.org/10.1039/C5CY01470K.

[44] Ameen M, Azizan MT, Yusup S, Ramli A, Yasir M. Catalytic hydrodeoxygenation of triglycerides: an approach to clean diesel fuel production. Renew Sustain Energy Rev 2017;80:1072–88. https://doi.org/10.1016/j.rser.2017.05.268.

[45] Zuo H, Liu Q, Wang T, Ma L, Zhang Q, Zhang Q. Hydrodeoxygenation of methyl palmitate over supported Ni catalysts for diesel-like fuel production. Energy Fuels 2012;26:3747–55. https://doi.org/10.1021/ef300063b.

[46] Harnos S, Onyestyák G, Kalló D. Hydrocarbons from sunflower oil over partly reduced catalysts. React Kinet Mech Catal 2012;106:99–111. https://doi.org/10.1007/s11144-012-0424-6.

[47] Tiwari R, Rana BS, Kumar R, Verma D, Kumar R, Joshi RK, et al. Hydrotreating and hydrocracking catalysts for processing of waste soya-oil and refinery-oil mixtures. Catal Commun 2011;12:559–62. https://doi.org/10.1016/j.catcom.2010.12.008.

[48] Veriansyah B, Han JY, Kim SK, Hong SA, Kim YJ, Lim JS, et al. Production of renewable diesel by hydroprocessing of soybean oil: effect of catalysts. Fuel 2012;94:578–85. https://doi.org/10.1016/j.fuel.2011.10.057.

[49] Hoang-Van C, Kachaya Y, Teichner SJ, Arnaud Y, Dalmon JA. Characterization of nickel catalysts by chemisorption techniques, x-ray diffraction and magnetic measurements. Effects of support, precursor and hydrogen pretreatment. Appl Catal 1989;46:281–96. https://doi.org/10.1016/S0166-9834(00)81123-9.

[50] Kalnes TN, Marker T, Shonnard DR, Koers KP. Green diesel production by hydrorefining renewable feedstocks. Biofuels Technol 2008;7–11.

[51] Hahn-Hägerdal B, Galbe M, Gorwa-Grauslund MF, Lidén G, Zacchi G. Bio-ethanol – the fuel of tomorrow from the residues of today. Trends Biotechnol 2006;24:549–56. https://doi.org/10.1016/j.tibtech.2006.10.004.

[52] Liu K, Atiyeh HK, Stevenson BS, Tanner RS, Wilkins MR, Huhnke RL. Continuous syngas fermentation for the production of ethanol, n-propanol and n-butanol. Bioresour Technol 2014;151:69–77. https://doi.org/10.1016/j.biortech.2013.10.059.

[53] Costa E, Ugulna A, Aguado J, Hernéndez PJ. Ethanol to gasoline process: effect of variables, mechanism, and kinetics. Ind Eng Chem Process Des Dev 1985;24:239–44. https://doi.org/10.1021/i200029a003.

[54] Tret'yakov VF, Makarfi YI, Tret'yakov KV, Frantsuzova NA, Talyshinskii RM. The catalytic conversion of bioethanol to hydrocarbon fuel: a review and study. Catal Ind 2010;2:402–20. https://doi.org/10.1134/S2070050410040161.

[55] Oujedans JC, Van Den OPF, Van BH. Conversion of ethanol over Zeoilite H-ZSM-5 in the presence of water. Catal Appl 1982;3:109–15. https://doi.org/10.1016/0166-9834(82)80084-5.

[56] Butter SA, Jurewicz AT, Kaeding WW. Conversion of alcohols, mercaptans, sulfides, halides and/or amines. US 3,894,107, 1975.

[57] Whitcraft DR, Verykios XE, Mutharasan R. Recovery of ethanol from fermentation broths by catalytic conversion to gasoline. Ind Eng Chem Process Des Dev 1983;22:452–7. https://doi.org/10.1021/i200022a019.

[58] Aldridge GA, Veryklos XE, Mutharasan R. Recovery of ethanol from fermentation broths by catalytic conversion to gasoline. 2. Energy analysis. Ind Eng Chem Process Des Dev 1984;23:733–7. https://doi.org/10.1021/i200027a018.

[59] Cobalt Technologies. Cobalt and the naval air warfare center team up to produce a renewable jet fuel from bio N-butanol; 2012.

[60] Varvarin AM, Khomenko KM, Brei VV. Conversion of n-butanol to hydrocarbons over H-ZSM-5, H-ZSM-11, H-L and H-Y zeolites. Fuel 2013;106:617–20. https://doi.org/10.1016/j.fuel.2012.10.032.

[61] Palla VCS, Shee D, Maity SK. Conversion of n-butanol to gasoline range hydrocarbons, butylenes and aromatics. Appl Catal Gen 2016;526:28–36. https://doi.org/10.1016/j.apcata.2016.07.026.

[62] Werpy T, Petersen G. Top value added chemicals from biomass volume I—results of screening for potential candidates from sugars and synthesis gas; 2004. https://doi.org/10.2172/15008859.

[63] Bozell JJ, Petersen GR. Technology development for the production of biobased products from biorefinery carbohydrates—the US Department of Energy's "top 10" revisited. Green Chem 2010;12:539–54. https://doi.org/10.1039/b922014c.

[64] Xing R, Subrahmanyam AV, Olcay H, Qi W, Van Walsum GP, Pendse H, et al. Production of jet and diesel fuel range alkanes from waste hemicellulose-derived aqueous solutions. Green Chem 2010;12:1933–46. https://doi.org/10.1039/c0gc00263a.

[65] Zapata PA, Faria J, Pilar Ruiz M, Resasco DE. Condensation/hydrogenation of biomass-derived oxygenates in water/oil emulsions stabilized by nanohybrid catalysts. Top Catal 2012;55:38–52. https://doi.org/10.1007/s11244-012-9768-4.

[66] Barrett CJ, Chheda JN, Huber GW, Dumesic JA. Single-reactor process for sequential aldol-condensation and hydrogenation of biomass-derived compounds in water. Appl Catal Environ 2006;66:111–8. https://doi.org/10.1016/j.apcatb.2006.03.001.

[67] Debecker DP, Gaigneaux EM, Busca G. Exploring, tuning, and exploiting the basicity of hydrotalcites for applications in heterogeneous catalysis. Chem Eur J 2009;15:3920–35. https://doi.org/10.1002/chem.200900060.

[68] Faba L, Díaz E, Ordóñez S. Aqueous-phase furfural-acetone aldol condensation over basic mixed oxides. Appl Catal Environ 2012;113–114:201–11. https://doi.org/10.1016/j.apcatb.2011.11.039.

[69] Xu C, Gao Y, Liu X, Xin R, Wang Z. Hydrotalcite reconstructed by in situ rehydration as a highly active solid base catalyst and its application in aldol condensations. RSC Adv 2013;3:793–801. https://doi.org/10.1039/C2RA21762G.

[70] Bohre A, Saha B, Abu-Omar MM. Catalytic upgrading of 5-hydroxymethylfurfural to drop-in biofuels by solid base and bifunctional metal-acid catalysts. ChemSusChem 2015;8:4022–9. https://doi.org/10.1002/cssc.201501136.

[71] Kitano M, Nakajima K, Kondo JN, Hayashi S, Hara M. Protonated titanate nanotubes as solid acid catalyst. J Am Chem Soc 2010;132:6622–3.

[72] Kikhtyanin O, Kelbichova V, Vitvarova D, Kubu M, Kubička D. Aldol condensation of furfural and acetone on zeolites. Catal Today 2014;227:154–62. https://doi.org/10.1016/j.cattod.2013.10.059.

[73] West RM, Liu ZY, Peter M, Grtner CA, Dumesic JA. Carbon-carbon bond formation for biomass-derived furfurals and ketones by aldol condensation in a biphasic system. J Mol Catal A Chem 2008;296:18–27. https://doi.org/10.1016/j.molcata.2008.09.001.

[74] Deng Q, Xu J, Han P, Pan L, Wang L, Zhang X, et al. Efficient synthesis of high-density aviation biofuel via solvent-free aldol condensation of cyclic ketones and furanic aldehydes. Fuel Process Technol 2016;148:361–6. https://doi.org/10.1016/j.fuproc.2016.03.016.

[75] Bui TV, Sooknoi T, Resasco DE. Simultaneous upgrading of furanics and phenolics through hydroxyalkylation/aldol condensation reactions. ChemSusChem 2017; 10:1631–9. https://doi.org/10.1002/cssc.201601251.

[76] Liu H, Xu W, Liu X, Guo Y, Guo Y, Lu G, et al. Aldol condensation of furfural and acetone on layered double hydroxides. Kinet Catal 2010;51:75–80. https://doi.org/10.1134/S0023158410010131.

[77] Chheda JN, Dumesic JA. An overview of dehydration, aldol-condensation and hydrogenation processes for production of liquid alkanes from biomass-derived carbohydrates. Catal Today 2007;123:59–70. https://doi.org/10.1016/j.cattod.2006.12.006.

[78] Olson ES, Heide C. Multiproduct biorefinery for the synthesis of fuel components and chemicals from lignocellulosics via Levulinate condensations. EP 2438144 A2; 2012.

[79] Corma A, de la TO, Renz M. High-quality diesel from hexose- and pentose-derived biomass platform molecules. ChemSusChem 2011;4:1574–7. https://doi.org/10.1002/cssc.201100296.

[80] Corma A, De TO, Renz M, Villandier N. Production of high-quality diesel from biomass waste products. Angew Chem 2011;123:2423–6. https://doi.org/10.1002/ange.201007508.

[81] Li G, Li N, Wang Z, Li C, Wang A, Wang X, et al. Synthesis of high-quality diesel with furfural and 2-methylfuran from hemicellulose. ChemSusChem 2012;5:1958–66. https://doi.org/10.1002/cssc.201200228.

[82] Li G, Li N, Wang XX, Sheng X, Li S, Wang A, et al. Synthesis of diesel or jet fuel range cycloalkanes with 2-methylfuran and cyclopentanone from lignocellulose. Energy Fuels 2014;28:5112–8. https://doi.org/10.1021/ef500676z.

[83] Zhu C, Shen T, Liu D, Wu J, Chen Y, Wang L, et al. Production of liquid hydrocarbon fuels with acetone and platform molecules derived from lignocellulose. Green Chem 2016;18:2165–74. https://doi.org/10.1039/C5GC02414E.

[84] Li G, Li N, Yang J, Wang A, Wang X, Cong Y, et al. Synthesis of renewable diesel with the 2-methylfuran, butanal and acetone derived from lignocellulose. Bioresour Technol 2013;134:66–72. https://doi.org/10.1016/j.biortech.2013.01.116.

[85] Wang W, Li N, Li S, Li G, Chen F, Sheng X, et al. Synthesis of renewable diesel with 2-methylfuran and angelica lactone derived from carbohydrates. Green Chem 2016;18:1218–23. https://doi.org/10.1039/c5gc02333e.

[86] Li S, Li N, Li G, Li L, Wang A, Cong Y, et al. Protonated titanate nanotubes as a highly active catalyst for the synthesis of renewable diesel and jet fuel range alkanes. Appl Catal Environ 2015;124–34. https://doi.org/10.1016/j.apcatb.2015.01.022.

[87] Xia Q, Xia Y, Xi J, Liu X, Zhang Y, Guo Y, et al. Selective one-pot production of high-grade diesel-range alkanes from furfural and 2-methylfuran over Pd/NbOPO$_4$. ChemSusChem 2017;10:747–53. https://doi.org/10.1002/cssc.201601522.

[88] O'Connor CT, Kojima M, Schumann WK. The oligomerisation of C$_4$ alkenes over cationic exchange resins. Appl Catal 1985;16:193–207. https://doi.org/10.1016/S0166-9834(00)84472-3.

[89] Sarkar A, Seth D, Ng FTT, Rempel GL. Selective oligomerization of isobutene on Lewis acid catalyst: kinetic modeling. Ind Eng Chem Res 2014;53:18982–92. https://doi.org/10.1021/ie501173z.

[90] Bond JQ, Alonso DM, Wang D, West RM, Dumesic JA. Integrated catalytic conversion of γ-valerolactone to liquid alkenes for transportation fuels. Science 2010;327:1110–4. https://doi.org/10.1126/science.1184362.

[91] Zhang X, Zhong J, Wang J, Zhang L, Gao J, Liu A. Catalytic performance and characterization of Ni-doped HZSM-5 catalysts for selective trimerization of n-butene. Fuel Process Technol 2009;90:863–70. https://doi.org/10.1016/j.fuproc.2009.04.011.

[92] Gary JH, Handwerk GE. Petroleum refining technology and economics. 4th ed. New York: Marcel Dekker, Inc; 2001.

[93] Fang Q, Hanna MA. Experimental studies for levulinic acid production from whole kernel grain sorghum. Bioresour Technol 2002;81:187–92. https://doi.org/10.1016/S0960-8524(01)00144-4.

[94] Girisuta B, Janssen LPBM, Heeres HJ. Kinetic study on the acid-catalyzed hydrolysis of cellulose to levulinic acid. Ind Eng Chem Res 2007;46:1696–708. https://doi.org/10.1021/ie061186z.

[95] Fitzpatrick SW. Production of levulinic acid from carbohydrate-containing materials. US 5608105, A; 1997.

[96] Lane J. GFBiochemicals reaches commercial-scale in renewable levulinic acid project. Biofuels Dig 2015;22.

[97] Tang X, Zeng X, Li Z, Hu L, Sun Y, Liu S, et al. Production of γ-valerolactone from lignocellulosic biomass for sustainable fuels and chemicals supply. Renew Sustain Energy Rev 2014;40:608–20. https://doi.org/10.1016/j.rser.2014.07.209.

[98] Mohan V, Raghavendra C, Pramod CV, Raju BD, Rama Rao KS. Ni/H-ZSM-5 as a promising catalyst for vapour phase hydrogenation of levulinic acid at atmospheric pressure. RSC Adv 2014;4:9660–8. https://doi.org/10.1039/c3ra46485g.

[99] Mohan V, Raju BD, Seetha K, Rao R. Vapour phase hydrogenation of Levulinic acid over carbon coated HZSM-5 supported Ni catalysts. J Catal Catal 2015;2:33–8.

[100] Mohan V, Venkateshwarlu V, Pramod CV, Raju BD, Rao KSR. Vapour phase hydrocyclisation of levulinic acid to γ-valerolactone over supported Ni catalysts. Cat Sci Technol 2014;4:1253–9. https://doi.org/10.1039/c3cy01072d.

[101] Varkolu M, Velpula V, Ganji S, Burri DR, Rao Kamaraju SR. Ni nanoparticles supported on mesoporous silica (2D, 3D) architectures: highly efficient catalysts for the hydrocyclization of biomass-derived levulinic acid. RSC Adv 2015;5:57201–10. https://doi.org/10.1039/C5RA10857H.

[102] Upare PP, Lee JM, Hwang YK, Hwang DW, Lee JH, Halligudi SB, et al. Direct hydrocyclization of biomass-derived levulinic acid to 2-methyltetrahydrofuran over nanocomposite copper/silica catalysts. ChemSusChem 2011;4:1749–52. https://doi.org/10.1002/cssc.201100380.

[103] Sudhakar M, Kumar VV, Naresh G, Kantam ML, Bhargava SK, Venugopal A. Vapor phase hydrogenation of aqueous levulinic acid over hydroxyapatite supported metal (M=Pd, Pt, Ru, cu, Ni) catalysts. Appl Catal Environ 2016;180:113–20. https://doi.org/10.1016/j.apcatb.2015.05.050.

[104] Balla P, Perupogu V, Vanama PK, Komandur VRC. Hydrogenation of biomass-derived levulinic acid to γ-valerolactone over copper catalysts supported on ZrO2. J Chem Technol Biotechnol 2016;91:769–76. https://doi.org/10.1002/jctb.4643.

[105] Kumar VV, Naresh G, Sudhakar M, Anjaneyulu C, Bhargava SK, Tardio J, et al. An investigation on the influence of support type for Ni catalysed vapour phase hydrogenation of aqueous levulinic acid to γ-valerolactone. RSC Adv 2016;6:9872–9. https://doi.org/10.1039/C5RA24199E.

[106] Varkolu M, Velpula V, Burri DR, Kamaraju SRR. Gas phase hydrogenation of levulinic acid to γ-valerolactone over supported Ni catalysts with formic acid as hydrogen source. New J Chem 2016;40:3261–7. https://doi.org/10.1039/c5nj02655e.

[107] Varkolu M, Raju Burri D, Rao Kamaraju SR, Jonnalagadda SB, Van Zyl WE. Hydrogenation of Levulinic acid using formic acid as a hydrogen source over Ni/SiO2 catalysts. Chem Eng Technol 2017;40:719–26. https://doi.org/10.1002/ceat.201600429.

[108] Upare PP, Jeong M, Hwang YK, kim HD, Kim Dok Y, Hwang DW, et al. Nickel-promoted copper – silica nanocomposite catalysts for hydrogenation of levulinic acid to lactones using formic acid as a hydrogen feeder. Appl Catal Gen 2015;491:127–35. https://doi.org/10.1016/j.apcata.2014.12.007.

[109] Ahmad E, Alam MI, Pant KK, Haider MA. Catalytic and mechanistic insights into the production of ethyl levulinate from biorenewable feedstocks. Green Chem 2016;18:4804–23. https://doi.org/10.1039/C6GC01523A.

[110] Siva Sankar E, Mohan V, Suresh M, Saidulu G, David Raju B, Rama Rao KS. Vapor phase esterification of levulinic acid over ZrO_2/SBA-15 catalyst. Catal Commun 2016;75:1–5. https://doi.org/10.1016/j.catcom.2015.10.013.

[111] Varkolu M, Moodley V, Potwana FSW, Jonnalagadda SB, Van Zyl WE. Esterification of levulinic acid with ethanol over bio-glycerol derived carbon–sulfonic-acid. React Kinet Mech Catal 2017;120:69–80. https://doi.org/10.1007/s11144-016-1105-7.

[112] Omoruyi U, Page S, Hallett J, Miller PW. Homogeneous catalyzed reactions of levulinic acid: to γ-valerolactone and beyond. ChemSusChem 2016;9:2037–47. https://doi.org/10.1002/cssc.201600517.

[113] Yan K, Wu G, Lafleur T, Jarvis C. Production, properties and catalytic hydrogenation of furfural to fuel additives and value-added chemicals. Renew Sustain Energy Rev 2014;38:663–76. https://doi.org/10.1016/j.rser.2014.07.003.

[114] Zheng J, Zhu J, Xu X, Wang W, Li J, Zhao Y, et al. Continuous hydrogenation of ethyl levulinate to γ-valerolactone and 2-methyl tetrahydrofuran over alumina doped Cu/SiO_2 catalyst: the potential of commercialization. Sci Rep 2016;6:28898https://doi.org/10.1038/srep28898.

[115] IEA Bioenergy. Direct thermochemical liquefaction, http://task34.ieabioenergy.com/pyrolysis-reactors/; 2017.

[116] Yan K, Yang Y, Chai J, Lu Y. Catalytic reactions of gamma-valerolactone: a platform to fuels and value-added chemicals. Appl Catal Environ 2015;179:292–304. https://doi.org/10.1016/j.apcatb.2015.04.030.

CHAPTER 13

Energy-Water Nexus: Renewable-Integrated Hybridized Desalination Systems

Hesamoddin Rabiee[‡], Kaveh Rajab Khalilpour*, John M. Betts[‡], Nigel Tapper[†]

*Faculty of Engineering and Information Technology, Monash University, Melbourne, VIC, Australia
[†]School of Earth Atmosphere & Environment, Monash University, Melbourne, VIC, Australia
[‡]Faculty of Information Technology, Monash University, Melbourne, VIC, Australia

Abstract

Increased water scarcity across increasing world populations has led to a greater demand for desalination. However, the energy intensity and subsequent high costs of desalination remain the main barrier for widespread deployment of desalination systems. Add to this, the sustainability concerns of fossil fuel energy sources. This challenge has led to focused international research on the energy-water nexus. In recent years, several types of renewable energy have been integrated with a variety of desalination processes. Various large-capacity, renewable-desalination (RE-desalination) plants have been built across the world, especially in Middle Eastern countries, where water is relatively scarce and renewable resources are abundant and accessible. In addition, the reduction in the cost of photovoltaic (PV) panels by almost 80% over the last decade has contributed to their greater economy and wide deployment worldwide. For remote areas, it is now reasonable to consider offgrid, small-capacity RE-desalination systems, since in these regions transportation of fuel or water and connection to the grid are prohibitively expensive or impractical. Various renewable energies—such as solar, wind, and geothermal—can be coupled with many desalination methods, based on the availability of these resources in different locations, and also on other factors such as reliability required or the capital cost of establishment. This chapter reviews these various methods of desalination and configurations of RE-desalination systems currently in use, or under development. In addition, the issues relating to grid connectivity of RE-desalination systems and the economy of grid connection versus complete or partial energy independence are explained.

Keywords: Water-energy nexus, Energy-water nexus, Desalination, Water, Renewable energy, Membrane, Solar, Wind

Polygeneration with Polystorage
https://doi.org/10.1016/B978-0-12-813306-4.00013-6

1 INTRODUCTION

Water is one of the essential needs of humans and all living organisms. Approximately 70% of the earth's surface area is covered with water. However, >97% of this water is salty [1]. Saline water includes brackish water with total dissolved solids (TDS) up to 10,000 ppm and seawater with TDS up to 50,000 ppm. Desalination aims at reducing the salinity to the range of potable water (500–1000 ppm). In addition to the need for fresh water for drinking, food production, and other basic requirements, increased industrialization over the last century has led to increased demand for fresh water to be used in manufacturing processes [2].

Human history shows that major civilizations have developed near rivers or lakes [3]. Indeed, major wars and migrations have occurred in searching for freshwater or fertilized land. However, population growth and competing demand for land have tended to push human communities further away from natural sources of potable water. In addition, climate change over recent decades has caused severe, worldwide water security problems. Having passed the threshold of 400 ppm atmospheric CO_2, and now moving toward 450 ppm, a warmer world is expected, which will foster the global trend of natural water cycle imbalances, the consequences of which are already observable [4]. According to the United Nations, approximately 700 million people in 43 countries currently suffer from water scarcity (defined as annual water supplies below $1000 \, m^3$ per person). With current climate change trends, the UN projects that by 2030 around 40% of the world's population will face a water deficit [5].

These concerns have necessitated the commissioning of desalination plants. These are typically located in coastal areas in order to have ready access to seawater [6, 7]. As of 2015, according to the International Desalination Association (IDA), there were 18,426 desalination units across 150 countries, supplying a daily volume of >86.8 million cubic meters to around 300 million people [8]. In fewer than 15 years, this value has increased more than six-fold from 14 million m^3/day in 2002. However, the number of desalination units has increased by only 50%, showing that the unit capacity of desalination units has been increasing over time and the economies of scale have improved. The fundamental issue with this technology is its energy intensity, which makes production expensive and contributes to global warming through carbon emissions [9].

Water and energy are closely associated in the process of desalination. To reduce the effect of desalination on climate change, sustainable/green energy resources should be employed to reduce carbon emissions. It is estimated that almost 25 kg CO_2 are released for one cubic meter of desalinated water [10]. Therefore, a sustainable solution requires the consideration of a "water-energy-carbon" nexus, leading to the application of renewable energies (RE)—such as solar and wind—as substitutes for conventional fuel sources [10]. Until the recent decade, RE desalination was mainly practiced only in countries with rich solar irradiation or reliable wind energy, such as the Persian Gulf [11]; but, with the recent revolution in the cost of some renewable technologies, widespread interest in utilizing renewable energies for water desalination is emerging [12–14]. Nonetheless, RE resource water desalination costs are not competitive with conventional energies in many jurisdictions, especially those with fossil fuel subsidies. Consequently, extensive research is now being carried out to make RE technologies cheaper [15].

Various forms of renewable energy have been coupled with different desalination methods, and it has been predicted that this trend will continue over coming decades. Fig. 1 shows the breakdown of renewable-powered desalination [16].

This chapter investigates the technoeconomic developments across a range of desalination technologies (thermal and nonthermal). Various

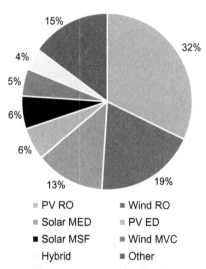

Fig. 1 Breakdown of renewable energy sources for desalination systems. *(From Ma Q, Lu H. Wind energy technologies integrated with desalination systems: review and state-of-the-art. Desalination 2011;277:274–80.)*

configurations of coupling RE sources with desalination methods are reviewed, and the technoeconomic aspects of their integration are illustrated. Finally, the impact of connectivity to the grid is analyzed.

2 DESALINATION METHODS

There are various desalination methods, as shown in Fig. 2. In this section, the most current desalination processes are discussed, which include membrane-based, mechanical, and thermal processes.

2.1 Membrane-Based Technologies

Membrane-based separation technologies are more popular than other conventional and thermal methods because they are more energy effective and compact. Here, four main membrane-based desalination technologies are discussed. Fig. 3 shows different types of membrane based on their pore size. Microfiltration (MF) [17] and ultrafiltration (UF) [18–21] are used in the pretreatment stages, as their pore size is not small enough to remove all the impurities from the water to be desalinated. Nanofiltration (NF) [22] and reverse osmosis (RO) [23] work at higher operating pressures and are

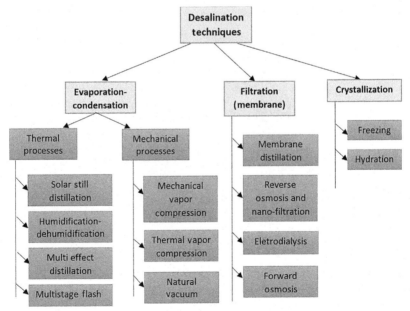

Fig. 2 Various types of desalination technologies.

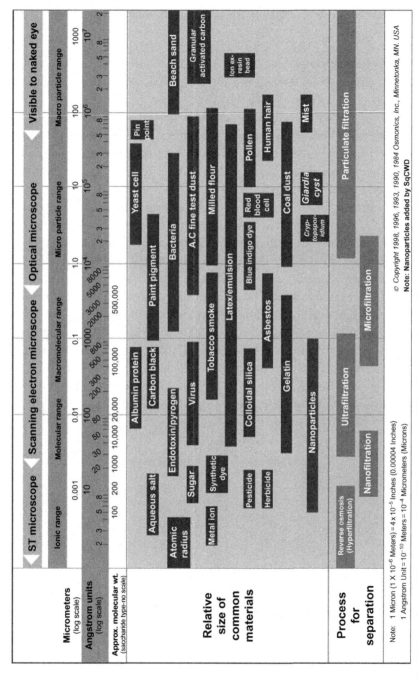

Fig. 3 Size of materials removed by different types of filtration membrane. *(Image: Courtesy of Osmonics, Inc.)*

able to remove dissolved ions (salts), decrease total dissolved solids, and produce water of acceptable quality for human consumption.

2.1.1 Reverse Osmosis

Reverse osmosis (RO) is a pressure-driven membrane desalination method that has become increasingly popular over recent decades (Fig. 4); it currently dominates the desalination market. In this process, a semipermeable membrane allows water to permeate and rejects salts and other dissolved ions. The pressure needed for this method varies with the salinity or osmotic pressure of the feed. Thus, higher pressure is required to filter seawater than is required for brackish water.

Pretreatment, including filtration and chemical addition, is a necessary step in RO systems to prevent biofouling of the membranes [26]. The three types of membranes for RO are: cellulosic, aromatic polyamide, and thin-film composite (TFC) [27]. TFC provides the highest rejection of organics and the highest water flux as well as having a better tolerance at very low or high pH, compared with the other membranes. However, its installation costs are higher than for the other membranes. RO needs high-pressure pumps to work at approximately 15–27 bar for brackish water desalination and at 50–80 bar for seawater desalination [27, 28]. Large-scale RO desalination units currently operating typically have a water flux up to $400,000 \, \text{m}^3/\text{day}$. As Fig. 1 shows, RO is a popular technology for coupling desalination with renewable energy, with >50% of RE-desalination plants

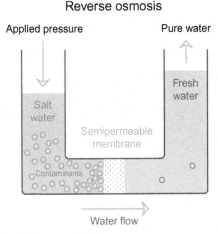

Reverse osmosis

Fig. 4 Schematic of the RO process [25].

using RO. In common with other membrane-based filtration methods, the most critical challenge for RO is fouling of the membrane, which hinders long-term performance and increases operational and maintenance costs [29].

2.1.2 Membrane Distillation

Membrane distillation (MD) is a thermal process that uses a hydrophobic membrane. Thus, MD combines thermal and membrane-based techniques. In this process, salty water is heated, and the vapor produced passes through a superhydrophobic membrane, which is only permeable by vapor (not water) [30]. The other side of the membrane is at low temperature, and this temperature gradient is the driving force for water vapor to pass through the membrane. The water vapor is further condensed after passing through the membrane through contact with a cooling liquid to form the condensate. Fig. 5 shows four different MD configurations: direct contact (DMCD), air gap (AGMD), sweeping gas (SGMD), and vacuum (VMD) [31]. Among these configurations, DCMD has undergone the greatest research, due to its potential for application on a large scale and potential for coupling with renewable energy resources [32, 33].

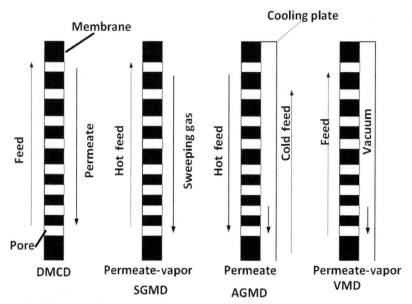

Fig. 5 Different configurations of membrane distillation.

MD has advantages over conventional distillation processes, as systems can be formed from compact modules, having low operating temperatures. The desalinated water from these systems has a high purity (theoretically 100% rejection of ions, macromolecules, and colloids). Given the possibility of using plastic equipment in this process, corrosion risks can also be reduced. In addition to water treatment, MD can also be used for a wide range of applications such as within the food industry [34] (concentration of juice or milk) or ethanol extraction from aqueous solutions [35]. Given the reliance of MD on a heating source, renewable thermal energies are suitable candidates to supply the required energy [36]. MD can be coupled with low-grade heats (waste heat), solar ponds, or geothermal energy [37, 38]. The membranes required in these types of systems for water desalination are microfiltration (polymeric or inorganic) with maximum possible hydrophobicity to allow only the passage of steam. Consequently, several research groups are currently focusing on the development of superhydrophobic materials for MD membranes to further improve the efficiency of these systems [39].

2.1.3 Electrodialysis

Electrodialysis (ED) is an electrochemical process that operates at normal atmospheric pressure. In contrast to RO and MD systems, in which water is separated from impurities, ED eliminates salt ions from the feed through an ion exchange membrane [40]. Unlike RO systems, ED only needs a low-pressure pump for circulation. In this process, water passes through a channel between two ion exchange membranes. A cation exchange membrane is only permeable to cations, and an anion exchange membrane only allows the passage of anions. The driving force for this movement is an electrical field created by two electrodes: the negative electrode collects the cations whereas the anions are attracted by the positive electrode. The flow passing through the ion exchange membranes has a high concentration of ions, so the feed salinity decreases and desalination occurs. Fig. 6 shows a schematic diagram of this process.

ED requires direct-current power supply, and therefore it is usually coupled with PV panels. These systems also require both pretreatment and posttreatment. ED is an appropriate option to produce drinking water from brackish water in small- to medium-sized quantities, as typical systems have a capacity ranging from 100 to 20,000 m^3/day [40]. This method is not very suitable for feed waters with TDS above 5000 mg L^{-1}. For systems operating under high salinity, RO seems to be the competitive

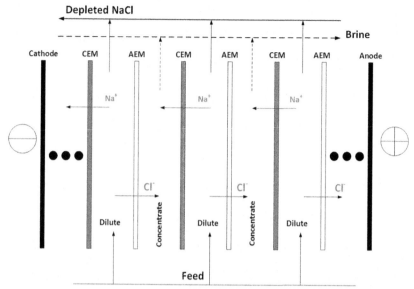

Fig. 6 Schematic diagram of an electrodialysis desalination process.

method [41]. A cost comparison RO and ED systems, based on salinity between, reveals that for salinity up to 5000 ppm, ED is cheaper than RO, with RO becoming more cost effective above 5000 ppm. The cost of desalinated water increases directly with feed salinity under both systems. However, costs for ED increase more steeply with salinity, since both operational and capital costs are affected by salinity, whereas for RO it is the operational cost only that primarily depends on water salinity [41].

The hybridization of RO with ED presents an interesting method to treat high-salinity feeds, above the operating range of RO [42, 43]. In this configuration, ED is used to dilute the feed to within the range of the RO operation (for example, a reduction from 120,000 ppm to 35,000 ppm as per the regular feed of RO for seawater desalination). This combination is also a very good solution to dilute high salinity brine remaining after RO [44], using the configuration shown in Fig. 7. In this process, ED1 is used to adjust a portion of the feed water to the desired concentration while diluting the remainder toward the concentration of the feed to the RO unit. ED2 dilutes the concentrate from the RO system (along with the dilute of ED1) to a concentration amenable to treatment with RO. ED2 also produces the concentrate as saline as the feed for ED1 to dilute. As well as treating high saline brines, an ED-RO configuration can also increase the recovery rate of RO working alone. For example, saline water with TDS of 4000 ppm can be

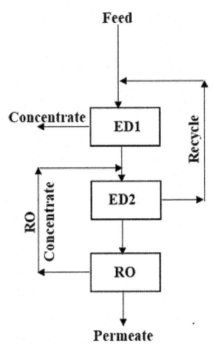

Fig. 7 Schematic diagram of the hybrid ED-RO system for the treatment of highly saline brine. *(From Ronan SMZ, McGovern K, John H, Lienhard V. Design and optimization of hybrid ED-RO systems for the treatment of highly saline brines. In: The international desalination association world congress on desalination and water reuse 2013/Tianjin, China, 2013.)*

purified to half its TDS using ED, after which it can be treated by RO to the purity of drinking water. Using this method, RO recovery can be as high as 60%, a good recovery rate for RO systems [43].

ED offers higher water recovery rate and longer lifespan for membranes than RO. However, ED is incapable of eliminating toxic materials like bacteria and viruses from feed water, so the outlet stream cannot be used without suitable posttreatment [45]. electrodialysis reversal (EDR) has been developed from ED technology. In this process, the periodic reversal of the direct current power supply creates to a self-cleaning process, whereby the salts deposited on the surface of the membranes, making them inefficient, are removed [46].

2.1.4 Forward Osmosis
Forward osmosis, FO, is a promising technology, rapidly gaining research interest. Unlike RO, FO does not need any external energy to push the feed

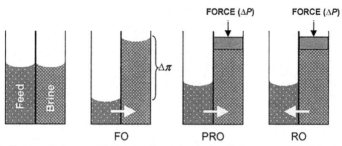

Fig. 8 Difference between FO, RO, and pressure retarded osmosis (PRO) [48].

through the membrane. Fig. 8 shows the schematic of the FO method compared with other osmotic processes. In FO, water is adsorbed through a semipermeable membrane due to the natural osmotic pressure differences between the feed and the draw solution.

Selecting the optimal draw solution is the key factor in developing FO systems since it is required to provide sufficient driving force for the FO system to operate [48]. A proper draw solution should provide enough osmotic pressure to be able to offer adequate water flux, be nontoxic, and have reasonable water recovery. To address this, several studies are ongoing for the development of new draw solutions [49, 50]. Consequently, FO is still mainly in the research scale, with systems coupled with other pressure-driven or thermal separation techniques to recover the draw solution. For instance, Fig. 9 shows an FO process integrated with RO (using RO to regenerate the draw solution).

2.2 Thermal Desalination

Thermal desalination methods use thermal energy to increase the temperature of the saline water feed. This energy can be supplied directly by some RE sources—especially solar. The main thermal desalination technologies, using nondirect thermal processes, are discussed in the following section.

2.2.1 Multistage Flash Desalination

Multistage flash desalination (MSF) is an energy-intensive distillation process requiring both thermal and electrical energy. Fig. 10 shows this process, in which saline water is pumped preheated and enters the stage under vacuum, a flash separation then takes place, and the water vapor is condensed and collected as fresh water. The feed seawater passes through the heat exchangers after the last stage to increase its temperature and reduce the energy needed for heating it up. The number of stages can vary from 4 to 40, enabling MSF systems to produce water volumes of the order of 10,000 to 40,000 m^3/day

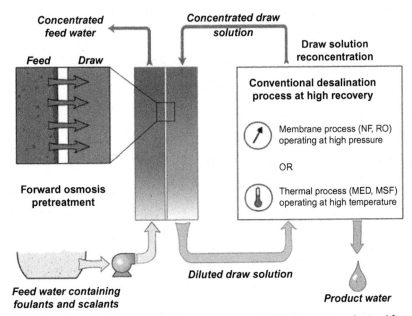

Fig. 9 Forward osmosis integrated with reverse osmosis [50]. *(Permission obtained from Elsevier.)*

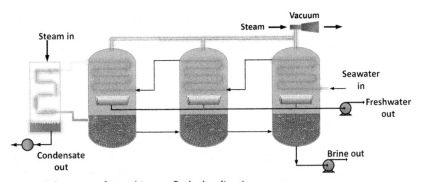

Fig. 10 Schematic of a multistage flash desalination process.

[2]. The stages have lower temperatures and pressures successively. The heating of the feed is from low-pressure steam (~2.5 bar and ~200° C), and the vacuums required in the stages are provided by generators working with medium-pressure steam. In addition to thermal energy, MSF requires separate pumps for seawater input and a freshwater output; hence, electrical energy sources are required. Around 26% of the worldwide desalination processes are MSF, making it the most popular

nonmembrane desalination technology [51]. Most of the energy is used to preheat seawater up to 90–120 °C, which can be supplied from fossil fuels, RE, or any other type of energy making it suitable for many different installation locations.

2.2.2 Multieffect Distillation

Multieffect distillation (MED) systems have a series of stages with a decreasing pressure gradient. A heat source is used to increase the temperature of the feed to 110 °C for the first stage. This heat might be supplied from a boiler working with fossil fuels, waste heat, or RE resources. The water vapor is produced in a series pattern, and in the first stage, it is transferred via a tube to the next stages to boil more seawater. Compared to MSF, MED is cheaper to install, being more compact, and having a smaller heat transfer. MED systems have an average of about 10 MED individual units, giving them a capacity of the order of 600 to 30,000 m^3/day for water production. Compared with MSF, MED systems consume lower electricity for the pumps (1.5–2 kW/ton of water) [10]. MED can work with waste heat or solar energy at a temperature of about 70°C (which is called low-temperature MED). However, in these systems, the maximum number of stages is around 10. Solar is the preferred RE source for MED. A schematic of a MED desalination process is shown in Fig. 11.

2.2.3 Vapor Compression

There are basically two types of vapor compression (VC) systems: mechanical vapor compression (MVC) and thermal vapor compression (TVC). These processes are shown schematically in Fig. 12A and B. For vapor

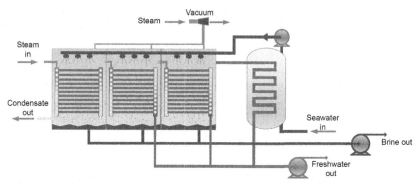

Fig. 11 MED desalination.

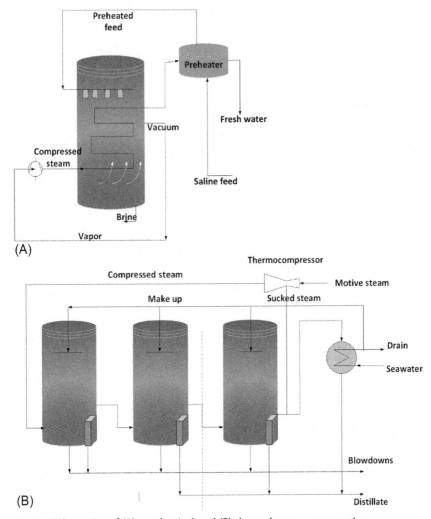

Fig. 12 Schematics of (A) mechanical and (B) thermal vapor compression.

compression, the feed, after passing through a heat exchanger, is converted to vapor and then compressed by mechanical or thermal means. The compressed vapor, as fresh water, has high temperature and pressure and is circulated through the initial heat exchanger to heat the incoming feed. The mechanical compressor works with electricity, and the thermal compressor uses a steam jet ejector to create a vacuum [52]. Typically, the capacity of MVC (100 to 3000 m³/day) is lower than that of TVC (10,000 to 30,000 m³/day).

2.2.4 Humidification-Dehumidification Desalination

The humidification-dehumidification (HDH) method is principally the evaporation of water into a carrier gas and then vapor stripping of that carrier gas [53]. A simple HDH desalination is shown schematically in Fig. 13. This method has attracted the attention of desalination designers since 2006 because it has a lower operating temperature than other thermal distillation processes ($<80°C$), and its operating pressure is near ambient air pressure [54]. In this approach, a carrier gas, usually air, is loaded with water vapor until it is saturated. The temperature of the humid air is then reduced, which dehumidifies the air, producing fresh water [55]. In the humidifier, air is saturated through interaction with hot seawater, which has been sprayed onto a tower to increase the contact area. The saturated air then passes over cooling coils, where dehumidification occurs. The main energy consumption for this method is through heating the saline seawater to produce vapor (for saturation of the carrier gas), making solar energy a good choice to couple with HDH systems [56].

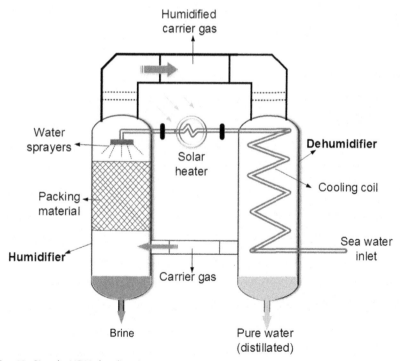

Fig. 13 Simple HDH desalination.

2.3 Innovative Desalination Methods

In this section, new and state-of-the-art methods of desalination are discussed. Capacitive deionization, ion concentration polarization, and adsorption desalination are three key research areas.

2.3.1 Capacitive Deionization

Capacitive deionization (CDI) is an emerging desalination method based on ion electrosorption. In this method, saline water passes between a pair of porous electrodes, which can be static or flow electrodes, and which are usually carbon-based. A differential voltage of between 1 and 1.4 V is applied to this pair of porous electrodes; thereby, the salt ions in the feed flow migrate to the electrical double layers (EDLs) at the carbon/water interface, which performs the desalination. Over the past decade, there have been several studies addressing different aspects of this process, including cell architecture, materials, applications and theory, discussed in [57].

Capacitive deionization offers several advantages over conventional pressure-driven systems (such as RO) or thermal technologies because there is no need for high-pressure pumps or heat sources [58]. Similar to electrodialysis, salt ions are separated from water, making it suitable for low salinity feeds. Consequently, this method has been considered a good replacement for RO in brackish water desalination [59]. However, one major drawback of this method is its inability to remove impurities other than ions (such as viruses, bacteria, etc.). Capacitive deionization can be coupled with two ion-exchange membranes (anion and cation) to create membrane capacitive deionization. In this process, the two membranes selectively pass the ions to the electrodes. The ion-exchange membranes can be coated on the porous electrodes, creating a thin layer membrane on the electrode instead of a standalone membrane [60]. A simple schematic of CDI and membrane CDI is shown in Fig. 14.

2.3.2 Ion Concentration Polarization

Ion concentration polarization (ICP) is similar in function to capacitive deionization by removing salt ions from the feed. However, ICP is an electrokinetic phenomenon, which occurs when a voltage is applied across a nanojunction or an ion permselective membrane [61]. The basis of this method is ion depletion and ion enrichment, which are dynamic changes in ion concentration near the membrane to maintain electroneutrality [62]. By contrast, in other electrochemical desalination methods such as electrodialysis and capacitive deionization, both cations

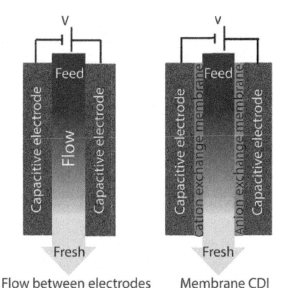

Flow between electrodes   Membrane CDI

Fig. 14 The schematic of CDI and membrane CDI [57], Published by The Royal Society of Chemistry.

and anions are removed from the feed via ion–selective electrodes or ion exchange membranes. In an ICP process, when ion concentration becomes polarized across the membrane, a boundary layer is formed on desalting the surface of the membrane, and the salt concentration in the layer is depleted, owing to the concentration polarization [63]. ICP is a suitable option to decrease the salinity of brines with very high salt concentration. For TDS >45,000 mg/L, RO is not cost effective, and other thermal distillation processes are expensive. However, ICP can be used to decrease the salinity of the feed water, after which RO and other thermal methods (designed for complete salt removal) can be used [64]. Desalination using ICP is shown schematically in Fig. 15. In this process, water is purified within a channel. Toward the end of the channel, a high salinity film is formed near the electrode and the ion exchange membrane, while desalinated water is discharged from the side of the channel [64].

2.3.3 Adsorption Desalination

Adsorption desalination (AD) is a thermally driven, low–cost desalination system, requiring low–grade heating (<85°C), which can be supplied by waste heat or renewable solar energy [65]. This process includes two adsorption and desorption beds that operate in a cycle (Fig. 16). The first step is

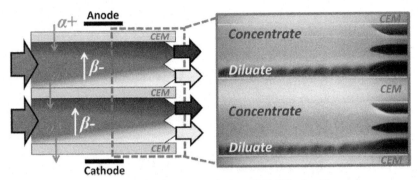

Fig. 15 Schematic of the ICP desalination technique [64]. *(Obtained from an open-access article in* Nature.*)*

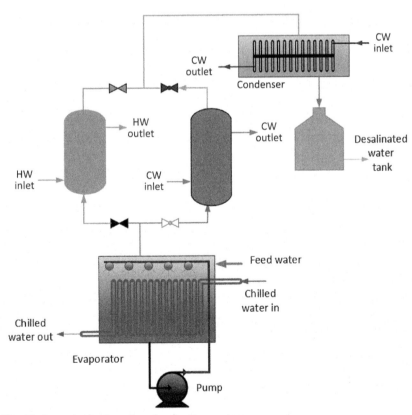

Fig. 16 A two-bed adsorption desalination system.

adsorption-evaporation. The suction effect from the adsorbent leads to the evaporation of water, which takes place in the evaporator under low pressure and temperature. The next step is *desorption-condensation,* where a low-temperature heat source is supplied to the bed, which is connected to a condenser. This results in the water vapor (evaporated by the low-grade heat in the second step) being condensed and collected as desalinated water [66].

AD technology can be integrated with a geothermal source for carbon-free desalination [65]. Because the heat source from geothermal is continuous, it can be designed for cogeneration systems (generation of electricity, potable water, and cooling or heating needs). Designs have already been considered in some countries like Saudi Arabia, where dry rock geothermal sources are in abundance, and the temperature gradient in the initial depths is of the order of 50 to 80°C per km and leveling off after reaching an equilibrium temperature.

3 RENEWABLE-ASSISTED DESALINATION SYSTEMS

The desalination technologies discussed in the previous section can be coupled with a variety of renewable energy sources, the two most popular being solar thermal and photovoltaic (PV). This section briefly introduces PV and solar thermal energy technologies and then reviews the more popular systems coupling desalination technologies with renewable energy.

PV cells are direct current (DC) electricity generators, of which there are three main types: *crystalline silicon (c-Si),* which are the most commonly used and reliable PV cells (approximately94% of cells produced in 2016 were this type); *amorphous silicon thin-film PV cells,* which are currently the cheapest PV cells to manufacture, but have a shorter lifespan and lower efficiency compared to crystalline silicon. They accounted for approximately 6% of total PV production in 2016; and *NanoPV technologies* [67]. Over recent years, there has been a pronounced increase in the efficiency of solar cells aimed at the mass market. In particular, there has been an increase from 16% in 2012 to almost 29% in 2016 [68, 69]. Although PV cells with higher efficiency exist, for example, 46% for high-concentration, multijunction solar cells, these are made from expensive materials and are unsuitable for large-scale fabrication [70].

Solar thermal desalination comprises two main methods: direct and indirect. In direct methods, evaporation and condensation happen in the same device. These techniques are usually able to produce water for small communities and households. Solar stills, water cones, and water pyramids are different varieties of direct solar thermal desalination [71]. Indirect methods have two steps: a solar collector and desalination unit. Different types of solar collectors can be coupled with various desalination systems (for example, membrane distillation (MD), multiple effect evaporation). Some thermal collectors, like solar ponds, can provide low-grade heat up to 90°C, which is not enough for conventional desalination systems like multieffect distillation (MED), but it is sufficient for membrane distillation. Other collectors, such as concentrating solar power (CSP), can go up to 200 °C and provide the required heat for thermal desalination systems [72]. In this section, different configurations of coupling RE with desalination are discussed.

3.1 Renewable-Assisted RO Desalination

3.1.1 Solar-Reverse Osmosis

Reverse Osmosis, RO, is the most popular and common desalination technology studied for coupling with RE sources. Fig. 17A and B shows two RO integrations with solar thermal and PV, respectively. RO requires high-pressure pumps to overcome the osmotic pressure of the saline water; thereby, the amount of energy needed by the system is related to the salt concentration of the feed. PV-powered RO systems have been favored for locations such as the Egyptian desert [73], remote places in Australia [74, 75] and rural areas of Jordan [76], due to their favorable geographical features. The first attempt to employ solar energy for RO was performed by Bowman et al. [77] to produce $80 \, m^3/day$ desalinated water from a feed with TDS of 5400 ppm. Their system used two hollow-fiber modules with an operating pressure of approximately 28 bar.

Thomson et al. [78] employed an RO system powered by 2.4-kWp PV cells to produce $3 \, m^3/day$ of fresh water. Their test location was Eritrea using water with TDS of 40,000 ppm as the feed. The estimated total cost of this system, including pumps, inverter, controller, PV arrays, and other components, was around €23,000. Their system connected PV cells directly to the pumps without using batteries. The quality of the water produced, and salt concentration was reported as being generally acceptable (from 200 to 500 mg/L). However, membrane aging led to a gradual increase in dissolved salt in the desalinated water over time.

Until the last decade, PV was an expensive technology. Herold et al. in 1998 estimated that water production via an RO-PV system on the Canary Islands would cost $16\$/m^3$, which is quite expensive, but they also predicted that it would become cheaper in the future. As Suleimani et al. [79] concluded, although the installation cost of PV might be higher than for a diesel

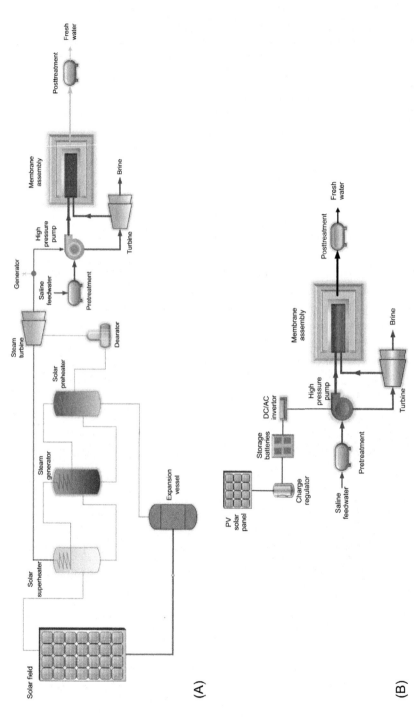

Fig. 17 (A) and (B) Schematic diagram of solarthermal-RO and PV-RO systems. *(From Al-Karaghouli A, Kazmerski LL. Energy consumption and water production cost of conventional and renewable-energy-powered desalination processes. Renew Sustain Energy Rev 2013;24:343–56.)*

system, in the long run, the cost of having a PV system for desalination is lower as its maintenance, and operating costs are relatively small.

In the past, the main problems to commercializing RO-PV systems were the cost of PV arrays, and also the drawback that the economic feasibility of having a PV-RO system, compared to conventional methods, highly depended on the distance to the electricity grid, the salt concentration of the feed and the plant capacity [80]. However, for several reasons, there was a disruption in the PV market and the PV price has declined almost 80% (from >$5/W to <$1/W [81]). Also, as mentioned before, PV efficiencies have increased significantly in the last few years, and PV panels have now been commercialized in many countries [68].

Richards et al. developed a prototype for using an RO/NF system powered by solar energy to produce drinking water for remote areas in Australia, where, in many regions, water is scarce or of poor quality [74]. They employed a submerged UF setup as the pretreatment stage before the RO/NF component (Fig. 18). This configuration was chosen because UF has better performance in fouling prevention than microfiltration (MF) or conventional technologies such as sand filtration [75]. Batteries were not used for energy storage. Consequently, power generated fluctuated, since solar radiation is not constant. Water produced showed that feed pressure for RO/NF treatment setup considerably influences water flux and recovery. In particular, when pump pressure is low (due to low solar radiation), water flux is low, and the contaminants diffuse further into the membrane. Hence, product water has a higher salinity in this scenario [82].

The main challenge of PV, in the absence of battery support, is power fluctuations due to dependency on solar irradiation, which affects the energy

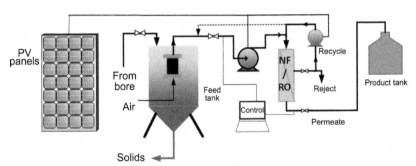

Fig. 18 Schematic diagram of the PV-powered UF/RO desalination system. *(From Richards BS, Schäfer AI. Design considerations for a solar-powered desalination system for remote communities in Australia. Desalination 2002;144:193–99.)*

provided for pumps, and hence, the pressure required for desalination. This issue was studied for an RO-PV system [83]. Desalinating brackish water (5000 mg/L NaCl), it was observed that when solar irradiance was in the range of 400–1200 W m^{-2}, the system had a stable performance with above 5 L m^{-2} h^{-1} water flux, and >80% salt retention for a permeate salinity target of 1 g L^{-1}. For solar irradiation in this range, fluctuations did not have a substantial effect on water quality and flux. However, for solar irradiation below 300 W m^{-2}, the permeate had a very low flux with high salt concentration. For these systems, when solar radiation is relatively high, the system can be shut down to perform backwashing (which disrupts the concentration of polarization layers on the membrane, degrading performance of the membranes).

3.1.2 Wind-Reverse Osmosis

In many coastal areas around the world, wind power is combined with RO for desalination. Initial studies have concluded that RO-wind is appropriate for regions with high wind speeds [84, 85]. However, over recent years, the development of membrane science, reductions in wind turbine cost, increased costs of fossil fuels, and growing climate change concerns have resulted in increased interest in this configuration, which can now compete with conventional desalination methods in regions with a sufficient wind resource and expensive fuels [86].

Liu et al. [87] employed the wind power directly, instead of generating power from wind and then using that power, to provide energy for RO brackish water desalination in Coconut Island, Australia. Wind energy was used as the energy source to create feed water pressure behind the membrane module. The authors concluded that this system could perform desalination if the wind speed was >5 m/s. Wind speeds in this region ranged from 4.5 to 9 m/s, making wind-RO feasible. At the average wind speed of 5 m/s, a salt rejection of 97% for brackish water with TDS of 3000 mg/L and a flow rate of 13 L/min was achieved.

Subiela et al. [88] described over 10 years' experience using renewable energy for desalination in the Canary Islands. They reported that several types of renewable desalination are now operating and that these are no longer novel technologies. Their experiences show that for small demands (up to 100 m^3/day), PV-powered RO is a reliable option. For medium demand (1000–5000 m^3/day), they reported the best option was using an offgrid windfarm and an RO plant since at the time (year) PV panels were too expensive. More recently, Loutatidou et al. [89] undertook a wide technoeconomic analysis for water desalination using wind power and an RO system in different regions in the UAE with wind speeds above 5 m/s. They mentioned ground slope as another important factor for a wind turbine installation, and so ground surfaces with slopes over 30 degrees were not considered. The levelized costs of water in their study were USD$1.57–1.63; 1.83–1.96; and 2.09–2.11 per m^3 of product water for RO-plant capacities of 7000; 10,500; and 14,000 m^3/day, respectively. This study claimed that wind-powered, variable-flow RO could be an appropriate replacement for current thermal desalination in the UAE; but, as the wind resources are unpredictable and low quality, it is better considered as complementary to the existing infrastructure for possible high demand in the future.

The logistic modeling of an autonomous wind-driven RO desalination system having lead-acid battery storage was undertaken to provide insights into the equipment selection criteria, size of the plant, size of wind turbines, and battery capacity decision [90]. This study showed that the most effective and economical way to increase the water production level was by using a larger desalination plant rather than a larger wind turbine. In addition, the authors observed that wind forecasting would benefit the system by avoiding excessive start/stop operations. The authors also found that, as a general rule, the size of the wind turbine should optimally be two to three times the desalination plant's base-level demand. Additionally, the battery should have a minimum capacity to provide the required energy for a proper shut down when there is insufficient wind energy. Moreover, the cost analysis of the autonomous wind-RO system showed that almost 75% of the cost is related to the equipment, with the wind turbine being responsible for over 40% of the total cost.

3.1.3 Solar-Wind-Reverse Osmosis (Hybrid System)
Analysis of using wind and solar as an alternative sources of energy for desalination has shown that these two RE resources are complementary. RO desalination coupled with PV-wind have been designed for regions such as Israel, Saudi Arabia, Mexico, and Tunis, where both solar and wind energy are available because of local geography: hot climate with sufficiently high winds and coastal location where access to seawater is possible [14, 91].

Mokheimer et al. [91] optimized the RO water desalination powered by hybrid wind and solar with a system comprising PV arrays, wind turbines, and batteries. In their methodology, they considered solar irradiation, wind speed, and ambient temperature as the variables. Their sizing and optimization results demonstrated that having multiple wind turbines reduced the cost of energy production. For example, the optimal hybrid system for producing 1-kW load for 12 h/day is 2 wind turbines, 40 PV modules, and 6 batteries, which result in 0.62 $/kWh cost of energy production. In summary, the hybrid system of RO desalination costs between $3.693 to $3.812/m^3.

Cherif et al. [92] modeled water and energy production for a large-scale PV-wind hybrid system connected to RO desalination in southern Tunisia (Fig. 19). This system was composed of 400 m^2 photovoltaic (PV) panels and a 10-kW wind turbine interconnected by a DC bus through converters. Based on the feed salinity, the pumps required sufficient energy to create 15.5 bar pressure. This system was able to provide between 57 and 111 m^3/d water, depending on the monthly climate and weather. A dynamic analysis showed the system was able to produce a stable potable water supply. Results over 3 months showed that no significant drop in water production efficiency occurred, indicating the system should work well for larger water desalination plants.

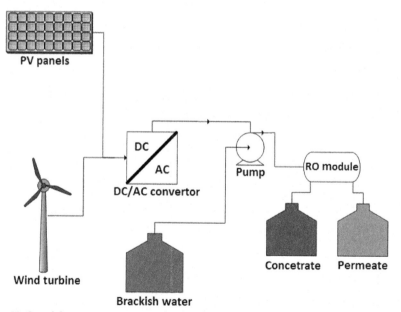

Fig. 19 Standalone reverse osmosis desalination unit powered by a standalone PV-wind hybrid system. *(From Cherif H, Belhadj J. Large-scale time evaluation for energy estimation of stand-alone hybrid photovoltaic–wind system feeding a reverse osmosis desalination unit. Energy 2011;36:6058–67.)*

Another standalone hybrid (solar-wind) coupled with RO desalination was investigated for use in remote coastal areas like the Greek Islands [93]. The researchers developed a numerical algorithm to inspect the operation and performance of the system in detail. They claimed that the specific water production cost was 1.5–3 €/m³, which is attractive as the current water supply for those areas is 5–8 €/m³ when considering transportation pricing. This system has two reservoirs (upper and lower) to be used as energy storage, in case the system is down (Fig. 20). When excessive energy is produced by the solar-wind system, this energy will be used to operate the pumps to store it as hydraulic energy in the upper reservoir. When the energy available for desalination is insufficient, a hydro turbine will transform the stored hydraulic energy into electricity.

A solar-wind hybrid system coupled with RO is one of the most reliable hybrid systems for energy production [94]. The investigation and sensitivity analysis of flexible, potable water production for Jordan revealed a very strong relationship between the type and cost of the fuel to perform desalination and the total annual cost of the system [95]. Thereby, integrated water-energy systems can benefit water production, both economically

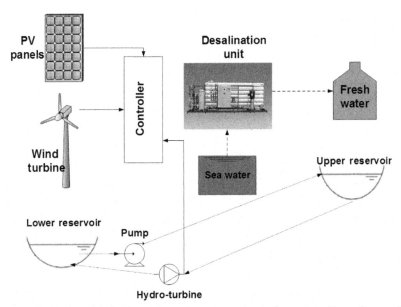

Fig. 20 RO-PV/wind hybrid system using reservoirs as batteries. *(From Spyrou ID, Anagnostopoulos JS. Design study of a stand-alone desalination system powered by renewable energy sources and a pumped storage unit. Desalination 2010;257:137–49.)*

and ecologically, as RE sources do not contribute to CO_2 emissions. In sum, hybrid solar-wind REs for RO desalination present considerable advantages with 1.3 times lower total costs and 2.24 times lower CO_2 emissions [95].

3.2 Renewable-Assisted MD Desalination

The integration of MD with renewable energies and waste heat sources has drawn particular attention since MD needs low-temperature heat to perform [96, 97]. The first attempt was reported in 1991 by Hogan et al. [98], who coupled hollow-fiber MD with solar energy. Their system could produce $0.05\,m^3/day$ desalinated water with an energy consumption of about $55.6\,kWh/m^3$. Suarez et al. [32] coupled membrane distillation with a solar pond and showed that the pond (a Lab-scale, salt gradient-solar pond with a depth of 1 m) was able to provide sufficient thermal energy for an MD system to produce $1\,L\,m^{-2}\,h^{-1}$ of desalinated water. Although 70% of the energy extracted from the solar pond was used for thermal desalination, only half was effectively used to transport water across the membrane, and the rest was lost by conduction in the membrane; thus, more efficient membranes and modules should be developed. Another MD–solar pond system has recently been studied by Nakoa et al. [38, 99], who aimed to implement a zero-liquid discharge desalination (ZLDD) system (Fig. 21). Their system was able to deliver $52\,L/day$ fresh water per $1\,m^2$ of membrane coupled with a solar pond with consumption of around $11\,kW/m^2$ of instant thermal energy.

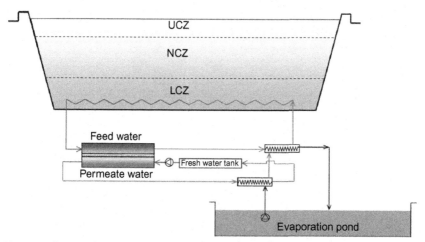

Fig. 21 Solar pond—MD desalination system [38]. *(Permission obtained from Elsevier.)*

Baghbanzadeh et al. [100] utilized the temperature gradient of the sea as a heating source. They used water from the sea surface as a heating source and water from beneath the sea as a cooling source. They concluded that the product water could be desalinated at the cost of $0.61/m^3, which is comparable to the commercial seawater RO process. Further to the economics, the key advantage of this technique is its reduced impact on aquatic ecosystems due to the avoidance of large volume discharge of concentrated brine.

The possibility of using solar energy for MD has also been considered by other researchers around the world [36, 101–105]. Fig. 22 shows the schematic of coupling MD with PV solar cells [105]. Most of the energy needed for a PV-MD unit is used for heating up the feed and the rest for pumps. However, PV-RO is more economically feasible as RO modules are cheaper, more commercial, and longer lasting than MD. Solar-powered (solar collectors) MD also shows higher specific thermal energy consumption compared to other mature thermal desalination technologies [106]. There may be applications where MD can fill utilization gaps of the MED process; for example, for the treatment of hot brine after an

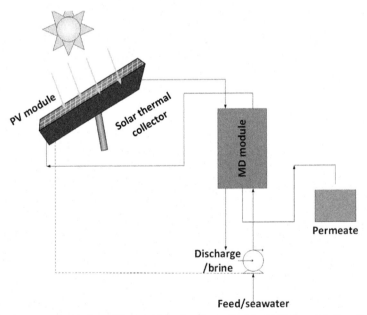

Fig. 22 Layout of coupling MD with PV cells. *(From Qtaishat MR, Banat F. Desalination by solar powered membrane distillation systems. Desalination 2013;308:186–97.)*

MED process, because MD can provide high distillate quality regardless of feed salinity, hence brines after regular thermal desalination can be treated by MD.

In addition to solar energy, Sarbatly et al. [37] used geothermal energy (GE) to heat feed water into the MD system. They showed that for the plant operating with GE—compared to the plant without GE utilization—the water production costs were less than $0.50/m^3 when the water fluxes exceeded 6.6 kg/m^2 h. Their results demonstrated that the prepared membrane from a PVDF and geothermal energy source was able to provide potable water with TDS below approximately 100 ppm (a decrease from 900 ppm). They also emphasized that using a geothermal energy source led to a 59% lower cost of water production for Malaysia, the country of study ($0.50/m^3 and $1.22/m^3 for the plant operating with and without geothermal energy, respectively).

3.3 Renewable-Assisted ED Desalination

As mentioned in Section 2.1.3, similar to other electrochemical methods of desalination, ED is an appropriate technique for brackish water where salinity is not very high. For example, when water salinity is 1000 and 3000 mg L^{-1}, ED consumes energy 75% and 30% less than RO, respectively [41]. ED is a preferred technology to RO for underground water desalination with relatively low salinity. As such, Wright et al. [41] used PV-powered ED for community-scale desalination in rural villages in India. Their analysis showed that their system is able to desalinate underground water with salinities up to 5000 mg L^{-1}, at which RO becomes more economical than ED (they claimed that 5000 mg L^{-1} is the threshold for this comparison between RO and ED, Fig. 23). This system showed suitable performance for Indian villages. In summary, considering an off-grid desalination system depends on the proper desalination method and the quality of the feed water.

In 2003, Al Madani et al. [107] integrated a small-scale, commercial-type ED with a PV cell. The ED system consisted of 24 cell pairs in four hydraulic stages and two electrical stages. They reported 95% salt removal from underground water with a TDS value of 1500 ppm. Fig. 24 schematically shows how ED couples with PV cells to provide the required electricity and the voltage difference.

Ortiz et al. [108, 109] also studied the possibility of using ED powered by PV (without any battery) for underground water desalination. Water salinity

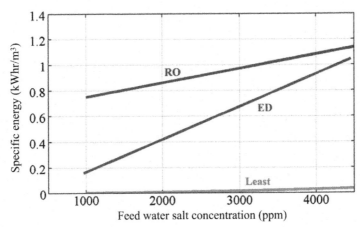

Fig. 23 Dependence of specific energy on feed water salinity [41]. *(Permission obtained from Elsevier.)*

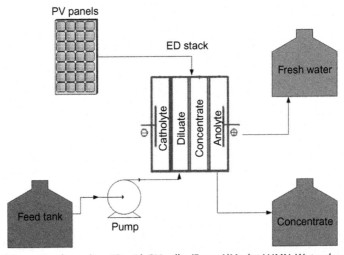

Fig. 24 Schematic of coupling ED with PV cells. *(From AlMadani HMN. Water desalination by solar powered electrodialysis process. Renew Energy 2003;28:1915–24.)*

in this study was between 2300 and 5000 ppm. Their results had good accordance with mathematical modeling and simulation. Zhang et al. [110] used PV-powered ED with FO for wastewater and brackish water desalination (Fig. 25). They used the concentrated brine from ED as a draw solution for FO, then subsequently used ED for desalination of the draw solution in

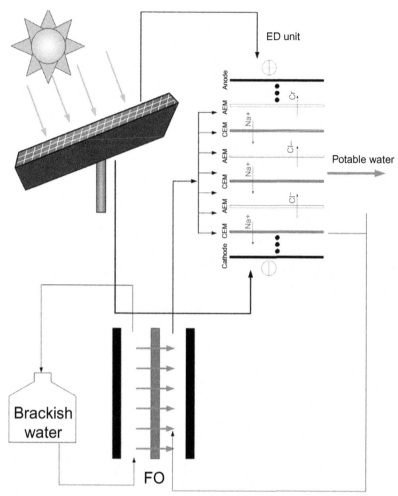

Fig. 25 Hybrid ED-FO desalination system coupled with PV cells. *(From Zhang Y, Pinoy L, Meesschaert B, Van der Bruggen B. A natural driven membrane process for brackish and wastewater treatment: photovoltaic powered ED and FO hybrid system. Environ Sci Technol 2013;47:10548–55.)*

a closed loop. The draw solution in FO was diluted from 0.5 M to 0.2 M, and then it was sent to ED for desalination and draw solution recovery.

3.4 Renewable-Assisted FO Desalination

Studies on renewable energy integration with FO desalination are limited. Recently, Khayet et al. [111] optimized FO desalination with solar thermal

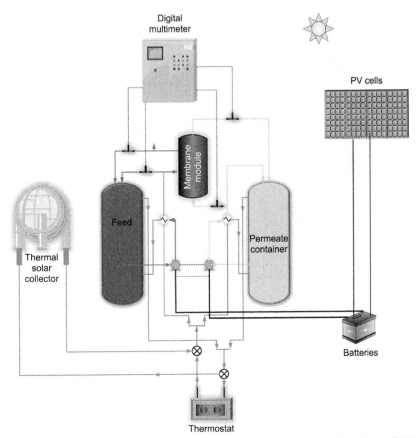

Fig. 26 Solar-assisted FO desalination setup. *(From Khayet M, Sanmartino JA, Essalhi M, García-Payo MC, Hilal N. Modeling and optimization of a solar forward osmosis pilot plant by response surface methodology. Solar Energy 2016;137:290–302.)*

and PV. They used the renewable energy for circulation pumps and also for an RO unit, which was used to regenerate the FO draw solution (Fig. 26). Their observations showed that most of the energy consumption was related to the reconcentration of draw solution and not the FO process. The draw solution used in this study was 35 g/L NaCl solution. The optimization revealed that for the highest FO performance the optimum flow rates of the feed and draw solutions were 0.83 L/min and 0.31 L/min, respectively. FO performance is optimized when water flux is as high as possible while reverse draw solution flux and energy consumption are low.

Another study focused on using the FO process for ethanol concentration (dilute ethanol was used as the feed), and the draw solution was regenerated by solar energy [112]. In their study, a gradient solar pond provided the required energy for draw solution regeneration. The researchers demonstrated that their system was successful to dehydrate ethanol to produce 50% w/w ethanol. However, due to the limited solubility of salts to create the draw solution and also the reduced evaporation rate of concentrated brine, production of 95% w/w ethanol with such a system was infeasible unless additional thermal distillation after the solar pond is considered.

3.5 Renewable-Assisted MED Desalination

The thermal energy needed for a MED process can be supplied by solar energy, particularly solar CSP collectors. Fig. 27 shows how CSP collectors are able to be coupled with a MED process to provide the required heat for the feed. First, CSP-MED units for desalination were built in Spain (Plataforma Solarde Almerı'a (PSA)), during the 1990s. Sagie et al. in 2000 did a study in Israel, and their results showed that at electricity tariffs above 7.1 solar–MED is more feasible than RO [113]. However, this conclusion cannot be trusted completely. Today, RO-PV as a popular RE-desalination process is more economical, since RO efficiency has improved greatly since 2000, and the PV price has decreased around 80% in the last decade (as mentioned earlier). A combination of MED with CSP is only suitable for medium-to-large ($>$12,000 m^3/day) production capacities. In smaller capacities, large evaporators are incompatible with the usual size and capacity of renewable resources (unless a huge solar field can be built) [10]. MED and RO are currently the best candidates for CSP coupling.

3.6 Renewable-Assisted MSF Desalination

Like MED, MSF can be coupled with different renewable energies. Fig. 28 shows schematics of an MSF integrated with a CSP system. An MSF system can also be coupled with a solar pond, which acts as the collector and storage of solar energy and provides this energy for MSF [14]. The capital cost of MSF is relatively lower than solar stills with a performance ratio 3 to 10 times higher [114]. The results of coupling MSF with a solar pond revealed that a solar pond (1500 m^2) at 70°C could supply the energy for a 10-unit MSF operating at 0.9 bar with 15 m^3/day capacity [115].

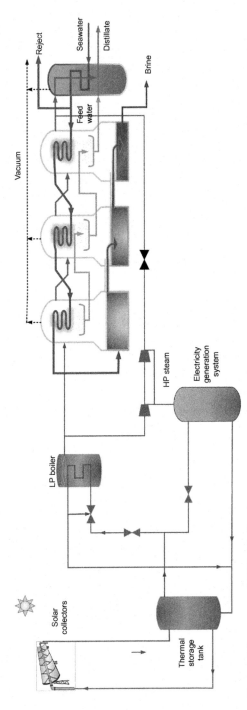

Fig. 27 Solar-powered MED desalination.

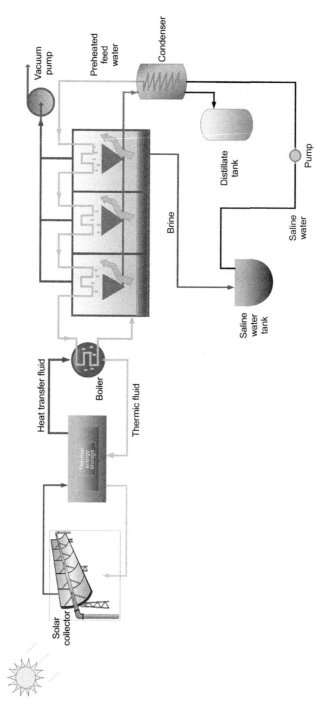

Fig. 28 Multistage flash desalination coupled with solar collectors.

3.7 Renewable-Assisted HDH Desalination

Similar to other thermal desalination systems, HDH can be coupled with solar energy to supply the required heat to increase the feed water temperature [116]. Usually, solar collectors heat up the feed water before it enters the humidifier (Fig. 29). In fact, solar collectors are responsible for the highest level of exergy loss in a solar system, and it has been suggested that improvements in the performance of solar collectors will result in more efficiency [117, 118]. Recently, Zubair et al. studied the performance and cost assessment of an HDH desalination system integrated with solar evacuated tubes [56]. Their cost analysis showed that the cost of water production varied from $0.032 to $0.038 per liter for the different locations that they considered. Their experiments took place during daytime (from 8 am to 3 pm).

4 ONGRID VS. OFFGRID (GRID-CONNECTED VS. STANDALONE)

One of the most detrimental factors in the feasibility of RE-desalination is the intended installation location. RE-desalination becomes a more viable option for remote and offgrid locations with high costs of electricity supply (generally diesel) and favorable renewable resources. However, daily and seasonal variations of renewable resources, such as solar and wind, are a critical concern for continuous operation of any process including desalination.

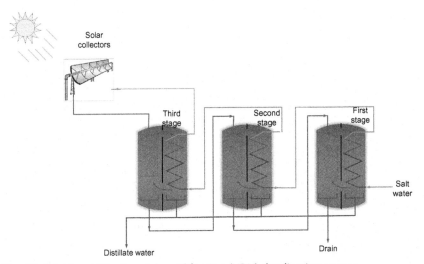

Fig. 29 Solar humidification-dehumidification (HDH) desalination system.

There are many factors and challenges that can affect the technical development of standalone systems and this section aims to discuss them.

Standalone (offgrid): This option is an excellent choice for isolated regions where transmission lines or transporting fuel are neither easily obtainable nor economical. In this scenario, it is recommended to combine a solar system and a wind system, as these two sources complement each other, and energy production will be more reliable. Some challenges about having a standalone solar-wind energy production are now explained.

High storage cost: It is necessary for standalone systems to consider energy storage, and it has been observed that the energy storage unit is the most expensive part of a system. The main solution to deal with this issue is combining a few renewable resources (for example, solar and wind) to bring greater reliability and reduce the fluctuations related to the energy output [119]. In addition, increasing the capacity of wind turbines and solar panels is a better option than having more batteries, because batteries have shorter lifespans and are more expensive [120]. Another factor that can improve energy production is considering more wind turbines with smaller sizes instead of less turbines with larger sizes. This is due to small turbines being able to work at lower wind speeds, which is suitable for small energy production targets like residential areas [121, 122].

Backup generator: Given weather condition unpredictability, it is possible that under certain extreme conditions renewable energy production becomes minimal for a few days and energy storage also runs out. This can put the fresh water supply at risk; integrating fuel cells with solar-wind can be a backup plan for such conditions [119, 121]. The main feed for a fuel cell is hydrogen, which can easily be shipped to the site. Also, fuel cells produce clean energy with a smaller carbon footprint, and the cost of increasing the size of storage is reasonable. Thereby, although considering a fuel cell as a substitution for the battery is not economical, it has other advantages mentioned here [121]. In total, even if energy storage is available for the system, it is better to consider other reliable backups in case the storage cannot provide the required energy. The backups can be a diesel generator system or fuel cells [67] and, generally, preference will be given to cheaper options ($/kW) even if the fuel cost might be higher.

Standalone desalination strongly depends on the reliability of the renewable sources; thereby, site selection plays a very important role in this regard. Offgrid systems are not possible in many countries, where sufficient solar irradiance and wind power is absent. An investigation into an offgrid, PV-powered RO (a photovoltaic generator consisting of 22 monocrystalline silicon modules, each at 50-W peak power) for different sites in Jordan revealed that the amount of energy production by solar irradiance in many places was enough to produce over 1600 m³/year water from the feed with TDS of 7000 mg/L [123]. Using wind power for offgrid RO seawater desalination in isolated coastal areas was widely studied by Petate et al. [124]. They concluded that water storage was crucial for such a system to meet water demands in periods of lack of wind or system breakdowns. They also tested offgrid and ongrid ED systems for brackish water desalination using wind as the source of energy.

A cost comparison of RE-desalination with standalone and grid-connected systems indicates that standalone systems are more expensive to produce drinking water. The reason is that for standalone systems, energy storage units and oversize desalination are necessary to guarantee the continuous production required by the plant.

5 ENERGY IMPLICATIONS AND THE COST OF DESALINATION PROCESSES INTEGRATED WITH RENEWABLE ENERGIES

Desalination intrinsically requires a considerable amount of energy. Consequently, all desalination methods are energy intensive. The minimum energy requirement for desalination follows the laws of thermodynamics and is independent of the desalination method and the number of stages of the system. When water salinity increases, the minimum energy required for desalination increases. For example, the theoretical minimum energy required for the desalination of seawater at 35000 ppm salt with a typical recovery of 50% is 1.06 kWh/m³ [6]. However, the actual energy requirement is higher than the theoretical minimum because desalination plants are finite in size and do not operate as a reversible thermodynamic process. The actual energy required for a typical RO desalination system (50% recovery from a feed with 35,000 ppm) has decreased dramatically during the last decades after the development of more efficient membranes, and it has not changed significantly since the [6, 125].

Among the major desalination approaches, RO requires electricity for pumps, but MSF and MED require both electricity and thermal energy. The energy costs for both thermal and membrane-based desalination processes cannot be underestimated. For example, a study on two identically sized desalination units in Libya, one with RO and another with MSF, showed that energy is responsible for 26% and 41% of the total desalination cost for RO and MSF, respectively [126]. Table 1 summarizes the amount of energy consumption by major desalination technologies within the existing literature [10, 114, 127–142].

Desalination costs mainly depend on the local availability of fuels and the cost of energy. In addition, site-specific aspects have effects on the cost of water production; for example, feed and water transportation, fresh water delivery to users, and plant size. Comparison of the two most used desalination technologies (RO and MSF) shows that MSF requires higher capital cost, while operational and maintenance costs are higher for RO [10]. Typically, the investment cost for a newly installed desalination plant is from 800 to 1500 USD per unit (m^3/d), and this value depends on a few factors like labor cost [143]. Usually, operational and maintenance costs are around 2%–2.5% of the annual investment cost [10]. Desalination is mainly an appropriate option for middle-income and rich countries and is still unsuitable for poor countries. Conventional desalination units running with fossil fuels usually cost 1–2 USD/m^3 (although, for the best conditions for large-scale and modern plants, it can be as low as 0.5 USD/m^3) [10]. Recently, there has been a great reduction in the cost of renewable energy, installation, and equipment; consequently, it is becoming more economical to integrate renewable energies with desalination. Table 2 summarizes the costs of some available RE-desalination systems from [10, 14, 31, 78, 144–149].

The cost of renewable desalination, like any other technology, follows the economy of scale. Currently, it is not economical to consider building a desalination plant for communities with low populations (<10 m^3/d consumption) [155], although there are extensive studies in this direction. There are suitable methods like solar still, solar membrane distillation, and solar multiple-effect humidification that can be used in areas where solar energy is abundant, like the Mediterranean islands and areas like Northern Africa and Oceania.

For medium-scale water usage (10–1000 m^3/d), the market is also large, and can address any area facing water shortage with a permanent or seasonal population from 500 up to 50,000 people (for example, villages, hotels, and island resorts). In addition, for remote areas, the cost of electricity is usually high; in such cases, considering desalination powered directly by wind can be a proper choice. However, the production of energy from wind is unstable and depends on the existence of adequate wind resources. For large-scale applications (>1000 m^3/d), RE-desalination is a replacement for any conventional desalination technology; however, using RE-desalination on this scale is still feasible where governmental supports or obligations exist.

Table 1 Energy consumption and economy of different desalination methods [10]

	RO [131, 136, 137]	RO [133, 138]	ED [141]	MSF [114]	MED [128, 130]	VC [10]
Feed water type	Seawater	Brackish	Brackish	Seawater	Seawater	Seawater
Typical unit size (m³/day)	Up to 128,000	98,000	2–145,000	23,000–528,000	5000–15,000	10,000–30,000 (TVC); 100–3000 (MVC)
Quality of product water (ppm)	400–500	200–500	150–500	10	10	10
Electricity requirement (kWh/m³)	4–6	1.5–2.5	1.5–4	2.5–5	2–2.5	1.8–1.6 (TVC); 7–12 (MVC)
Thermal energy demand (kWh/m³)	–	–	–	15.83–23.5	12.2–19.1	14.5 (TVC)
Total energy consumption (kWh/m³)	4–6	1.5–2.5	1.5–4	19.58–27.25	14.45–21.35	15.5–16.5 (TVC); 7–12 (MVC)
Operating temperature (°C)	Ambient	Ambient	Ambient	90–110	70–75	70–90
Cost of water (USD/m³)	0.45–0.66 (for 100–320 10³ m³/day); 0.48–1.62 (for 15–60 10³ m³/day); 0.7–1.72 (for 1–4.8 10³ m³/day)	0.26–0.54 (for 40,000 m³/day); 0.78–1.33 (for 20–1200 m³/day); 0.56–12.99 (for few m³/day)	0.6 (large capacity); 1.05 (small capacity)	0.56–1.75	0.52–1.01 (for 91–320 10³ m³/day); 0.95–1.5 (for 12–55 10³ m³/day); 2.0–8.0 (for less than 100 m³/day)	0.87–0.95 (for TVC with 30,000 m³/day); 2.0–2.6 (for MVC with 1000 m³/day)

Permission obtained from Elsevier.

Table 2 Energy implications of different RE-desalination methods [10]

	Design capacity (m³/day)	Energy demand (kWh/m³)	Water cost (USD/m³)
PV-RO	<100	Electrical: 0.5–1.5 (BW); 45 (SW) [149]	6.5–9.1 (BW); 11.7–15.6 (SW) [78]
Wind-RO [150–152]	50–2000	Electrical: 0.5–1.5 (BW); 4–5 (SW)	Units under 100 m³/d: 3.9–6.5 (BW); 6.5–9.1 (SW); Over 1000 m³/d: 2–5.2
PV-EDR	<100	Electrical: 3–4 [149]; 0.6–1 [153] (brackish water)	10.4–11.7 [149]; 3–16 [153]
Wind-ED	–	–	2–3.5 (BW)
Solar-MD	0.15–10	Thermal: 150–200 [149]; 436 [154]; 180–2200 [106]; 100–600 [153]	10.4–19.5 [149] 13–18 [153]
Geothermal-MED [146]	–	–	3.8–5.7 (SW)
CSP-MED [149]	>5000	Thermal: 60–70 Electrical: 1.5–2	2.3–3
Wind-MVC [150]	<100	10–14 electrical for seawater	5.2–7.8
Solar stills	<0.1	Solar passive	1.3–6.5
Solar AD	8	Electrical: 1.38 Thermal: 39.8 [65]	0.7 (electrical cost only) [65]
Geothermal-MD	20,000 [37]	66.03 kW/kg h⁻¹	0.5
Solar-multieffect humidification [149]	1–100	31.1 kWh/m³	2.6–6.5

The quality of desalinated water after thermal desalination technologies like MSF or MED is similar to distilled water with negligible salt, as can be seen in Table 1. In thermal desalination approaches, the product water is collected from the steam, and the dissolved salt is left behind. However, for membrane-based cases, the mechanism of salt removal is different and works based on the size of the dissolved substances. As can be seen from Table 1, the product of RO has a TDS up to 500 ppm, which is considered healthy water based on the regulations of the World Health Organization (WHO). ED is efficient in removing salts and other ions as it is an electrochemical technique. However, ED cannot remove viruses and bacteria, while RO is able to remove most of the impurities. Thereby, considering the quality of the feed water, it would then be necessary to consider disinfection pre- or posttreatment steps [156].

6 CONCLUSION

Different desalination methods and their integration with renewable energies were discussed in this chapter. Desalination technologies are mainly divided into two types: membrane-based and nonmembrane-based (thermal). In recent years, membrane-based methods have received greater attention as they are compact and energy effective. However, as desalination is very energy consuming by nature, about 50% of the cost of desalination is related to energy requirements. This means that providing cheaper energy for desalination is key to its development and to making it practical, which is why green, renewable sources of energy have been taken into account. Renewable resources, mostly solar and wind, have unreliable nature and their availability depends on weather and location. Therefore, designing systems that consider hybrid systems (which are complementary systems) is of great importance. In addition, renewable resources need to be provided with batteries to deal with issues related to the unreliability of renewable resources. Another discussed matter is the ongrid and offgrid structures of RE-desalination systems. For those areas where the local grid is distant, and the connections are quite expensive, it is recommended to design an offgrid system in which desalination is completely independent of the grid. However, offgrid systems have some requirements, and batteries would seem necessary.

REFERENCES

[1] Goosen MFA, Mahmoudi H, Ghaffour N. Today's and future challenges in applications of renewable energy technologies for desalination. Crit Rev Environ Sci Technol 2014;44:929–99.
[2] Sharon H, Reddy KS. A review of solar energy driven desalination technologies. Renew Sustain Energy Rev 2015;41:1080–118.

[3] Durant W, Durant A, Rouben Mamoulian Collection (Library of Congress). The story of civilization. New York: Simon and Schuster; 1935.

[4] IPCC. Climate change 2014: Impacts, adaptation, and vulnerability. Part A: Global and Sectoral Aspects. In: Field CB, Barros VR, Dokken DJ, Mach KJ, Mastrandrea MD, Bilir TE, Chatterjee M, Ebi KL, Estrada YO, Genova RC, Girma B, Kissel ES, Levy AN, MacCracken S, Mastrandrea PR, White LL, editors. The contribution of Working Group II to the fifth assessment report of the intergovernmental panel on climate change. Cambridge, United Kingdom and New York, NY, USA: Cambridge University Press; 2014.

[5] Connor R. The United Nations world water development report 2015: water for a sustainable world. Paris: UNESCO Publishing; 2015.

[6] Elimelech M, Phillip WA. The future of seawater desalination: energy, technology, and the environment. Science 2011;333:712–7.

[7] Shatat M, Worall M, Riffat S. Opportunities for solar water desalination worldwide: review. SCS 2013;9:67–80.

[8] International Desalination Association (IDA). Desalination by the numbers. 2015.

[9] Subramani A, Badruzzaman M, Oppenheimer J, Jacangelo JG. Energy minimization strategies and renewable energy utilization for desalination: a review. Water Res 2011;45:1907–20.

[10] Al-Karaghouli A, Kazmerski LL. Energy consumption and water production cost of conventional and renewable-energy-powered desalination processes. Renew Sustain Energy Rev 2013;24:343–56.

[11] Lindemann JH. Wind and solar powered seawater desalination applied solutions for the Mediterranean, the Middle East and the Gulf countries. Desalination 2004;168:73–80.

[12] Bahrami M, Abbaszadeh P. An overview of renewable energies in Iran. Renew Sustain Energy Rev 2013;24:198–208.

[13] Alamdari P, Nematollahi O, Alemrajabi AA. Solar energy potentials in Iran: a review. Renew Sustain Energy Rev 2013;21:778–88.

[14] Al-Karaghouli A, Renne D, Kazmerski LL. Solar and wind opportunities for water desalination in the Arab regions. Renew Sustain Energy Rev 2009;13:2397–407.

[15] Wittholz MK, O'Neill BK, Colby CB, Lewis D. Estimating the cost of desalination plants using a cost database. Desalination 2008;229:10–20.

[16] Ma Q, Lu H. Wind energy technologies integrated with desalination systems: review and state-of-the-art. Desalination 2011;277:274–80.

[17] Hwang K-J, Chiang Y-C. Comparisons of membrane fouling and separation efficiency in protein/polysaccharide cross-flow microfiltration using membranes with different morphologies. Sep Purif Technol 2014;125:74–82.

[18] Rabiee H, Mojtaba Seyedi S, Rabiei H, Arya A, Alvandifar N. Preparation and characterization of PVC/PAN blend ultrafiltration membranes: effect of PAN co. Desalin Water Treat 2017;58:1–11.

[19] Rabiee H, Shahabadi SMS, Mokhtare A, Rabiei H, Alvandifar N. Enhancement in permeation and antifouling properties of PVC ultrafiltration membranes with addition of hydrophilic surfactant additives: Tween-20 and Tween-80. J Environ Chem Eng 2016;4:4050–61.

[20] Rabiee H, Seyedi SM, Rabiei H, Alvandifar N. Improvements in permeation and fouling resistance of PVC ultrafiltration membranes via addition of Tetronic-1107 and Triton X-100 as two non-ionic and hydrophilic surfactants. Water Sci Technol 2016;74:1469–83.

[21] Rabiee H, Vatanpour V, Farahani MHDA, Zarrabi H. Improvement in flux and antifouling properties of PVC ultrafiltration membranes by incorporation of zinc oxide (ZnO) nanoparticles. Sep Purif Technol 2015;156:299–310.

[22] Fujioka T, Khan SJ, McDonald JA, Nghiem LD. Nanofiltration of trace organic chemicals: a comparison between ceramic and polymeric membranes. Sep Purif Technol 2014;136:258–64.

[23] Chian EK, Chen JP, Sheng P-X, Ting Y-P, Wang L. Reverse osmosis technology for desalination. In: Wang L, Hung Y-T, Shammas N, editors. Advanced physicochemical treatment technologies. Totowa, NJ: Humana Press; 2007. p. 329–66.

[24] Deleted in Review.

[25] www.puretecwater.com.

[26] Fritzmann C, Löwenberg J, Wintgens T, Melin T. State-of-the-art of reverse osmosis desalination. Desalination 2007;216:1–76.

[27] Greenlee LF, Lawler DF, Freeman BD, Marrot B, Moulin P. Reverse osmosis desalination: water sources, technology, and today's challenges. Water Res 2009; 43:2317–48.

[28] Bourouni K, Ben M'Barek T, Al Taee A. Design and optimization of desalination reverse osmosis plants driven by renewable energies using genetic algorithms. Renew Energy 2011;36:936–50.

[29] Lee KP, Arnot TC, Mattia D. A review of reverse osmosis membrane materials for desalination—development to date and future potential. J Membr Sci 2011;370:1–22.

[30] Drioli E, Ali A, Macedonio F. Membrane distillation: recent developments and perspectives. Desalination 2015;356:56–84.

[31] Alkhudhiri A, Darwish N, Hilal N. Membrane distillation: a comprehensive review. Desalination 2012;287:2–18.

[32] Suárez F, Ruskowitz JA, Tyler SW, Childress AE. Renewable water: direct contact membrane distillation coupled with solar ponds. Appl Energy 2015;158:532–9.

[33] Camacho L, Dumée L, Zhang J, Li J-d, Duke M, Gomez J, Gray S. Advances in membrane distillation for water desalination and purification applications. Water 2013;5:94–196.

[34] Calabro V, Jiao BL, Drioli E. Theoretical and experimental study on membrane distillation in the concentration of orange juice. Ind Eng Chem Res 1994;33:1803–8.

[35] Udriot H, Ampuero S, Marison IW, von Stockar U. Extractive fermentation of ethanol using membrane distillation. Biotechnol Lett 1989;11:509–14.

[36] Kim Y-D, Thu K, Choi S-H. Solar-assisted multi-stage vacuum membrane distillation system with heat recovery unit. Desalination 2015;367:161–71.

[37] Sarbatly R, Chiam C-K. Evaluation of geothermal energy in desalination by vacuum membrane distillation. Appl Energy 2013;112:737–46.

[38] Nakoa K, Rahaoui K, Date A, Akbarzadeh A. An experimental review on coupling of solar pond with membrane distillation. Solar Energy 2015;119:319–31.

[39] Seyed Shahabadi SM, Rabiee H, Seyedi SM, Mokhtare A, Brant JA. Superhydrophobic dual layer functionalized titanium dioxide/polyvinylidene fluoride-co-hexafluoropropylene (TiO$_2$/PH) nanofibrous membrane for high flux membrane distillation. J Membr Sci 2017;537:140–50.

[40] Tanaka Y. 12: Electrodialysis. In: Ion exchange membranes. 2nd ed. Amsterdam: Elsevier; 2015. p. 255–93.

[41] Wright NC, Winter AG. Justification for community-scale photovoltaic-powered electrodialysis desalination systems for inland rural villages in India. Desalination 2014;352:82–91.

[42] Oren Y, Korngold E, Daltrophe N, Messalem R, Volkman Y, Aronov L, Weismann M, Bouriakov N, Glueckstern P, Gilron J. Pilot studies on high recovery BWRO-EDR for near zero liquid discharge approach. Desalination 2010; 261:321–30.

[43] Thampy S, Desale GR, Shahi VK, Makwana BS, Ghosh PK. Development of hybrid electrodialysis-reverse osmosis domestic desalination unit for high recovery of product water. Desalination 2011;282:104–8.

[44] Ronan SMZ, McGovern K, John H, Lienhard V. In: Design and optimization of hybrid ed-ro systems for the treatment of highly saline brines. The international desalination association world congress on desalination and water reuse 2013/Tianjin, China; 2013.

[45] Strathmann H. Electrodialysis, a mature technology with a multitude of new applications. Desalination 2010;264:268–88.

[46] Tanaka Y. 16: Electrodialysis reversal. In: Ion exchange membranes. 2nd ed. Amsterdam: Elsevier; 2015. p. 345–67.

[47] Deleted in Review.

[48] Cath TY, Childress AE, Elimelech M. Forward osmosis: principles, applications, and recent developments. J Membr Sci 2006;281:70–87.

[49] Chung T-S, Zhang S, Wang KY, Su J, Ling MM. Forward osmosis processes: yesterday, today and tomorrow. Desalination 2012;287:78–81.

[50] Shaffer DL, Werber JR, Jaramillo H, Lin S, Elimelech M. Forward osmosis: where are we now? Desalination 2015;356:271–84.

[51] Li C, Goswami Y, Stefanakos E. Solar assisted sea water desalination: a review. Renew Sustain Energy Rev 2013;19:136–63.

[52] Belessiotis V, Kalogirou S, Delyannis E. Chapter six: Indirect solar desalination (MSF, MED, MVC, TVC). In: Thermal solar desalination. London, UK: Academic Press; 2016. p. 283–326.

[53] Kabeel AE, Hamed MH, Omara ZM, Sharshir SW. Water desalination using a humidification-dehumidification technique—a detailed review. Nat Resour 2013;04:286–305.

[54] Zheng H. Chapter 6 - Humidification–dehumidification solar desalination systems. In: Solar energy desalination technology. Amsterdam: Elsevier; 2017. p. 447–535.

[55] Belessiotis V, Kalogirou S, Delyannis E. Chapter 5: Humidification–dehumidification. In: Thermal solar desalination. London, UK: Academic Press; 2016. p. 253–81.

[56] Zubair MI, Al-Sulaiman FA, Antar MA, Al-Dini SA, Ibrahim NI. Performance and cost assessment of solar driven humidification dehumidification desalination system. Energ Conver Manage 2017;132:28–39.

[57] Suss ME, Porada S, Sun X, Biesheuvel PM, Yoon J, Presser V. Water desalination via capacitive deionization: what is it and what can we expect from it? Energ Environ Sci 2015;8:2296–319.

[58] Jeon S-i, Park H-r, Yeo J-g, Yang S, Cho CH, Han MH, Kim DK. Desalination via a new membrane capacitive deionization process utilizing flow-electrodes. Energ Environ Sci 2013;6:1471.

[59] Hatzell KB, Iwama E, Ferris A, Daffos B, Urita K, Tzedakis T, Chauvet F, Taberna P-L, Gogotsi Y, Simon P. Capacitive deionization concept based on suspension electrodes without ion exchange membranes. Electrochem Commun 2014;43:18–21.

[60] Simon P, Gogotsi Y. Materials for electrochemical capacitors. Nat Mater 2008;7:845–54.

[61] Li M, Anand RK. Recent advancements in ion concentration polarization. Analyst 2016;141:3496–510.

[62] Kwak R, Pham VS, Kim B, Chen L, Han J. Enhanced salt removal by unipolar ion conduction in ion concentration polarization desalination. Sci Rep 2016;6:25349.

[63] Tanaka Y. 6: Concentration polarization. In: Ion exchange membranes. 2nd ed. Amsterdam: Elsevier; 2015. p. 101–21.

[64] Kim B, Kwak R, Kwon HJ, Pham VS, Kim M, Al-Anzi B, Lim G, Han J. Purification of high salinity brine by multi-stage ion concentration polarization desalination. Sci Rep 2016;6:31850.

[65] Ng KC, Thu K, Kim Y, Chakraborty A, Amy G. Adsorption desalination: an emerging low-cost thermal desalination method. Desalination 2013;308:161–79.

[66] Thu K, Kim Y-D, Amy G, Chun WG, Ng KC. A hybrid multi-effect distillation and adsorption cycle. Appl Energy 2013;104:810–21.

[67] Badwawi RA, Abusara M, Mallick T. A review of hybrid solar PV and wind energy system. Smart Sci 2016;3:127–38.

[68] www.pv-magazine.com/2014/09/11/irena-pv-prices-have-declined-80-since-2008_ 100016383.

[69] www.solarchoice.net.au/blog/solar-power-system-prices.

[70] Dimroth F, Grave M, Beutel P, Fiedeler U, Karcher C, Tibbits TND, Oliva E, Siefer G, Schachtner M, Wekkeli A, Bett AW, Krause R, Piccin M, Blanc N, Drazek C, Guiot E, Ghyselen B, Salvetat T, Tauzin A, Signamarcheix T, Dobrich A, Hannappel T, Schwarzburg K. Wafer bonded four-junction GaInP/ GaAs//GaInAsP/GaInAs concentrator solar cells with 44.7% efficiency. Prog Photovolt Res Appl 2014;22:277–82.

[71] M C, Yadav A. Water desalination system using solar heat: a review. Renew Sustain Energy Rev 2017;67:1308–30.

[72] Kalogirou S. Seawater desalination using renewable energy sources. Prog Energy Combust Sci 2005;31:242–81.

[73] Ahmad GE, Schmid J. Feasibility study of brackish water desalination in the Egyptian deserts and rural regions using PV systems. Energ Conver Manage 2002;43:2641–9.

[74] Richards BS, Schäfer AI. Design considerations for a solar-powered desalination system for remote communities in Australia. Desalination 2002;144:193–9.

[75] Schäfer AB, Andreas, Richards BS. Membranes and renewable energy – a new era of sustainable development for developing countries. Membr Technol 2005;6–10.

[76] Gocht W, Sommerfeld A, Rautenbach R, Melin T, Eilers L, Neskakis A, Herold D, Horstmann V, Kabariti M, Muhaidat A. Decentralized desalination of brackish water by a directly coupled reverse-osmosis-photovoltaic-system – a pilot plant study in Jordan. Renew Energy 1998;14:287–92.

[77] Bowman TE, El-Nashar AM, Thrasher BH, Husseiny AA, Unione AJ. Design of a small solar-powered desalination system. Desalination 1981;39:71–81.

[78] Thomson M, Infield D. A photovoltaic-powered seawater reverse-osmosis system without batteries. Desalination 2003;153:1–8.

[79] Al Suleimani Z, Nair VR. Desalination by solar-powered reverse osmosis in a remote area of the Sultanate of Oman. Appl Energy 2000;65:367–80.

[80] García-Rodríguez L. Renewable energy applications in desalination: state of the art. Solar Energy 2003;75:381–93.

[81] Khalilpour KR, Vassallo A. PV-battery nanogrid systems. In: Community energy networks with storage. Green energy and technology. Singapore: Springer; 2016. p. 61–82.

[82] Schäfer AI, Richards BS. Testing of a hybrid membrane system for groundwater desalination in an Australian national park. Desalination 2005;183:55–62.

[83] Richards BS, Capão DPS, Früh WG, Schäfer AI. Renewable energy powered membrane technology: impact of solar irradiance fluctuations on performance of a brackish water reverse osmosis system. Sep Purif Technol 2015;156:379–90.

[84] Habali SM, Saleh IA. Design of stand-alone brackish-water desalination wind energy system for Jordan. Solar Energy 1994;52:525–32.

[85] Kiranoudis CT, Voros NG, Maroulis ZB. Wind energy exploitation for reverse osmosis desalination plants. Desalination 1997;109:195–209.

[86] Forstmeier M, Mannerheim F, D'Amato F, Shah M, Liu Y, Baldea M, Stella A. Feasibility study on wind-powered desalination. Desalination 2007;203:463–70.

[87] Liu CCK, Jae-Woo P, Migita R, Gang Q. Experiments of a prototype wind-driven reverse osmosis desalination system with feedback control. Desalination 2002; 150:277–87.

[88] Subiela VJ, de la Fuente JA, Piernavieja G, Peñate B. Canary Islands institute of technology (ITC) experiences in desalination with renewable energies (1996–2008). Desalin Water Treat 2012;7:220–35.

[89] Loutatidou S, Liosis N, Pohl R, Ouarda TBMJ, Arafat HA. Wind-powered desalination for strategic water storage: techno-economic assessment of concept. Desalination 2017;408:36–51.

[90] Koklas PA, Papathanassiou SA. Component sizing for an autonomous wind-driven desalination plant. Renew Energy 2006;31:2122–39.

[91] Mokheimer EMA, Sahin AZ, Al-Sharafi A, Ali AI. Modeling and optimization of hybrid wind–solar-powered reverse osmosis water desalination system in Saudi Arabia. Energ Conver Manage 2013;75:86–97.

[92] Cherif H, Belhadj J. Large-scale time evaluation for energy estimation of stand-alone hybrid photovoltaic–wind system feeding a reverse osmosis desalination unit. Energy 2011;36:6058–67.

[93] Spyrou ID, Anagnostopoulos JS. Design study of a stand-alone desalination system powered by renewable energy sources and a pumped storage unit. Desalination 2010;257:137–49.

[94] Koutroulis E, Kolokotsa D. Design optimization of desalination systems power-supplied by PV and W/G energy sources. Desalination 2010;258:171–81.

[95] Novosel T, Ćosić B, Pukšec T, Krajačić G, Duić N, Mathiesen BV, Lund H, Mustafa M. Integration of renewables and reverse osmosis desalination – case study for the Jordanian energy system with a high share of wind and photovoltaics. Energy 2015;92:270–8.

[96] Moudjeber DE, Ruiz-Aguirre A, Ugarte-Judge D, Mahmoudi H, Zaragoza G. Solar desalination by air-gap membrane distillation: a case study from Algeria. Desalin Water Treat 2016;57:22718–25.

[97] Dow N, Gray S, Li J-d, Zhang J, Ostarcevic E, Liubinas A, Atherton P, Roeszler G, Gibbs A, Duke M. Pilot trial of membrane distillation driven by low grade waste heat: membrane fouling and energy assessment. Desalination 2016;391:30–42.

[98] Hogan PA, Sudjito AG, Fane GLM. Desalination by solar heated membrane distillation. Desalination 1991;81:81–90.

[99] Nakoa K, Rahaoui K, Date A, Akbarzadeh A. Sustainable zero liquid discharge desalination (SZLDD). Solar Energy 2016;135:337–47.

[100] Baghbanzadeh M, Rana D, Lan CQ, Matsuura T. Zero thermal input membrane distillation, a zero-waste and sustainable solution for freshwater shortage. Appl Energy 2017;187:910–28.

[101] Zaragoza G, Ruiz-Aguirre A, Guillén-Burrieza E. Efficiency in the use of solar thermal energy of small membrane desalination systems for decentralized water production. Appl Energy 2014;130:491–9.

[102] Cipollina A, Di Sparti MG, Tamburini A, Micale G. Development of a membrane distillation module for solar energy seawater desalination. Chem Eng Res Des 2012;90:2101–21.

[103] Koschikowski J, Wieghaus M, Rommel M, Ortin VS, Suarez BP, Betancort Rodríguez JR. Experimental investigations on solar driven stand-alone membrane distillation systems for remote areas. Desalination 2009;248:125–31.

[104] Koschikowski J, Wieghaus M, Rommel M. Solar thermal-driven desalination plants based on membrane distillation. Desalination 2003;156:295–304.

[105] Qtaishat MR, Banat F. Desalination by solar powered membrane distillation systems. Desalination 2013;308:186–97.

[106] Saffarini RB, Summers EK, Arafat HA, Lienhard V JH. Technical evaluation of stand-alone solar powered membrane distillation systems. Desalination 2012;286:332–41.

[107] AlMadani HMN. Water desalination by solar powered electrodialysis process. Renew Energy 2003;28:1915–24.

[108] Ortiz J, Exposito E, Gallud F, Garciagarcia V, Montiel V, Aldaz A. Desalination of underground brackish waters using an electrodialysis system powered directly by photovoltaic energy. Solar Energy Mater Sol Cells 2008;92:1677–88.

[109] Ortiz JM, Expósito E, Gallud F, García-García V, Montiel V, Aldaz A. Photovoltaic electrodialysis system for brackish water desalination: modeling of global process. J Membr Sci 2006;274:138–49.

[110] Zhang Y, Pinoy L, Meesschaert B, Van der Bruggen B. A natural driven membrane process for brackish and wastewater treatment: photovoltaic powered ED and FO hybrid system. Environ Sci Technol 2013;47:10548–55.

[111] Khayet M, Sanmartino JA, Essalhi M, García-Payo MC, Hilal N. Modeling and optimization of a solar forward osmosis pilot plant by response surface methodology. Solar Energy 2016;137:290–302.

[112] Schrier J. Ethanol concentration by forward osmosis with solar-regenerated draw solution. Solar Energy 2012;86:1351–8.

[113] Sagie D, Feinerman E, Aharoni E. Potential of solar desalination in Israel and in its close vicinity. Desalination 2001;139:21–33.

[114] Borsani R, Rebagliati S. Fundamentals and costing of MSF desalination plants and comparison with other technologies. Desalination 2005;182:29–37.

[115] Safi MJ. Performance of a flash desalination unit intended to be coupled to a solar pond. Renew Energy 1998;14:339–43.

[116] Hou S, Zhang H. A hybrid solar desalination process of the multi-effect humidification dehumidification and basin-type unit. Desalination 2008;220:552–7.

[117] Hou S, Zeng D, Ye S, Zhang H. Exergy analysis of the solar multi-effect humidification–dehumidification desalination process. Desalination 2007;203: 403–9.

[118] Parekh S, Farid M, Selman J, Alhallaj S. Solar desalination with a humidification-dehumidification technique—a comprehensive technical review. Desalination 2004;160:167–86.

[119] Ahmed NA, Miyatake M, Al-Othman AK. Power fluctuations suppression of stand-alone hybrid generation combining solar photovoltaic/wind turbine and fuel cell systems. Energ Conver Manage 2008;49:2711–9.

[120] Chedid R, Saliba Y. Optimization and control of autonomous renewable energy systems. Int J Energy Res 1996;20:609–24.

[121] Nelson DB, Nehrir MH, Wang C. Unit sizing and cost analysis of stand-alone hybrid wind/PV/fuel cell power generation systems. Renew Energy 2006;31:1641–56.

[122] Huang Q, Shi Y, Wang Y, Lu L, Cui Y. Multi-turbine wind-solar hybrid system. Renew Energy 2015;76:401–7.

[123] Hrayshat ES. Brackish water desalination by a stand alone reverse osmosis desalination unit powered by photovoltaic solar energy. Renew Energy 2008;33:1784–90.

[124] Peñate B, Castellano F, Bello A, García-Rodríguez L. Assessment of a stand-alone gradual capacity reverse osmosis desalination plant to adapt to wind power availability: a case study. Energy 2011;36:4372–84.

[125] MacHarg TFSJ. B. Sessions. In: Desalination and water reuse. vol. 18. South Croydon, Surrey, UK: Faversham House Group; 2008. p. 30–9.

[126] Zotalis K, Dialynas E, Mamassis N, Angelakis A. Desalination technologies: Hellenic experience. Water 2014;6:1134–50.

[127] Al-Karaghouli A, Renne D, Kazmerski LL. Solar and wind opportunities for water desalination in the Arab regions. Renew Sustain Energy Rev 2009;13:2397–407.

[128] Wu SR. Analysis of water production costs of a nuclear desalination plant with a nuclear heating reactor coupled with MED processes. Desalination 2006; 190:287–94.

[129] Sambrailo D, Ivic J, Krstulovic A. Economic evaluation of the first desalination plant in Croatia. Desalination 2005;179:339–44.

[130] Ophir A, Lokiec F. Advanced MED process for most economical sea water desalination. Desalination 2005;182:187–98.

[131] Mohamed ES, Papadakis G, Mathioulakis E, Belessiotis V. The effect of hydraulic energy recovery in a small sea water reverse osmosis desalination system; experimental and economical evaluation. Desalination 2005;184:241–6.

[132] Agashichev SP, El-Nashar AM. Systemic approach for techno-economic evaluation of triple hybrid (RO, MSF and power generation) scheme including accounting of CO_2 emission. Energy 2005;30:1283–303.

[133] Jaber IS, Ahmed MR. Technical and economic evaluation of brackish groundwater desalination by reverse osmosis (RO) process. Desalination 2004;165:209–13.

[134] Agashichev SP. Analysis of integrated co-generative schemes including MSF, RO and power generating systems (present value of expenses and "levelised" cost of water). Desalination 2004;164:281–302.

[135] Wu SR, Zhang ZY. An approach to improve the economy of desalination plants with a nuclear heating reactor by coupling with hybrid technologies. Desalination 2003;155:179–85.

[136] Avlonitis SA, Kouroumbas K, Vlachakis N. Energy consumption and membrane replacement cost for seawater RO desalination plants. Desalination 2003;157:151–8.

[137] Avlonitis SA. Operational water cost and productivity improvements for small-size RO desalination plants. Desalination 2002;142:295–304.

[138] Al-Wazzan Y, Safar M, Ebrahim S, Burney N, Mesri A. Desalting of subsurface water using spiral-wound reverse osmosis (RO) system: technical and economic assessment. Desalination 2002;143:21–8.

[139] Wade NM. Distillation plant development and cost update. Desalination 2001; 136:3–12.

[140] Tian JF, Shi G, Zhao ZY, Cao DX. Economic analyses of a nuclear desalination system using deep pool reactors. Desalination 1999;123:25–31.

[141] Adiga MR, Adhikary SK, Narayanan PK, Harkare WP, Gomkale SD, Govindan KP. Performance analysis of photovoltaic electrodialysis desalination plant at tanote in THAR desert. Desalination 1987;67:59–66.

[142] Alkaisi A, Mossad R, Sharifian-Barforoush A. A review of the water desalination systems integrated with renewable energy. Energy Procedia 2017;110:268–74.

[143] Karagiannis IC, Soldatos PG. Water desalination cost literature: review and assessment. Desalination 2008;223:448–56.

[144] Qiblawey HM, Banat F. Solar thermal desalination technologies. Desalination 2008;220:633–44.

[145] El-Sebaii AA, Aboul-Enein S, Ramadan MRI, Khallaf AM. Thermal performance of an active single basin solar still (ASBS) coupled to shallow solar pond (SSP). Desalination 2011;280:183–90.

[146] Barbier E. Geothermal energy technology and current status: an overview. Renew Sustain Energy Rev 2002;6:3–65.

[147] Garcia-Rodriguez L, Romero-Ternero V, Gomez-Camacho C. Economic analysis of wind-powered desalination. Desalination 2001;137:259–65.

[148] Saffarini RB, Summers EK, Arafat HA, Lienhard V JH. Economic evaluation of stand-alone solar powered membrane distillation systems. Desalination 2012;299:55–62.

[149] Papapetrou M, Wieghaus M, Biercamp C. Roadmap for the development of desalination powered by renewable energy: promotion for renewable energy for water production through desalination. ProDes project by the Intelligent Energy for Europe Programme; 2010.

[150] Miranda MS, Infield D. A wind-powered seawater reverse-osmosis system without batteries. Desalination 2003;153:9–16.

[151] Kershman SA, Rheinlander R, Neumann T, Goebel O. Hybrid wind/PV and conventional power for desalination in Libya – GECOL's facility for medium and small scale research at Ras Ejder. Desalination 2005;183:1–12.

[152] de la Nuez Pestana I, Javier García Latorre F, Argudo Espinoza C, Gómez Gotor A. Optimization of RO desalination systems powered by renewable energies. Part I: Wind energy. Desalination 2004;160:293–9.

[153] Ali MT, Fath HES, Armstrong PR. A comprehensive techno-economical review of indirect solar desalination. Renew Sustain Energy Rev 2011;15:4187–99.

[154] Kim Y-D, Thu K, Ghaffour N, Choon Ng K. Performance investigation of a solar-assisted direct contact membrane distillation system. J Membr Sci 2013;427: 345–64.

[155] Eltawil MA, Zhengming Z, Yuan L. A review of renewable energy technologies integrated with desalination systems. Renew Sustain Energy Rev 2009;13:2245–62.

[156] Nayar KG, Sundararaman P, O'Connor CL, Schacherl JD, Heath ML, Gabriel MO, Shah SR, Wright NC, Winter VAG. Feasibility study of an electrodialysis system for in-home water desalination in urban India. Dev Eng 2017;2:38–46.

CHAPTER 14

Renewable Energy Integration in Combined Cooling, Heating, and Power (CCHP) Processes

Dia Milani
CSIRO Energy Centre, Newcastle, NSW, Australia

Abstract

A combined cooling, heating, and power (CCHP) system with distributed cogeneration units and renewable energy integration provides effective solutions to many energy-related problems.

This chapter explains the fundamental of energy conversion and shows the advantages of CCHP in primary energy saving, exergy efficiency, emission reduction, and annual total energy saving. The renewable energy integration with CCHP systems is categorized based on their primary service as in electric harvest or thermal harvest domains. It is emphasized that due to high fluctuation in supply-demand dynamics, large-scale projects benefit most from diversifying energy supply options along with different demand profiles of end-users. This tangible advantage would require sophisticated optimization techniques based on consistent and reliable communication between various stakeholders, which lead to the introduction of CCHP microgrid concept. CCHP microgrids feature many advantages over the individual CCHP systems in better volatility management, flexibility operation, waste heat utilization, automated response, and technoeconomic superiority.

Keywords: Trigeneration, Dynamics, Distributed energy system, Electric demand management, Thermal demand management, CCHP

Abbreviation

ANN	artificial neural network
ATCS	annual total cost saving
BLM	base load management
CCHP	combined cooling, heating, and power
CHP	combined heat and power
EDM	electric demand management
ER	emission reduction
GHG	greenhouse gas
HVAC	heating, ventilation, and air-conditioning
KPI	key performance indicator
ORC	organic Rankine cycle

Polygeneration with Polystorage
https://doi.org/10.1016/B978-0-12-813306-4.00014-8

PES   primary energy savings
PGU   power generation unit
PVT   photovoltaic thermal
TDM   thermal demand management
TES   thermal energy storage

1 INTRODUCTION

As a promising remedy to the emerging global energy-related problems, combined cooling, heating, and power (CCHP) generation systems may provide an effective alleviation to reduce fossil fuel consumption and subsequently greenhouse gas (GHG) emissions and also provide economic benefits. The concept of CCHP was initially proposed to encounter the considerable deficiency related to the bulk movement of electrical energy in pronged transmission networks, where the electricity eventually and substantially consumed in heating and cooling applications by end-users. The concept of CCHP, also known as trigeneration, offers a distributed energy supply approach with the flexibility to incorporate renewable energy and energy storage technologies. The key objective of utilizing CCHP is to lessen our dominant reliance on the centralized power infrastructure and utilize small-scale decentralized hybrid power systems. These tailor-designed distributed power systems can be a step forward in delivering more efficient and cost effective solutions. They will not only be site-flexible and community-friendly self-sustainable energy systems, but they will also be highly capable of renewable energy integration. They can also provide a reliable support to the centralized electric power plants particularly in shaving up the peak loads [1–3], reducing the electric market turbulences [4,5], and in emergency shutdowns.

One proven approach of distributed energy generation is the use of CCHP systems, which combine the use of traditional technology with newly developed techniques to satisfy the modern lifestyles and complex interactions between the energy demand, economic viability, and environmental constrains. Compared to the centralized approach that is generating and distributing bulk electrical power traveling for long distances, CCHP systems are also able to utilize the waste heat energy from the nearby fossil-fuel-fired power plants and/or other industrial plants for heating, ventilation, and air-conditioning (HVAC) purposes. It is estimated in building sector alone, 46% of the total worldwide energy demand is attributed to various HVAC applications [6]. Since HVAC systems account for a considerable portion of the energy consumption in residential and industry sectors, this approach may conserve and utilize unwanted/unused thermal energy from the vicinity infrastructures. Alternatively, CCHP can easily be integrated and hybridized with a variety of renewable energy sources, such as solar and geothermal. This concept has already been deployed in a wide range of buildings such as universities (Penn State University, USA), airports (Shanghai Pudong International Airport, China), seawater desalination plants (Iran), or for district heating (Finland, Denmark, Germany), and in many hospitals and supermarkets worldwide [7,8].

For an economic, efficient, and low GHG emission, it is important to tailor-design the CCHP application in respect to full consideration of the energy demands and power capacity of that specific building/entity. The capacity of CCHPs is often categorized in four representative groups depending on the application scale/area as shown in Table 1 [7,9].

In this chapter, efficiency improvement measures in thermodynamic power cycles are explored. The role of utilizing waste heat in combination with power generation and the expected improvement in thermal efficiencies are mathematically demonstrated. The concept of heat integration in the form of combined heat and power (CHP) is developed and the most common operation strategies are introduced. This combination has led to introduce the comprehensive concept of CCHP that is not only proven by the enhanced thermal efficiency viability, but also supported by tangible benefits in other key performance indicators (KPIs) such as exergy efficiency, emission reduction (ER), and annual total cost savings (ATCSs). Next, innovative renewable energy integration methods are probed in the literature to recognize the recent advances in various renewable energy integrated systems. The wide adaptability and scalability of these renewable sources in combination with a better understanding of power market dynamics have led to pursue the advanced frontiers of CCHP microgrids. This chapter highlights the key elements of modern CCHP microgrids and showcase the characteristics of intelligent energy networks.

Table 1 Power system scale in respect to the power capacity and application area

Configuration	Capacity	Application
Microscale	<20 kW	Distributed energy system
Small scale	20 kW–1 MW	Supermarkets, retail stores, hospitals, office buildings, university campuses
Medium scale	1–10 MW	Large factories, hospitals, schools
Large scale	>10 MW	Large industries. Waste heat used in universities, district heating

2 EFFICIENCY IMPROVEMENT DESIGN STRATEGIES

The main objective for the design of any power conversion plant is to achieve the highest work output from a given heat/fuel supply. The power plant must be tailor-designed at every aspect for that common objective. The eminent Carnot cycle, consisting of two isothermal processes and two adiabatic processes, is the most efficient heat engine cycle allowed by physical laws (Eq. 1):

$$\zeta_{carnot} = \frac{W}{Q} = 1 - \frac{T_{min}}{T_{max}} \tag{1}$$

In a conventional thermal power plant, typically about one-third of the input fuel energy ends up as a useful electric power; and the rest of the energy is often wasted in the form of lukewarm water either in the cooling towers, or to the adjacent rivers/sea. Therefore, in order to raise the thermal efficiency, other design modifications are necessary in order to utilize the exhaust heat. The heat exiting the turbine can be reused by one or a combination of the following methods: (i) regeneration [10–12], (ii) combined gas and steam cycles [13], (iii) cogeneration of heat and power [14,15], or (iv) the trigeneration (CCHP) systems [9,16].

The T-S diagram of the basic Rankine and Joule-Brayton power cycles is presented in Fig. 1 showing a few possible modifications that can be implemented in order to increase the thermal efficiency of those conventional cycles. For example, regeneration method would reuse the waste heat within the cycle by bleeding steam back to the boiler or by using effective heat exchangers in the thermodynamic circuit. However, the scope of these modifications in improving the thermal efficiency is limited.

Alternatively, in a combined cycle plant (Fig. 2), two power cycles are combined in such a way that the heat rejected from the higher (topping) cycle (i.e., Joule-Brayton power cycle) Q_{HL} of efficiency ζ_H, is used as an input to the lower (bottoming) cycle (i.e., Rankine steam cycle) of efficiency ζ_L. The two plants are cyclic and use two different working fluids. A brief analysis of this plant for determining thermal efficiency is given as follows:

$$W_H = \zeta_H \, Q_H \tag{2}$$

$$W_L = \zeta_L \, Q_{HL} \tag{3}$$

$$Q_{HL} = Q_H \left(1 - \zeta_H\right) \tag{4}$$

$$\zeta_{th} = \frac{W_H + W_L}{Q_H} = \zeta_H + \zeta_L - \zeta_H \zeta_L \tag{5}$$

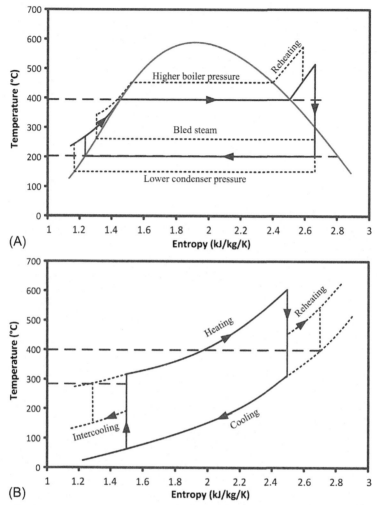

Fig. 1 Temperature-entropy diagram showing possible modifications to basic (A) Rankine cycle and (B) Joule-Brayton cycle in order to increase thermal efficiency.

Thus, the efficiency of the single topping cycle is increased by $\zeta_L(1 - \zeta_H)$ by utilizing the rejected heat to generate power in the lower cycle [17].

Furthermore, a CHP unit provides the ability to utilize the otherwise wasted steam/heat in a combined heat-power cycle plant. The simplified configuration in a CHP facility integrated with the conventional separated heat and electric generation is presented in Fig. 3. The industrial plant requires E_1 and Q_1 quantities of electricity and heat, respectively. The CHP unit supplies E and Q quantities of electricity and heat. If the CHP

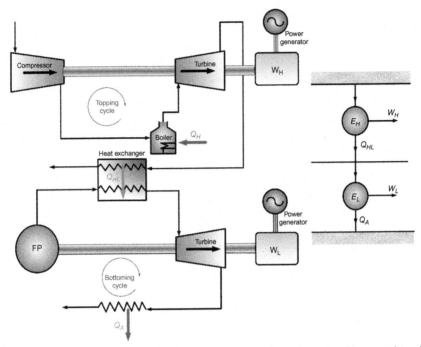

Fig. 2 A schematic and block diagram of a gas turbine/Steam turbine combined power plant.

unit cannot meet the entire electricity/heat demand to satisfy the industrial actions, an additional electricity and/or fuel may be purchased to satisfy the requirements of the plant. The overall efficiency of the CHP unit may be written as:

$$\zeta_{CHP} = \frac{Q + E}{Q_o} \tag{6}$$

where ζ_{CHP} is also named the energy utilization factor. Since all of the heat that is not used for electric generation is utilized to supply heat, the overall efficiency of a CHP unit can reach as high as 60%–80% of total primary energy input compared to the 35%–40% for the conventional separated production (SP) plants [9].

Moving forward particularly for industrial plants and commercial/residential buildings that have ongoing cooling demand will introduce the concept of a more advanced cycle of CCHP (Tri-generation). A slight difference between CCHP and CHP is that thermal or electrical/mechanical energy is further utilized to provide space or process cooling capacity in a CCHP application. In winter, many CCHP systems in commercial/residential

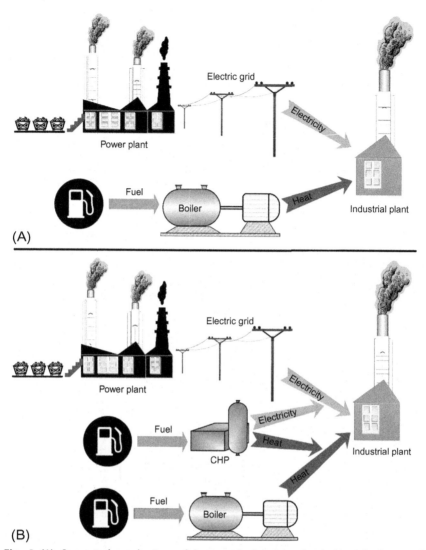

Fig. 3 (A) Separated production of heat and electricity for industrial plant and (B) cogeneration of heat and electricity via utilizing a CHP unit.

buildings may simply turn to CHP units when there is no cooling demand of the building HVAC system. In other words, the CHP unit is in fact a CCHP without any thermally activated equipment for generating cooling power, though this difference may considerably change the structure of these systems and affect the economics. A typical CCHP is illustrated in Fig. 4 comprising a gas turbine, a generator, and an absorption/adsorption chiller. The turbine is running by natural gas combustion and the mechanical energy is converted

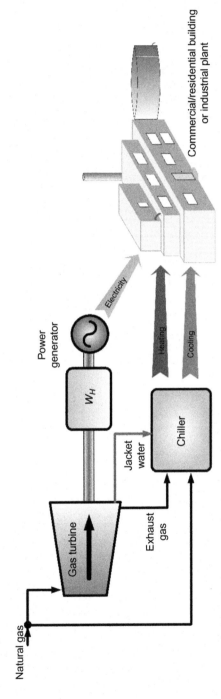

Fig. 4 A typical CCHP cycle tailor-designed for a specific building or industrial plant.

into electric power by the generator. At the same time, the absorption/adsorption chiller would generate cooling power in summer and heating power in winter by utilizing the exhaust gas and the jacket water derived from the gas turbine engine. If the waste heat from the gas turbine does not satisfy the cooling/heating target, a furnace in the chiller can be used to burn a supplementary of natural gas. Thus, the energy demands of the cooling, heating and electric power in a building or a district can always be met by this system configuration.

Compared with the conventional energy supply from the large centralized power plant that is eventually to some extent consumed in the local air-conditioning systems, the locally distributed CCHP systems show massive thermal efficiency improvement in the range of 70%–90% [9]. This is because of less primary energy is needed to obtain the same amount of electricity and thermal energy. In addition to the saving in primary energy, vast reductions in net fuel costs, power transmission, and distribution savings can be achieved [9].

3 OPERATION STRATEGIES

The mathematical principles for analysis and optimization of the CCHP plants compared to the conventional SP are presented in this section. The design of CCHP systems should consider the best strategies in respect to the load demand, the tradeoffs among capital cost saving, energy saving, and also the net GHG emissions. For this purpose, several performance criteria and operation strategies are discussed here. The CCHP operation strategy is often designed based on the load objectives and/or fuel consumption targets. There are a number of effective strategies that may be used to control the performance of CCHP systems [7]:

I. Electric demand management (EDM):

The priority in this strategy is to generate all the electricity needed to satisfy the electric load and the waste heat generated as a byproduct in the power generation unit (PGU) is used to satisfy as much of the thermal load as possible. If the recovered heat is not sufficient, an auxiliary boiler is used to supplement the extra heat needed by the facility.

II. Thermal demand management (TDM):

The priority in this strategy is to satisfy all the thermal demand in the facility, while the electricity generated in the process is used to satisfy a portion of the electric power demand. Additional electricity may need to be purchased from the grid if necessary.

III. Base load management (BLM):

The CCHP system is designed to cover only a constant amount of the electric and thermal loads for the facility. Any additional requirement of electricity/heat needs to be individually maintained from the electric grid and/or an auxiliary boiler, respectively.

The EDM and TDM strategies may not guarantee the optimum technoeconomic performance of the CCHP system. This is due to factors like variations in load demands and fuel/electricity prices and availabilities. Hence, in order to maximize the performance of the system, certain optimization techniques are necessary. Next, a more detailed CCHP system optimization model in comparison with the conventional SP model is presented [18].

3.1 SP System Model

The typical SP system is anticipated to provide full services for cooling, heating, and electric power for a particular building as shown in Fig. 5A. Except for some heat energy provided by burning fuel (F_b^{SP}) in the boiler, all the energy demand is assumed to be satisfied by the electric grid (E_{grid}^{SP}). The total energy from the grid is given as:

$$E_{grid}^{SP} = E_{bld} + E_c + E_p \tag{7}$$

where E_{bld} is the building electric load; E_c is the electric power required for the cooling system; and E_p is the electric power needed for the supplementary of heating/cooling systems. The electric power for the energy consumed by the chiller can be calculated by:

$$E_c = \frac{Q_c}{COP_{ch}} \tag{8}$$

where COP_{ch} is the coefficient of performance for the chiller. The fuel energy used to provide the grid electric power E_{grid}^{SP} is:

$$F_p^{SP} = \frac{E_{grid}^{SP}}{\zeta_e^{SP} \zeta_{grid}} \tag{9}$$

where F_p^{SP} is the combusted fuel in the central power plant to provide the E_{grid}^{SP}; while ζ_e^{SP} is the efficiency of power generation; and ζ_{grid} is the efficiency of power transmission and distribution, respectively [18]. Similarly, the fuel energy used to provide Q_h for the boiler is:

$$F_b^{SP} = \frac{Q_h}{\zeta_b^{SP} \zeta_h^{SP}} \tag{10}$$

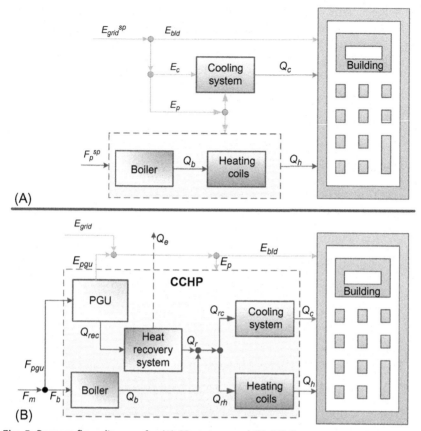

Fig. 5 Process flow diagram for (A) SP system and (B) CCHP system.

where F_b^{SP} is the combusted fuel in the boiler; ζ_b^{SP} is the thermal efficiency of the boiler; and ζ_h^{SP} is the thermal efficiency of the heating coils in the SP system model. Finally, the total fuel consumed in SP system can be estimated as:

$$F^{SP} = \frac{E_{bld}}{\zeta_e^{SP}\,\zeta_{grid}} + \frac{E_p}{\zeta_e^{SP}\,\zeta_{grid}} + \frac{Q_c}{COP_{ch}\,\zeta_e^{SP}\,\zeta_{grid}} + \frac{Q_h}{\zeta_b^{SP}\,\zeta_h^{SP}} \qquad (11)$$

3.2 CCHP System Model

The schematic for a representative CCHP system is presented in Fig. 5B. From the figure, the electric energy balance for the CCHP system can be expressed in Eq. (12):

$$E_{grid} + E_{pgu} = E_{bld} + E_p \qquad (12)$$

where the electric power from the grid (E_{grid}) and the power generation unit (E_{pgu}) are used for the building (E_{bld}) and also for the auxiliary power (E_p) within the CCHP. The fuel energy consumed to provide E_{pgu} is:

$$F_{pgu} = \frac{E_{pgu}}{\zeta_e} \qquad (13)$$

The recovered waste heat is given as:

$$Q_r = F_{pgu}\,\zeta_{rec}\,(1 - \zeta_e) \qquad (14)$$

The heat balance equation between the cooling system and heating coils yields:

$$Q_r + Q_b = Q_{rc} + Q_{rh} \qquad (15)$$

The heat consumed by the cooling system and the heating coils is estimated as:

$$Q_{rc} = \frac{Q_c}{COP_{ch}} \qquad (16)$$

$$Q_{rh} = \frac{Q_h}{\zeta_h} \qquad (17)$$

The additional fuel energy required in the auxiliary boiler is given as:

$$F_b = \frac{Q_{rc} + Q_{rh} - Q_r}{\zeta_b} \qquad (18)$$

Finally, the local fuel energy consumed in the CCHP system is:

$$F_m = F_{pgu} + F_b \qquad (19)$$

While the total fuel (F) consumed in CCHP scenario is:

$$F = F_{grid} + F_m = \frac{E_{grid}}{\zeta_e^{SP}\,\zeta_{grid}} + \frac{E_{pgu}}{\zeta_e} + \frac{Q_{rc} + Q_{rh} - Q_r}{\zeta_b} \qquad (20)$$

For EDM operation strategy, the $E_{grid} = 0$ is substituted in these equations as all the electric demand is fulfilled by the CCHP system, which may end in excess heat production that could be wasted as Q_e. Similarly, for TDM operation strategy, all local fuel is used in PGU and $Q_b = 0$ as no heat is required from the boiler. This may lead to excess production of electric power that could be stored and/or sold back to the local grid, although this option might be uneconomical.

4 PERFORMANCE CRITERIA

In order to quantify the tangible benefits achieved by CCHP system over the conventional SP system, a number of KPIs have been expressed as the following:

4.1 Primary Energy Savings

The primary energy savings (*PES*) is the ratio of energy saved by CCHP system in comparison with an equivalent SP system divided by the original energy consumed in the SP system.

$$PES = \frac{F^{SP} - F}{F^{SP}} = 1 - \frac{F}{F^{SP}} \tag{21}$$

The *PES* is measured relative to a reference *SP* system. While the primary energy ratio (*PER*) is a criteria with an absolute value. It is defined as the ratio of the energy demand to the fuel energy required to satisfy the demand.

$$PER = \frac{E_{bld} + Q_c + Q_h}{F} \tag{22}$$

where E_{bld} is the electricity demand for the building, Q_h is the heat demand, and Q_c is the cooling demand to satisfy the HVAC of the building.

4.2 Exergy Efficiency

Exergy analysis is used to identify sources of possible irreversibility losses both internal and external to the system. The exergy of electricity, cooling, and heating is defined as:

$$EX_e = E \tag{23}$$

$$EX_c = \left(\frac{T_o}{T_c} - 1\right) Q_c \tag{24}$$

$$EX_h = \left(1 - \frac{T_o}{T_h}\right) Q_h \tag{25}$$

respectively, where T_o is the ambient temperature; T_c and T_h are the cold water and hot water temperatures, respectively. The exergy of fuel is given as:

$$EX_f = \gamma_f \, V_f \, HHV_f = \gamma_f \, V_f \, R_f \, LHV_f \tag{26}$$

where γ_f denotes the exergy grade function for the fuel, defined as the ratio of fuel chemical exergy to the fuel higher heating value HHV_f and V_f is the gas consumption. The LHV_f is the low heating value of gas and R_f is the ratio of HHV_f to LHV_f. For natural gas, the product of γ_f and R_f is 1.03 [18]. Based on these definitions of component exergies, the exergy efficiency of the CCHP system is defined as:

$$\zeta_{ex} = \frac{EX_e + EX_c + EX_h}{EX_f} \tag{27}$$

4.3 Emission Reduction

The amount of CO_2 emission (CO_2E) from a system can be determined using the emission conversion factor of the fuel combusted locally or that related to the grid purchase:

$$CO_2E = \mu_{CO_2,g}\, F + \mu_{CO_2,e}\, E_{grid} \tag{28}$$

where $\mu_{CO_2,\, g}$ and $\mu_{CO_2,\, e}$ are the CO_2 emission conversion factors for the local fuel and the electricity from the grid (E_{grid}), respectively. In comparison to the SP system, the CO_2 emission reduction (ER) by using CCHP is defined as:

$$ER = \frac{CO_2E^{SP} - CO_2E}{CO_2E^{SP}} = 1 - \frac{CO_2E}{CO_2E^{SP}} \tag{29}$$

where the ER factor shows the environmental benefits achieved by using the CCHP system compared to an equivalent SP system [18].

4.4 Annual Total Cost Saving

In this KPI, the economic performance of CCHP system is compared to a reference SP system. The annual total cost is a sum of the capital cost of the equipment (C_e) and the energy charge (C_m). The equations are defined as [7]:

$$ATCS = C_e + C_m \tag{30}$$

$$ATCS = \frac{ATC^{SP} - ATC}{ATC^{SP}} = 1 - \frac{ATC}{ATC^{SP}} \tag{31}$$

Several studies have combined different cycles and/or heat inputs aiming to improve the exergy and energy efficiencies by reducing energy losses and subsequently increasing the fuel exergy and $ATCS$ ratio. In two-part study, Balli et al. [19,20] investigated thermodynamic and thermoeconomic methodologies for a CCHP system with a gas-diesel engine. They applied the developed methodologies to an actual CCHP system with a rated output of 6.5-MW gas-diesel engine installed in the Eskisehir industry estate, Turkey. Abusoglu and Kanoglu [14] pointed to several thermodynamic KPIs, namely, energy and exergy efficiencies, equivalent electrical efficiency, fuel energy saving ratio, fuel exergy saving ratio, and others were determined for the desired system. Fang et al. [21] indicated that electricity-to-thermal-energy ratio is an important factor in measuring the performance of a CCHP. They proposed a complementary CCHP based on ORC system and optimized the operation strategy in order to solve the randomness of load demands. Mago and Luck [22] conducted an analysis to determine the economic, energetic, and environmental potential benefits that can be obtained from the implementation of a combined microturbine ORC versus a simple microturbine or a topping-cycle CHP system. Ghaebi et al. [23] considered the cost rate of product of a gas turbine CCHP system as an objective function and applied genetic algorithm technique to find the optimum operating of system. Guo et al. [24] presented a two-stage optimal planning and design method for CCHP microgrid system. NSGA-II method was applied to solve the optimal design problem including the optimization of equipment type and capacity and mixed-integer linear programming algorithm was used to solve the optimal dispatch problem. In most of these studies, it is explicitly agreed that renewable energy integration can significantly enhance one or more of these KPIs. The wide adaptability and scalability of renewable energy sources was lately associated with a remarkable declining trend in production and installation cost. In addition, a variety of renewable energy and storage technologies are now commercially available that can be perfectly hybridized with different sizes of CCHP systems. The following section highlights the main renewable integration potentials and reviews the most advanced strategies and case studies in the literature.

5 RENEWABLE INTEGRATION

As seen in Eq. (21), reducing the fuel consumption (F) in a CCHP system will proportionally increase the PES and subsequently ER factor. As mentioned earlier, 46% of the total worldwide energy demand is attributed to HVAC applications in buildings [6]. Renewable energy resources are sustainable alternatives to fossil fuels in driving traditional CHP/CCHP systems [25]. Depending on the renewable energy applications, two methods of energy harvesting and integration are possible: either electric energy harvesting or thermal energy harvesting. Although there are few renewable energy applications that can harvest both forms of energy (i.e., solar photovoltaic thermal (PVT) collectors), very limited work has addressed the use of PVT collectors particularly in integration with small-scale CHP systems [26–28]. Alternatively, most of the CHP/CCHP systems are integrated either electrically or thermally with renewable energy resources.

5.1 Electric Energy Harvest

Decentralized power generation systems have the flexibility of diversifying power production sources combined with various energy storage systems, which bring enormous resilience in matching the power supply and demand at low cost. This option has been progressively deployed worldwide, particularly in islands and remote areas where the accessibility to the national grid is often unfeasible or expensive. The renewable sources that match this type of distributed power generation systems would be solar PV fields, wind farms, or a combination of them. The integration model for these renewable applications with CCHP systems is most likely to follow the EDM strategy where the priority is in maintaining the electrical load of the building/entity (Fig. 6). The DC electric power generated by solar PV, wind turbine, or a combination of them is converted to AC power by an inverter. Beforehand, the surplus/deficit power is regulated by using a set of batteries. The primary objective of the renewable energy sources + energy storage + the power generated by the local PGU is to maintain the electric load and also run an

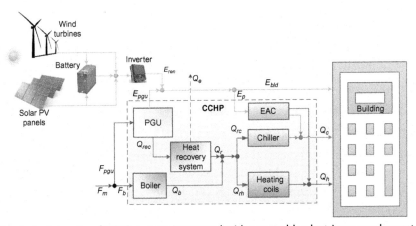

Fig. 6 A schematic of CCHP system integrated with renewable electric energy harvest applications.

electric air conditioner to balance and adjust the cooling/heating load for the building. The exhaust heat recovered from the PGU is used in the adjacent chiller or heating coils, accordingly (Fig. 6).

One of the key challenges of renewable integration particularly for EDM models is the predictability of power supply at least for short terms. Because of the intermittent nature of renewable energy output, many studies have focused on developing accurate forecasting models for the PV and wind power output. Solar irradiance, wind speed, humidity, and ambient temperature will not only affect the power output, but will also influence the power demand particularly for HVAC systems. This is a very important practice especially for EDM strategy. For a day ahead EDM, the power output from PV and/or wind power systems is often modeled by two main methods. The first and most common method is named the deterministic model, which forecasts the power output over a certain period, while the other method is called the probabilistic model, which considers the random nature of the PV and wind power sources, by using chance-constrained programming or an expectancy model [29]. The model predictability is often based on two stages. In the first stage, the resource availability (i.e., solar irradiance) is calculated based on various time scales using complex algorithms such as artificial neural networks (ANNs) [30,31], autoregression [32], or hybrid methods [33]. The deterministic model is used in a number of studies [34,35] based on the Hottel-Liu-Jordan formula to calculate the solar radiation, for which the latitude and longitude along with the typical meteorological year data are required. In the second stage, the forecasted solar irradiance and ambient temperature data are used as inputs for commercial PV simulation software, such as Trnsys, SAM, and Homer Energy. Unlike two-stage models, a single-stage approach predicts the PV power directly based on prior information or some readily accessible data [36,37]. For wind power, the power output is directly related to wind speed, which is calculated by a number of methods such as ANNs [38,39], Kalman filtering [40,41], and time series [42]. Similarly, these forecasting models can also be applicable to some extent in the thermal energy harvest from some of renewable sources, i.e., concentrated solar power technologies, but are not going to be elaborated further in next section.

5.2 Thermal Energy Harvest

In thermal energy harvest from renewable energy sources, the priority of energy harvest is to satisfy thermal energy demand for the CHP/CCHP systems and, therefore, it is most likely to follow the TDM strategy. Among the compatible renewable energy resources, biomass, geothermal, and solar thermal applications are the most attractive options for CHP/CCHP integration.

5.2.1 Biomass Integration

Several studies have summarized the state-of-the-art technologies for biomass-fueled CHP/CCHP systems in near-future perspectives [25, 43–45]. The focus of these studies was on conversion methods for biomass to fuel CHP/CCHP systems. The primary conversion methods include combustion, gasification, pyrolysis, biochemical, and chemical processes [25]. Based on the target temperatures and heat quality, the working fluid such as hot water, steam, or other gaseous/liquid fluids is determined in the secondary conversion technologies (e.g., steam turbine [46], organic Rankine cycle (ORC) [47], gas turbine [46], internal combustion engine (ICE) [48,49], among others). The main objective is optimizing power, heat, and cooling production to match the various loads of a specific building/entity. Different integration methods of biomass conversion and utilization in CHP/CCHP systems have been proposed and investigated. Mertzis et al. [48] studied the performance of a microbiomass CHP system and reported the effect of different types of biomass feedstock, gasification parameters, and engine intake mixtures on long-term operation and energy combination output. Fryda et al. [50] investigated three different microbiomass CHP configurations integrated with solid oxide fuel cells (SOFC) and microgas turbine (MGT), and they presented an exergy analysis to optimize these configurations. Bang-Møller et al. [51] analyzed the exergy destruction of a biomass steam gasification, SOFC, and MGT hybrid plant and optimized its configuration. Huang et al. [52] compared the technoeconomic performances of biomass gasification and ORC-based CHP systems, and they emphasized that both systems are economically unfeasible in generating the required power to satisfy the aimed loads. Al-Sulaiman et al. [53] reported that the energy and exergy analyses of a biomass CCHP system using an ORC and concluded that there is a significant improvement when CCHP is used in EDM scenario. Biomass combustion was the most common and straightforward utilization methods in CHP/CCHP systems. Ahmadi et al. [54,55] designed a biomass CCHP system integrated with ORC and proposed a thermoeconomic multiobjective optimization model to improve energy, exergy, and environmental impact. Lian et al. [56] evaluated the thermoeconomic potentials of four different biomass-driven

CCHP configurations and concluded that exergy destruction is the most extensive in the furnace, counting to nearly 60% of net exergy loss. Compared with biomass combustion, gasification is more efficient at producing electric power and heat. Wang et al. [57] presented a unique study in integrating biomass gasification process with a CCHP system and also a heat pipe heat exchanger to recover the waste heat from the high-temperature product gas and optimize energy utilizations corresponding to different operation/load conditions. Fig. 7 illustrates a simple integration between biomass gasification and CCHP system to satisfy the power and HVAC loads for a building. Puig-Arnavat et al. [58] developed a simple but rigorous CCHP model for designing, optimizing, and simulating small/medium-scale plants, including a realistic biomass gasification model, and achieved approximately 10% PES. Huang et al. [59] modeled a small-scale CCHP system integrated by a biomass downdraft gasifier, an internal combustion engine, and an absorption chiller. They concluded that this system would be beneficial to buildings that often demand an appropriate load characteristic.

5.2.2 Geothermal Integration
Geothermal energy can be withdrawn from the hot water circulating among the rocks below the Earth's surface, or by pumping cold water into the hot rocks and returning the heated water back to the surface. Unlike intermittent renewable energy sources, geothermal energy can be utilized almost continuously to provide a stable baseload energy for a particular building/district or an industrial plant. Applying the idea of integrating geothermal energy with the concept of CHP/CCHP has been addressed in a number of studies. The key objective of geothermal integrated CCHP systems is the diversification of energy sources, particularly in those promising geographical locations [60]. This system is widely studied and optimized to demonstrate thermodynamic and thermoeconomic benefits. Esen and co-workers [61] assessed the performance of a horizontal ground source heat pump using R22 as a working fluid and compared its cost effectiveness with conventional heating methods. They compared a ground-coupled heat pump with an air-coupled heat pump system from the thermoeconomic viewpoint and concluded that ground-coupled heat pump systems were economically preferable to air-coupled heat pump systems for the purpose of space cooling [62]. They also performed energetic and exergetic analyses on the ground-coupled heat pump system with two different horizontal ground heat pumps and assessed the influences of the buried depth of the heat exchanger on the energetic and exergetic efficiencies [63]. In another

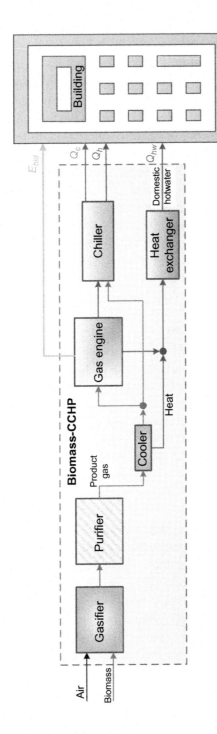

Fig. 7 A schematic of biomass gasification integrated with a CCHP system.

study, Esen and Yuksel hybridized geothermal with solar energy to heat up a greenhouse in eastern Turkey [64]. More attention is paid to geothermal integrated CCHP systems due to their renowned environmental and economic sustainability [9]. Zhu et al. [65] reviewed geothermal energy potential in China and considered its utilization, development roadmap, and legislative support. Yazici [66] performed energy and exergy analyses for a renovated Afyon geothermal heating system in Turkey, which uses geothermal water as the energy source. The author presented a detailed energy and exergy flow diagrams for the system. Kanoglu et al. [67] coupled ammonia-water absorption refrigeration system with Claude cycle for liquefaction of hydrogen. They used geothermal energy as the heat source in the generator of the absorption cycle and carried out energy and exergy analyses on the system. Using the geothermal energy, they reduced the required work in the cycle by 24%. A number of other research teams [68–74] have studied different fluids in Kalina cycle driven by geothermal heat and/or solar heat. In most of these studies, it was concluded that Kalina cycle is better for power generation from geothermal energy as its exergy efficiency was higher than the ORC cycle. Some other research teams [75–77] studied the integration of geothermal energy with Goswami cycle [78], a novel thermodynamic cycle that was proposed by Yogi Goswami in 1998. The concept of geothermal power integration with CHP/CCHP systems has already being demonstrated in a number of European countries. However, the scale of economy of these integrations is more feasible for large-scale projects, i.e., district CHP systems.

5.2.3 Solar Thermal Integration

Solar thermal collectors are probably the most common renewable energy technology that can be integrated with CHP/CCHP applications. This is because of their high scalability, affordability, and flexibility for various geographical locations. The collector type and quantity is usually determined by the required thermal load and the outlet temperature. Sizing a solar thermal field is highly flexible with/without thermal energy storage (TES) and merely related to the project technoeconomic feasibility. There are two main categories of solar collectors: stationary and tracking collectors. Tracking collectors are also subcategorized into single-axis tracking and two-axes tracking, which are often used for larger-scale projects that require a relatively greater heat input [79]. Fig. 8 presents a categorization of commercially available solar thermal collectors based on their sun-tracking motion. Historically, the utilization of solar thermal energy began with

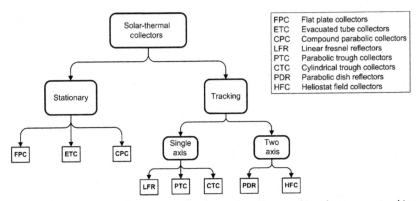

Fig. 8 Categorization of various solar-thermal collectors based on sun tracking motion [83].

the concept of solar heating and later expanded to solar cooling and refrigeration. With the increasing emphasis on decentralized applications such as CHP/CCHP, solar thermal technologies were in the spotlight of many process hybridization/integration studies. The main challenge for the use of solar energy in the CHP/CCHP is the intermittence and time lag between the solar irradiance and the peak heat demand [80]. Hence, this technology needs to be coupled with a proper design of TES system, being often a vessel/tank filled with a high thermal density fluid such as water or a phase change material (PCM) [81,82].

Among those studies, Wang et al. [84] proposed a novel solar CCHP based on a Rankine cycle and an ejector refrigeration cycle. Zhai et al. [85] proposed and extensively investigated a hybrid solar CCHP system integrated with parabolic trough collectors (PTCs). Annual energy and exergy efficiencies of the system were evaluated under the climate of North Western region of China. Meng et al. [86] presented a new CCHP system driven by a combination of solar thermal energy and industrial waste heat. Two pairs of metal hydrides were selected and the working principle of the system was discussed for the proposed system. Wang et al. [87] proposed a new solar-driven CCHP system combining a Brayton cycle and a transcritical CO_2 refrigeration cycle with ejector-expansion device. A parametric study was carried out to assess the effect of major parameters on the exergy efficiency. In two-part study, Al-Sulaiman et al. [88,89] examined the thermoeconomic optimization formulations of three new ORC-CCHP systems, i.e., SOFC-CCHP, biomass-CCHP, and solar-CCHP systems. Results showed the better performance for SOFC-CCHP from the exergy viewpoint while the cost per exergy unit for solar-CCHP was obtained less than that for other systems. Sanaye and Hajabdollahi [90] modeled and optimized a solar-assisted CCHP plant. Both the genetic algorithm and particle swarm optimization were used to maximize the actual annual benefit. Boyaghchi and Heidarnejad [91,92] proposed and optimized a new CCHP system integrated with evacuated tube collectors (TECs) based on ORC for both summer and winter seasons. Thermal and exergy efficiencies as well as total product cost rate were selected as the objective functions. In another study, Boyaghchi and Chavoshi [93] presented a combined solar-geothermal heat input for a proposed CCHP system and performed a comprehensive thermodynamic, economic, and environmental analyses. In solar part, they used flat plate collectors (FPTs) circulating water/copper oxide (CuO) nanofluid as the heat transfer fluid (HTF) and compared four different working fluids from the exergy, exergoeconomic, and exergoenviromental points of view. Earlier, Faninger [80] proposed a solar-biomass combination to overcome the mismatch between solar energy supply and energy demand. Wang et al. [94] thermodynamically optimized a solar-driven CCHP system for three modes, i.e., power, cooling + power, and heating + power. The average useful output and the total heat transfer area were selected as the two objective functions in the functionality of NSGA-II technique to find the optimal operation of that system. Ezzat and Dincer [95] combined geothermal and solar energy as heat source for a multigeneration system, which produces five output commodities, including refrigeration, heated air, hot water, dried food, and electricity. They used energy and exergy analyses to calculate system efficiency. Mehrpooya and co-authors [96] also studied the performance of a hybrid solar collector-geothermal heat pump system. This system was used to provide the required heat load for a greenhouse. Ghasemi et al. [97] optimized power production in a geothermal ORC cycle and then coupled the plant with a solar PTC. They showed that produced power by the proposed system increases by 5% compared to optimized geothermal system. Ayub et al. [98] further studied the proposed system by Ghasemi et al. by incorporating economic aspect. They calculated levelized cost of electricity (*LCOE*) for the system. It was shown that the *LCOE* of the new cycle is 2% lower than the standalone geothermal system. Since the integration of most renewable energy sources has shown greater technoeconomic feasibility in larger-scale projects, the research has rapidly moved forward for a more comprehensive approach. Integrating more than one renewable energy source and optimizing the diversity of various thermal and electric loads of end-users have shown enormous sweet-spots.

However, the optimization of supply-demand dynamics would require a sophisticated control system that is based on consistent and reliable communication between various stakeholders, which leads to the introduction of broader term named "CCHP Microgrid."

6 CCHP MICROGRIDS

From the technical, economical, and environmental viewpoints, many complications may affect the performance of a CCHP for a wider community or large buildings/districts. Solving these complications should lead to an efficient energy distribution system providing major opportunities for those liberated/decentralized networks. The development of CCHP microgrid utilizing distributed cogeneration equipment and renewable energy sources has recently caught considerable attention in the research community. Compared with conventional CCHP systems, the CCHP microgrids show a novel and greater functionality. In this context, CCHP microgrid not only satisfies the cooling, heating and power demands for a particular consumption profile (i.e., residential buildings, schools, department stores, and industrial plants), but also interacts with the main grid to provide reserve, peak shaving, demand response services, and improved capability for integration with renewable energy sources. A typical CCHP microgrid consists of four main elements: the prime mover, the storage system, the load, and the renewable energy sources. For technoeconomic purposes, CCHP microgrids can be categorized into two groups: (i) large-scale CCHP applications that can serve for large loads, or (ii) distributed CCHP units with relatively smaller capacity and advanced control strategies designed to meet diverse demands in the commercial, residential, institutional, and small industrial sectors [99–101]. The advantage of using distributed CCHP units is in custom-designed bridge between supply and demand, which highly contributes in an improved energy efficiency and ER. This provides a flexible option in using new technologies of prime movers such as fuel cells and microturbines that pose much lower exhaust emissions than the traditional technologies used in centralized large-scale CCHP. Therefore, CCHP microgrid structures vary from site to site, with diverse prime movers, cooling options, rated size ranges, load characteristics, and operational objectives. Fig. 9 illustrates an advanced CCHP microgrid that consists of a renewable energy integration, energy storage, and a prime

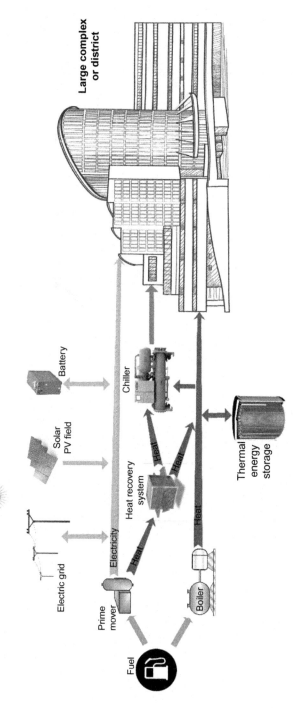

Fig. 9 Illustration of CCHP microgrid and its key components custom-designed for large buildings or residential complexes.

mover. The cooling and heating demands are satisfied by an absorption chiller and a thermal boiler, respectively. By utilizing the waste heat from the prime mover, heat storage is used to store the excess heat. A natural-gas-fueled boiler is used as an auxiliary heating equipment when the waste heat is insufficient. The proportion of renewable energy integration (e.g., solar PV systems) in the CCHP microgrid is scalable and is often oversized to respond for afterhours demands (i.e., night loads). The electric energy storage systems (batteries) are therefore used to mitigate fluctuations caused by renewable energy intermittence and also the loads.

In summary, CCHP microgrids feature the following advantages over traditional CCHP systems [29]: (1) better management of the volatility in the cooling/heating loads and the intermittent nature of renewable energy; (2) better flexibility of the prime movers at different operational modes and under different operation conditions resulting in improved power and heating efficiencies, which varies with the capacity and the load rate; (3) the waste heat can be utilized not only for heating, but also for cooling, resulting in enhanced integration between the three types of energy in the system; (4) the opportunity for automated and rapid response to load variation with the lowest time lags; and (5) the potential of technoeconomic optimization in complex energy balances at a variety of scheduling strategies. For example, waste heat from the prime mover has the priority in meeting heating demands vs cooling demands; or the power production in the prime mover can be assigned based on the different operating modes, etc.

One way of solving complex distribution system problems is in redesigning the distribution system to include the integration of high levels of distributed energy resources, using CCHP microgrid concept. The basic objectives of a CCHP microgrid are to improve local network reliability and promote high penetration of renewable sources, dynamic supply/demand interconnection, and improved efficiencies via waste heat reuse. Managing significant levels of distributed energy resources with a wide and dynamic set of resources and control points may show overwhelming results. The best way to manage such a complex system is to break the distribution into small clusters of microgrids, aiming for a distributed optimizing controls coordinating multimicrogrids [102]. The concept of "intelligent energy networks" (IENs) has recently been proposed to represent an intelligent management system for a complete set of energy sources, including electricity, heat, hydrogen, biofuels, and non-biofuels [103]. The CCHP microgrid is set to play a vital role in this type of energy synthesis system. In order to effectively incorporate all the components and ensure optimal operation, energy management systems (EMS)

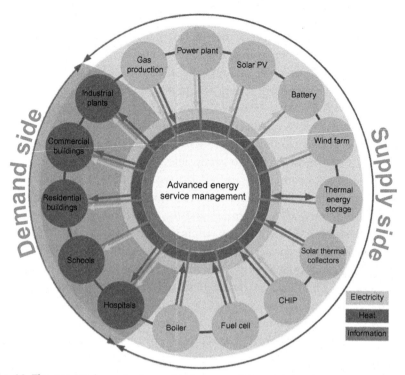

Fig. 10 The potential supply-demand interconnectivity and the electric/heat traffic in this advanced energy management system.

are used, which utilizes optimization algorithms and real-time control schemes for scheduling the CCHP microgrids. Fig. 10 shows a complete set of different power/heat suppliers and consumers, where a large amount of communication data is exchanged and analyzed in a central data-processing center to compute the optimum technoeconomic interaction between the supply and demand sides.

7 CONCLUSIONS

The deficits of bulk movement of electric power from traditional centralized power plants are now well acknowledged and the concept of utilizing the renewable-integrated decentralizing CHP/CCHP systems is progressively deployed in many locations worldwide. Advanced CHP/CCHP systems can provide a reliable power and heat for HVAC systems at different scales and at much higher exergy and thermal efficiencies. They also play a vital role in energy supply security, emission control, peak shaving, reducing the electric market turbulences, and in emergency shutdowns. In this work, the persuading segments of energy conversion were discussed and the boosting role of a typical CCHP on PES, exergy efficiency, ER, and annual total energy saving was highlighted. The renewable energy integration with CCHP systems was categorized based on their primary service as in electric harvest or thermal harvest domains, which may support the EDM or TDM schemes, respectively. For EDM scheme, solar PV and wind turbines are the most accessible renewable applications, While for TDM scheme, biomass, geothermal, and solar thermal application are considered the most appropriate applications. The literature was intensively probed for the promising integration potentials and few of hybrid renewable integrations were also highlighted. It was emphasized that due to high fluctuation in supply–demand dynamics, large-scale projects would benefit from diversifying energy supply options along with the randomness of the lifestyles of the end-users. This tangible advantage would require sophisticated optimization techniques based on consistent and reliable communication between various stakeholders, which leads to the introduction of a more advanced concept "CCHP Microgrid." CCHP microgrids feature many advantages over the individual CCHP systems in better volatility management, flexibility operation, waste heat utilization, automated response, and technoeconomic feasibility.

REFERENCES

[1] El-Khattam W, Salama MM. Distributed generation technologies, definitions and benefits. Electr Pow Syst Res 2004;71(2):119–28.

[2] Wang H-C, et al. Techno-economic analysis of a coal-fired CHP based combined heating system with gas-fired boilers for peak load compensation. Energy Policy 2011;39(12):7950–62.

[3] Oudalov A, Cherkaoui R, Beguin A. Sizing and optimal operation of battery energy storage system for peak shaving application. In: Power tech, 2007 IEEE Lausanne. IEEE; 2007.

[4] Furaiji F, Łatuszyńska M, Wawrzyniak A. An empirical study of the factors influencing consumer behaviour in the electric appliances market. Contemp Econ 2012;6 (3):76–86

[5] Khalilpour KR, Vassallo A. Community energy networks with storage: modeling frameworks for distributed generation. Singapore: Springer; 2016.

[6] Olsthoorn D, Haghighat F, Mirzaei PA. Integration of storage and renewable energy into district heating systems: a review of modelling and optimization. Sol Energy 2016;136:49–64.

[7] Cho H, Smith AD, Mago P. Combined cooling, heating and power: a review of performance improvement and optimization. Appl Energy 2014;136:168–85.

[8] Wiltshire R. Advanced district heating and cooling (DHC) systems. Cambridge, UK: Woodhead Publishing; 2015.

[9] Wu D, Wang R. Combined cooling, heating and power: a review. Prog Energy Combust Sci 2006;32(5):459–95.

[10] Habib M, Zubair S. Second-law-based thermodynamic analysis of regenerative-reheat Rankine-cycle power plants. Energy 1992;17(3):295–301.

[11] Ahmadi MH, Mohammadi AH, Dehghani S. Evaluation of the maximized power of a regenerative endoreversible Stirling cycle using the thermodynamic analysis. Energ Convers Manage 2013;76:561–70.

[12] Ying Y, Hu EJ. Thermodynamic advantages of using solar energy in the regenerative Rankine power plant. Appl Therm Eng 1999;19(11):1173–80.

[13] Kehlhofer R, et al. Combined-cycle gas & steam turbine power plants. Oklahoma, USA: Pennwell Books; 2009.

[14] Abusoglu A, Kanoglu M. Exergoeconomic analysis and optimization of combined heat and power production: a review. Renew Sustain Energy Rev 2009;13(9):2295–308.

[15] Vélez F, et al. A technical, economical and market review of organic Rankine cycles for the conversion of low-grade heat for power generation. Renew Sustain Energy Rev 2012;16(6):4175–89.

[16] Deng J, Wang R, Han G. A review of thermally activated cooling technologies for combined cooling, heating and power systems. Prog Energy Combust Sci 2011;37 (2):172–203.

[17] Horlock J, Heat C-C. Power (CHP). New York: Pergamon Press; 1987.

[18] Wang J-J, et al. Performance comparison of combined cooling heating and power system in different operation modes. Appl Energy 2011;88(12):4621–31.

[19] Balli O, Aras H, Hepbasli A. Thermodynamic and thermoeconomic analyses of a trigeneration (TRIGEN) system with a gas–diesel engine: part I—methodology. Energ Convers Manage 2010;51(11):2252–9.

[20] Balli O, Aras H, Hepbasli A. Thermodynamic and thermoeconomic analyses of a trigeneration (TRIGEN) system with a gas–diesel engine: part II—an application. Energ Convers Manage 2010;51(11):2260–71.

[21] Fang F, et al. Complementary configuration and operation of a CCHP-ORC system. Energy 2012;46(1):211–20.

[22] Mago PJ, Luck R. Evaluation of the potential use of a combined micro-turbine organic Rankine cycle for different geographic locations. Appl Energy 2013;102:1324–33.

[23] Ghaebi H, Saidi M, Ahmadi P. Exergoeconomic optimization of a trigeneration system for heating, cooling and power production purpose based on TRR method and using evolutionary algorithm. Appl Therm Eng 2012;36:113–25.

[24] Guo L, et al. A two-stage optimal planning and design method for combined cooling, heat and power microgrid system. Energ Convers Manage 2013;74:433–45.

[25] Dong L, Liu H, Riffat S. Development of small-scale and micro-scale biomass-fuelled CHP systems—a literature review. Appl Therm Eng 2009;29(11):2119–26.

[26] Tonekaboni N, et al. Energy and exergy economic analysis of cogeneration cycle of homemade CCHP with PVT collector. Can J Basic Appl Sci 2015;03(08):224–33.

[27] Zheng C, Yang G. Integration of CCHP with renewable energy. In: Wang R, Zhai X, editors. Handbook of energy systems in green buildings. Berlin, Germany: Springer; 2018. p. 1449–84.

[28] Milani D, et al. Experimentally validated model for atmospheric water generation using a solar assisted desiccant dehumidification system. Energ Buildings 2014;77:236–46.

[29] Gu W, et al. Modeling, planning and optimal energy management of combined cooling, heating and power microgrid: a review. Int J Electr Power Energy Syst 2014;54:26–37.

[30] Yona A, et al. In: Application of neural network to one-day-ahead 24 hours generating power forecasting for photovoltaic system. International conference on intelligent systems applications to power systems, 2007 (ISAP 2007)IEEE; 2007.

[31] Mellit A, et al. A simplified model for generating sequences of global solar radiation data for isolated sites: using artificial neural network and a library of Markov transition matrices approach. Sol Energy 2005;79(5):469–82.

[32] Bacher P, Madsen H, Nielsen HA. Online short-term solar power forecasting. Sol Energy 2009;83(10):1772–83.

[33] Cao JC, Cao S. Study of forecasting solar irradiance using neural networks with pre-processing sample data by wavelet analysis. Energy 2006;31(15):3435–45.

[34] Rahman M, Nakamura K, Yamashiro S. In: A grid-connected PV-ECS system with load leveling function taking into account solar energy estimation. Proceedings of the 2004 IEEE International conference on electric utility deregulation, restructuring and power technologies, 2004 (DRPT 2004). IEEE; 2004.

[35] Rahman MH, Yamashiro S. Novel distributed power generating system of PV-ECaSS using solar energy estimation. IEEE Trans Energy Conver 2007;22(2):358–67.

[36] Chen C, et al. Online 24-h solar power forecasting based on weather type classification using artificial neural network. Sol Energy 2011;85(11):2856–70.

[37] Shi J, et al. Forecasting power output of photovoltaic systems based on weather classification and support vector machines. IEEE Trans Ind Appl 2012;48(3):1064–9.

[38] Varshney K, Poddar K. Prediction of wind properties in urban environments using artificial neural network. Theor Appl Climatol 2012;107(3–4):579–90.

[39] Hossain R, Ooa AMT, Alia AS. Historical weather data supported hybrid renewable energy forecasting using artificial neural network (ANN). Energy Procedia 2012;14:1035–40.

[40] Louka P, et al. Improvements in wind speed forecasts for wind power prediction purposes using Kalman filtering. J Wind Eng Ind Aerodyn 2008;96(12):2348–62.

[41] Wei Z, Weimin W. In: Wind speed forecasting via ensemble Kalman filter. 2nd International conference on advanced computer control (ICACC)IEEE; 2010.

[42] An X, et al. Wind farm power prediction based on wavelet decomposition and chaotic time series. Exp Syst Appl 2011;38(9):11280–5.

[43] Maraver D, et al. Assessment of CCHP systems based on biomass combustion for small-scale applications through a review of the technology and analysis of energy efficiency parameters. Appl Energy 2013;102:1303–13.

[44] Raj NT, Iniyan S, Goic R. A review of renewable energy based cogeneration technologies. Renew Sustain Energy Rev 2011;15(8):3640–8.

[45] Ahrenfeldt J, et al. Biomass gasification cogeneration—a review of state of the art technology and near future perspectives. Appl Therm Eng 2013;50(2):1407–17.

[46] Bagdanavicius A, Jenkins N, Hammond GP. Assessment of community energy supply systems using energy, exergy and exergoeconomic analysis. Energy 2012;45(1):247–55.

[47] Taljan G, et al. Optimal sizing of biomass-fired organic Rankine cycle CHP system with heat storage. Renew Energy 2012;41:29–38.

[48] Mertzis D, et al. Performance analysis of a small-scale combined heat and power system using agricultural biomass residues: the SMARt-CHP demonstration project. Energy 2014;64:367–74.

[49] Hossain A, Davies P. Pyrolysis liquids and gases as alternative fuels in internal combustion engines—a review. Renew Sustain Energy Rev 2013;21:165–89.

[50] Fryda L, Panopoulos K, Kakaras E. Integrated CHP with autothermal biomass gasification and SOFC–MGT. Energ Convers Manage 2008;49(2):281–90.

[51] Bang-Møller C, Rokni M, Elmegaard B. Exergy analysis and optimization of a biomass gasification, solid oxide fuel cell and micro gas turbine hybrid system. Energy 2011;36(8):4740–52.

[52] Huang Y, et al. Comparative techno-economic analysis of biomass fuelled combined heat and power for commercial buildings. Appl Energy 2013;112:518–25.

[53] Al-Sulaiman FA, Dincer I, Hamdullahpur F. Energy and exergy analyses of a biomass trigeneration system using an organic Rankine cycle. Energy 2012;45(1):975–85.

[54] Ahmadi P, Dincer I, Rosen MA. Development and assessment of an integrated biomass-based multi-generation energy system. Energy 2013;56:155–66.

[55] Ahmadi P, Dincer I, Rosen MA. Thermoeconomic multi-objective optimization of a novel biomass-based integrated energy system. Energy 2014;68:958–70.

[56] Lian Z, Chua K, Chou S. A thermoeconomic analysis of biomass energy for trigeneration. Appl Energy 2010;87(1):84–95.

[57] Wang J-J, et al. Energy and exergy analyses of an integrated CCHP system with biomass air gasification. Appl Energy 2015;142:317–27.

[58] Puig-Arnavat M, Bruno JC, Coronas A. Modeling of trigeneration configurations based on biomass gasification and comparison of performance. Appl Energy 2014;114:845–56.

[59] Huang Y, et al. Biomass fuelled trigeneration system in selected buildings. Energ Convers Manage 2011;52(6):2448–54.

[60] Frunzulica R, Damian A. Small-scale cogeneration-a viable alternative for Romania, In: 4th IASME/WSEAS International Conference on Energy, Environment, Ecosystems and Sustainable Development; 2008 Algarve, Portugal.

[61] Esen H, Inalli M, Esen M. Technoeconomic appraisal of a ground source heat pump system for a heating season in eastern Turkey. Energ Convers Manage 2006;47 (9):1281–97.

[62] Esen H, Inalli M, Esen M. A techno-economic comparison of ground-coupled and air-coupled heat pump system for space cooling. Build Environ 2007;42(5):1955–65.

[63] Esen H, et al. Energy and exergy analysis of a ground-coupled heat pump system with two horizontal ground heat exchangers. Build Environ 2007;42(10):3606–15.

[64] Esen M, Yuksel T. Experimental evaluation of using various renewable energy sources for heating a greenhouse. Energ Buildings 2013;65:340–51.

[65] Zhu J, et al. A review of geothermal energy resources, development, and applications in China: current status and prospects. Energy 2015;93:466–83.

[66] Yazici H. Energy and exergy based evaluation of the renovated Afyon geothermal district heating system. Energ Buildings 2016;127:794–804.

[67] Kanoglu M, Yilmaz C, Abusoglu A. Geothermal energy use in absorption precooling for Claude hydrogen liquefaction cycle. Int J Hydrogen Energy 2016;41 (26):11185–200.

[68] Ashouri M, et al. Techno-economic assessment of a Kalina cycle driven by a parabolic trough solar collector. Energ Convers Manage 2015;105:1328–39.

[69] Modi A, et al. Thermoeconomic optimization of a Kalina cycle for a central receiver concentrating solar power plant. Energ Convers Manage 2016;115:276–87.

[70] Shokati N, Ranjbar F, Yari M. Exergoeconomic analysis and optimization of basic, dual-pressure and dual-fluid ORCs and Kalina geothermal power plants: a comparative study. Renew Energy 2015;83:527–42.

[71] Yari M, et al. Exergoeconomic comparison of TLC (trilateral Rankine cycle), ORC (organic Rankine cycle) and Kalina cycle using a low grade heat source. Energy 2015;83:712–22.

[72] Vatani A, Mehrpooya M, Tirandazi B. A novel process configuration for co-production of NGL and LNG with low energy requirement. Chem Eng Process Process Intensif 2013;63:16–24.

[73] Zare V. A comparative thermodynamic analysis of two tri-generation systems utilizing low-grade geothermal energy. Energ Convers Manage 2016;118:264–74.

[74] Mohammadi A, Mehrpooya M. Energy and exergy analyses of a combined desalination and CCHP system driven by geothermal energy. Appl Therm Eng 2017;116:685–94.

[75] Tamm G, et al. Theoretical and experimental investigation of an ammonia–water power and refrigeration thermodynamic cycle. Sol Energy 2004;76(1):217–28.

[76] Tamm G, et al. Novel combined power and cooling thermodynamic cycle for low temperature heat sources, part I: theoretical investigation. J Sol Energy Eng 2003;125(2):218–22.

[77] Padilla RV, et al. Analysis of power and cooling cogeneration using ammonia-water mixture. Energy 2010;35(12):4649–57.

[78] Yogi Goswami D. Solar thermal power technology: present status and ideas for the future. Energy Source 1998;20(2):137–45.

[79] Milani D. Modelling framework of solar assisted dehumidification system to generate freshwater from "thin air" Sydney, Australia: The University of Sydney; 2012. 274.

[80] Faninger G. Combined solar–biomass district heating in Austria. Sol Energy 2000;69(6):425–35.

[81] Zalba B, et al. Review on thermal energy storage with phase change: materials, heat transfer analysis and applications. Appl Therm Eng 2003;23(3):251–83.

[82] Luu MT, et al. Dynamic modelling and analysis of a novel latent heat battery in tankless domestic solar water heating. Energ Buildings 2017;152:227–42.

[83] Khalilpour R, et al. A novel process for direct solvent regeneration via solar thermal energy for carbon capture. Renew Energy 2017;104:60–75.

[84] Wang J, et al. A new combined cooling, heating and power system driven by solar energy. Renew Energy 2009;34(12):2780–8.

[85] Zhai H, et al. Energy and exergy analyses on a novel hybrid solar heating, cooling and power generation system for remote areas. Appl Energy 2009;86(9):1395–404.

[86] Meng X, et al. Theoretical study of a novel solar trigeneration system based on metal hydrides. Appl Energy 2010;87(6):2050–61.

[87] Wang J, et al. Parametric analysis of a new combined cooling, heating and power system with transcritical CO_2 driven by solar energy. Appl Energy 2012;94:58–64.

[88] Al-Sulaiman FA, Dincer I, Hamdullahpur F. Thermoeconomic optimization of three trigeneration systems using organic Rankine cycles: part I—formulations. Energ Convers Manage 2013;69:199–208.

[89] Al-Sulaiman FA, Dincer I, Hamdullahpur F. Thermoeconomic optimization of three trigeneration systems using organic Rankine cycles: part II—applications. Energ Convers Manage 2013;69:209–16.

[90] Sanaye S, Hajabdollahi H. Thermo-economic optimization of solar CCHP using both genetic and particle swarm algorithms. J Sol Energy Eng 2015;137(1). 011001.

[91] Boyaghchi FA, Heidarnejad P. Thermodynamic analysis and optimisation of a solar combined cooling, heating and power system for a domestic application. Int J Exergy 2015;16(2):139–68.

[92] Boyaghchi FA, Heidarnejad P. Thermoeconomic assessment and multi objective optimization of a solar micro CCHP based on organic Rankine cycle for domestic application. Energ Convers Manage 2015;97:224–34.

[93] Boyaghchi FA, Chavoshi M. Multi-criteria optimization of a micro solar-geothermal CCHP system applying water/CuO nanofluid based on exergy, exergoeconomic and exergoenvironmental concepts. Appl Therm Eng 2017;112:660–75.

[94] Wang M, et al. Multi-objective optimization of a combined cooling, heating and power system driven by solar energy. Energ Convers Manage 2015;89:289–97.

[95] Ezzat M, Dincer I. Energy and exergy analyses of a new geothermal–solar energy based system. Sol Energy 2016;134:95–106.

[96] Mehrpooya M, Hemmatabady H, Ahmadi MH. Optimization of performance of combined solar collector-geothermal heat pump systems to supply thermal load needed for heating greenhouses. Energ Convers Manage 2015;97:382–92.

[97] Ghasemi H, et al. Hybrid solar–geothermal power generation: optimal retrofitting. Appl Energy 2014;131:158–70.

[98] Ayub M, Mitsos A, Ghasemi H. Thermo-economic analysis of a hybrid solar-binary geothermal power plant. Energy 2015;87:326–35.

[99] Tse LKC, et al. Solid oxide fuel cell/gas turbine trigeneration system for marine applications. J Power Sources 2011;196(6):3149–62.

[100] Maeda K, Masumoto K, Hayano A. A study on energy saving in residential PEFC cogeneration systems. J Power Sources 2010;195(12):3779–84.

[101] Angrisani G, et al. Experimental results of a micro-trigeneration installation. Appl Therm Eng 2012;38:78–90.

[102] Lasseter RH. Smart distribution: coupled microgrids. Proc IEEE 2011;99(6):1074–82.

[103] Orecchini F, Santiangeli A. Beyond smart grids—the need of intelligent energy networks for a higher global efficiency through energy vectors integration. Int J Hydrogen Energy 2011;36(13):8126–33.

CHAPTER 15

Design and Operational Management of Energy Hubs: A DS4S (Screening, Selection, Sizing, and Scheduling) Framework

Kaveh Rajab Khalilpour
Faculty of Engineering and Information Technology, Monash University, Melbourne, VIC, Australia

Abstract

The contemporary climate change crisis has motivated the need for decentralization of energy networks. Over recent years, several often closely defined concepts have been introduced for decentralized energy networks. These include "microgrid," "mesogrid," "nanogrid," "energy internet," "community energy network," "social energy network," "peer-to-peer energy network," and "virtual power plant."

The critical challenge in this context is determining the optimal extent of decentralization and the design and operation of new local energy networks. With the rapid developments in distributed generation and storage (DGS) technologies and the introduction of various products with diverse specifications, the selection of a DGS system, with decisions as to its appropriate size, has become a complex problem.

DS4S is a decision support tool based on discrete optimization algorithms. It supports energy hub developers in the concurrent screening, selection, sizing, and scheduling of DGS systems based on various types of objective (minimum initial investment, minimum levelized cost of energy, highest reliability, etc.). It supports both on-network and off-network conditions.

Discussion is also provided in this chapter about the development of cooperative energy hubs and associated challenges. Two mechanisms are considered for the operation of multiuser energy hubs, namely, "DSO-out-of-the-loop" and "DSO-in-the-loop" systems. Although for a single energy hub DSO-out-of-the-loop might be an effective approach, networks of energy hubs may require a "DSO-in-the-loop" system due to complexities involved.

Keywords: Future energy systems, Microgrid, Nanogrid, Energy internet, Community energy network, Social energy network, Virtual power plant (VPP), Peer-to-peer energy network, Cooperation, Flexible manufacturing systems, DSO-out-of-the-loop, DSO-in-the-loop

Polygeneration with Polystorage
https://doi.org/10.1016/B978-0-12-813306-4.00015-X
493

1 BACKGROUND TO CENTRALIZED ENERGY NETWORK AND CHALLENGES

It was a few years after the invention of electricity generators that Edison invented the light bulb in 1879. With a gifted business intelligence, Edison began to think about a centralized power plant offering electricity to the wealthy shops of Manhattan, New York City's financial district, for lighting and keeping customers around the streets even in dark evenings. He turned this imagination into reality in 1882 by building the world's first commercial power plant. This was the ignition of a new industry and a critical infrastructure that today somehow links all aspects of our life.

Power generation began with scattered early adopters, and therefore initially the electricity network was decentralized (and disconnected). But it could be imagined that the ambition of the time was centralization of the network, with the goal of utilizing "economy of scale" for increased revenue for suppliers and affordable electricity for the society. Centralization was also linked to social welfare, as district power generation was often noisy, fuel supply (e.g., coal) was a dirty and tedious task, and operation of the generators needed skilled people. Centralization could move all the noise, pollution, and operation responsibilities away from the users. It could also improve the reliability of supply. In the mindset of the time, therefore, centralization was a positive action. But today, more than a century since the centralization wave, we have learned that overcentralization, at least in energy and electricity networks, has drawn the overall system into suboptimality. Not only do electricity networks account for a nontrivial amount of energy loss during transmission and distribution, but also the development capital of huge network infrastructure accounts for a major part of the energy bill. Fig. 1 shows the decomposition of electricity price (c/kWh) for several jurisdictions [1]. The price is composed of energy volume price, network cost, environmental taxes, and goods-and-services tax. The figure clearly shows the notable fraction of network cost in the electricity price. Indeed, in some countries, that fraction accounts for half of the price.

The other concern about centralized networks is linked to climate change. Grid centralization has produced considerable distances between the supplier (generator) and the consumers. The transmission of electricity over the network, at various voltages, results in the loss of a nontrivial amount of energy as well as CO_2 emissions that could otherwise be avoided. A similar concern arises from the transmission of high-pressure natural gas over long pipelines, where pressure drop leads to notable energy losses requiring costly recompression stations.

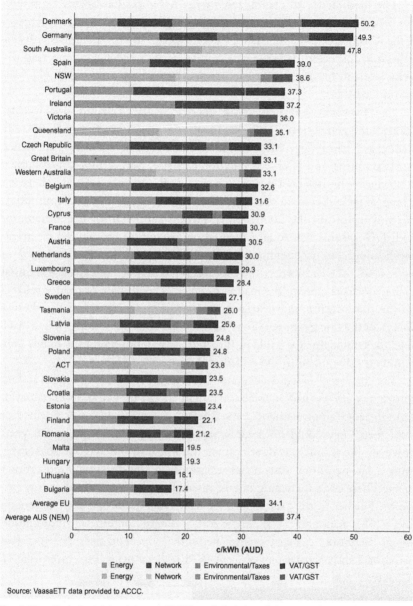

Fig. 1 Nominal electricity prices (c/kWh) and their breakdown in the European region and Australian states. *(Image: Courtesy of Australian Competition and Consumer Commission (ACCC), Restoring electricity affordability and Australia's competitive advantage: retail electricity pricing inquiry—final report. Canberra: Australian Competition and Consumer Commission; 2018.)*

The inefficiency of centralized energy networks has been long known but, in the absence of practical solutions, less discussed. After the emergence of the Kyoto Protocol in the late 1990s, however, interest in network decentralization increased. It was just a few years after the Kyoto meeting that terms relevant to the decentralized networks, such as "microgrid" [2], were introduced [3].

The possibility of generating energy at the demand side, termed "distributed generation" (DG), has many advantages in terms of energy efficiency, as it can reduce energy losses due to network transmission, the network footprint, and reserve generation capacity. Furthermore, it can utilize coproduced thermal energies, which would otherwise be wasted if there was a long distance between the thermal energy generation and consumption. For this reason, in the 2000s, a mainstream approach for decentralization with DG systems was sought through cogeneration (combined electricity and heating) and trigeneration (combined electricity, heating, and cooling) using fossil fuels or biomass and biofuels [4]. The key concept was to install such generators close to the load in order to (1) eliminate or reduce reliance on the transmission and distribution network, and (2) utilize the thermal energy of the flue gas in buildings' heating and cooling systems. When available, the cooled flue gas could be directed to greenhouses for the highly concentrated CO_2 usage in the photosynthesis process [5].

The fossil-fuel-based cogenerator and trigenerator (simply cogens and trigens) approach was just one step forward in energy networks decentralization and efficiency improvement. Nevertheless, due to their major reliance on fossil fuels, they could not address sustainability concerns. In recent years, however, the technology development of some distributed renewable energy sources has permitted sustainable realization to the decentralization movement. There is an increasing rate of research into decentralized energy networks. Further to the "microgrid," several similar concepts such as "energy hub" [6], "nanogrid" [7], "mesogrid," "energy internet," "community energy network," "social energy network," "peer-to-peer (P2P) energy network," and "virtual power plant (VPP)" have been introduced (Fig. 2) (Chapter 3 and [8]).

2 DESIGN AND OPERATION OF ENERGY HUBS

The most critical aspect of the decentralized energy networks is the rise of prosumers. Over the last century, centralization converted the demand-side users to mere consumers. With DG, however, consumers are taking an

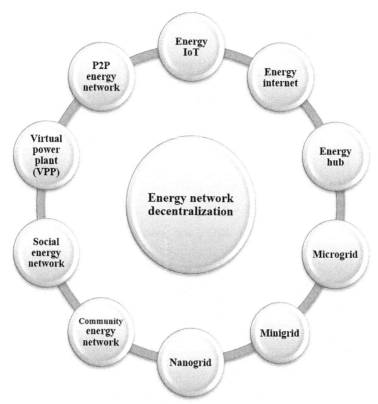

Fig. 2 New concepts in decentralized energy network context (*IoT*, internet of things; *P2P*, peer to peer).

active role in the energy supply chain—they are being transformed into prosumers who are partially consumers and partially producers. In such a context, there are critical challenges for the prosumers, including but not limited to those listed in Fig. 3.

In just the first step, complex decision making is involved in technology screening and selection, along with overall system configuration. This decision-making problem includes various generation and storage options, along with different energy demands such as DC/AC electricity, and heating energies at various temperature ranges. Deciding on the interconnection to a nearby network is another challenging problem [9].

One interesting issue in the network decentralization movement is the degree of decentralization. The extreme condition is network atomization, in which every household is disconnected from the grid with the formation

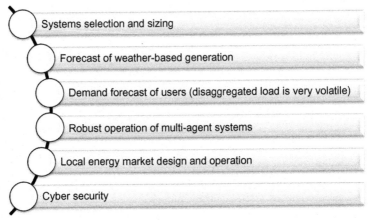

Fig. 3 Some investment decision making and operation management challenges for distributed prosumers.

of its own nanogrid. At another level, several consumers (or prosumers) build an energy hub with or without shared generation and storage technologies. One level beyond that is when multiple energy hubs are integrated. Each of those scenarios requires a different system and market design. Here, we briefly discuss the decision-making challenges associated with the establishment of such networks.

From a wider perspective, we should also differentiate between decentralization and network disconnection (atomization). Decentralizing the network seems to be an environmentally and economically viable option at resourced locations, as the cost of building networks in rural areas with low population density is generally much greater than that in high-density urban areas. Therefore, decentralization and the development of microgrids for locations with suitable resources might be a viable option, one that is also supported by utility companies due to economic feasibility [10].

2.1 Single-User Energy Hubs

When an energy consumer decides to develop its own energy system (generation and/or storage), the decision must be made whether to be self-sustaining or to remain with a connection to available networks in the vicinity (Fig. 5).

Various combinations of energy generation and storage technology have been studied for such applications, with solar systems being of the highest interest. The earliest simple configurations were photovoltaic (PV)-grid,

PV-diesel [11], and PV-battery. The configurations have diversified over time with the addition of various hybrid DGS systems such as PV-hydrogen, PV-diesel-battery, PV-wind-battery [12], PV-wind-diesel [13], PV-wind-diesel-battery [14], and PV-wind-diesel-hydrogen-battery [15]. The list of configurations could be much longer if other generation types depicted in Fig. 4 (e.g., bioenergy, hydro, gas turbine) and storage (e.g., hydro, compressed air, flywheel, capacitance, chemical conversions) are included [16]. Here, we provide a superstructure framework as a decision support for screening, selection, sizing, and scheduling (DS4S) of DGS systems for personalized energy hubs.

Now, consider an energy consumer analyzing energy usage (electricity and thermal) for a planning horizon of H segments (weeks, months, years) with P' multiple periods of a given fixed length (minute, hour, etc.).

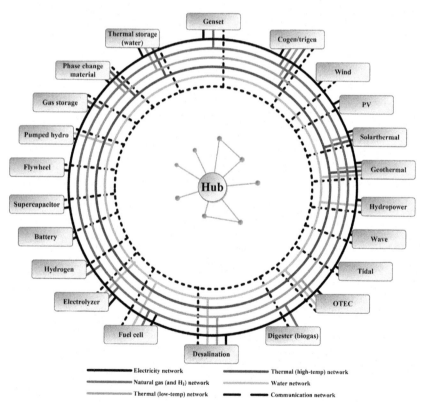

Fig. 4 Illustration of an energy hub with multiple generators, multiple storage systems, and interconnection to multiple supply networks (*OTEC*, ocean thermal energy conversion).

Thus, the planning horizon consists of a total of $P = H \times P'$ periods (p: 1, 2, ..., P). The current optimization study is occurring in the base period ($p = 0$).

Fig. 5 provides a schematic of the decision problem. The consumer has K (k: 1, 2, ..., K) different energy demands, with the quantity of L_{kp} for energy type k during period p. There are N (n: 1, 2, ..., N) different supply networks in the vicinity (e.g., gas and electricity). We also assume wind and solar irradiation as potential supply networks.

The consumer is interested to investigate the feasibility of DG systems and/or storage systems to reduce energy costs over the planning horizon. There are many DGS suppliers in the market, with a wide range of costs, sizes, efficiency, and operational performance. The consumer is considering I (i: 1, 2, ..., I) number of DG systems, each with a capital cost of CX_i^{DG} (per unit power), to ultimately select the best one(s).

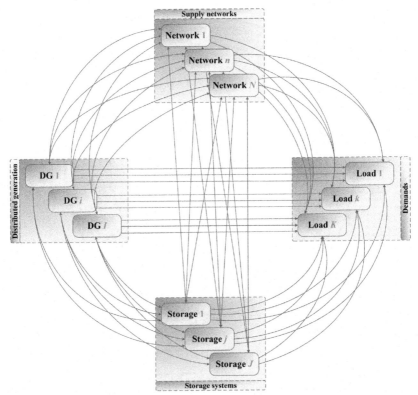

Fig. 5 The superstructure of a DS4S (decision support for screening, selection, sizing, and scheduling) problem for an energy hub.

The DG systems i can receive feed F_{nip}^{DG} from supply network n and generate one or more of the K energy types to use locally, to sell to the network, or to store in the storage system(s) for a later use.

Each DG has a design specification of S_i^{DG} (power unit) with the nominal (standard) design efficiency of η_i^{DG} and occupying an area of A_i with volume V_i. The real DG efficiency in any period p is taken as η_{ip}^{DG}, which might be function of several parameters (weather condition, aging, etc.). The term "performance ratio" is sometimes used to represent the real efficiency [17]. The performance ratio is obtained by dividing the real DG efficiency by the nominal efficiency.

Likewise, the consumer considers J (j: 1, 2, ..., J) number of storage systems with a capital cost of CX_i^S (per unit energy) to select the optimal one(s). The storage can receive energy k from the DG system and/or the supply network n. The storage system capacity can be used locally and/or sold to the network. The consumer may be indifferent to the two options of off-network or on-network connectivity and may leave it to the program to select the best option with the highest benefit. Each storage system has a nominal size of S_j^S (an energy unit), with the nominal charge and discharge efficiency of η_j^C and η_j^D, respectively. Storage j occupies an area of A_j with the volume of V_j. The real storage charge and discharge efficiency are functions of many parameters and are taken as η_{jp}^C and η_{jp}^D, respectively, during period p with consideration of influential factors such as ambient temperature and aging. The storage system also self-discharges at the rate of β_{jp} in any period p. Each storage system has a lower bound and upper bound to its state of charge (SOC), SOC_j^L, and SOC_j^U. As such, the storage system needs a charge controller with the efficiency of η_j^{CC} for regulation of the input/output power. Storage systems also have limitations to their rate of charge/discharge. We take CR_j and DR_j as the maximum possible charge and discharge rates of the storage system, respectively, per period. The nominal efficiency of the energy convertor (or invertor) is taken as η_i^{DGin} and η_j^{Sin} for DG and storage system, respectively. When the nominal efficiency is nonlinear (e.g., in Ref. [18]), it can be taken as a variable (a function of input power flow) in each period p, for DG and storage system, η_{ip}^{DGin} and η_{jp}^{Sin}, respectively. However, this will convert the linear program formulation into a nonlinear program.

It is possible that the energy hub has a space limitation that precludes installation of DGS systems with total area or volume greater than A^m and V^m, respectively. Also, one system might be considered as more than one agent. For instance, an electric vehicle could be considered as one storage unit and one load unit or a hydro plant could be considered as one generation unit and one storage unit.

The feed-in-tariff (FiT) for selling the energy to the network n is highly policy-related and the consumer projects the value of FiT_{np} during period p over the planning horizon. The market price of input feed to the generators is a function of the time of use, with occasional price modifications. The long-term supply price can be a function of various parameters (economic growth, carbon tax, etc.). Given the current retail prices and all other possible parameters, the consumer anticipates that the price and the connection fee (or supply charge) will be EP_{np} and CF_{np}, respectively, for supply network n in period p. Each DG and storage technology has periodical fixed operation and maintenance costs given by FOM_{ip}^{DG} and FOM_{jp}^{S}, respectively, during period p.

With these definitions, the decision support for screening, selection, sizing, and scheduling (DS4S) problem can now be formulated as identifying the best investment plan in the DGS system to build a personalized energy hub and minimize the energy cost over the planning horizon. The definition of objective function could be a personal choice. For instance, it could be the minimum net present value of costs (NPV_C) or the maximum net present value of overall savings in energy costs (NPV_S) over the planning horizon (h). It could be also the internal rate of return, which is, in fact, another representation of NPV_S. The formulations of NPV_C and NPV_S along with all governing constraints are provided in Table 1.

2.2 Multiuser Energy Hub

2.2.1 The Issue of DGS Underutilization

Obviously, a key issue in the success of energy market decentralization and the development of energy hubs is the market price parity of energy produced by distributed energy technologies versus centralized generators. Traditionally, energy hubs, in simple forms such as PV-diesel or PV-battery, have been used in remote locations without network access. But the recent rapid decline in some renewable technology prices has brought their price to network parity in several jurisdictions [19]. As a result, there has been an unexpected rate of increase in residential level uptake of DG systems (e.g., PV) around the world. The prices of storage systems have also shown a notable declining trend and it is anticipated that battery technology may follow the price trajectory of PV [20]. In brief, the concept of energy hubs is not novel: Edison was thinking in that direction one and a half centuries

Table 1 The key formulas governing the DS4S (decision support for screening, selection, sizing, and scheduling) process for a personalized energy hub development

Formula	Formula description	Note
$\sum_{i=1}^{I} \gamma_i \leq N^{DG}$	Limit on the number of selected DG systems	$\gamma_i = \begin{cases} 1, & \text{if DG system } i \text{ is selected} \\ 0, & \text{otherwise} \end{cases}$
$\sum_{j=1}^{J} \gamma'_j \leq N^S$	Limit on the number of selected storage systems	$\gamma'_j = \begin{cases} 1, & \text{if storage system } j \text{ is selected} \\ 0, & \text{otherwise} \end{cases}$
$\sum_{i=1}^{I} \gamma_i A_i + \sum_{j=1}^{J} \gamma'_j A_j \leq A^m$	Limit on the total area occupied by the DGS system	Total area limit: A^m
$\sum_{i=1}^{I} \gamma_i V_i + \sum_{j=1}^{J} \gamma'_j V_j \leq V^m$	Limit on the total volume occupied by the DGS system	Total volume limit: V^m
$FC_{ip} = \sum_{n=1}^{N} X_{ip}^{DG} F_{nip} / \eta_{ip}^{DG}$	Feed supply cost of DG unit i at period p	F_{nip}: Feed price per unit supply from supply network n for DG unit i at period p
$E_{ip} = \sum_{n=1}^{N} X_{ip}^{DG} CI_{nip} / \eta_{ip}^{DG}$	CO_2-equivalent GHG emission from DG i at period p	CI_{nip}: Carbon intensity of DG unit i per unit of feed energy from network n at period p
$GHC_{ip} = E_{ip} CP_p$	Incurred GHG cost for DG i at period p	CP_p: CO_2-equivalent GHG emission penalty at period p
$X_{ip}^{DG} = \sum_{n=1}^{N} X_{nip}^{DG.N} + \sum_{k=1}^{K} X_{kip}^{DG.K} + \sum_{j=1}^{J} X_{ijp}^{DG.S} \leq \gamma_i C_{ip}^{DG}$	Limit on the total energy production from DG i at period p	C_{ip}^{DG}: The maximum "generatable" capacity of a DG unit i at period p; $X_{nip}^{DG.N}$: Export from DG i to network n at period p; $X_{kip}^{DG.K}$: Supply from DG i to load k at period p; $X_{ijp}^{DG.S}$: Supply from DG i to storage j at period p

Continued

Table 1 The key formulas governing the DS4S (decision support for screening, selection, sizing, and scheduling) process for a personalized energy hub development—cont'd

Formula	Formula description	Note
$$X_{ip}^{DG}/\eta_{ip}^{DG} = \sum_{n=1}^{N} X_{nip}^{N.DG} + \sum_{j=1}^{J} X_{jip}^{S.DG}$$	Quantity of feed received by DG i at period p	$X_{nip}^{N.DG}$: Supply from network n to DG i at period p $X_{jip}^{S.DG}$: Supply from storage j to DG i at period p
$$\sum_{n=1}^{N} X_{nkp}^{N.K} + \sum_{i=1}^{I} \eta_{ip}^{DGin} X_{kip}^{DG.K} + \sum_{j=1}^{J} X_{kjp}^{S.K} \le L_{kp}$$	Local load k limit in any period p	$X_{kip}^{S.DG}$: Supply from network n to load k at period p $X_{kip}^{DG.K}$: Supply from DG i to load k at period p $X_{kjp}^{S.K}$: Supply from storage j to load k at period p
$$B_{jp} = \left(1 - \beta_{jp}\right)\left(\eta_j^{CC}\eta_{jp}^{C}\sum_{i=1}^{I} X_{ijp}^{DG.S} + \eta_{jp}^{Sin}\eta_j^{CC}\eta_{jp}^{C}\sum_{n=1}^{N} X_{njp}^{N.S}\right)$$ $$- \sum_{n=1}^{N} X_{njp}^{S.N}/\left(\eta_{jp}^{Sin}\eta_j^{CC}\eta_{jp}^{D}\right) - \sum_{k=1}^{K} X_{kjp}^{S.K}/\left(\eta_{jp}^{Sin}\eta_j^{CC}\eta_{jp}^{D}\right)$$	Input-output balance of storage system j in period p	β_{jp}: Self-discharges of storage system j during period p $X_{njp}^{N.S}$: Supply from network n to storage j at period p $X_{njp}^{S.N}$: Supply from storage j to network n at period p
$$SOC_{jp} = \sum_{p'=1}^{p} B_{jp'}$$	SOC of storage system j at period p	
$$\gamma_j' SOC_j^L \le SOC_{jp} \le \gamma_j' SOC_j^U$$	Lower and upper limit on SOC of storage system j	SOC_j^L: Lower bound on SOC SOC_j^U: Upper bound on SOC
$$NC_{nh} = \sum_{p=(h-1)P'+1}^{hP'}\left(\sum_{k=1}^{K}\left(X_{nkp}^{N.K}\right)\right)$$ $$+ \sum_{n=1}^{N}\sum_{i=1}^{I}\left(X_{nip}^{DG.N} + X_{nip}^{N.DG}\right) + \sum_{j=1}^{J}\left(X_{njp}^{S.N} + X_{njp}^{N.S}\right)$$	Magnitude of connectivity to network n during horizon h	$X_{nkp}^{N.K}$: Supply from network n to load k at period p

$$NC_{nh} \leq M \cdot \gamma''_{nh}$$

$$NPV_c = \sum_{i=1}^{I} CX_i^{DG} + \sum_{j=1}^{J} CX_j^{S}$$
$$+ \sum_{h=1}^{H} \left[\sum_{p=(h-1)P'+1}^{hP'} \left(\sum_{i=1}^{I} \gamma_i FOM_{ip}^{DG} + \sum_{i=1}^{I} \gamma_j^t FOM_{ip}^{S} \right) \right] /$$
$$(1+r)^h + \sum_{h=1}^{H} \left[\sum_{p=(h-1)P'+1}^{hP'} \right.$$
$$\left(\sum_{n=1}^{N} \sum_{k=1}^{K} X_{nkp}^{N.K} EP_{np} \right.$$
$$+ \sum_{n=1}^{N} \gamma''_{nh} CF_{np} + \sum_{j=1}^{J} \left(\sum_{n=1}^{N} EP_{np} X_{njp}^{N.S} - \sum_{n=1}^{N} FiT_{np} X_{njp}^{S.N} \right)$$
$$- \sum_{i=1}^{I} \left(\eta_{ip}^{DGin} \sum_{n=1}^{N} FiT_{np} X_{nip}^{DG.N} \right)$$
$$+ \sum_{i=1}^{I} (FC_{ip} + GHG_{ip}) \right) \right] / (1+r)^h$$

$$NPV_S = \sum_{h=1}^{H} \left[\sum_{p=(h-1)P'+1}^{hP'} \sum_{n=1}^{N} \left(CF_{np} + \sum_{k=1}^{K} EP_{np} L_{kp} \right) \right] /$$
$$(1+r)^h - NPV_c$$

If connected to network n during horizon h

DS4S objective for the net present value of costs

DS4S objective for the net present value of savings

$$\gamma''_{nh} = \begin{cases} 1, & \text{if } NC_{nh} > 0 \\ 0, & \text{if } NC_{nh} = 0 \end{cases}$$

r: Discount rate over h

ago. The difference is that Edison's customers were early adopters, whereas today the customers are becoming the majority.

While the cumulative global installed PV capacity was <10 GW in 2007, it increased 10-fold to 100 GW by 2012 and surpassed 200 GW in early 2015. Over the last three years, there has been around 100 GW new capacity annually, and it is expected that by end of 2018 the global installed PV capacity will exceed 500 GW. An almost similar trend has occurred for wind turbine installations. As a result of this increase in production scale and also the shift of global production to China, PV system prices have decreased dramatically since 2010.

As an example, Fig. 6 shows Australia's PV installation trend in terms of cumulative installations and also the average installed capacity. By 2010, there were only around 93000 PV installations totalling around 150 MW. On average, the size of PV systems on each customer's rooftop was only 1.43 kW. Four years later, by the end of 2014, the number of installations had increased by around 14 times (1,366,264), with the total installed capacity jumping to 4083 MW. This meant that the average PV size on each now-prosumer's rooftop had increased more than three times, reaching 4.84 kW. As of March 2018, the number of installations had exceeded 1,845,000 (equal to ~1/5 of total Australian households), with the total installed capacity of 7804 MW. Now, the average installation size has reached 6.36 kW.

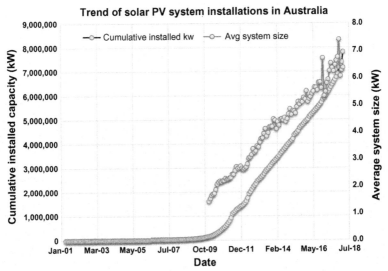

Fig. 6 An exemplary case of Australia with fast-growing photovoltaic (PV) cell installation while the average size per installation has also increased thanks to the continuous decline in PV market prices.

Fig. 6 clearly shows the trend of the demand-side behavior change. The growing size of DG systems on prosumers' location is the origin of a new problem and opportunity discussed here. Even with a small PV system (say 1.0 kW), there might be a time during the day (especially midday) when a house's PV system generates electricity surplus to its demand, which should be either curtailed or dispatched to the grid with a given feed-in tariff. With increased PV (or any DG) system size, the quantity of this unused energy will further elevate. Therefore, while a house without a PV system is fully reliant on the retailer for purchasing electricity, a house with PV will be doubly reliant on the retailer, for both purchasing and selling the electricity. The public concern is that over time, the retailer feed-in tariff might unrealistically decline against the prosumers' benefit. This is the main driver of the public interest in storage systems, which provide prosumers with extra flexibility for their energy management and load shifting.

2.2.2 Advantages of Cooperative Energy Hubs

Some years ago, observation of the initial growth in the size of rooftop PV systems, and the projection of future large underutilized distributed generation and storage (DGS) systems drew our attention to the analogy of high-performance computing (HPC). Imagine a person with occasional computation-intensive tasks for which a normal PC does not satisfy all the requirements within a given timeframe. Procurement of an expensive HPC machine will solve the problem. But this expensive resource will be underutilized with only occasional use. In fact, such situations resulted in the birth of the distributed or grid computing concept in the late 1990s [21], for clustering a collection of computer resources from multiple locations to reach a common goal. The more advanced and publicized form of grid computing is today's cloud computing or cloud storage. Thus, the concept of distributed computing relies on sharing resources to achieve higher efficiency and economy of scale over a network, and all of us are receiving the benefits of this concept today.

We draw such an analogy in the context of DGS systems. A small energy generation system (similar to a normal PC) might not satisfy all the requirements of a consumer, whereas a large energy generation system (similar to an HPC) would be underutilized. On this basis, we envisaged the future in the development of localized cooperative communities in which, unlike centralized microgrids, all members are interlinked and exchange energy among themselves (Fig. 7). The two key requirements of a successful energy hub are

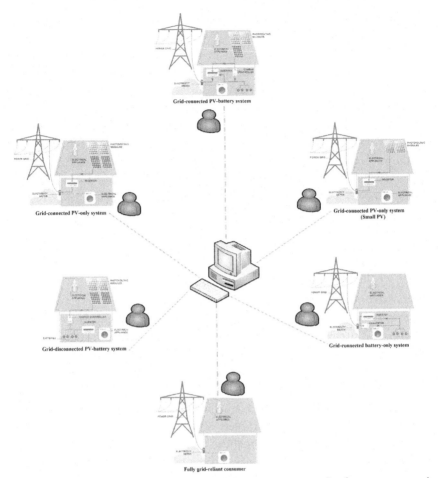

Fig. 7 Schematic of a cooperative community energy network of consumers and prosumers with various distributed generation and storage configurations, with or without connection to a nearby network.

(a) diverse DGS systems (type and size) in the local network, and (b) diverse consumption patterns.

Diverse DGS systems (type and size) in the local network: Some consumers might have only a storage system, but not a DG (e.g., those living in apartments); some may have only DG; some will install both. In the absence of an appealing offer (feed-in tariffs) from an existing network retailer, the local energy hub will enable the prosumers with unused energy to exchange their surplus energies with their neighbours rather than curtailing or exporting to the network at an unattractive tariff.

Diverse consumption patterns: Generally, the consumption patterns of all users are unique. Families, with children staying at home, have their main peak in the mornings, while all-working families have the main peak in the early evenings. Seniors have almost flat curves during a typical day, whereas youth residencies might have main peaks in late evenings. Therefore, if neighbors with mixed load patterns cooperate with energy exchange, they could achieve better utilization of their installed resources and reduce the costs of their peak demands.

In summary, the success of local energy hubs is founded on the "dissimilarities" or heterogeneity of the hub members' DGSs and loads. For instance, a cooperative energy hub might be less successful in a retirement village than in a village with diverse populations.

It is worth highlighting the critical role of local energy hubs in network stability. The growing size of DG systems at every prosumer's location can cause network stability challenges if all prosumers export their surplus generation at some specific timeframes (e.g., PV output at midday). Local energy exchange systems can play a positive role in damping such positive or negative disturbances on the network and improve the network robustness.

2.2.3 Two Multi-user Energy Hubs: DSO-out-of-the-Loop and DSO-in-the-Loop

There could be two main energy hubs, *direct cooperation* and *indirect cooperation*, based on the involvement of distribution system operators (DSOs) or distribution service providers (DSPs). The direct hub is a "DSO-out-of-the-loop" peers-to-peers system, in which the community members build their own local energy market (LEM) based on an agreement achieved in periodical meetings of the members' committee. The indirect scenario is "DSO-in-the-loop," in which a third-party *aggregator* business takes control of the members' DGS systems and operates all equipment with its preferred centralized market model [22]. Members receive their financial benefits based on individual agreements with the DSO (or DSP), distinct from the other network members. In this scenario, the community members might not know each other. They are not cooperating directly, though there is an indirect cooperation.

Already several such initiatives have existed around the world, though majorly focused on indirect LEM models. Examples from Australia include the Ausnet's Mooroolbark Mini Grid trial with 14 homes [23], Reposit Power-led 250-home project in Canberra [24], and AGL's 5 MW programme with 1000 homes [25]. The most recent example is the revolutionary 50000-home and 250-MW South Australia-Tesla program, which is the world's largest virtual power plant (VPP) [26]. These examples reveal that less attention has been given to direct LEMs in which community members are directly involved in their network operation. Zhang et al. [27] recently reviewed existing P2P energy-trading projects and concluded that the majority of these trials considered the members to have the conventional "supplier role," overlooking the development of "necessary communication and control networks that could enable P2P energy trading in or among local microgrids."

It is also noteworthy that, following the recent academic and commercial excitement about blockchains (i.e., a dynamic list of records or blocks linked using cryptography) [28], some initiatives have arisen regarding Bitcoin-inspired and blockchain-based P2P energy trading [29, 30]. Although these might be effective platforms for easy and transparent energy exchange, they do not still address the network operation and control challenges. There is a need for an integrated framework, which concurrently addresses both the technical issues of the network development and the market mechanisms.

To address this challenge, we have developed a methodology for dynamic management of networks with a LEM mechanism so that all community members benefit fairly from the cooperation. The detailed concept and methodology were published as a book [3], and we have carried out several case studies and evaluated the system's performance.

The results have shown that the highest benefit for a community can be achieved only when all generators and storage of the community members become an integral part of the community network [31]. Members can develop a bidding-based LEM or, alternatively, set some fair local tariffs and then operate the LEM centrally. Our examples show clearly that fair and cooperative operation of the energy network not only reduces all community members' costs but also helps the macrogrid with reduction of the load during peak demand periods. This result implies the economic advantage of the community network for members of the community as well as improvement in the efficiency of the macrogrid and, from a larger perspective, enhancement of global sustainability through the optimized use of renewable power generation.

2.2.4 Networks of Energy Hubs

Another level of decentralized energy networks is created when multiple energy hubs are integrated to build a LEM. Probably, in such conditions when multiple communities build a larger community, an individual acquaintance is relatively out of context. Consequently, the network operation mechanism may converge into a bidding and game-based LEM, which can be best operated with DSO-in-the-loop.

3 CONCLUSION

One concurrently interesting and challenging issue in the movement toward network decentralization is the extent of decentralization. The extreme condition is network atomization, in which every household is disconnected from the grid with the formation of its own nanogrid. Another level is when several consumers (or prosumers) build an energy hub with or without shared generation and storage technologies. A further level beyond this is when multiple energy hubs are integrated. Each of these scenarios requires a different system and market design.

In this chapter, we have briefly discussed the challenges and opportunities associated with the establishment of such networks. We introduced the DS4S (decision support for screening, selection, sizing, and scheduling) framework for optimal design and operation of energy hubs. We also discussed the two scenarios of "DSO-out-of-the-loop" and "DSO-in-the-loop" systems. While for a single energy hub, DSO-out-of-the-loop might be an effective approach, a network of energy hubs may require a DSO-in-the-loop system.

REFERENCES

[1] ACCC. Restoring electricity affordability and Australia's competitive advantage: retail electricity pricing inquiry—final report. Canberra: Australian Competition and Consumer Commission; 2018.
[2] Jopp K. Energy forum. ABB bets on alternative energies. "The future belongs to decentralized electricity grids." Brennstoff-Waerme-Kraft 2000;52(9):21.
[3] Khalilpour, R. and A. Vassallo, Community energy networks with storage: modeling frameworks for distributed generation. Energy policy, economics, management & transport, 2016, Singapore: Springer.
[4] Onovwiona HI, Ugursal VI. Residential cogeneration systems: review of the current technology. Renew Sustain Energy Rev 2006;10(5):389–431.
[5] Compernolle T, et al. Analyzing a self-managed CHP system for greenhouse cultivation as a profitable way to reduce CO2-emissions. Energy 2011;36(4):1940–7.
[6] Geidl M, et al. Energy hubs for the future. IEEE Power Energ Mag 2007;5(1):24–30.
[7] Khalilpour KR, Vassallo A. A generic framework for DGS nanogrids. In: Community energy networks with storage: modeling frameworks for distributed generation. Singapore: Springer; 2016. p. 41–59.
[8] Khalilpour KR, Vassallo A. Community energy networks with storage: modeling frameworks for distributed generation. Singapore: Springer; 2016.

[9] Khalilpour R, Vassallo A. Leaving the grid: an ambition or a real choice? Energy Policy 2015;82:207–21.

[10] Stobbe R. In: Parkinson G, editor. SA network operator: rural communities could quit the grid. RenewEconomy; 2014.

[11] Yamegueu D, et al. Experimental study of electricity generation by solar PV/diesel hybrid systems without battery storage for off-grid areas. Renew Energy 2011; 36(6):1780–7.

[12] Nema P, Nema RK, Rangnekar S. A current and future state of art development of hybrid energy system using wind and PV-solar: a review. Renew Sustain Energy Rev 2009;13(8):2096–103.

[13] McGowan JG, Manwell JF. Hybrid wind/PV/diesel system experiences. Renew Energy 1999;16(1–4):928–33.

[14] Merei G, Berger C, Sauer DU. Optimization of an off-grid hybrid PV–wind–diesel system with different battery technologies using genetic algorithm. Sol Energy 2013;97:460–73.

[15] Dufo-López R, Bernal-Agustín JL. Multi-objective design of PV–wind–diesel–hydrogen–battery systems. Renew Energy 2008;33(12):2559–72.

[16] Neves D, Silva CA, Connors S. Design and implementation of hybrid renewable energy systems on micro-communities: a review on case studies. Renew Sustain Energy Rev 2014;31:935–46.

[17] Dierauf T, et al. Weather-corrected performance ratio. Colorado: NREL; 2013.

[18] Velasco G, et al. In: Power sizing factor design of central inverter PV grid-connected systems: a simulation approach. Proceedings of 14th international power electronics and motion control conference (EPE-PEMC 2010); 2010.

[19] Khalilpour KR, Vassallo A. Technoeconomic parametric analysis of PV-battery systems. Renew Energy 2016;97:757–68.

[20] Szatow T, et al. What happens when we un-plug? Exploring the consumer and market implications of viable, off-grid energy supply; 2014.

[21] Foster I, Kesselman C. In: Ian F, Carl K, editors. The grid: blueprint for a new computing infrastructure. Morgan Kaufmann Publishers; 1999. p. 677.

[22] Khalilpour KR, Vassallo A. Noncooperative community energy networks. In: Khalilpour KR, Vassallo A, editors. Community energy networks with storage: modeling frameworks for distributed generation. Singapore: Springer; 2016. p. 131–49.

[23] Ausnet. In: Services A, editor. moorebank mini grid project: exploring our energy future. Melbourne: Ausnet; 2017. p. 1–4.

[24] Trask S. Canberra businesses and homes trial 'world's largest virtual power plant; 2017, https://www.canberratimes.com.au/national/act/canberra-businesses-and-homes-trial-worlds-largest-virtual-power-plant-20171130-gzvwiu.html.

[25] Parkinson G. AGL hits pause on virtual power plant in technology "rethink"; 2017, https://reneweconomy.com.au/agl-hits-pause-on-virtual-power-plant-in-technology-rethink-57487/.

[26] Parkinson G. Tesla to build 250MW "virtual power plant" in South Australia; 2018, https://reneweconomy.com.au/tesla-to-build-250mw-virtual-power-plant-in-south-australia-44339/.

[27] Zhang C, et al. Review of existing peer-to-peer energy trading projects. Energy Procedia 2017;105:2563–8.

[28] Marsal-Llacuna M-L. Future living framework: is blockchain the next enabling network? Technol Forecast Soc Chang 2018;128:226–34.

[29] Powell D. Aussie blockchain startup power ledger forms US partnership to roll out "hundreds" of energy trading projects. Startup News Analysis 2018.

[30] Tushar W, et al. Transforming energy networks via peer-to-peer energy trading: the potential of game-theoretic approaches. IEEE Signal Process Mag 2018;35(4):90–111.

[31] Khalilpour KR, Vassallo A. Cooperative community energy networks. In: Community Energy Networks With Storage: Modeling Frameworks for Distributed Generation. Singapore: Springer; 2016. p. 151–82.

CHAPTER 16

The Transition From X% to 100% Renewable Future: Perspective and Prospective

Kaveh Rajab Khalilpour
Faculty of Engineering and Information Technology, Monash University, Melbourne, VIC, Australia

Abstract

Until very recent times, the topic of a "100% renewable energy system" was a type of intellectual and mostly academic discussion. But some revolutionary movements in the renewable energy markets, e.g., photovoltaic (PV) cells and wind technologies, have raised hopes and converted "100% renewables" to an attainable scenario in policy and planning analyses with several successful deployment stories already around the world. Here, we investigate some historical background of this new field from the academic literature and the ongoing research directions. We also elaborate on the associated energy system planning challenges in this context.

Keywords: Future energy systems, X% renewables, 100% renewables, Net-zero emission system, Net-negative emission systems, Flexible manufacturing systems, Flexibility, Climate change mitigation, Scenarios of scenarios, Superscenarios, Social network analysis, Network theory

1 INTRODUCTION

The topic of 100% renewable energy system was a type of intellectual and mostly academic discussion until very recent times. But some revolutionary movements in the renewable energy markets, e.g., photovoltaic (PV) cells and wind technologies, have converted "100% renewable" to an attainable scenario in policy and planning analyses, with several successful deployment stories already around the world. Here, we investigate some historical background of this new field and discuss the way forward.

Until the Industrial Revolution, life around the world was decentralized and relied mainly on the resources in the vicinity. In other words, the producers were the consumers of their products (farms and agriculture), and the redundancies were supplied to neighborhood community markets, and vice

Polygeneration with Polystorage
https://doi.org/10.1016/B978-0-12-813306-4.00016-1

versa. Energy resources were mostly renewable (e.g., wood, solar thermal, wind and water mills), and humans, as well as other creatures, had learned to adapt to the variability of renewable resource availability (e.g., working by daylight, and sleeping at night) (Chapter 1). The development of steam engines and the subsequent Industrial Revolution transformed the lifestyle by centralizing the energy and commodity production systems in order to benefit from the economy of scale. The new-born giant industries driven by steam turbines needed reliable and economic sources of energy when renewable energies were not sufficient.

Though renewable resources have several critical advantages, including abundance and relatively scattered geographic distribution, combinations of reasons, including their intermittency and limited availability, made them an unattractive source of energy for the booming industries. Thus, the need for a continuous source of energy led to the utilization of fossil fuels and gradually the global energy supply was dominated by cheap and "reliable" fossil fuels, which kept most renewable energy technologies out of the game.

Fig. 1 nicely shows this historical transition nicely. According to the figure, until the Industrial Revolution era, wood was the main energy source. Gradually, coal was explored and overtook wood by the early 1900s. The age of coal continued until oil and gas joined the mix. In the second half of the 20th century, oil was the dominant energy source. Along with fossil fuels, some cleaner energy sources such as hydro and nuclear have evolved and diversified the energy portfolio over the last century, but their share in the global energy mix has been marginal.

The impact of fossil-fuel-based emissions on global warming has been long discussed. The greenhouse effect was discovered in the early 19th century by Joseph Fourier, and by the end of that century, Svante Arrhenius predicted that emissions of carbon dioxide from the burning of fossil fuels were large enough to cause global warming [1]. Since then, there has been continuous evidence of the catastrophic consequences of greenhouse gas (GHG) emissions. However, in the absence of technoeconomically feasible alternatives, until the late 20th century, denial or ignorance were the two mainstream approaches to GHG emissions. Toward the late 20th century, observation of the environmental consequences of fossil fuels led to increased concerns being voiced by researchers (e.g., [2, 3]). The main global actions began in the 1980s. In 1988, the Intergovernmental Panel on Climate Change was formed to collate and assess evidence on climate change. Four years later, in 1992, the United Framework Convention on Climate Change (UNFCCC) was formed for "stabilization of greenhouse

Global primary energy consumption, world
Global primary energy consumption, measured in terawatt-hours (TWh) per year. Here "other renewables" are renewable technologies not including solar, wind, hydropower, and traditional biofuels.

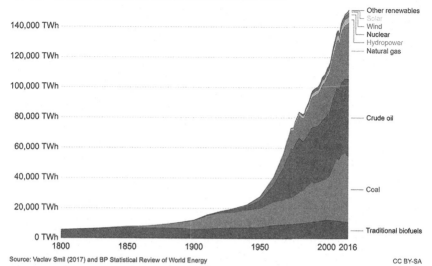

Source: Vaclav Smil (2017) and BP Statistical Review of World Energy   CC BY-SA

Global primary energy consumption, world
Global primary energy consumption, measured in terawatt-hours (TWh) per year. Here "other renewables" are renewable technologies not including solar, wind, hydropower, and traditional biofuels.

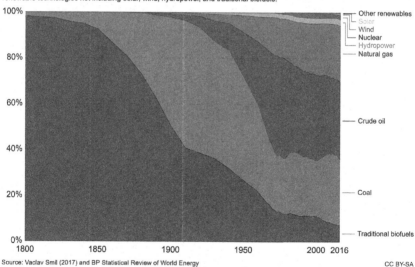

Source: Vaclav Smil (2017) and BP Statistical Review of World Energy   CC BY-SA

Fig. 1 Evolution of the global primary energy mix since the Industrial Revolution; left: quantity profile; right: percentage profile. (*Image: Courtesy of Our World in Data.*)

gas concentrations in the atmosphere at a level that would prevent dangerous anthropogenic interference with the climate system." The UNFCCC efforts led to the Kyoto Protocol of 1997, which pledged that the developed nations would reduce emissions by an average 5% by 2008–12.

With binding commitments from most OECD countries to reduce their GHG emissions, research and commercial interest in renewable energies increased. The identified technological solutions had three pillars: (1) efficiency improvement, (2) replacing fossil fuels with clean energy sources such as renewables, biofuels, and nuclear energy, and (3) CO_2 capture, storage, and utilization (CCSU) [4, 5] (Fig. 2).

From this period, research, development, demonstration, and commercialization (RD&DC) activities focused on the three given directions boomed [6]. National energy planning studies began to consider these three elements. Energy systems planning naturally involves investment in diverse energy technologies, with the goal of reducing GHG emissions as well as other environmental impacts, while imposing a marginal impact on the delivered energy costs. Fig. 3 illustrates such planning goals, starting with BAU (business as usual), which assumes no change in energy supply sources. Countries can adopt a combination of aims for each of these emission abatement plans (carbon capture and storage (CCS), energy efficiency, and renewables) based on their desired or committed emission reduction goals over a planning period. The extreme of the BAU could be 100% renewable

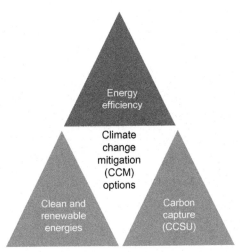

Fig. 2 The main technological options for climate change mitigation (*CCSU*, carbon capture, storage, and utilization).

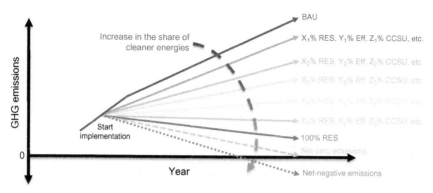

Fig. 3 Sustainable energy system planning scenarios (*BAU*, business as usual; *RES*, renewable energy sources; *Eff*, energy efficiency; *CCSU*, carbon capture, storage, and utilization).

energy systems. However, thinking about such a scenario was so distant from practicality that not only was it abandoned from commercial discussion but also only rare discussions appeared in the academic literature before the current millennium.

Renewable energy technologies had been long been under development but, except for some cases such as hydro and biofuel, the technologies were costly. In fact, there was a peak in attention to renewable technologies just a couple of decades before the Kyoto Protocol, in the 1970s. The energy crisis of that period brought renewable energies to the forefront of discussions in energy-dependent countries. But, as the motivation focus during this period was mostly on energy security and least on sustainability, the fall in oil prices in the 1980s further decreased the competitiveness of renewable energy technologies. Now, in the post-Kyoto era with countries committed to emission reduction, some mechanisms had to be introduced to improve the competitiveness of renewables technology.

The gradual introduction of carbon taxing mechanisms and renewable energy subsidies improved the prospect of clean energy technology development. As such, renewable energy technologies found a stronger share in countries' energy policy-making and planning activities. Two decades after the Kyoto meeting, today a few more renewable technologies, including photovoltaic cells and wind, have reached price parity with fossil fuels in most jurisdictions around the world [7, 8].

In those days, CCS was considered a bridging option over the next several decades while transitioning to future clean energy systems. Today, however, with the rapid growth in renewables technologies and their markets,

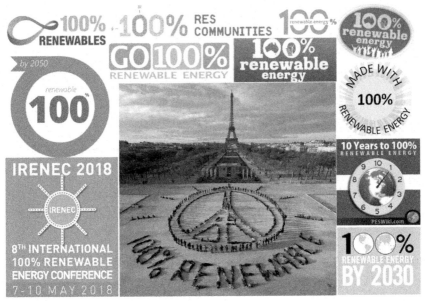

Fig. 4 Examples of social, academic, and industrial movements toward 100% renewables.

100% renewable energy systems have emerged among the possible scenarios in countries' national energy planning. There are even discussions about beyond 100% renewables and toward a "net-zero emission" or "net-negative emission" future (see Fig. 3).

Gradually, the renewables movement, particularly the "100% renewables," has evolved from a technical, academic, commercial, or policy-related topic into a social demand. There are ongoing social activities around the world encouraging policy makers toward regulations that lead to 100% renewable societies (Fig. 4). History has shown that a bipartisan movement involving both industry and society has almost always succeeded.

In the next section, we analyze the historical development of the emerging research field of "100% renewables" and investigate the direction the field is taking.

2 LITERATURE HISTORY OF 100% RENEWABLES

To review the literature on 100% renewables, we extracted publication records from Scopus. This bibliometric website was searched for publications having any of the following keywords: "100%renewable," "100% renewable,"

"100% renewables," "100%renewables," "100 percent renewable," "100 percent renewables," "100-percent renewables," "100%-renewables," "100%-renewable," "hundred percent renewable," "hundred percent renewables," "hundred-percent renewable," "hundred-percent-renewable," "hundred-percent-renewable" in their titles, abstracts, or keywords.

The search in mid-2018 produced 465 records, of which 414 items had been published by the end of 2017. Fig. 5 shows the histogram of the publications during 2001–17. It is evident from the number of the references and also from the publication trends that this is an emerging topic with an accelerating growth rate.

It is noteworthy that this study investigated a literature set, which is both indexed and has reference to "100% renewable," or its associated terms, within the articles' keywords, titles, or abstracts. It is expected that other publications exist, which are either not indexed in the scholarly publication databases or do not explicitly refer to the topic of the study, though they

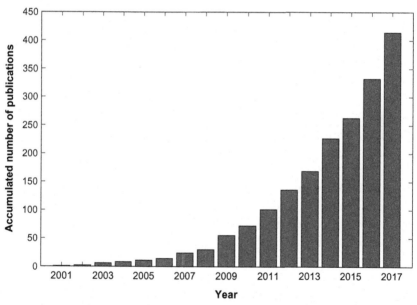

Fig. 5 Trend of publications with explicit relevance to 100% renewables. (Keywords searched: *"100%renewable,"* "100% renewable," "100% renewables," "100% renewables," "100 percent renewable," "100 percent renewables," "100-percent renewables," "100%-renewables," "100%-renewable," "hundred percent renewable," "hundred percent renewables," "hundred-percent renewable," "hundred-percent-renewable," "hundred-percent-renewable.")

have addressed this problem indirectly. For instance, John Burdon Sanderson Haldane in 1923 [9] sketched a plan for the future energy systems of England relying on "intermittent but inexhaustible sources of power, the wind and the sunlight" acknowledging that water power would not be sufficient. According to him, "four hundred years hence the power question in England may be solved somewhat as follows: The country will be covered with rows of metallic windmills working electric motors which in their turn supply current at a very high voltage to great electric mains. At suitable distances, there will be great power stations where during windy weather the surplus power will be used for the electrolytic decomposition of water into oxygen and hydrogen. These gasses will be liquefied, and stored in vast vacuum jacketed reservoirs, probably sunk in the ground... In times of calm, the gasses will be recombined in explosion motors working dynamos which produce electrical energy once more, or more probably in oxidation cells." It is clear that Haldane was sketching the 100% renewable future of England.

As another example, two years after the world energy crisis of 1973, Bent Sørensen [10] wrote an article in *Science* magazine describing his energy system planning for Denmark's energy demand in 2050. His article, though it did not specify 100% renewables, came to the conclusion that a combination of wind and solar panels would be economically feasible to supply the 2050 energy demand, though it required substantial initial investment, which was not practical for the socioeconomic context of Denmark of the time.

Returning to the literature set of this study (see Fig. 5), the first explicit use of the "100% renewable" keyword was by the US Environmental Protection Agency (EPA) in 2001, which reported the EPA's agreement with a renewable energy marketing company to power EPA Cincinnati research facilities with "100 percent renewable energy" (wind and biomass energy) [11]. The EPA declared this contract as "an attempt to pave the way for the green power market" and "as an example for other companies hoping to implement clean energy in the future."

The term "100% renewable" is applicable to various scales (see Fig. 6). Further to its application for energy systems, the term "100% renewables" has also been used for renewable materials and fuels, which are either derived from renewable resources or are recyclable (renewable in nature). For instance, in 2003, BASF introduced a new fiber, Savant, which is fully recyclable and 100% renewable [12]. The *Textile Asia* journal also discussed Ingeo (ingredients from the earth), another fiber, which is produced from 100% renewable resources with biodegradability under the right conditions [13]. Another instance of a 100% renewable was the *Petroleum Review*'s [14] brief report in 2003 on Shell's strategy for "evolution from improved

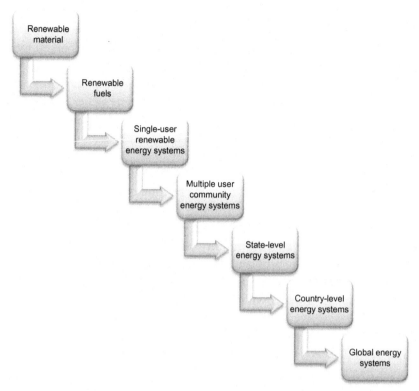

Fig. 6 Multiple scales in 100% renewable energy research (spanning from 100% renewable materials and fuels to energy systems from a single house, up to national and global levels); a single user could be a household or commercial, which attempts to satisfy its energy demands with available distributed generation and storage. Community scale is a group of users who build a so-called microgrid or energy hub to satisfy their demands. Macrolevel could be the energy management of a town, state, or country.

conventional gasoline and diesel through biodiesel and bioethanol blends to arrive at 100% renewable blends."

As time went on, however, the dominant direction in "100% renewables" became energy systems. Generally, energy system planning can be conducted at various scales, including single user, community, city, state, country, or global levels (Fig. 6). Initial research activities pertained to small-scale applications. For instance, Van Helden [15] discussed the optimal design of photovoltaic and thermal (PVT) systems, which could supply buildings with 100% renewable electricity and heat, and with improved efficiency compared to separate PV and solar thermal systems.

But slowly, the view increased to 100% renewables in larger systems. Kassels [16] studied energy transition in the Galápagos islands, Ecuador. More than 97% of the islands are UNESCO-protected national parks. Even the nearby ocean is protected as a Marine Resources Reserve. As a concern of growing population and consequences of fossil fuels pollution, the government supported converting the island to 100% renewables mostly with PV technology. The Fraunhofer Institute for Environmental Safety & Energy Technology (UMSICHT) in Germany integrated a sewage treatment plant with a molten carbonate fuel cell (MCFC), the use of which allowed 100% renewable power generation using the sewage off-gas [17].

Moving closer to the Kyoto Protocol's commitment period of 2008–12, research into climate change mitigation (CCM) options, including renewable energy, also accelerated (evident in Fig. 5). Also, discussions expanded to large-scale applications at community and state levels. For instance, Ashok [18] discussed issues encompassing the electrification of remote rural areas, with high relevance to diesel gens, and presented methodologies for designing hybrid energy systems with 100% renewable energy contribution for communities. In 2005, a thesis was published by Gregor Czisch from Kassel University, studying a renewable electricity system for Europe and North Africa [19]. This was a rigorous study considering hourly supply/demand profiles and assessing renewable electricity shares up to 100% renewable.

As discussed earlier, the scholarly field of "100% renewables" has blossomed from Denmark, starting with the works of Bent Sørensen [10]. The most significant contribution to the field also comes from Denmark, with the works of Henrik Lund and his colleagues. They published a paper in 2006 [20] for the analysis of integrating wind, PV, and wave power into a Danish electricity system to supply 0%–100% of the electricity demand. In the following year, 2007, Lund published a paper [21], discussing the challenges and opportunities of converting Denmark's energy systems into a 100% renewable energy system. The positive conclusion of the manuscript may be considered the main igniter of the emergence of the "100% renewable scenario" as one of the scenarios in state-level planning practices. This study was later supported with more detailed studies from Lund and his colleagues [22–24]. Furthermore, the publication of a textbook along with the EnergyPlan software, with the capability of 100% renewable scenario analysis, helped the growth of the field (*Ed. 1* [25], and *Ed. 2* [22]).

Closer to the Copenhagen meeting of 2009, global warming was at the core of attention. By this time, the "100% renewable" research had already been scaled up to national level energy system planning. That year, Jacobson and Delucchi [26] published a controversial paper in *Scientific American* claiming that wind, water, and solar (WWS) technologies could eliminate all fossil fuels, providing 100% of the world's energy in 2030. According to the authors, the utilizable WWS resources were over 600 TW, while the demand in 2030 would be only 11.5 TW. This demand could be optimally supplied by a mix of 9% water, 51% wind, and 40% solar. The paper was later backed up with more details by the authors [27, 28]. They also studied country-level cases. For instance, they debated the stability and economic feasibility of 100% renewables for the US electricity grid [29]. More recently, the same team discussed the feasibility of 100% WWS for 139 countries in 20 regions across the world [30].

Given the variability of renewable energies, there is another research category, which works on the interconnection of multiple energy networks (e.g., crosscountry import-export) to further improve the feasibility and technical stability of 100% renewable energy systems. For instance, there is increasing interest in supergrids or supersmart grids (Chapter 6) [31].

The growing academic developments (see Fig. 5) have gradually motivated policy makers to bring the 100% renewables proposition into policy analysis. In recent years, "100% renewables" has become one of the main benchmarks in several countries, and almost all countries have set a goal of "X% renewables" in their energy policy and planning to be achieved within a certain timeframe. The "X% renewables" scenario is, in fact, a measure to assess a country's distance from the ultimate "100% renewable" ambition. For instance, Fig. 7 shows the plan of 28 European countries for the share of renewable energies in gross-final demand by 2020 with values ranging from ∼10% (Malta) to >50% (Sweden) [32]. Details of some countries' policy actions toward 100% renewables are provided elsewhere [33, 34]. In fact, as already stated, the 100% renewables journey mainly requires the evolution of economic, legal, and regulatory frameworks without the need for significant technological breakthroughs, as they are fairly available today [35].

3 RESEARCH DIRECTIONS IN 100% RENEWABLES

3.1 The Analysis Methodology

To investigate the research directions in the field of 100% renewables, we conducted a keyword network analysis. The keywords, which had been carefully assigned to scientific papers by their authors, describe the main topics and research foci of the papers. Exploring these keywords and

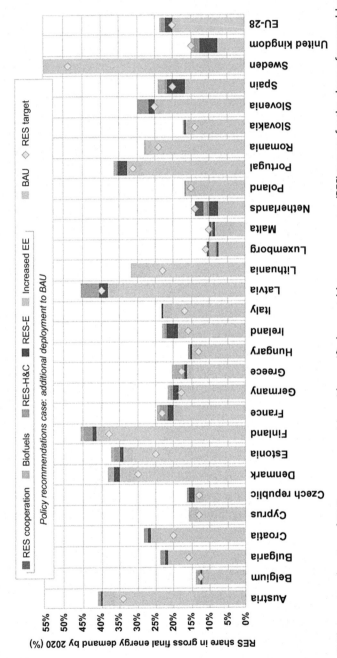

Fig. 7 The energy planning of 28 European countries with specified renewable energy sources (RES) targets for the share of renewables in their energy mix by 2020 [32]. (Image courtesy of EUFORES—European Forum for Renewable Energy Sources.)

identifying the commonest terms and their associations in a discipline can unfold its knowledge structure [36, 37]. Coword analysis is a technique, which uses keywords to build a network establishing relations between concepts and ideas in a corpus. Two keywords are linked in the network if they appear together in a scientific paper [36, 38]. This approach has been used in different fields including service innovation [39], hydrogen energy and fuel cells [40], ecology [41], climate change [6], polymer chemistry [42], zoonotic disease propagation [43], simulation-optimization of supply chains [44], and information retrieval [45].

We examined keyword co-occurrence networks (keywords networks for short). Each network consists of a set of nodes and ties connecting certain nodes together. In the network, nodes are keywords, and a tie between two keyword nodes is an indication of the two keywords appearing in the list of keywords in a paper. Details of the analysis are provided elsewhere (Chapter 3) [6].

The computed network data (adjacency matrix) are visualized using VOSviewer software, which provides different views of network maps. In the label view, each node is represented by a circle and its corresponding label. The sizes of the circles and labels vary based on the weight of the nodes. The weight of a node is assigned as the total strength of all the links of the node. To avoid overlapping, the set of all labels is shown partially. In the density view (or heat map view), the colors of the labels are in the blue-red range, determined based on the density of the nodes. Colors closer to red represent higher density, while colors closer to blue represent lower density. The density of a node is defined based on the number of its neighboring nodes and their weights. This view can be used to identify important areas of the map [46, 47].

3.2 Analysis of Publications Until 2010

From the annual publication profile (Fig. 5), it can be seen that there was an insignificant number of publications from the first publication incidence (i.e., 2001) to the beginning of 2010. The database contains only 55 publications by the year 2010. An increasing trend of publication starts from this period. Therefore, it would not be expected to notice a meaningful research network by studying the literature up to 2010. Fig. 8 meets this expectation by revealing the field's very limited keyword network. Nevertheless, the figure highlights the role of hydrogen, energy vector, and system planning. H2RES and Balmorel are two energy system planning tools, which seem to have been used during this period for 100% renewable planning.

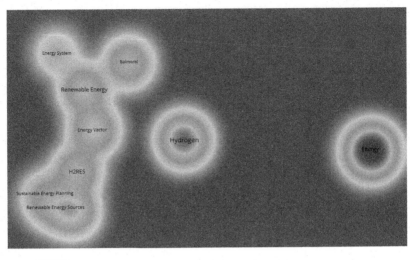

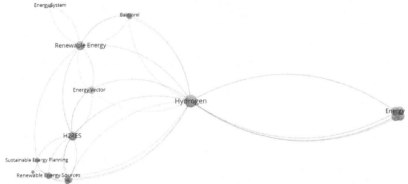

Fig. 8 Keywords network maps for all the records of "100% renewables" until 2010; upper: heat map, and lower: label network map.

3.3 Publications Until 2015

While only 55 records for "100% renewables" were found until 2010, the value increased more than threefold and reached 227 by the beginning of 2015 (Fig. 5). The density map of the keywords network until 2015 (Fig. 9) clearly shows the development of a distinct cluster with "renewable energy (system)" as the overarching keyword. Fig. 10 shows the zoomed map. As evident from this figure, the other strong keywords in this cluster include "smart grid," "district heating," "energy storage," and "hydrogen." These imply that by this time the "100% renewables" research community has directed its focus onto the challenge, i.e., the key reliability of future "smart grids," which require "energy storage," with "hydrogen" being one

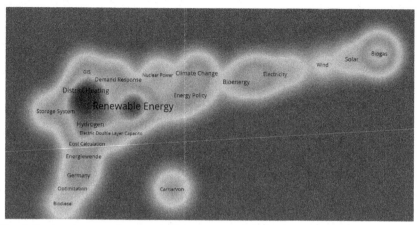

Fig. 9 Keywords network map of all the records for "100% renewables" until 2015.

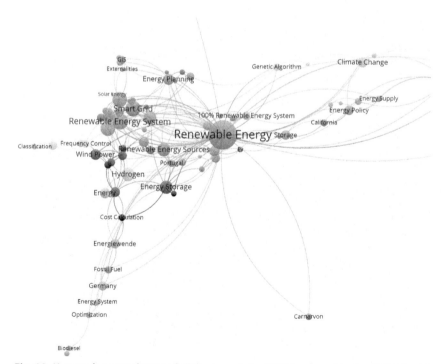

Fig. 10 Keywords network map of all the records for "100% renewables" until 2015, with focus on strong clusters.

of the options. Also, there is attention to renewable energy applications for district heating (and cooling), of particular interest for the European region.

3.4 Publications Until Mid-2018

In the few years from 2015 to mid-2018, the number of publication records more than doubled (from 227 to 465). The most recent density map of the keywords network is shown in Fig. 11, illustrating the emergence of "100% renewable energy" as a key cluster. Fig. 12 also shows the map with a focus on the key cluster region.

Further to the "smart grid," "energy storage," and "hydrogen" keywords, which were top terms in the previous period, some other terms also have become popular by 2018, the two top being "energy transition" and "power-to-gas." "Solar" and "wind are also common keywords. It is also noteworthy that topics related to "energy systems analysis" have been used frequently. These include "smart energy system," "renewable energy

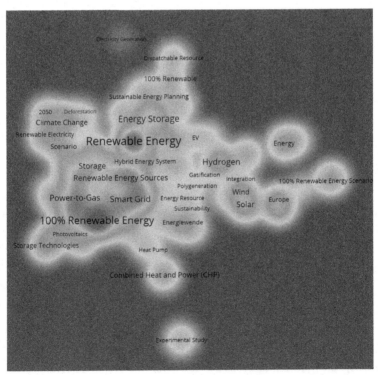

Fig. 11 Keywords network map of all the records for "100% renewables" until mid-2018.

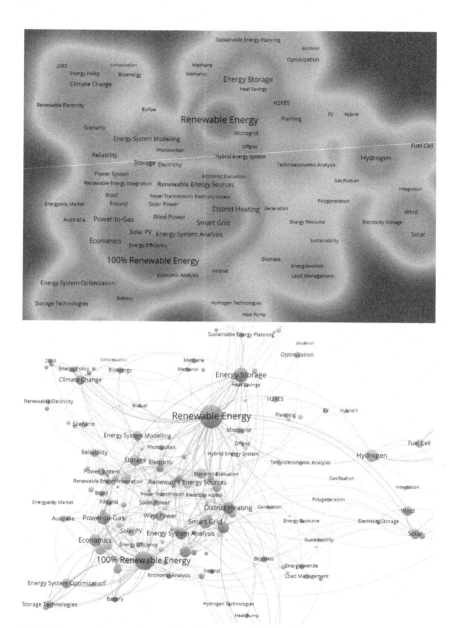

Fig. 12 Keywords network map for all the records all the records on "100% renewables" until mid-2018: focus on strong clusters.

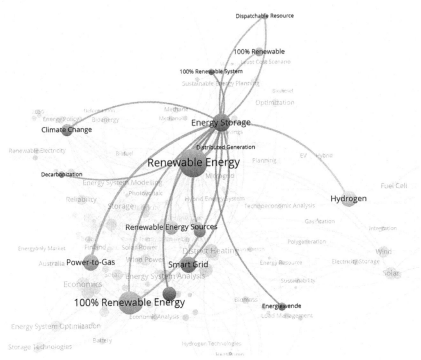

Fig. 13 The key connection ties of "energy storage" in the "100% renewables" research network.

system," "energy system modelling," "optimization," "grid integration," "demand response," and "scenario."

As shown in both Figs. 11 and 12, "energy storage" is an emerging cluster. Different relevant keywords such as "storage," "storage system," "storage technology," "electricity storage," "storage demand" are used.

Undoubtedly, energy storage is a complement to variable renewable energy technologies. Fig. 13 illustrates the key connection ties of "energy storage" in the research network. The figure clearly shows the strong tie of between "energy storage" and "renewable energy" and also "100% renewable energy." With the growing share of renewable technologies and over the transition to 100% renewables, the role of energy storage will become prominent.

Given the diverse geographic availability of renewable resources, a sustainable future energy system is expected to be composed of various energy generation and storage options. As such, like the "polygeneration" term, which represents the diversity of generation options, polystorage will overarch diverse energy storage options [48].

4 FUTURE ENERGY SYSTEMS PLANNING

The centrality of energy system analysis in 100% renewable research was elaborated in the research network analysis of the previous section. Any decision making in this space requires access to decision support tools. Balmorel and H2RES (see Fig. 8) are two such decision support tools, which were used in the early 100% renewable studies, thanks also to being open source. There were also applications of some commercial tools such as PLEXOS [49]. An elaborated list of existing energy system analysis tools, with the pros and cons of each one, is available elsewhere [50].

In the next section, the overall decision-making framework and the associated challenges are discussed. Also, the modeling structure of some case studies is assessed.

4.1 Special Requirements of Decentralized Energy Network Planning

4.1.1 Integration of Energy Storage: A Necessary Feature

Increasing attention has been paid to the structure of future energy networks, considering wide ranges of renewable energies in response to market trends and climate change policies. However, relatively little work exists, which addresses the integration of future networks (with a wide range of intermittent renewable generations) with energy storage, especially at the distribution level. Thanks to the revolutions in PV and wind technologies, and their consequences, the large-scale and small-scale deployment of energy storage into future energy networks is now widely accepted [51–56] and the key research questions relate to the size, location, ownership/operation, and storage device cost trajectories, which are still evolving [57]. Many applications in existing electricity networks could benefit from storage (schematic in Fig. 14).

That said, most conventional energy system planning tools cannot accommodate energy storage features. One immediate reason is that most such models are coarse, with the time granularity of a year. Even tools with finer time frequency (such as hourly profiles) are mostly dispatch models, which cannot accommodate all features of energy storage. In fact, dispatch models noticeably underestimate the value of energy storage. These constraints leave us with finger-count models suitable to future energy system planning.

4.1.2 Consideration of Distribution Network and the Prosumer-Side

Mainstream studies of future energy networks are limited to generation-transmission levels of the network and avoid modeling the distribution level.

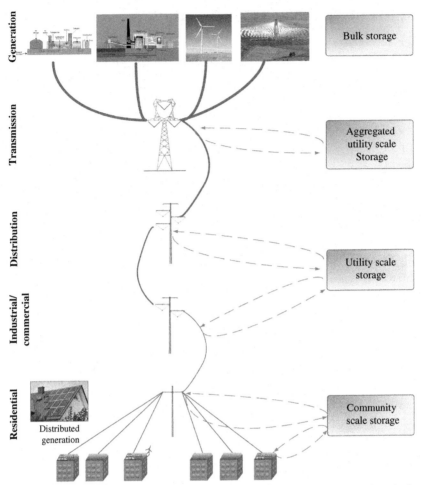

Fig. 14 Schematic of energy storage integration with various levels of the electricity grid.

They generally make some lumpsum assumptions of the uptake of distributed generation/storage and electric vehicles and develop demand forecasts on that basis. In fact, in future prosumer-oriented energy networks, generation and storage can be deployed at all scales and locations on the grid—from large-scale hydroelectric energy storage, through transmission and distribution (T&D) assets (for example, underground compressed air for multihour energy storage, or lithium battery storage for ancillary services), associated with generation (for example at wind farms), throughout the distribution network (for example at substations, and at major load centres such as shopping centres), all the way to households. Even onboard storage of electric vehicles can be considered if this capacity is made available when the vehicle is plugged in.

4.2 A Possible Framework for a Future Energy Network Planning Tool

A proper model for future energy system planning should enable energy storage allocation at all locations within a network, including transmission, distribution, and demand side. One possible way of modeling such a multi-scale planning problem (from supplier to user) is a two-stage framework (see Fig. 15). At the first stage, individual and aggregated electricity generation and a load of the distribution network are modeled. The results of Stage 1 are then used as inputs for the overall grid modeling, including a centralized generation and transmission network (Stage 2). The optimal result of the model would be the technology mix (generation and storage), the operation schedule, and the topology of the new transmission network.

Stage 1 (microlevel): The conventional electricity network models are based at the transmission level. But in the decentralized network era, generation and storage systems can be also located at the distribution level. A possible approach to accommodate this is to aggregate distribution level demand up to the transmission level, using estimates of residential, commercial, and

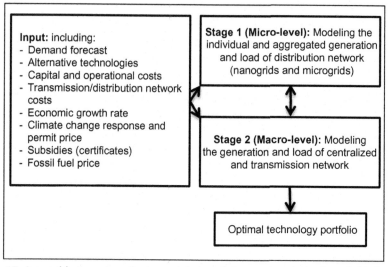

Fig. 15 A possible two-stage framework for modeling and optimization of the future multiscale electricity network.

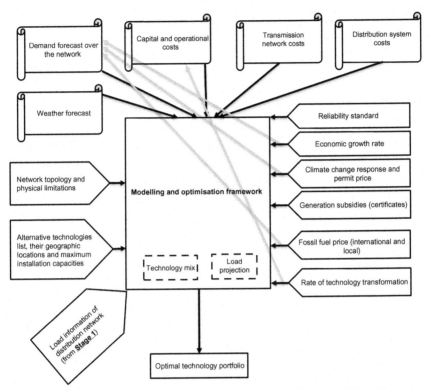

Fig. 16 Modeling structure of electricity network planning (Stage 2 of Fig. 15).

industrial loads. It is necessary to use load data with high granularity (hourly or less) because of observed strong seasonal demand patterns.

Stage 2 (macrolevel): The modeling structure of the macrolevel network (Stage 2) is illustrated in Fig. 16. Given the technoeconomic input parameters and the information from the distribution network (from Stage 1), the geographic regions under study will be meshed into polygons with the assumption of unique weather profile, demand, and costs.

4.3 Planning Optimization Framework

A planning tool is, in fact, an optimization program. Any optimization process consists of a few key components including assumptions, input parameters, variables, constraints, and an objective function.

4.3.1 Assumptions

It is almost impossible to model a phenomenon or a system precisely. This was ironically articulated by George Box, who wrote, "all models are wrong but some are useful" [58]. Any model is associated with some assumptions; the following are some assumptions for future energy systems planning:

- The existing transmission system will be available.
- The future energy network modeling is carried out within a geographically limited location with known boundaries.
- Associated social and political changes may or may not be considered.
- Some generation or storage technologies might be excluded due to local constraints.
- Transmission and distribution (T&D) connection is possible between all components unless otherwise specified.

4.3.2 Input Parameters

The term "parameter" here refers to all those values that are given as "inputs" to the model. Inputs are technical, socioenvironmental, microeconomic, or macroeconomic parameters, which are supplied by decision makers for a given problem. Figs. 15 and 16 show some such parameters for energy network modeling, including but not limited to weather forecasts, demand forecasts, alternative technologies, capital and operational costs, transmission/distribution network costs, economic growth rate, climate change responses and CO_2 permit price, subsidies (certificates), and fossil fuel price.

4.3.3 Variables

The purpose of the entire optimization practice is to identify the values of unknowns. These can be binary (yes/no) decisions, or quantities (integer or continuous). Once the decision maker supplies the input parameters, the optimization initiates to find the best combinations of all the variables to achieve the optimal goal.

Energy generation technologies and storage options are the main variables in an energy system planning optimization. A list of generation and storage options (e.g., Fig. 17), together with their technoeconomic parameters, must be supplied to the decision support tool. Obviously, some options might be geographically unavailable or not recommended by expert inputs for a given jurisdiction. They could be removed from the list in priori.

4.3.4 Constraints

In any technoeconomic assessment, constraints are involved. They could be high-level economic constraints such as the maximum available investment

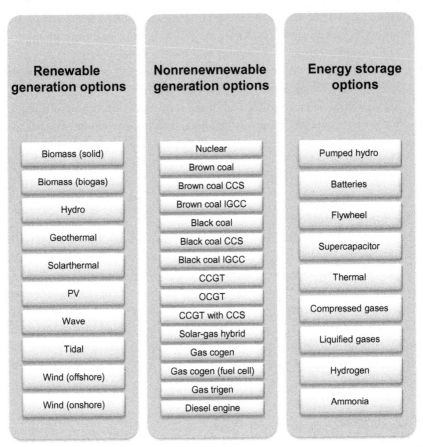

Fig. 17 A typical list of energy generation and storage options (*CCS*, carbon capture and storage; *IGCC*, integrated gasification combined cycle; *CCGT*, closed-cycle gas turbine; *OCGT*, open-cycle gas turbine).

budget, or they could be technical limitations enforced by the physics or chemistry of the system.

Constraints are those equations, which can direct the optimization to operate within physical and logical boundaries. Some constraints are enforced by assumptions; others are imposed by technoeconomic limitations of generation, storage, and T&D networks. The followings are some key constraints of energy system planning:

- The total emission by the selected portfolio of technologies must meet the given national emission reduction target.

- Generation, storage, and T&D components have lower and upper bounds of operational capacity. Their transition rates (ramping up and ramping down) are also constrained.
- The annual reliability standard of every technology and/or region must be met.
- There is a limit to any technology at a certain location and time (e.g., due to non/renewable resource limitation.
- For overall stability, voltage and frequency must be operated within a given range.

4.3.5 Objective

Selection of the objective function is the most critical step in any technoeconomic assessment and is entirely related to project finances, policy inputs, and long-term vision. A correct objective function can guarantee a sustainable decision. An objective function might be merely minimization of total capital investment, simply capital expenditure (CAPEX), or operating expenditure (OPEX). It could also be the minimization of the levelized cost of energy (LCOE), maximization of the internal rate of return (IRR), maximization of the net present value (NPV) of cash flow, and so forth.

One proper objective function in energy system planning is to find the least-cost electricity generation portfolio considering storage, transmission, and distribution network that meet the given constraints (e.g., required reliability standards) while considering generation and storage.

Having defined all variables, uncertain parameters, constraints, and assumptions the optimization task will be carried out to find the least-cost technology mix to satisfy the projected demand (under the given scenario) over the planning horizon complying with the reliability standards.

4.4 Superscenarios (Scenarios of Scenarios): To Tackle Uncertainty and Computation Intensity Challenges

Inputs are technical, socioenvironmental, microeconomic, or macroeconomic parameters, which are supplied by decision makers for a given problem. In the majority of conditions, especially in strategic decision making with a wide planning horizon, these parameters are not available with certainty. These uncertainties in the parameters are the most crucial threats to the accuracy of decisions. In the absence of certain parameters, the usual practice is to consider numerous scenarios with various probabilistic values for each parameter, ranging from lowest to the highest extreme values.

Table 1 Some uncertain parameters in future energy network planning

Uncertain parameter	Scenarios
Carbon reduction target	X% reduction (below 2000 level) by 2030, and Y% reduction (below 2000 level) by 2050
Carbon price	Low, medium, high
Energy subsidies (certificates)	Low, medium, high
Economic growth	Low, medium, high
Productivity growth	Low, medium, high
Population growth	Low, medium, high
Commodity prices	Low, medium, high
International energy (oil, gas, and coal) prices	Low, medium, high
Local gas price	Low, medium, high
Local coal price	Low, medium, high
Local gas production	Low, medium, high
Weather data profiles	Low, medium, high
Penetration of electric vehicles (EVs)	Low, medium, high
Development of EV charging infrastructure	Low, medium, high
Development of hydrogen infrastructure	Low, medium, high
Penetration of distributed generation	Low, medium, high
Demand-side participation	Low, medium, high
Demand-side penetration of energy storage	Low, medium, high
Disconnections from the grid	Low, medium, high
Energy efficiency	Low, medium, high
Wholesale electricity price	Low, medium, high
Retail electricity price	Low, medium, high
Electricity demand increase	Low, medium, high
Natural gas demand increase	Low, medium, high
Costs of large-scale renewables generation and storage technologies	Grubb curve data for given jurisdiction from a preferred source
Costs of distributed generation and storage technologies	Grubb curve data for given jurisdiction from a preferred source

For instance, Table 1 lists the uncertain parameters in the context of energy systems planning.

From Table 1, which reflects only a fraction of the uncertain parameters in planning energy systems, the main challenge arises. Deterministic

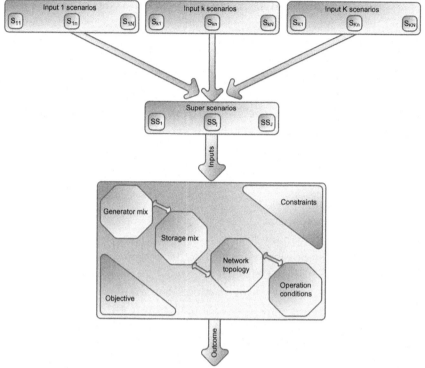

Fig. 18 Illustration of bundling uncertain input parameters into limitted number of superscenarios to be introduced to the planning optimization program.

optimization of energy systems planning is itself a challenge due to computational intensity. Now, let us think of a small problem, which has only 10 uncertain parameters, each with three possible values (low, medium, high). This translates to a total number of $3^{10} = 59049$ optimization problems to be solved. Considering the scale of energy system planning, especially at the national level, it is almost impossible to run so many scenarios. One alternative option is stochastic and parametric programming [59], which is a growing field. But all such models are still under development, with limited achievements in overcoming the computation time. Another alternative is "scenarios of scenarios" or simply "superscenarios." This approach, which is receiving increasing attention due to its practicality, is based on the bundling of input scenarios into some limited case studies (Fig. 18).

As such, the superscenario or scenarios-of-scenario (SS) analysis approach is the most efficient (and sometimes the only) way forward for large systems entailing multiple input parameters with a wide range of uncertainties. In this approach, combinations of a few parameters with relevant behaviors are bundled into a group.

4.5 Superscenario Analysis Examples From Australia

4.5.1 Studies of the Australian Energy Market Operator (AEMO)

The AEMO conducts an annual National Transmission Network Development Plan (NTNDP) study, which is supported by computational rigor. One of the reports considers four key uncertainties confronting the future development of the energy industry in Australia:

economic growth

response to climate change

technology development

fuel price and availability [60]

Considering the possibility of change (from weak/low to strong/high) for each of these factors, AEMO has developed six different superscenarios (SSs). In terms of climate change action, they employed the Australian Treasury's three SSs of zero, core (5% reduction from 2000 emissions by 2020) and high (25% reduction from 2000 emissions by 2020) carbon prices. Fig. 19 illustrates the matrix of superscenarios. SSs 1 and 5 consider two extreme conditions of fast economic growth together with a fast response to climate change (SS 1) and slow economic growth together with a slow response to climate change (SS 5). The other four SSs assume a Treasury core response to climate change with various economic growth values, the lowest being for SS 6 and the highest for SS 2. SSs 3 and 4 are similar in all parameters except for technology development, where the former assumes moderate penetration of distributed generation and EVs in contrast with the latter assuming a strong transformation and penetration.

After the announcement of the Clean Energy Future plan by the Australian Government in July 2011 and passing of the legislation in November 2011, AEMO expanded its planning study to include the case for 100% renewables [61]. AEMO considered four different SSs including rapid or moderate transformation, of which two are for the year 2030 and the other two are for the year 2050. The specifications of these four SSs are summarized in Table 2. The key elements in defining the SSs were the availability and costs of generation and storage, demand forecasts, load profiles, and reliability standard assumptions. Worley Parsons prepared the initial technology costs for the Australian Bureau of Resources and Energy Economics (BREE)'s AETA (Australian Energy Technology Assessment)

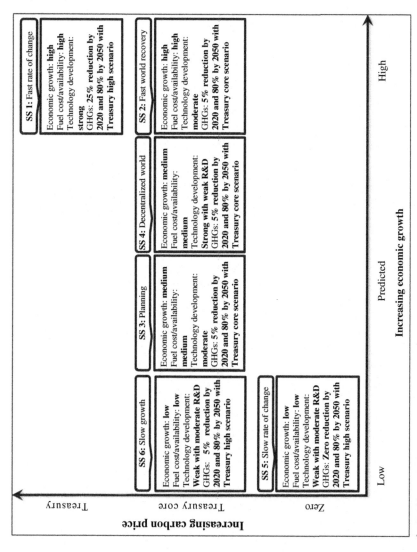

Fig. 19 Matrix of AEMO's superscenarios [60].

Table 2 Summary of AEMO's superscenarios for a 100% renewable electricity network (information source: [61])

SS number	SS 1	SS 2	SS 3	SS 4
Key features				
Timeframe	2030	2050	2030	2050
Transformation of the electricity sector	Rapid	Rapid	Moderate	Moderate
Economic and electricity demand growth	Moderate	Moderate	Robust	Robust
Demand-side participation	Strong	Strong	Weak	Weak
Objective	Least-cost modeling to determine an optimal combination of generation, storage, and transmission investments for the given forecast customer demand			
Key general assumptions	➢ Electricity supply beyond the planning regions (NEM: national energy market) is not considered ➢ The associated social, political, and economic changes are not considered ➢ The transition path from the current power system to 100% renewable power systems is not considered ➢ Distribution system costs are not included ➢ The existing transmission system will be available ➢ The five NEM regions are divided into 43 locational polygons ➢ Fossil fuels, CCS, and nuclear are excluded ➢ Except for EVs, there is no other switching away from fossil fuels toward renewables–based electricity across the rest of the Australian economy ➢ Half of the demand side participation (DSP) is curtailable load (demand can be reduced at a given cost) and half is "movable demand" or flexible demand (reduction in demand is possible at no cost provided that it is consumed at an alternative time that day) ➢ The long-term average unserved energy (USE) over a year is less than 0.002% of annual energy consumption (at least 99.998% of energy requirements are met) ➢ CST costing is based on central receiver technology ➢ Utility PV costing is based on single axis technology ➢ Transmission losses across the NEM are 5%			

	Scenario 1	Scenario 2	Scenario 3	Scenario 4
	⋏ Eight renewable technologies are considered: biomass, geothermal (hot sedimentary aquifer), geothermal (deep hot rock), utility PV, solar thermal, wave, onshore wind, and offshore wind ⋏ Five categories of large utility-scale energy storage technologies are considered (batteries; biomass, as solid matter and as biogas; compressed air; pumped hydroelectricity; and molten salt thermal energy storage associated with CST energy collection) ⋏ Existing pumped hydro in the NEM will remain ⋏ Historical data are generated based on 2009–10 demand			
Key specific assumptions	⋏ Up to 10% of demand in any hour is available for DSP ⋏ Capital costs are based on 2012 costs given by AETA 2012 [62] and CSIRO's GALLM ⋏ Rapid technology transformation will drive real reductions in operation and maintenance (O&M) costs outweighing any increases projected in AETA 2012 leading to O&M costs reduction by 12.5% by 2030	⋏ Up to 10% of demand in any hour is available for DSP ⋏ Capital costs are based on 2012 costs given by AETA 2012 [62] and CSIRO's GALLM ⋏ Rapid technology transformation will drive real reductions in O&M costs outweighing any increases projected in AETA 2012 leading to O&M costs reduction by 25% by 2050	⋏ Up to 5% of demand in any hour is available for DSP ⋏ Capital costs are based on midpoint costs given by AETA 2012s Grubb curve [62] ⋏ O&M costs escalate at around 150% of the Consumer Price Index (CPI), leading to cost increases of 27% by 2030	⋏ Up to 5% of demand in any hour is available for DSP ⋏ Capital costs are based on midpoint costs given by AETA 2012s Grubb curve [62] ⋏ O&M costs escalate at around 150% of CPI, leading to cost increases of 46% by 2050
Input data sources	⋏ Capital and operational costs from BREE's Australian Energy Technology Assessment 2012 (AETA 2012) ⋏ Demand forecasts from AEMO's 2012 National Energy Forecasting Report (NEFR 2012) [63] ⋏ Transmission costs are based on AEMO's electricity transmission cost assumptions			

report [62], and those values were used for the SSs with a moderate technology transformation assumption. Australia's Commonwealth Scientific and Industrial Research Organisation (CSIRO) projected a "trajectory of technology costs" using its global and local learning model (GALLM). The midpoint of the so-called Grubb curve was used for the SSs with the assumption of rapid technological transformation. AEMO has categorized power generating technologies into three groups:

- *Nondispatchable (PV, wind, and wave):* These technologies have low operating costs. But their load is weather-dependent although forecastable to some extent. Their output load and response to demand are controllable only by curtailment or energy storage.
- *Baseload (geothermal, biomass (wood), bagasse):* These technologies have no intermittency limitations and their output load can be controlled at a cost. They are, however, relatively slow to respond to volatile demand changes.
- *Peak dispatchable (hydro, pumped hydro, concentrated solar thermal (CST), biogas):* These are flexible, fast-to-respond technologies with the limitation of low annual energy potential requiring costly energy storage.

According to AEMO, "all three categories must be optimally combined to reliably meet supply for the lowest cost" [61]. Thus, five different large utility-scale energy storage technologies were studied [batteries; biomass, as solid matter and as biogas; compressed air; pumped hydroelectricity; and molten salt thermal energy storage associated with (CST energy collection)], with the objective of determining an optimal combination of generation, storage, and transmission investments for the given forecast customer demand in the years 2030 and 2050.

Although the study results discourage the application of batteries and compressed air, as well as adopting the new addition of pumped hydro, they confirm that the combined dispatch of all three technologies is sufficient to match demand under all four SSs, even with the unavailability of PV load at peak-hour evenings.

4.5.2 CSIRO Study With a Focus on Demand-Side Participation

The Future Grid Forum led by the CSIRO studied four different SSs with the main focus on consumer uptake of distributed energy systems. Fig. 20 illustrates the four SSs and their levels with respect to consumer response. SS 1 considers the lowest possibility of consumer uptake. Under this SS, the increase in retail electricity price will motivate a modest number of customers to take up demand management. They will prefer just to "set and

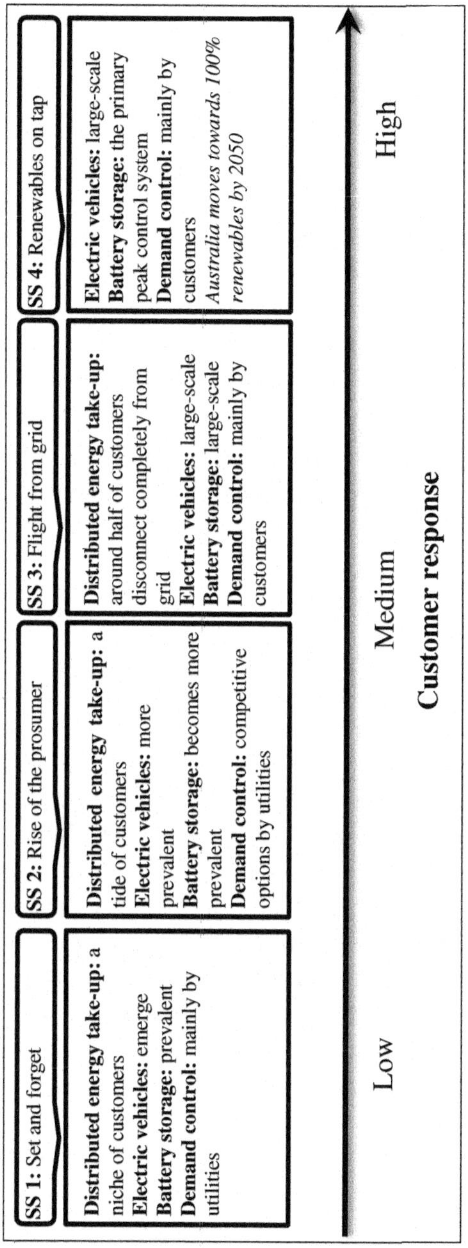

SS 1: Set and forget	SS 2: Rise of the prosumer	SS 3: Flight from grid	SS 4: Renewables on tap
Distributed energy take-up: a niche of customers **Electric vehicles:** emerge **Battery storage:** prevalent **Demand control:** mainly by utilities	**Distributed energy take-up:** a tide of customers **Electric vehicles:** more prevalent **Battery storage:** becomes more prevalent **Demand control:** competitive options by utilities	**Distributed energy take-up:** around half of customers disconnect completely from grid **Electric vehicles:** large-scale **Battery storage:** large-scale **Demand control:** mainly by customers	**Electric vehicles:** large-scale **Battery storage:** the primary peak control system **Demand control:** mainly by customers *Australia moves towards 100% renewables by 2050*

Low   Medium   High

Customer response

Fig. 20 Example of superscenarios based on demand-side participation.

forget" a level of demand control system and pass the responsibility of its management to the utility companies. Electric vehicles will be prevalent under this SS but liquid-fueled transport and centralized distribution will still be dominant. Under SS 2, the higher electricity price together with reduced costs of PV panels will motivate more residential customers to install distributed generation systems. Moreover, distribution service becomes competitive to the benefit of customers. Electric vehicles become more prevalent, resulting in a notable reduction in Australia's oil demand. SS 3 assumes that customers will prefer to manage their own energy generation, together with battery storage, rather than receiving service from utilities. The continued decline of battery prices will motivate many customers to totally disconnect from the grid, resulting in escalation of retail electricity prices as the fixed costs are apportioned to a smaller customer base. SS 4 assumes that the reduction of renewable generating technologies' costs will result in the introduction of a 100% renewable target by 2050. Under this SS, storage will be implemented for both large-scale and onsite generation.

5 CONCLUSION

The topic of "100% renewable" energy system is almost a new research field. There were trivial relevant publication records until the year 2000, and even by 2010 there were fewer than 100 total records. But during the 2010s, the field had grown noticeably. Our keyword network analysis showed the development of a field with a clear research direction. As shown in Fig. 11, along with obvious solar and wind power, "smart grid," "energy storage," "hydrogen," and "power-to-gas" were the key components of this research field. Another critical element in this rising field was "energy systems analysis," which has been referred to by various terms such as "smart energy system," "renewable energy system," "energy system modelling," "optimization," "grid integration," and "demand response."

To address the challenges of energy system analysis, we discussed the requirement of a proper energy system planning tool in the context of a decentralized energy network, which requires (1) high time resolution; (2) consideration of energy storage; (3) consideration of system dynamics (ramping ups and downs); and (4) uncertainties. Some discussion was provided on the consideration of input uncertainties, for which the practical K suggested was to bundle the uncertain input parameters into a "superscenario" or "scenarios of scenarios."

REFERENCES

[1] Maslin M. Global warming: a very short introduction. Oxford: OUP; 2008.

[2] Hansen J, et al. Climate impact of increasing atmospheric carbon dioxide. Science 1981;213(4511):957–66.

[3] Kerr RA. Hansen vs. the World on the Greenhouse Threat. Scientists like the attention the greenhouse effect is getting on Capitol Hill, but they shun the reputedly unscientific way their colleague James Hansen went about getting that attention. Science 1989;244(4908):1041–3.

[4] Damm DL, Fedorov AG. Conceptual study of distributed $CO2$ capture and the sustainable carbon economy. Energy Convers Manag 2008;49(6):1674–83.

[5] IEA. World energy outlook 2009 edition-climate change excerpt. Paris: IEA; 2009.

[6] Karimi F, Khalilpour R. Evolution of carbon capture and storage research: trends of international collaborations and knowledge maps. Int J Greenhouse Gas Control 2015;37:362–76.

[7] Khalilpour KR, Vassallo A. Technoeconomic parametric analysis of PV-battery systems. Renew Energy 2016;97:757–68.

[8] Khalilpour RK, Vassallo A. Grid revolution with distributed generation and storage. In: Community energy networks with storage: modeling frameworks for distributed generation. Singapore: Springer; 2016. p. 19–40.

[9] Haldane JBS. Daedalus: or, science and the future; a paper read to the Heretics. Cambridge: E.P. Dutton; 1924 (on February 4th, 1923).

[10] Sørensen B. Energy and resources. A plan is outlined according to which solar and wind energy would supply Denmark's needs by the year 2050. Science 1975;189(4199):255–60.

[11] EPA. Industry news: EPA leader in use of renewable energy. Pollut Eng 2001;33(10):9.

[12] Norberg K. BASF bets on renewable smart fiber. Int Fiber J 2002;17(4):38–40.

[13] Textile-Asia. Ingeo—a new fibre. Knitting Technol 2003;25(4):31.

[14] Petroleum-Review. News downstream: challenge posed by global warming. Pet Rev 2003;57(683):10.

[15] van Helden WGJ, van Zolingen RJC, Zondag HA. PV thermal systems: PV panels supplying renewable electricity and heat. Prog Photovolt Res Appl 2004;12(6):415–26.

[16] Kassels S. Energy evolution—renewable energy in the Galápagos Islands. Refocus 2003;4(5):36–8.

[17] Fuel-Cells-Bulletin. German project to operate Ansaldo MCFC on sewage off-gas. Fuel Cells Bull 2004;2004(2):6–7.

[18] Ashok S. Optimised model for community-based hybrid energy system. Renew Energy 2007;32(7):1155–64.

[19] Czisch G. Szenarien zur zukünftigen Stromversorgung: kostenoptimierte Variationen zur Versorgung Europas und seiner Nachbarn mit Strom aus erneuerbaren Energien; 2005, https://kobra.bibliothek.uni-kassel.de/bitstream/urn:nbn:de:hebis:34-200604119596/1/DissVersion0502.pdf.

[20] Lund H. Large-scale integration of optimal combinations of PV, wind and wave power into the electricity supply. Renew Energy 2006;31(4):503–15.

[21] Lund H. Renewable energy strategies for sustainable development. Energy 2007;32(6):912–9.

[22] Lund H. Renewable energy systems: the choice and modeling of 100% renewable solutions. New York: Elsevier Science; 2009.

[23] Lund H, Mathiesen BV. Energy system analysis of 100% renewable energy systems-the case of Denmark in years 2030 and 2050. Energy 2009;34(5):524–31.

[24] Lund H, Ostergaard PA. Sustainable towns: the case of Frederikshavn–100% renewable energy. In: Sustainable communities. New York: Springer; 2009. p. 155–68.

[25] Lund H. Renewable energy systems: a smart energy systems approach to the choice and modeling of 100% renewable solutions. New York: Elsevier Science; 2014.

[26] Jacobson M, Delucchi M. A path to sustainable energy by 2030. Sci Am 2009; 301:58–65.

[27] Jacobson MZ, Delucchi MA. Providing all global energy with wind, water, and solar power, part I: technologies, energy resources, quantities and areas of infrastructure, and materials. Energy Policy 2011;39(3):1154–69.

[28] Delucchi MA, Jacobson MZ. Providing all global energy with wind, water, and solar power, part II: reliability, system and transmission costs, and policies. Energy Policy 2011;39(3):1170–90.

[29] Jacobson MZ, et al. The United States can keep the grid stable at low cost with 100% clean, renewable energy in all sectors despite inaccurate claims. Proc Natl Acad Sci 2017;114(26):E5021–3.

[30] Jacobson MZ, et al. Matching demand with supply at low cost in 139 countries among 20 world regions with 100% intermittent wind, water, and sunlight (WWS) for all purposes. Renew Energy 2018;123:236–48.

[31] Battaglini A, et al. Development of SuperSmart Grids for a more efficient utilisation of electricity from renewable sources. J Clean Prod 2009;17(10):911–8.

[32] Resch G, et al. Keeping track of renewable energy targets towards 2020: EU tracking roadmap 2015. Brussels: European Forum for Renewable Energy Sources (EUFORES); 2015. p. 105.

[33] Gipe P. 100 percent renewable vision building, in renewable energy world; 2013, https://www.renewableenergyworld.com/articles/2013/04/100-percent-renewable-vision-building.html.

[34] Hohmeyer OH, Bohm S. Trends toward 100% renewable electricity supply in Germany and Europe: a paradigm shift in energy policies. Wiley Interdiscip Rev Energy Environ 2015;4(1):74–97.

[35] Schellekens G, et al. 100% renewable electricity: a roadmap to 2050 for Europe and North Africa; 2010, http://pure.iiasa.ac.at/id/eprint/9383/.

[36] Romo-Fernandez LM, Guerrero-Bote VP, Moya-Anegon F. Co-word based thematic analysis of renewable energy (1990-2010). Scientometrics 2013;97(3):743–65.

[37] Yi S, Choi J. The organization of scientific knowledge: the structural characteristics of keyword networks. Scientometrics 2012;90(3):1015–26.

[38] Ravikumar S, Agrahari A, Singh SN. Mapping the intellectual structure of scientometrics: a co-word analysis of the journal Scientometrics (2005–2010). Scientometrics 2014;102:1–27.

[39] Zhu WJ, Guan JC. A bibliometric study of service innovation research: based on complex network analysis. Scientometrics 2013;94(3):1195–216.

[40] Chen YH, Chen CY, Lee SC. Technology forecasting of new clean energy: the example of hydrogen energy and fuel cell. Afr J Bus Manag 2010;4(7):1372–80.

[41] Budilova EV, Drogalina JA, Teriokhin AT. Principal trends in modern ecology and its mathematical tools: an analysis of publications. Scientometrics 1997;39(2):147–57.

[42] Callon M, Courtial JP, Laville F. Co-word analysis as a tool for describing the network of interactions between basic and technological research–the case of polymer chemistry. Scientometrics 1991;22(1):155–205.

[43] Hossain L, et al. Evolutionary longitudinal network dynamics of global zoonotic research. Scientometrics 2015;103(2):337–53.

[44] Huerta-Barrientos A, Elizondo-Cortes M, de la Mota IF. Analysis of scientific collaboration patterns in the co-authorship network of simulation-optimization of supply chains. Simul Model Pract Theory 2014;46:135–48.

[45] Ding Y, Chowdhury GG, Foo S. Bibliometric cartography of information retrieval research by using co-word analysis. Inf Process Manag 2001;37(6):817–42.

[46] van Eck NJ, Waltman L. In: VOSviewer: a computer program for bibliometric mapping. Proceedings of ISSI 2009–12th international conference of the international society for scientometrics and informetrics, vol. 2; 2009. p. 886–97.

[47] van Eck NJ, Waltman L. Software survey: VOSviewer, a computer program for bibliometric mapping. Scientometrics 2010;84(2):523–38.

[48] Khalilpour KR. Polygeneration with polystorage: for chemical and energy hubs. New York: Elsevier Science; 2018.

[49] Global-Change-Institute. Delivering a competitive Australian power system—part 2: the challenges, the scenarios. St Lucia: University of Queensland; 2013.

[50] Lund H, et al. Tool: the EnergyPLAN energy system analysis model. In: Renewable energy systems—a smart energy systems approach to the choice and modelling of 100 % renewable solutions, 2nd ed. New York: Academic Press; 2014. p. 1–6.

[51] Hadjipaschalis I, Poullikkas A, Efthimiou V. Overview of current and future energy storage technologies for electric power applications. Renew Sust Energ Rev 2009; 13(6–7):1513–22.

[52] Lemaire E, et al. European white book on grid-connected storage. France: European Distributed Energy Resources Laboratories; 2011.

[53] Battke B, et al. A review and probabilistic model of lifecycle costs of stationary batteries in multiple applications. Renew Sust Energ Rev 2013;25(0):240–50.

[54] Koohi-Kamali S, et al. Emergence of energy storage technologies as the solution for reliable operation of smart power systems: a review. Renew Sust Energ Rev 2013;25(0):135–65.

[55] Denholm P, et al. The value of energy storage for grid applications. Colorado: National Renewable Energy Laboratory; 2013.

[56] Directorate-general-for-energy. DG ENER working paper—the future role and challenges of energy storage, European Commission; 2013. https://ec.europa.eu/energy/sites/ener/files/energy_storage.pdf.

[57] Hoffman MG, et al. Analysis tools for sizing and placement of energy storage in grid applications: a literature review. Washington: Pacific Northwest National Laboratory; 2010.

[58] Box GEP. Science and statistics. J Am Stat Assoc 1976;71(356):791–9.

[59] Khalilpour R, Karimi IA. Parametric optimization with uncertainty on the left hand side of linear programs. Comput Chem Eng 2014;60:31–40.

[60] AEMO. 2012 Scenarios descriptions (NTNDP). Australia: Australian Energy Market Operator; 2012.

[61] AEMO. 100 percent renewables study—modelling outcomes. Melbourne: Australian Energy Market Operator; 2013.

[62] BREE. Australian energy technology assessments (AETA) 2012. Canberra: Bureau of Resources and Energy Economics; 2012.

[63] AEMO. National electricity forecasting report (NEFR) 2012. Melbourne: Australian Energy Market Operator; 2012.

INDEX

Note: Page numbers followed by *f* indicate figures and *t* indicate tables.

Printed in the United States
By Bookmasters